Biological Markers in Sediments and Petroleum

A Tribute to Wolfgang K. Seifert

J. Michael Moldowan

Chevron Oil Field Research Company
Richmond, California

Pierre Albrecht

Institut de Chimie
Université Louis Pasteur
Strasbourg, France

R. Paul Philip

School of Geology and Geophysics
University of Oklahoma
Norman, Oklahoma

Editors

Prentice Hall, Englewood Cliffs, New Jersey 07632

Library of Congress Cataloging-in-Publication Data

Biological markers in sediments and petroleum : a tribute to Wolfgang K. Seifert / J. Michael Moldowan, Pierre Albrecht, R. Paul Philp, editors.
p. cm.
Based on contributions to the American Chemical Society Meeting, held in Dallas, Tex., Apr. 9–14, 1989.
Includes bibliographical references and index.
ISBN 0-13-083742-3
1. Petroleum—Prospecting—Congresses. 2. Biogeochemical prospecting—Congresses. 3. Petroleum—Geology—Congresses. I. Seifert, Wolfgang. II. Moldowan, J. M. (J. Michael) III. Albrecht, Pierre. IV. Philp, R. P. V. American Chemical Society. Meeting (1989 : Dallas, Tex.)
TN271.P4B48 1992
622′.1828—dc20 91-16205
CIP

Editorial/production supervision/interior design: *Jean Lapidus*
Cover design: *Joe Di Dominico*
Copy editor: *Maria Caruso*
Acquisitions editor: *Michael Hays*
Editorial assistant: *Dana Mercure*
Prepress buyer: *Mary E. McCartney*
Manufacturing buyer: *Susan Brunke*

The publisher offers discounts on this book when ordered in bulk quantities. For more information, write: Special Sales/Professional Marketing, Prentice Hall, Professional & Technical Reference Division, Englewood Cliffs, NJ 07632.

ISBN 0-13-083742-3

90000>

9 780130 837424

Printed in the United States of America

10 9 8 7 6 5 4 3 2 1

ISBN 0-13-083742-3

PRENTICE-HALL INTERNATIONAL (UK) LIMITED, *London*
PRENTICE-HALL OF AUSTRALIA PTY. LIMITED, *Sydney*
PRENTICE-HALL CANADA INC., *Toronto*
PRENTICE-HALL HISPANOAMERICANA, S.A., *Mexico*
PRENTICE-HALL OF INDIA PRIVATE LIMITED, *New Delhi*
PRENTICE-HALL OF JAPAN, INC., *Tokyo*
SIMON & SCHUSTER ASIA PTE. LTD., *Singapore*
EDITORA PRENTICE-HALL DO BRASIL, LTDA., *Rio de Janeiro*

Contents

Dedication in Honor of Wolfgang K. Seifert

1931–1985

We stand at the point where two periods of geochemical biomarker research overlap. In the last two decades, fundamental research progressed to the stage where biomarkers have been proven applicable to oil exploration problem solving. A sharpening of skills continues. In the coming decade, there will be a melding of many areas of organic geochemistry into a more unified science.

We recognize Professor Alfred Treibs as the founder of organic geochemistry in the 1930s with his theory on the origin of porphyrins from chlorophyll. Professor Treibs was also the mentor for Wolfgang Seifert in his doctoral studies on pyrrole synthesis. Structure elucidation of a few biomarker compounds proceeded again in the late 1960s and early 1970s.

Development of computerized gas chromatography-mass spectrometry (GC-MS) systems allowed an exponential advancement of this science in the late 1970s to early 1980s. Biomarker oil/oil and oil/source rock correlations have become progressively less empirical as a body of chemical knowledge developed yielding chemical natural product precursor relationships. We honor Dr. Wolfgang K. Seifert in this book because his work was the fundamental catalyst for this rapid growth period.

During the early 1970s, Dr. Seifert's interests evolved into the biological marker specialty in Organic Geochemistry. His work on petroleum carboxylic acids led to two landmark publications: One, an article on structure elucidation of steroid carboxylic acids in petroleum (1972), which received the Organic Geochemistry Division's Best Paper Award, and the other, a review of "Carboxylic Acids in Petroleum and Sediments" published in "Progress in the Chemistry or Organic Natural Products" in 1975. It was about this time that Wolfgang started participating actively and most creatively in the biannual Gordon Research Conferences on Organic Geochemistry. However, his goal was to create a niche in the field of biomarker applications to show that they could be used in petroleum exploration, and he published several important articles to demonstrate this application. This work relied heavily on GC-MS, but also on precise structural elucidation of individual biomarkers that was essential from his point of view as an organic chemist. It resulted in the development of applications that helped establish a growth in petroleum-related organic geochemistry research and development throughout the oil industry. Wolfgang expressed great satisfaction at having helped the profession in this way.

The next decade will see a greater cross-linking of various geochemical disciplines. A more unified, more powerful science will result. For example, vitrinite reflectance, a maturity parameter measured only on kerogen can be linked with biomarker maturity parameters measured in bitumen and oil. The link may be enhanced by pyrolysis of the kerogen and comparison of bulk pyrolysate maturity parameters with biomarker maturity ratios in the pyrolysate. Integrated with kerogen kinetics, geology, and geophysics, refined basin models are already beginning to emerge.

Our knowledge of biomarker relationships to biochemical and diagenetic pathways will improve, allowing more precise paleoreconstruction of depositional environments. Study of the relationships between biomarkers in immature recent sediments and mature ancient sediments continues to be important and ongoing research in this area will be enhanced by the measurement of stable isotope ratios on biomarkers. The future payout from this research will be a quantum increase in understanding at the organism class-specific level. Thus, the evolution of some species will be traceable through chemical markers and, once established, this chemical fossil history will be applicable to age-dating petroleum.

These research targets are being realized, in part, through developments in high-technology analytical instruments. Tandem mass spectrometers, high-resolution NMR instruments, improved chromatographic science, and isotope ratio monitoring GC-MS systems head this list.

The recent advances and continued progress in the field of biomarker organic geochemistry, exemplified by the contributions to this book, are a tribute to the legacy of Wolfgang K. Seifert. Most of the contributions to this book were presented at the "Biomarkers in Petroleum-Memorial Symposium" for Wolfgang K. Seifert presented before the Divisions of Petroleum Chemistry and Geochemistry at the National Meeting of the American Chemical Society, Dallas, April 9–14, 1989.

Contributors

ANDJA ALAJBEG, INA, Proleterskih Brigada 78, 41000 Zagreb, Yugoslavia.

PIERRE ALBRECHT, Institut de Chimie, Université Louis Pasteur, 1, rue Blaise Pascal, 67008 Strasbourg, France.

ROBERT ALEXANDER, Centre for Petroleum and Environmental Organic Geochemistry, Curtin University of Technology, Box U1987 GPO, Perth, Western Australia 6001.

DONALD E. ANDERS, U.S. Geological Survey, Box 25046, MS 977, Denver, Colorado 80225, USA.

PHILIPPE BISSERET, Ecole Nationale Supérieure de Chimie de Mulhouse, 3 rue Alfred Werner, F68093 Mulhouse Cedex, France.

CHRISTOPHER J. BOREHAM, Division of Continental Geology, Bureau of Mineral Resources, Geology and Geophysics, GPO Box 378, Canberra, ACT, 2601, Australia.

ROBERT M. K. CARLSON, Chevron Oil Field Research Company, P.O. Box 1627, Richmond, California 94802-0627, USA.

DANIEL E. CHAMBERLAIN, Chevron Oil Field Research Company, P.O. Box 1627, Richmond, California 94802-0627, USA.

PATRICIA CHOSSON, Sanofi-Elf Biorecherches, 31328 Labege Cedex, France.

JACQUES CONNAN, Elf Aquitaine, Cst Jean Feger, 64018 Pau Cedex, France.

HENRICUS CATHARINA COX, Organic Geochemistry Unit, Faculty of Chemistry and Materials Science, Delft University of Technology, De Vries van Heystplantsoen 2, 2628 RZ Delft, The Netherlands.

WILLIAM R. CROASMUN, Kraft Technology Center, 801 Waukegan Road, Glenview, Illinois 60025, USA.

JOSEPH A. CURIALE, Unocal, Inc., P.O. Box 76, Brea, California 92621, USA.

GERARD J. DEMAISON, Chevron Overseas Petroleum, Inc., P.O. Box 5046, San Ramon, California 94583-0946, USA. *Current Address:* P.O. Box 1877, Capitola, California 95010, USA.

DANIEL DESSORT, Elf Aquitaine, Cst Jean Feger, 64018 Pau Cedex, France.

EMILIO J. GALLEGOS, Chevron Research Company, P.O. Box 1627, Richmond, California 94802-0627, USA.

BOGDAN GJUKIĆ, INA, Proleterskih Brigada 78, 41000 Zagreb, Yugoslavia.

HANS LODEWIJK TEN HAVEN, Institute of Petroleum and Organic Geochemistry (ICH-5), KFA Jülich, P.O. Box 1913, D-5170 Jülich 1, Federal Republic of Germany. *Current address:* TOTAL Route de Versallies, 78470 St. Rémy Les Chevreuse, France.

LINDA L. JAHNKE, Planetary Biology Branch, NASA Ames Research Center, Moffett Field, California 94035, USA.

ANTHONY JERVOISE, Organic Geochemistry Unit, School of Chemistry, University of Bristol, Cantock's Close, Bristol BS8 1TS, UK.

ROBERT I. KAGI, Centre for Petroleum and Environmental Organic Geochemistry, Curtin University of Technology, Box U1987 GPO, Perth, Western Australia 6001.

OREST E. KAWKA, Petroleum Research Group, College of Oceanography, Oregon State University, Corvallis, Oregon 97331, USA.

COLETTE LANAU, Sanofi-Elf Biorecherches, 31328 Labege Cedex, France.

ALFONS V. LARCHER, Centre for Petroleum and Environmental Organic Geochemistry, Curtin University of Technology, Box U1987 GPO, Perth, Western Australia 6001.

CATHY Y. LEE, Chevron Oil Field Research Company, P.O. Box 1627, Richmond, California 94802-0627, USA. *Current address:* Chevron Overseas Petroleum, Inc., P.O. Box 5046, San Ramon, California 94583-0946, USA.

JAN WILLEM DE LEEUW, Organic Geochemistry Unit, Faculty of Chemistry and Materials Science, Delft University of Technology, De Vries van Heystplantsoen 2, 2628 RZ Delft, The Netherlands.

RALF LITTKE, Institute of Petroleum and Organic Geochemistry (ICH-5), KFA Jülich, P.O. Box 1913, D-5170 Jülich 1, Federal Republic of Germany.

LESLIE B. MAGOON, U.S. Geological Survey, 345 Middlefield Road/MS 999, Menlo Park, California 94025, USA.

ROGER MARZI, Institute of Petroleum and Organic Geochemistry at the Research Centre (KFA) Jülich, P.O. Box 1913, D-5170 Jülich, Federal Republic of Germany. *Current address:* Centre for Petroleum and Environmental Organic Geochemistry, School of Applied Chemistry, Curtin University of Technology, GPO Box U 1987, Perth, Western Australia, 6001, Australia.

JAMES R. MAXWELL, Organic Geochemistry Unit, School of Chemistry, University of Bristol, Cantock's Close, Bristol BS8 1TS, UK.

J. MICHAEL MOLDOWAN, Chevron Oil Field Research Company, P.O. Box 1627, Richmond, California, 94802-0627, USA.

SERGE NEUNLIST, Ecole Natíonale Supérieure de Chimie de Mulhouse, 3 rue Alfred Werner, F68093 Mulhouse Cedex, France.

JUNG-N. OUNG, School of Geology and Geophysics, University of Oklahoma, Norman, Oklahoma 73019, USA.

TORREN M. PEAKMAN, Organic Geochemistry Unit, Faculty of Chemistry and Materials Science, Delft University of Technology, De Vries van Heystplantsoen 2, 2628 RZ Delft, The Netherlands. *Current address:* Institute of Petroleum and

Organic Geochemistry (ICH-5), KFA-Jülich, P.O. Box 1913, D-5170 Jülich, Federal Republic of Germany.

R. PAUL PHILP, School of Geology and Geophysics, University of Oklahoma, Norman, Oklahoma 73019, USA.

PETER L. PRICE, CSR Oil and Gas, GPO Box 880, Brisbane, Queensland 4001.

JOSEF A. RECHKA, Organic Geochemistry Unit, School of Chemistry, University of Bristol, Cantock's Close, Bristol BS8 1TS, UK.

ADOLFO G. REQUEJO, ARCO Oil & Gas Company, 2300 W. Plano Parkway, Plano, Texas 75075, USA. *Current address:* Exxon Production Research Company, P.O. Box 2189, Houston, Texas 77001, USA.

MICHEL ROHMER, Ecole Natíonale Supérieure de Chimie de Mulhouse, 3 rue Alfred Werner, F68093 Mulhouse Cedex, France.

JÜRGEN RULLKÖTTER, Institute of Petroleum and Organic Geochemistry at the Research Centre (ICH-5), (KFA) Jülich, P.O. Box 1913, D-5170 Jülich, Federal Republic of Germany.

TITO SALVATORI, ENIRICERCHE 20097 S. Donato, Milanese, Italy.

ODETTE SIESKIND, Institut de Chimie, Université Louis Pasteur, 1, rue Blaise Pascal, 67008 Strasbourg, France.

BERND R. T. SIMONEIT, Petroleum Research Group, College of Oceanography, Oregon State University, Corvallis, Oregon 97331, USA.

NACER-EDDINE SLOUGUI, University of Kentucky, Department of Chemistry, Lexington, Kentucky, USA.

ROGER E. SUMMONS, Division of Continental Geology, Bureau of Mineral Resources, Canberra, ACT, 2601, Australia.

PADMANABHAN SUNDARARAMAN, Chevron Oil Field Research Company, P.O. Box 446, La Habra, California 90633-0466, USA.

SYLVIE TRIFILIEFF, Institut de Chimie, Université Louis Pasteur, 1, rue Blaise Pascal, 67008 Strasbourg, France.

GONG-MING WANG, Petroleum Research Group, College of Oceanography, Oregon State University, Corvallis, Oregon 97331, USA.

DAVID S. WATT, University of Kentucky, Department of Chemistry, Lexington, Kentucky, USA.

GEORGE A. WOLFF, Organic Geochemistry Unit, School of Chemistry, University of Bristol, Cantock's Close, Bristol BS8 1TS, UK. *Current address:* Department of Earth Sciences, Oceanography Laboratories, Bedford Street North, P.O. Box 147, Liverpool L69 3BX, UK.

Acknowledgments

The editors would like to thank the following people for their help in reviewing chapters for this book:

Robert Alexander
Christopher J. Boreham
Robert M. K. Carlson
Jacques Connan
Joseph A. Curiale
Jeremy E. Dahl
Dick Drozd
Emilio J. Gallegos
Peter Grantham
Francis Xavier de las Heras
William B. Hughes
Bradley J. Huizinga
Ian Kaplan
Jean-Pierre Kintzinger
Steve Larter
James Maxwell
Beverly L. McFarland
Walter Michaelis
Richard Patience
Kenneth E. Peters
J. Martin E. Quirke
Michel Rohmer
Jürgen Rullkötter
Bernd Simoneit
Padmanabhan Sundararaman
Clifford Walters
David S. Watt
John E. Zumberge

JMM thanks Alison L. Whitlock for handling many details that made this book feasible.

1

The Hopanoids, Prokaryotic Triterpenoids and Precursors of Ubiquitous Molecular Fossils

Michel Rohmer, Philippe Bisseret, and Serge Neunlist

Abstract. Although hopanoids were known first as simple C_{30} pentacyclic triterpenoids from few higher plants and cryptogams, they were more recently recognized as a wider fascinating family of prokaryotic lipids. Indeed the C_{35} bacteriohopane derivatives presenting the unique feature of a carbon/carbon bond between a triterpene and a D-ribose unit characterize solely bacteria and cyanobacteria and include triterpenoids linked to all major families of natural products (amino acids, sugars, nucleosides). These compounds are apparently essential metabolites for the bacteria synthesizing them, acting basically as membrane reinforcers. According to the distribution and the structures of the geohopanoids found in every sedimentary rock, it is clear that these compounds are the molecular fossils of the bacterial hopanoids. The narrow gap between bio- and geohopanoids is under investigation and could be already partially filled using a geomimetic chemical oxidation of the hopane framework.

INTRODUCTION

Triterpenoids of the hopane series are probably the most abundant complex organic compounds of the geosphere. Indeed, their global stock could be estimated at least at 10^{13} or 10^{14} tons (Ourisson et al., 1984). Most of these compounds are buried in the organic matter of the sedimentary rocks, whatever their age, nature, or origin. Extensive research started around the early 1970s led to the identification of more than 150 geohopanoids, covering a broad spectrum of organic compounds with various functional groups (Ourisson et al., 1979; Brassel et al., 1983; Ourisson et al., 1984). The most striking structural feature of these geohopanoids is the presence of an additional side chain having up to five carbon atoms linked to the isopropyl group of the hopane skeleton. Thus, the long known C_{30} hopanoids isolated from scattered higher plant taxa and numerous cryptogams such as ferns, mosses, lichens, and a few filamentous fungi (Ourisson et al., 1987) could not account for the widespread occurrence of these geohopanoids. According to their ubiquity, Bird and Reid on the one hand (Bird et al., *Tetrahedron Lett.*, 1971) and Ourisson and co-workers on the other hand (Ensminger et al., 1972) proposed that these triterpenoids were indeed molecular fossils of yet unknown lipids from microorganisms widespread in the biosphere and without specific distribution. This assertion was already supported by findings of hop-22(29)-ene **2** (Fig. 1.1), a C_{30} olefin, in three cyanobacteria (Gelpi et al., 1970), in the thermoacidophilic bacterium *Bacillus acidocaldarius* (De Rosa et al., 1971) and in the obligate methylotroph *Methylococcus capsulatus* (Bird et al., 1971) and proved later on to be correct after isolation of the C_{35} bacteriohopane derivatives from numerous bacteria. The developments of the chemistry and the biochemistry of the biohopanoids, a long completely overlooked family of important prokaryotic lipids are presented emphasizing whenever possible the links between bio- and geohopanoids.

THE STRUCTURAL VARIETY OF PROKARYOTIC HOPANOIDS

The C_{30} Hopanoids: Diploptene and Diplopterol

Two C_{30} hopanoids, diploptene **2** and diplopterol **3** (Fig. 1.1), already isolated a long time ago from several ferns, are always present, usually as minor compounds, in all hopanoid producing bacteria (Rohmer et al., 1984). Next to diploptene or diplopterol several other pentacyclic triterpenes have been repeatedly reported as minor compounds: for example, diploptene isomers or their C_{31} homologs from *Bacillus acidocaldarius* (De Rosa et al., 1973), *Rhodomicrobium vannielii* (Howard and Chapman, 1981), and *Zymomonas mobilis* (Barrow et al., 1983) or even olefins belonging to other triterpenic series such as those described from *Zymomonas mobilis* (Tornabene et al., 1982). The amounts were so low or the mixtures so complex that they could not be excluded in the absence of any blank experiment that some of these triterpenes could arise from chemical alteration of diploptene or diplopterol during the culture of the microorganisms or the isolation procedure or from accidental contaminations. However, they might deserve more attention since some of them might represent important intermediates in the biosynthetic pathway leading to bacterial hopanoids.

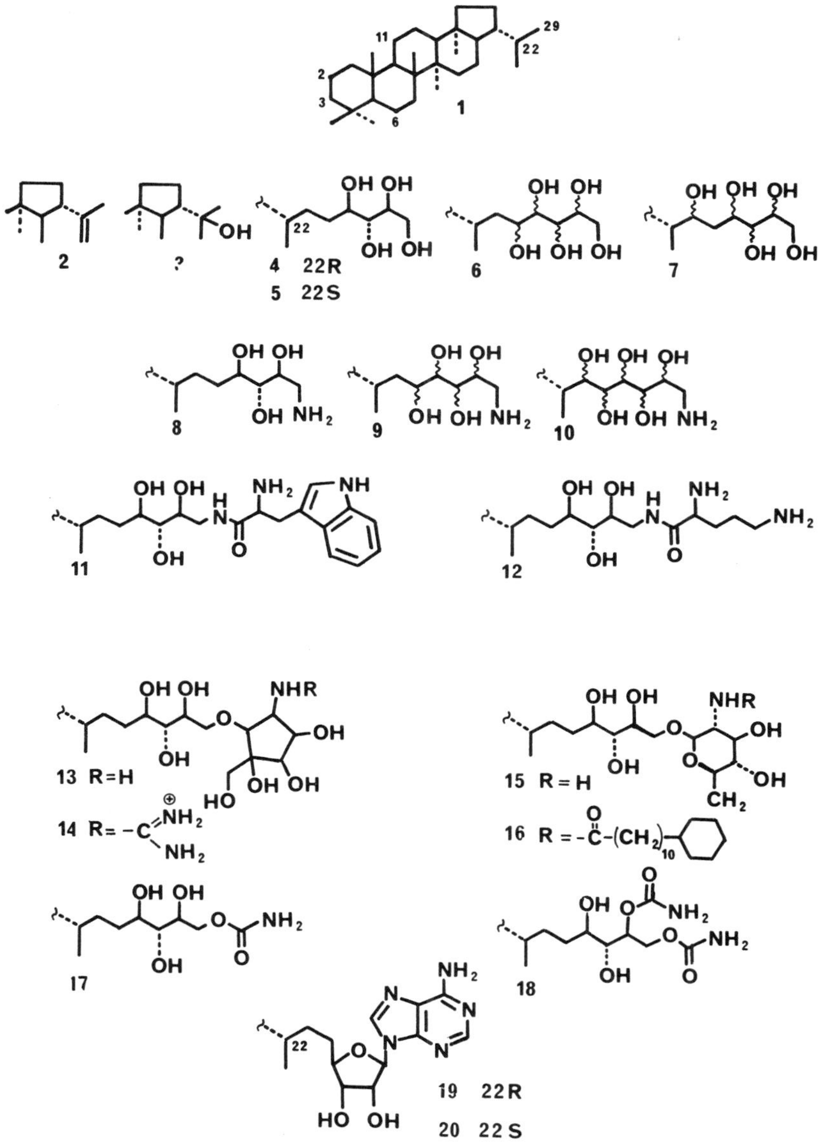

Figure 1.1 Hopane **1** and prokaryotic hopanoids **2** to **20**.

Polyols and Aminopolyols

Trying to find out the compound responsible for the orientation of the cellulose microfibrils produced by the bacterium *Acetobacter aceti* ssp. *xylinum*, the groups of Biemann and Colvin isolated the first hopanoids with an extended side chain (Förster et al., 1973, Haigh et al., 1973). Their basic C_{35} framework was formed by the C_{30} pentacyclic ring system linked to a polyhydroxylated *n*-alkyl side chain and was called bacteriohopane by the first authors. Two potential configurations were proposed. However, one of them could not account for the role of bacteriohopane derivatives as precursors for geohopanoids. We showed by chemical correlation that the other possible structure which was compatible with those of the geohopanoids was the correct one (Rohmer, 1975; Rohmer and Ourisson, 1976a), and that the *Acetobacter* hopanoids were indeed a complex mixture of triterpenoids which were the first ones from a long series of new natural products (Rohmer, 1975; Rohmer and Ourisson, 1976b, 1976c). The most important structural variations are listed as follows:

1. In all hopanoids isolated so far from bacteria, the stereochemistry is always 17β(H) and 21β(H) indicating that the molecular fossils with 17β(H), 21α(H) or 17α(H), 21β(H) configurations arise most probably from the isomerization of biohopanoids via diagenetic processes as emphasized (Ensminger et al., 1976; Dastillung and Albrecht, 1976; Seifert, 1978; Mackenzie et al., 1980).
2. The side-chain stereochemistry at C-22 is in most cases unique and identical in nearly all bacteriohopane derivatives, that is, (22*R*) for bacteriohopanetetrol **4** or aminobacteriohopanetriol **8** (Fig. 1.1) (Rohmer and Ourisson, 1976a; Zundel and Rohmer, 1985; Neunlist et al., 1988) and (22*S*) for aminobacteriohopanepentol **10** (Neunlist and Rohmer, 1985b). The three former hopanoids have all the same C-22 stereochemistry, the nomenclatural difference resulting only from different priorities of the substituents at the asymmetric center according to the Cahn-Ingold-Prelog rules. Significant amounts of the other epimer, (22*S*)-bacteriohopanetetrol **5**, are only present in the acetic acid bacteria (Rohmer and Ourisson, 1976a; Rohmer et al., 1984).
3. Bacteriohopanetetrols possessing Δ^6 and/or Δ^{11} double-bonds are regularly present in all *Acetobacter* spp. as well as in a few other scattered taxa (cyanobacteria, methylotrophs) (Rohmer and Ourisson, 1976b; Rohmer et al., 1984). However, the presence of hopanoids tentatively identified on the basis of their mass spectra as Δ^6-trisnorhopan-21-one (ten Haven et al., 1987) in eastern Mediterranean late Quaternary sediments and of a Δ^6-C_{32} hopanoid carboxylic acid in a peat (Quirk et al., 1984) along with their saturated counterparts might reflect a far more common occurrence of these unsaturated hopanoids in prokaryotes than deduced from the few existing reports.
4. An additional methyl group may be present on ring A, either at C-2β leading to compounds with gas chromatography (GC) retention times nearly identical

on most columns to those of their nonmethylated analogs (Bisseret et al., 1985; Babadjamian et al., 1984) or at C-3β resulting in homologs with longer retention times (Rohmer and Ourisson, 1976c; Zundel and Rohmer, 1985a, 1985b). Both methylated series have been detected in sediments (Dastillung et al., 1980; Summons and Capon, 1988).

5. Concerning the structure of the polar side-chain, tetrol **4** and aminotriol **8** are the most common bacteriohopane moieties found until now (Rohmer et al., 1984). Synthesis of the eight side-chain stereoisomers of bacteriohopanetetrol (Bisseret and Rohmer, 1989), chemical correlation with adenosylhopane **19** (Neunlist et al., 1988), stereochemical analysis of aminotriol derivatives (Neunlist and Rohmer, 1988) as well as incorporation of ^{13}C labeled acetate and glucose into the hopanoids of several bacteria (Flesch and Rohmer, 1988; Rohmer et al., 1989) showed that the additional five carbon atoms derive from D-ribose which is linked via its C-5 carbon atom to the hopane framework by a carbon/carbon bond. More oxidized side chains corresponding to pentols **6** and **7** or aminotetrol **9** and aminopentol **10** (Fig. 1.1) accompany often at least as minor compounds the two former hopanoids (Rohmer et al., 1984) or are even the major compounds in some cyanobacteria or methylotrophs (Rohmer and Ourisson, 1976b; Bisseret et al., 1985; Neunlist and Rohmer, 1985a, 1985b).

Composite Hopanoids: Glycosides, Ethers, Peptides, Nucleosides

In recent years, we have carefully studied several strains of hopanoid producing bacteria. Free polyols or aminopolyols were only present in two *Acetobacter* species (Rohmer et al., unpublished results), in methylotrophs (Neunlist and Rohmer, 1985a, 1985b) and in *Rhodopseudomonas palustris* (Neunlist et al., 1988). In all other bacteria, the bacteriohopanepolyols are linked to polar moieties. As early as 1976, a N-acylglycoside of bacteriohopanetetrol **16** has been isolated from *Bacillus acidocaldarius* (Langworthy and Mayberry, 1976; Langworthy et al., 1976). During the last few years, we have identified novel groups of composite hopanoids where the bacteriohopane skeleton is linked to moieties belonging to the major groups of natural products: amino acids, sugar derivatives, and nucleosides.

Thus in *Rhodomicrobium vannielii* aminotriol is linked via a peptide bond to tryptophane or ornithine (**11**,**12**) (Neunlist, et al., 1985). In *Methylobacterium organophilum* (Renoux and Rohmer, 1985), *Zymomonas mobilis* (Renoux and Rohmer, 1985; Flesch 1987; Flesch and Rohmer, 1989) or *Rhodopseudomonas acidophila* (Neunlist et al., 1988) bacteriohopanetetrol is linked to glucosamine via a glycosidic bond (**15**) or to novel carbocyclic pentose analogues via an ether bond (**13**,**14**). Finally, in *Rhodopseudomonas acidophila* bacteriohopanetetrol forms mono-and dicarbamoyl derivatives (**17** and **18**), and the hopane framework is linked by a C-30/C-5′ bond to adenosine in the 22*R* and 22*S* series (Neunlist and Rohmer, 1985c; Neunlist et al., 1988).

HOPANOID DISTRIBUTION IN PROKARYOTES

Detection of Bacteriohopane Derivatives

In spite of the abundance of works devoted to bacterial metabolites, the hopanoids, and especially the composite derivatives, have been completely overlooked by nearly all authors. This raises clearly the problems involved in handling these highly amphipathic molecules. As free compounds, they are poorly soluble in most organic solvents and can not be recovered with satisfactory yields from silica gel after thin-layer chromatography (TLC) or column chromatography. Successful isolation could only be performed after acetylation (Neunlist and Rohmer, 1985a, 1985b, 1985c; Renoux and Rohmer, 1985). Indeed the acetylated derivatives are easy to handle and useful for spectroscopic identifications. They permitted, for instance, to identify by ^{1}H-NMR spectroscopy the hopanoid containing fractions obtained after chromatography by the characteristic methyl singlets of the hopane skeleton and of the acetoxy groups.

The presence of several different bacteriohopane derivatives in the same microorganism, for example, up to six in the purple bacterium *Rhodopseudomonas acidophila* (Neunlist et al., 1988), and the lack of any specific detection method of the native compounds or their acetylated derivatives led us to choose an analytical method which was fully independent of the structure of the side-chain (Rohmer,

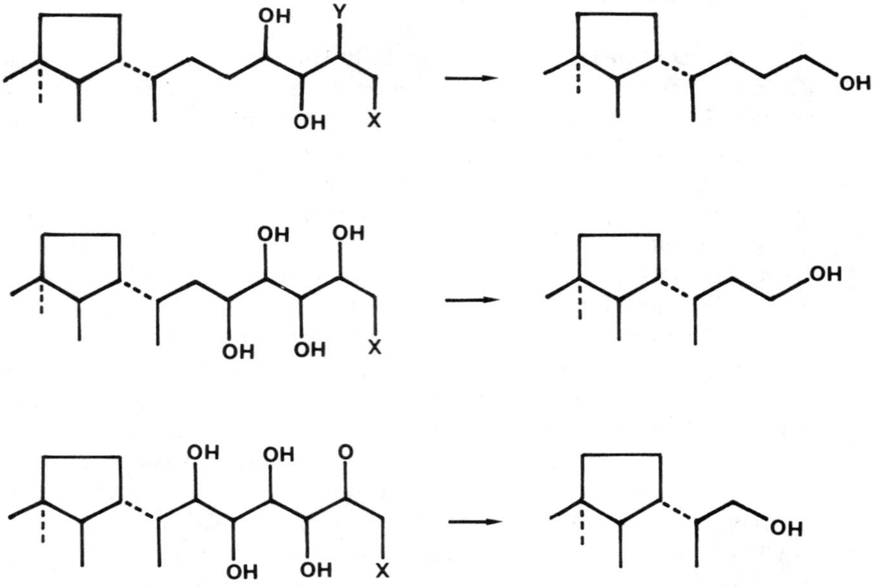

OH Y OH X OH OH OH OH X OH OH OH O OH OH X OH

Figure 1.2 Degradation ($H_5IO_6/NaBH_4$) of side chains of bacteriohopane derivatives.

1975; Rohmer et al., 1984). Thus the crude $CHCl_3$–CH_3OH (2:1) extract containing the hopanoids is successively treated with H_5IO_6 and $NaBH_4$ (Fig. 1.2). Provided that two vicinal hydroxy groups are present, all bacteriohopane derivatives known until now with the exception of adenosylhopanes **19** and **20** and bacteriohopanepentol **7** give the same primary alcohols easily isolated by TLC and analyzed by GC and GC-mass spectroscopy (MS). Although much information is lost concerning the structure of the side chain, this is the simplest and fastest method to prove or disprove the presence of most of the bacteriohopane derivatives in a prokaryote. This allows the simultaneous quantification of diploptene, diplopterol, as well as of bacteriohopanepolyols.

Distribution of Hopanoids

Although about 120 strains have been studied for the presence of hopanoids (Rohmer et al., 1984; Ourisson et al., 1987), no clear-cut trends in the distribution of this triterpenoid series among bacteria can be drawn. The selected species representing only a few among the multitude of strains already described and the present distribution (Table 1.1), revealing perhaps rather the preferences of the investigator and the availability of the strains in collections. This shows, however, that hopanoids are produced by numerous different prokaryotic taxa (Table 1.1). They can be found in Gram negative as well as in Gram positive bacteria, most often in soil and fresh water microorganisms living free in the environment and which are of geochemical interest (e.g., cyanobacteria, Rhodospirillaceae, methylotrophs, some *Streptomyces* spp.) or in bacteria tolerating high-ethanol concentrations (e.g., *Acetobacter* spp. or *Zymomonas mobilis*). They are apparently absent in all analyzed symbiotic or parasitic bacteria (*e.g.*, enterobacteria). This is perhaps related to the protected environment of these bacteria which might not require hopanoids as membrane reinforcers.

Concerning the negative results, it has to be pointed out on the one hand that they have been obtained with a single analytical method. Hopanoids possessing no vicinal hydroxy groups such as adenosylhopanes **19** and **20** or pentol **7** can not be detected as primary alcohols after $H_5IO_6/NaBH_4$ treatment, and hopanoids bound to polymer (peptides, polysaccharides) might not be extracted by $CHCl_3/CH_3OH$. On the other hand, these results have been obtained using specific growth conditions. Since it has been observed that the hopanoid content of the cells is highly dependent on pH, temperature, presence of alcohols, and age of the cultures, it might be possible that hopanoid biosynthesis was not expressed in some of our cultures.

Finally, it has to be emphasized that the presence of bacteriohopane derivatives is restricted to prokaryotes. These triterpenoids do not represent minor metabolites. Their concentration in bacterial cells is of the same order of magnitude as those reported for sterols in eukaryotes, that is, 0.1 to 3 mg per g (dry weight) (Rohmer et al., 1984). Sterols were absent in nearly all strains we have analyzed, or when they were present, this was in amounts of the same order of magnitude

Table 1.1 DISTRIBUTION OF HOPANOIDS IN PROKARYOTES

Cyanobacteria	Rhodospirillaceae	Methylotrophs	Gram negative chemoautotrophs	Gram negative chemoheterotrophs	Gram positive chemoheterotrophs
Anabaena sp.	*Rhodomicrobium vannielii* **8**,**11**,**12**	*Methylomonas albus*	*Nitrosomonas europaea*	*Acetobacter* spp. (11 strains) **4**,**5**	*Bacillus acidocaldarius* **4**,**15**,**16** (Langworthy, et al., 1976)
Calothrix sp.	*Rhodopseudomonas acidophila* **4**,**13**,**17**, **18**,**19**,**20**	*Methylomonas methanica* **9**,**10**		*Azotobacter* sp. (2 species)	*Bacillus acidoterrestris* (Deinhard, et al., 1987)
Fischerella sp.	*Rhodopseudomonas palustris* **8**	*Methylomonas* sp.		*Gluconobacter oxydans*	*Corynebacterium* sp. (Babadjamian, et al., 1984)
Lyngbia/Phormidium/ Plectonema group	*Rhodospirillum rubrum*	*Methylocystis parvus*		*Hyphomicrobium* sp.	*Streptomyces* sp.
Nostoc sp. **4**,**6**,**7**		*Methylosinus sporium*		*Methylobacterium organophilum* **4**,**13**,**14**,**15**	(5 species)
Scytonema sp.		*Methylosinus trichosporium* **8**,**9**		*Pseudomonas cepacia*	*Eubacterium limosum*
Synechocystis sp.		*Methylococcus capsulatus* **8**,**9**,**10**		*Pseudomonas syringae*	
				Pseudomonas C45 (Natori, et al., 1981)	
				Acetobacter methanolicus (Vier and Voigt, 1986)	

Unless otherwise indicated, all analyses by Rohmer and coworkers. Structures when known, are included, and refer to Figure 1.1. In all other cases, the presence of bacteriohopane derivatives was detected after degradation of the side-chain (Fig. 1.2).

as those found in blank experiments (Bouvier, 1978). We could show that in several strains reported to contain sterols, the greater the precautions against contaminations, the lower the sterol content in the analyses. Although sterols have been repeatedly described from prokaryotes, there is usually no proof concerning their *de novo* biosynthesis, no unambiguous characterization of the radiopurity in case of labeling experiments or no mention of blank experiments to eliminate the possible introduction of contaminants. There are until now only two bacteria for which *de novo* sterol biosynthesis has been clearly documented: the methylotroph *Methylococcus capsulatus* (Bird et al., 1971; Bouvier et al., 1976; Rohmer et al., 1980) and the gliding bacterium *Nannocystis exedens* (Kohl et al., 1983).

HOPANOIDS AS PROKARYOTIC MEMBRANE CONSTITUENTS

Are hopanoids essential metabolites for the bacteria producing them or are they just secondary metabolites without crucial importance? What happens when their biosynthesis is blocked? Do the bacteria grow further, or is their growth inhibited? Only one enzyme of the hopanoid biosynthetic pathway is fairly well known. We characterized the activity of the squalene cyclase catalyzing the cyclization of squalene **21** into diplopterol **3** and diploptene **2** (Fig. 1.3) in several cell-free systems (Anding et al., 1976; Rohmer et al., 1980a, 1980b; Bouvier et al., 1980). This enzyme has been recently purified and isolated independently from *Bacillus acidocaldarius* by two German groups (Neumann and Simon, 1986; Seckler and Poralla, 1986). It is strongly inhibited by squalene analogues such as 2,3-dihydro-2-azasqualene **22** or 2,3-dihydro-2,3-epiminosqualene **23** (Fig. 1.3), known as potent inhibitors of squalene oxide cyclases of eukaryotes (Corey et al., 1963; Duriatti et al., 1985; Flesch and Rohmer, 1987). At 0.1 μM inhibitor concentration in cell-

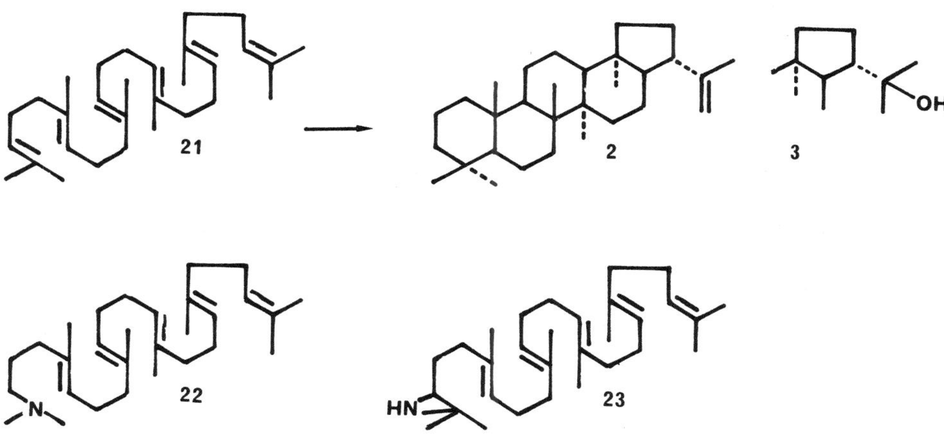

Figure 1.3 Cyclization of squalene **21** into diploptene **2** and diplopterol **3** and squalene cyclase inhibitors **22** and **23**.

free systems, we observed 50 percent inhibition of enzymatic activity. When these inhibitors were added to the culture medium, they inhibited specifically the growth of the hopanoid producers at low concentrations (around 1 μM), whereas the growth of bacteria containing no detectable hopanoids was not affected even at the highest concentration tested (e.g., 200 μM). This suggests that hopanoids are essential metabolites for the bacteria synthesizing them.

The structural analogies between bacteriohopanetetrol and cholesterol led us to postulate ten years ago similar functional roles for both polyterpenoid series (Rohmer et al., 1979). Indeed both are amphiphilic molecules, possessing each a rigid planar ring system and similar dimensions with a cross section enabling close packing with acyl chains of phospholipids and a length corresponding to the half of the section of a phospholipid bilayer, fulfilling the requirements for membrane stabilizers (Demel and De Kruyff, 1976). This assertion has been later on supported by experimental proofs obtained on artificial membrane models as well as by experiments on whole bacterial cells (Ourisson et al., 1987, and references cited therein). On the one hand, sterols as well as bacteriohopane derivatives modify much in the same way the properties of phospholipid mono- and bilayers, quenching the gel to liquid crystal phase transition and modulating their fluidity and permeability. On the other hand, evidences have been obtained concerning the role of hopanoids as membrane constituents. When the ciliate *Tetrahymena pyriformis* is grown in the absence of sterols, it synthesizes a quasi hopanoid, tetrahymanol, which is incorporated in the membranes. When grown in the presence of sterols, tetrahymanol biosynthesis is blocked and sterols are present in the membranes (Conner et al., 1968). This was the first evidence for the equivalence of two different triterpenoid series as membrane stabilizers. The sterol requirement of the parasitic prokaryote *Mycoplasma capricolum* can be fulfilled by a hopanoid, diplopterol **3** (Kannenberg and Poralla, 1982). The tolerance towards high temperature of *Bacillus acidocaldarius* (Poralla et al., 1980) or towards high ethanol concentrations for *Zymomonas mobilis* (Bringer et al., 1985; Schmidt et al., 1986) is modulated by bacteriohopanetetrol derivatives. The higher the growth temperature for the former bacterium or the higher the ethanol content of the culture medium for the latter, the higher the hopanoid proportion in the total lipids of these bacteria. This suggests that these bacteria are capable of counterbalancing the destabilizing influence of the temperature or the solvent by increasing the amounts of membrane reinforcers. Finally, it could be recently shown that hopanoids are apparently located in the former Gram negative bacterium *Zymomonas mobilis* in the inner as well as in the outer membrane (Tahara et al., 1988).

DEGRADATION OF BIOHOPANOIDS: THE ROUTE TO GEOHOPANOIDS

Free biohopanoids have been detected in several recent sediments: diploptene **2** and diplopterol **3** in quaternary freshwater and marine deposits (Rohmer et al., 1980; Philp, 1985; ten Haven et al., 1987; Venkatesan, 1988) and even bacteriohopanetetrol **4** in recent freshwater muds (Rohmer et al., 1980), cyanobacterial mats (Boon et al., 1983) or in an organic rich soil (Ries-Kautt, 1986; Ries-Kautt and Albrecht, 1989). Free tetrol has not been isolated from any older sediments. Even in the Eocene Messel oil shale which has undergone a very mild diagenesis, this hopanoid could not be detected in the organic solvent soluble fraction (Rohmer et al., 1980). However, intact bacteriohopanetetrol was released from the kerogen of this shale by hydrogenolysis on a rhodium/charcoal catalyst (Mycke et al., 1987), showing that the polymeric matrix of the kerogen could protect the polyol from further transformation. All these observations are in accordance with a probable rapid degradation of the polyfunctionalized side chains of all bacteriohopane derivatives, leaving nearly untouched the pentacyclic nucleus. Nothing is known concerning the catabolism or the abiotic degradation of the triterpenoids of the hopane series. However, some hints can be deduced from the structures of oxidized hopanoids isolated from living organisms or obtained by chemical oxidation (Fig. 1.4). This shows that oxidation reactions of the hopane skeleton do exist and might be relevant to explain the formation of geohopanoids (e.g., aromatization, demethylation at C-18, epimerization at C-17 and C-21). Thus, some bacteria (*Acetobacter* spp., *Nostoc* spp., *Pseudomonas* sp. and *Methylosinus* sp.) are capable of introducing Δ^6 or Δ^{11} double bonds on the hopane skeleton. Filamentous fungi and lichens (most probably their mycobionts) oxidize nonactivated carbon atoms (Fig. 1.4) leading to alcohols and carboxylic acids (Corbett and Young, 1966; Corbett and Cumming, 1971; Huneck, 1971; Yosioka et al., 1972; Ejiri and Shibata, 1974; Hveding-Bergseth et al., 1983; Van Eijk et al., 1986).

Finally the hopane framework can also be easily modified by mild abiotic reactions. Peracid treatment of saturated 17β(H), 21β(H)-hopanoids led to hydroxylations either at C-17 or at C-21 (Neunlist, 1987; Rohmer et al., unpublished results). These positions appear as specially fragile and correspond to the centers where epimerization occurs during diagenesis and maturation of the sediment. Indeed the former tertiary alcohols were readily dehydrated even under mild conditions (e.g., on standing in $CHCl_3$ solution) leading quantitatively to $\Delta^{17(21)}$-hopanoids which are already known from sediments (Ensminger, 1977; Ourisson et al., 1979; Meunier-Christmann, 1988) and are possible precursors after hydrogenation of the 17α(H), 21β(H)-geohopanoids.

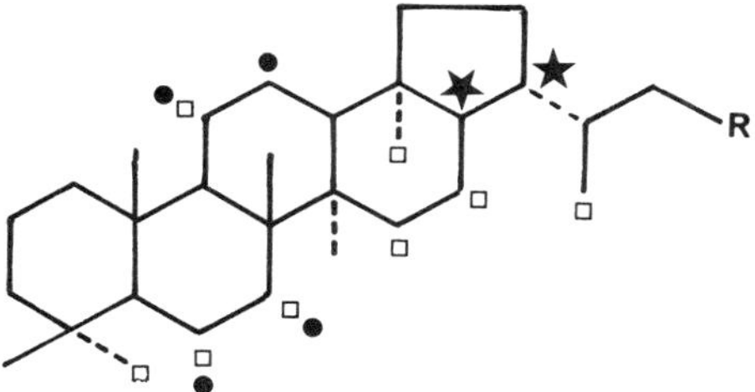

Figure 1.4 Oxidation of the hopane skeleton: □ by fungi and lichens (hydroxylation), • by bacteria (introduction of double bonds), and ★ by abiotic reagents (hydroxylation by peracids).

CONCLUSION

According to the structures and the wide distribution of triterpenoids of the hopane series in prokaryotes it is now clear that geohopanoids from the organic matter of sediments are molecular fossils of a long undisclosed class of microbial lipids. However, contrary to the first assumptions, bacteriohopanetetrol (4) which has been detected in a few sediments is not their only precursor. Indeed from our current knowledge of hopanoid distribution, aminobacteriohopanetriol (8) or composite bacteriohopane derivatives (Fig. 1.1) are as likely precursors as the former tetrol.

Since in most cases there is no clear cut correlation between the structure of a biohopanoid and the belonging of the bacterium which synthesized it to a determined taxonomic group, it is not possible to draw from the structures of fossil hopanoids any conclusion concerning the prokaryotes living in the environment where the sediment deposited, the more as all information contained in the structures of the complex side chains have been lost. C_{29} and C_{30} hopanoids with ethyl or isopropyl side chains might however represent an exception, arising perhaps predominantly either from aminobacteriohopanepentol (10) or from aminobacteriohopanetetrol (9) and revealing thus the presence of obligate methylotrophs, or eventually from bacteriohopanepentol (7) and indicating the intervention of cyanobacteria (Zundel and Rohmer, 1985).

Finally, in the many prokaryotes we have analyzed, no polycyclic terpenoids other than hopanoids (and in two cases sterols) could be detected. Thus, other polycyclic isoprenoids that are widespread in sediments, such as tricyclopolyprenol derivatives or isoarborinol, are still orphan molecular fossils and waiting for the identification of their precursors.

Acknowledgments

This work on biohopanoids has been supported by the Centre National de la Recherche Scientifique (Unité de Recherche Associée 135) and by the Ministère de l'Education Nationale (Réseau Européen de Laboratoires).

REFERENCES

Anding, C., Rohmer, M., and Ourisson, G. (1976) Non-specific biosynthesis of hopane triterpenes in a cell-free system from *Acetobacter rancens*. *J. Am. Chem. Soc. 98*, 1274–1275.

Babadjamian, A., Faure, R., Laget, M., Dumenil, G., and Padieu, P. (1984) Occurrence of triterpenoids in methanol-oxidizing bacteria. 2-Methyl-22-hydroxyhopane from *Corynebacterium*. *Chem. Commun.* 1657–1658.

Barrow, K.D., Collins, J.G., Rogers, P.L., and Smith, G.M. (1983) Lipid composition of ethanol-tolerant strain of *Zymomonas mobilis*. *Biochim. Biophys. Acta 753*, 324–330.

BIRD, C.W., LYNCH, J.M., PIRT, J.R., and REID, W.W. (1971) The identification of hop-22(29)-ene in prokaryotic organisms. *Tetrahedron Lett.* 3189–3190.

BIRD, C.W., LYNCH, J.M., PIRT, S.J., REID, W.W., BROOKS, C.J.W., and MIDDLEDITCH, B.S. (1971) Steroids and squalene in *Methylococcus capsulatus* grown on methane. *Nature 230*, 473–474.

BISSERET, P. and ROHMER, M. (1989) Bacterial sterol surrogates. Determination of the absolute configuration of bacteriohopanetetrol side-chain by hemisynthesis of its diastereoisomers. *J. Org. Chem.*, in press.

BISSERET, P., ZUNDEL, M., and ROHMER, M. (1985) Prokaryotic triterpenoids. 2. 2β-Methylhopanoids from *Methylobacterium organophilum* and *Nostoc muscorum*, a new series of prokaryotic triterpenoids. *Eur. J. Biochem. 150*, 29–34.

BOON, J.J., HINES, H., BURLINGAME, A.L., KLOK, J., RIJPSTRA, W.I.C., DE LEEUW, J.W., EDMUNDS, K.E., and EGLINTON, G. (1983) Organic geochemical studies of Solar lake laminated cyanobacterial mats. In *Advances in Organic Geochemistry* (eds. M. Bjorøy et al.), pp. 207–227, J. Wiley & Sons, Chichester.

BOUVIER, P. (1978) Biosynthèse de stéroïdes et de triterpénoïdes chez les procaryotes et les eucaryotes. Ph. D. dissertation, Université Louis Pasteur, Strasbourg, France.

BOUVIER, P., BERGER, Y., ROHMER, M., and OURISSON, G. (1980) Non-specific biosynthesis of gammacerane derivatives by a cell-free system from *Tetrahymena pyriformis*. *Eur. J. Biochem. 112*, 549–556.

BOUVIER, P., ROHMER, M., BENVENISTE, P., and OURISSON, G. (1976) $\Delta^{8(14)}$-Steroids in the bacterium *Methylococcus capsulatus*. *Biochem. J. 159*, 267–271.

BRASSEL, S.C., EGLINTON, G., and MAXWELL, J.R. (1983) The geochemistry of terpenoids and steroids. *Biochem. Soc. Trans. 11*, 575–586.

BRINGER, S., HÄRTNER, T., PORALLA, K., and SAHM, H. (1985) Influence of ethanol on the hopanoid content and the fatty acid pattern in batch and continuous cultures of *Zymomonas mobilis*. *Arch. Microbiol. 140*, 312–316.

CONNER, R.L., LANDREY, J.R., BURNS, C.H., and MALLORY, F.B. (1968) Cholesterol inhibition of pentacyclic triterpenoid biosynthesis in *Tetrahymena pyriformis*. *J. Protozool. 15*, 600–605.

CORBETT, R.E. and CUMMING, S.D. (1971) Lichens and fungi. Part VII. Extractives from the lichen *Sticta mougeotiana* var. *dissecta* Del. *J. Chem. Soc.* (C) 955–960.

CORBETT, R.E. and YOUNG, H. (1966) Lichens and fungi. Part II. Isolation and structural elucidation of 7β-acetoxy-22-hydroxyhopane from *Sticta billardierii* Del. *J. Chem. Soc.* (C) 1556–1563.

COREY, E.J., ORTIZ DE MONTELLANO, P.R., LIN, K., and DEAN, P.D.G. (1963) 2,3-Iminosqualene, a potent inhibitor of enzymic cyclization of 2,3-oxido-squalene to sterols. *J. Am. Chem. Soc. 89*, 2797–2798.

DASTILLUNG, M. and ALBRECHT, P. (1976) Molecular test for oil pollution in surface sediments. *Bull. Mar. Pollut. 7*, 13–15.

DASTILLUNG, M., ALBRECHT, P., and OURISSON, G. (1980) Aliphatic and polycyclic alcohols in sediments: Hydroxylated derivatives of hopane and of 3-methylhopane. *J. Chem. Res.* (S), 168–169.

DEINHARD, G., BLANZ, P., PORALLA, K., and ALTAN, E. (1987) *Bacillus acidoterrestris* sp. nov., a new thermotolerant acidophile isolated from different soils. *System. Appl. Microbiol. 10*, 47–53.

DEMEL, R.A. and DE KRUYFF, B. (1976) The function of sterols in membranes. *Biochim. Biophys. Acta 457*, 109–132.

DE ROSA, M., GAMBACORTA, A., MINALE, L., and BU'LOCK, J.D. (1971) Bacterial Triterpenes. *Chem. Commun.* 619–620.

DE ROSA, M., GAMBACORTA, A., MINALE, L., and BU'LOCK, J.D. (1973) Isoprenoids from *Bacillus acidocaldarius. Phytochemistry 12*, 1117–1123.

DURIATTI, A., BOUVIER-NAVE, P., BENVENISTE, P., SCHUBER, F., DELPRINO, L., BALLIANO, G., and CATTEL, L. (1985) In vitro inhibition of animal and higher plant 2,3-oxidosqualene-sterol cyclases by 2-aza-2,3-dihydrosqualene and derivatives and other ammonium containing molecules. *Biochem. Pharmacol. 34*, 2765–2777.

EJIRI, H. and SHIBATA, S. (1974) Zeorin from the mycobiont of *Anaptychia hypoleuca. Phytochemistry 13*, 2871.

ENSMINGER, A. (1977) Evolution de composés polycycliques sédimentaires. Ph. D. Dissertation, Université Louis Pasteur, Strasbourg, France.

ENSMINGER, A., ALBRECHT, P., OURISSON, G., and TISSOT, B. (1976) Evolution of polycyclic alkanes under the effect of burial (Early Toarcian shales, Paris Basin). In *Advances in Organic Geochemistry* (eds. R. Campos and J. Goni) pp. 42–45, Enadimsa, Madrid.

ENSMINGER, A., ALBRECHT, P., OURISSON, G., KIMBLE, B.J., MAXWELL, J.R. and EGLINTON, G. (1972) Homohopane in Messel oil shale: first identification of a C_{31} pentacyclic triterpane in nature. *Tetrahedron Lett.* 3861–3864.

FLESCH, G. (1987) Biosynthèse des bactériohopanepolyols, triterpénoïdes bactériens. Inhibiteurs de la squalène-cyclase de procaryotes. Ph. D. Dissertation, Université de Haute Alsace, Mulhouse, France.

FLESCH, G. and ROHMER, M. (1987) Growth inhibition of hopanoid synthesizing bacteria by squalene cyclase inhibitors. *Arch. Microbiol. 147*, 100–104, and references cited therein.

FLESCH, G. and ROHMER, M. (1988) Prokaryotic hopanoids: The biosynthesis of the bacteriohopane skeleton. Formation of isoprenic units from two different acetate pools and a novel type of carbon/carbon linkage between a triterpene and D-ribose. *Eur. J. Biochem. 175*, 405–411.

FLESCH, G. and ROHMER, M. (1989) A novel hopanoid from the ethanol-producing bacterium *Zymomonas mobilis. Biochem. J. 262*, 673–675.

FÖRSTER, H.J., BIEMANN, K., HAIGH, W.G., TATTRIE, N.H., and COLVIN, J.R. (1973) The structure of novel C_{35} pentacyclic terpenes from *Acetobacter xylinum. Biochem. J. 135*, 133–143.

GELPI, E., SCHNEIDER, H., MANN, J., and ORO, J. (1970) Hydrocarbons of geochemical significance in microscopic algae. *Phytochemistry 9*, 603–612.

HAIGH, W.G., FÖRSTER, H.J., BIEMANN, K., TATTRIE, N.H., and COLVIN, J.R. (1973) Induction of orientation of bacterial cellulose microfibrils by a novel terpenoid from *Acetobacter xylinum. Biochem. J. 135*, 145–149.

HOWARD, D.L. and CHAPMAN, D.J. (1981) Structural elucidation of two hopanoids from the photosynthetic bacterium *Rhodomicrobium vannielii. Chem. Commun.* 468–469.

HUNECK, S. (1971) Chemie und Biosynthese der Flechtenstoffe. In *Fortschritte der Chemie organischer Naturstoffe 29*, 209–306.

HVEDING-BERGSETH, N., BRUUN, T., and KJOSEN, H. (1983) Isolation of 30-nor-21α-hopan-22-one (isoadiantone) from the lichen *Plastimatia glauca. Phytochemistry 22*, 1826–1827.

KANNENBERG, E. and PORALLA, K. (1982) The influence of hopanoids on growth of *Mycoplasma mycoides*. *Arch. Microbiol. 133*, 100–102.

KOHL, N., GLOE, A., and REICHENBACH, H. (1983) Steroids from the myxobacterium *Nannocystis exedens*. *J. Gen. Microbiol. 129*, 1629–1635.

LANGWORTHY, T.A. and MAYBERRY, W.R. (1976) A 1,2,3,4-tetrahydroxypentane-substituted pentacyclic triterpene from *Bacillus acidocaldarius*. *Biochim. Biophys. Acta 431*, 570–577.

LANGWORTHY, T.A., MAYBERRY, W.R., and SMITH, P.F. (1976) A sulfonolipid and novel glucosaminyl glycolipids from the extreme thermoacidophile *Bacillus acidocaldarius*. *Biochim. Biophys. Acta 431*, 550–569.

MACKENZIE, A.S., PATIENCE, R.L., MAXWELL, J.R., VANDENBROUCKE, M., and DURAND, B. (1980) Molecular parameters of maturation in the Toarcian shales, Paris Basin—I. Changes in the configurations of acyclic isoprenoid alkanes, steranes and triterpanes. *Geochim. Cosmochim. Acta 44*, 1709–1721.

MEUNIER-CHRISTMANN, C. (1988) Géochimie organique de phosphates de schistes bitumineux marocains: étude du processus de phosphatogenèse. Ph. D. Dissertation, Université Louis Pasteur, Strasbourg, France.

MYCKE, B., NARJES, F., and MICHAELIS, W. (1987) Bacteriohopanetetrol from chemical degradation of an oil shale kerogen. *Nature 326*, 179–181.

NATORI, Y., KAMEI, T., and NAGASAKI, T. (1981) Occurrence of triterpenes and polyprenyl alcohols in *Pseudomonas* C-45, a mutant. *Agric. Biol. Chem. 45*, 2337–2338.

NEUMANN, S. and SIMON, H. (1986) Purification, partial characterization and substrate specificity of a squalene cyclase from *Bacillus acidocaldarius*. *Biol. Chem. Hoppe Seyler 367*, 723–729.

NEUNLIST, S. (1987) Hopanoïdes de bactéries méthylotrophes et de Rhodospirillacées. Ph.D. Dissertation, Université de Haute Alsace, Mulhouse, France.

NEUNLIST, S. and ROHMER, M. (1985a) The hopanoids of *Methylosinus trichosporium:* aminobacteriohopanetriol and aminobacteriohopanetetrol. *J. Gen. Microbiol. 131*, 1363–1367.

NEUNLIST, S. and ROHMER, M. (1985b) Novel hopanoids from the methylotrophic bacteria *Methylococcus capsulatus* and *Methylomonas methanica*. (22*S*)-35-Aminobacteriohopane-30,31,32,33,34-pentol and (22*S*)-35-amino-3β-methylbacteriohopane-30,31,32,33,34-pentol. *Biochem. J. 221*, 635–639.

NEUNLIST, S. and ROHMER, M. (1985c) A novel hopanoid, 30-(5′-adenosyl)-hopane, from the purple non-sulphur bacterium *Rhodopseudomonas acidophila*. *Biochem. J. 228*, 769–771.

NEUNLIST, S. and ROHMER, M. (1988) A convenient route to an acetylenic C_{35} hopanoid and the absolute configuration of the side chain of amino-bacteriohopanetriol. *Chem. Comm.* 830–832.

NEUNLIST, S., BISSERET, P., and ROHMER, M. (1988) The hopanoids of the purple nonsulfur bacteria *Rhodopseudomonas palustris* and *Rhodopseudomonas acidophila* and the absolute configuration of bacteriohopanetetrol. *Eur. J. Biochem. 171*, 245–252.

NEUNLIST, S., HOLST, O., and ROHMER, M. (1985) Prokaryotic triterpenoids. The hopanoids of the purple nonsulphur bacterium *Rhodomicrobium vannielli*: An aminotriol and its aminoacyl derivatives, *N*-tryptophanyl and *N*-ornithinyl aminotriol. *Eur. J. Biochem. 147*, 561–568.

OURISSON, G., ALBRECHT, P., and ROHMER, M. (1979) The hopanoids: Palaeochemistry and biochemistry of a group of natural products. *Pure Appl. Chem. 51*, 709–729, and references cited therein.

OURISSON, G., ALBRECHT, P., and ROHMER, M. (1984) The microbial origin of fossil fuels. *Scientific Amer. 251*, 44–51.

OURISSON, G., PORALLA, K., and ROHMER, M. (1987) Prokaryotic hopanoids and other polyterpenoid sterol surrogates. *Ann. Rev. Microbiol. 41*, 301–333, and references cited therein.

PHILP, R.P. (1985) Fossil fuel biomarkers: Applications and spectra. Elsevier, 294p. and references cited therein.

PORALLA, K., KANNENBERG, E., and BLUME, A. (1980) A glycolipid-containing hopane isolated from the acidophilic, thermophilic *Bacillus acidocaldarius* has a cholesterol-like function in membranes. *FEBS Lett. 113*, 107–110.

QUIRK, M.M., WARDROPER, A.M.K., WHEATLEY, R.E., and MAXWELL, J.R. (1984) Extended hopanoids in peat environments. *Chemical Geology 42*, 25–43.

RENOUX, J.M. and ROHMER, M. (1985) Prokaryotic triterpenoids. New bacteriohopane cyclitol ethers from the methylotropic bacterium *Methylobacterium organophilum. Eur. J. Biochem. 151*, 405–410.

RIES-KAUTT, M. (1986) Etude des lipides dans différents types de sol. Aspects moléculaires. Ph. D. Dissertation, Université Louis Pasteur, Strasbourg France.

RIES-KAUTT, M. and ALBRECHT, P. (1989) Hopane derived triterpenes in soils. *Chem. Geol. 76*, 143–151.

ROHMER, M. (1975) Triterpénoïdes de procaryotes. Ph. D. Dissertation, Université Louis Pasteur, Strasbourg, France.

ROHMER, M. and OURISSON, G. (1976a) Structures des bactériohopanetétrols d'*Acetobacter xylinum. Tetrahedron Lett.* 3633–3636.

ROHMER, M. and OURISSON, G. (1976b) Dérivés du bactériohopane: variations structurales et répartition. *Tetrahedron Lett.* 3637–3640.

ROHMER, M. and OURISSON, G. (1976c) Méthylhopanes d'*Acetobacter xylinum* et d'*Acetobacter rancens*: une nouvelle famille de composés triterpéniques. *Tetrahedron Lett.* 3641–3644.

ROHMER, M., ANDING, C., and OURISSON, G. (1980a) Non-specific biosynthesis of hopane triterpenes by a cell-free system from *Acetobacter pasteurianus. Eur. J. Biochem. 112*, 541–547.

ROHMER, M., BOUVIER, P., and OURISSON, G. (1979) Molecular evolution of biomembranes: structural equivalents and phylogenetic precursors of sterols. *Proc. Natl. Acad. Sc. USA 76*, 847–851.

ROHMER, M., BOUVIER, P., and OURISSON, G. (1980b) Non-specific lanosterol and hopanoid biosynthesis by a cell-free system from the bacterium *Methylococcus capsulatus. Eur. J. Biochem. 112*, 557–560.

ROHMER, M., BOUVIER-NAVE, P., and OURISSON, G. (1984) Distribution of hopanoid triterpenes in prokaryotes. *J. Gen. Microbiol. 130*, 1137–1150.

ROHMER, M., DASTILLUNG, M., and OURISSON, G. (1980) Hopanoids from C_{30} to C_{35} in recent muds: Chemical markers for bacterial activity. *Naturwissenschaften 67*, 456–458.

ROHMER, M., SUTTER, B., and SAHM, H. (1989) Bacterial sterol surrogates. Biosynthesis of the side chain of bacteriohopanetetrol and of a carbocyclic pseudopentose from ^{13}C labelled glucose in *Zymomonas mobilis. Chem. Commun.*, 1471–1472.

SCHMIDT, A., BRINGER-MEYER, S., PORALLA, K., and SAHM, H. (1986) Effect of alcohols and temperature on the hopanoid content of *Zymomonas mobilis. Appl. Microbiol. Biotechnol. 25*, 32–36.

SECKLER, B. and PORALLA, K. (1986) Characterization and partial purification of squalene hopene cyclase from *Bacillus acidocaldarius. Biochim. Biophys. Acta 881*, 356–363.

SEIFERT, W. (1978) Steranes and terpanes in kerogen pyrolysis for correlation of oils and source rocks. *Geochim. Cosmochim. Acta 42*, 473–484.

SUMMONS, R.E. and CAPON, R.J. (1988) Fossil steranes with unprecedented methylation in ring A. *Geochim. Cosmochim. Acta 55*, 2733–2736.

TAHARA, Y., YUMARA, H., and YAMADA, Y. (1988) Distribution of tetrahydroxybacteriohopane in the membrane fractions of *Zymomonas mobilis. Agric. Biol. Chem. 52*, 607–609.

TEN HAVEN, H.L., BAAS, M., DE LEEUW, J.W., MAASEN, J.M.,and SCHECK, P.A. (1987) Organic geochemical characteristics of sediments from the anoxic brine-filled Tyro basin (eastern Mediterranean). *Org. Geochem. 11*, 605–611.

TORNABENE, T.G., HOLZER, G., BITTNER, A.S., and GROHMANN, K. (1982) Characterization of the total extractable lipids of *Zymomonas mobilis* var. *mobilis. Can. J. Microbiol. 28*, 1107–1118.

VAN EIJK, G.W., ROEIJMANS, H.J., and SEYKENS, D. (1986) Hopanoids from the entomogenous fungus *Aschersonia aleyrodis. Tetrahedron Lett. 27*, 2533–2534.

VENKATESAN, M.J. (1988) Diploptene in Antarctic sediments. *Geochim. Cosmochim. Acta 52*, 217–222, and references cited therein.

VIER, B. and VOIGT, B. (1986) Zum Vorkommen von Hop-22(29)-en in *Acetobacter methanolicus. J. Basic Microbiol. 26*, 547–549.

YOSIOKA, I., NAKANISHI, T., YAMAKI, M., and KITAGAWA, I. (1972) Lichen triterpenoids. (IV) The structures of leucotylic acid and methyl isoleucotylate, an acid-induced isomer of methyl leucotylate. *Chem. Pharm. Bull. 70*, 487–501.

ZUNDEL, M. and ROHMER, M. (1985a) Prokaryotic triterpenoids. 1. 3β-Methylhopanoids from *Acetobacter* species and *Methylococcus capsulatus. Eur. J. Biochem. 150*, 23–27.

ZUNDEL, M. and ROHMER, M. (1985b) Prokaryotic triterpenoids. 3. The biosynthesis of 2β-methylhopanoids and 3β-methylhopanoids of *Methylobacterium organophilum* and *Acetobacter pasteurianus* spp. *pasteurianus. Eur. J. Biochem. 150*, 35–39.

ZUNDEL, M. and ROHMER, M. (1985) Hopanoids of the methylotropic bacteria *Methylococcus capsulatus* and *Methylomonas* sp. as possible precursors of C_{29} and C_{30} hopanoid chemical fossils. *FEMS Microbiol. Lett. 28*, 61–64.

2

Qualitative and Quantitative Evolution and Kinetics of Biological Marker Transformations—Laboratory Experiments and Application to the Michigan Basin

Roger Marzi and Jürgen Rullkötter

Abstract. The evolution of three biological marker compound ratios commonly observed in sediments was simulated by a series of laboratory hydrous pyrolysis experiments using immature Toarcian shale from northern Germany as starting material. The extent of isomerization of C_{29} steranes at C-20 as well as the mono- to triaromatic steroid hydrocarbon ratios show a reversal instead of a steady approach toward a maximum end value. This is not the case for the epimer ratio of 17α(H)-homohopanes at C-22. As in nature, the absolute concentrations of all these biological marker compound classes decrease by several orders of magnitude with increasing thermal stress in the experiments. A fractionation effect was observed for the release of tri- versus monoaromatic steroid hydrocarbons from the rock chips during hydrous pyrolysis. The pyrolysis results were suitable for the calculation of kinetic data for the progress of sterane and hopane isomerization. They were found to differ significantly from some of those determined on natural samples or from previous experiments. Due to the fractionation effect, kinetic data for aromatic steroid hydrocarbons are ambiguous. The sterane kinetic data were used to put constraints on the geothermal history of Paleozoic and Mesozoic sediments in the Michigan Basin. The sterane isomerization values observed in natural sediment samples can best be explained by additional subsidence and higher heat flow during the Upper Carboniferous. This significantly modifies the present understanding of the geological development in the Michigan Basin.

INTRODUCTION

It has been put forward that a number of geochemical reactions dependent on time and temperature involve specific transformations of organic molecules having characteristic biochemically controlled structures into well-defined geochemical products as determined from the progressive change of biological marker compound ratios in geological samples which have been exposed to increasing thermal stress (e.g., Seifert and Moldowan, 1978, 1980, 1981; Mackenzie, 1984). The precursor/product relationship of a number of these so-called biological markers was said to be sufficiently well established for the transformation reactions to be used as a tool for thermal maturity assessment on organic matter in sediments and on crude oils. Among the reactions are the isomerization of 24-ethyl-5α(H),14α(H),17α(H)-cholestane at C-20 and of 17α(H)-homohopane at C-22 (Mackenzie et al., 1980) as well as the transformation of C-ring monoaromatic into ABC-ring triaromatic steroid hydrocarbons (Mackenzie et al., 1981; Moldowan and Fago, 1986; Riolo et al., 1986). Recent studies (e.g., Peakman and Maxwell, 1988; Abbott et al., 1990) indicate, however, that direct interconversions of hydrocarbon epimers or aromatic species may be less important but that biological marker compound ratios are rather influenced by the combined effects of (1) the initial assemblage of a certain biomarker type, (2) the neoformation of such biomarkers (from macromolecular organic matter or polar precursors) with a compound ratio different from that of the initial assemblage, and (3) the preferential thermal decomposition of one of the biomarker species used for compound ratio calculation. Reference to transformation *reactions* in this study, thus, is a formal treatment and merely implies the observed changes in compound ratios; this approach nevertheless appears to be useful in practical applications.

Where the inferred reactions could be simulated by laboratory heating experiments or studied in geological environments whose geothermal histories were reasonably well understood, the kinetic parameters were determined (e.g., Mackenzie and McKenzie, 1983; Suzuki, 1984; Abbott et al., 1985; Alexander et al., 1986; Sajgo and Lefler, 1986; Rullkötter and Marzi, 1988; Strachan et al., 1989). Under the assumption that the observed changes follow pseudo-first-order kinetics, the Arrhenius equation can be applied, and the derived parameters of activation energy and frequency factor provide information on how the compound ratios respond to the influence of temperature and time during the geological history of a sedimentary layer (Mackenzie and McKenzie, 1983).

Unfortunately, the various kinetic studies yielded inconsistent results. For example, activation energy/frequency factor pairs for sterane isomerization were as different as 91 kJ $mol^{-1}/6 \times 10^{-3}s^{-1}$ (Mackenzie and McKenzie, 1983), 147 kJ $mol^{-1}/6.5 \times 10^{7}s^{-1}$ (Suzuki, 1984), 84 kJ $mol^{-1}/6 \times 10^{-3}s^{-1}$ (Alexander et al.,

1986), 92 kJ mol^{-1}/2.4 × $10^{-3}s^{-1}$ (Sajgo and Lefler, 1986), and 170 kJ mol^{-1}/6 × $10^{8}s^{-1}$ (preliminary results; Rullkötter and Marzi, 1988). Among these data, those of Mackenzie and McKenzie (1983), Alexander et al. (1986), and Sajgo and Lefler (1986) are closest to each other because the authors used similar methods and/or partly identical sample series. Kinetic data for steroid aromatization published so far vary in the following way: 200 kJ mol^{-1}/1.8 × $10^{14}s^{-1}$ (Mackenzie and McKenzie, 1983), 145 kJ mol^{-1}/6.7 × $10^{12}s^{-1}$ (Abbott et al., 1985), and 121 kJ mol^{-1}/1.1 × $10^{3}s^{-1}$ (Sajgo and Lefler, 1986), although the latter authors do not consider their steroid aromatization kinetics data to be applicable in practice.

The consequences of the different values for the progress of biological marker reactions as a function of temperature and time are hard to predict from the numbers directly (e.g., by just comparing activation energies), because activation energy and frequency factor can compensate each other to a large extent. A preliminary sensitivity analysis for sterane isomerization under an assumed geological heating rate of 2.6°C Ma^{-1} showed that the reaction progress is fairly similar when the kinetic data of Mackenzie and McKenzie (1983), Sajgo and Lefler (1986) or Rullkötter and Marzi (1988) are used, although the numerical values are largely different (Marzi, 1989). In contrast to this, the simulated reaction proceeds much faster under the same heating conditions, but with the kinetic data of Suzuki (1984) or Alexander et al. (1986).

The inconsistency of the kinetic data derived from laboratory heating experiments may be caused by the different conditions applied. That this may be the case is supported by the recent observation that the rock matrix (shale or coal) has a major influence on the rate of apparent sterane isomerization (Strachan et al., 1989). In our attempt to contribute to the resolution of this conflict, we chose hydrous pyrolysis (Winters et al., 1983) as a simulation method which is widely believed to represent reasonably well natural maturation conditions. Hydrous pyrolysis experiments have been used before to study the qualitative and quantitative aspects of sterane formation and isomerization or aromatization (Lewan et al., 1986; Eglinton and Douglas, 1988) but the kinetics of the reactions were not studied by these authors.

The kinetic data for steroid isomerization from our hydrous pyrolysis experiments are then applied to the Michigan Basin in order to help differentiate between the currently different interpretations of its geothermal history.

EXPERIMENTAL METHODS

Samples

Unextracted rock chips (5–15 mm) from a core of the Lower Toarcian Posidonia shale (Lias ϵ) from well Wenzen-1001 in the Hils syncline in northern Germany were used for hydrous pyrolysis experiments. The organic matter in this sediment is immature (e.g., 0.48 percent vitrinite reflectance). For more details on the sample material see Rullkötter et al. (1988), Littke et al. (1988), and Littke et al. (in press).

Twenty-five samples of rock, ranging in age from Late Cambrian to Pennsylvanian, were selected for an investigation of the biological marker composition in the Michigan Basin. Sample preparation and bulk parameters are described in Rullkötter et al. (in press). Selection of the rocks was biased toward clastic samples to avoid anomalous biological marker compound ratios known to occur in carbonates and other clay-free sediments (ten Haven et al., 1986). Contamination of a number of rock samples with diesel drilling mud additives presented serious analytical problems and appeared to have affected sterane isomerization values in the most heavily contaminated rocks.

Crude oils are of Ordovician origin (in Ordovician and Devonian reservoirs), Silurian origin (in Silurian reservoirs) and Devonian origin (in Devonian and early Carboniferous reservoirs). Aspects of oil/source correlation and bulk properties have been dealt with by Rullkötter et al. (1986; in press).

Hydrous Pyrolysis

Sealed vessel hydrous pyrolysis experiments under isothermal conditions were carried out following the method of Winters et al. (1983). Details can be found in the work of Marzi (1989). Briefly, the 250 ml autoclaves were filled with 50 g of crushed rock and 100 ml of deionized water. The headspace was purged with helium. Heating to the desired temperatures was achieved within two hours. Pyrolysis temperatures in the 43 experiments varied from 210°C to 350°C and were held constant for time intervals between 2 and 90 days. During the work-up procedures, the expelled pyrolysate floating free on the water phase or adsorbed on the rock surfaces (removed by pipetting or rinsing the rock chips with dichloromethane, respectively) and the residual bitumen in the rock chips (removed by extraction with dichlormethane after grinding the rock) were treated separately. Hydrocarbon fractions were separated by medium pressure liquid chromatography (Radke et al., 1980).

Gas Chromatography-Mass Spectrometry

Gas chromatography-mass spectrometry (GC-MS) measurements were carried out on a VG 7070E mass spectrometer linked directly to a Carlo Erba Model 4160 gas chromatograph. Samples were injected in the splitless mode onto a 25 m × 0.3 mm ID fused silica column with a chemically bound DB 5 stationary phase; the column was introduced directly into the mass spectrometer ion source. Helium was used as carrier gas, and the temperature was programmed from 70 to 300°C at a rate of 4° per minute. The mass spectrometer was operated at an ionisation energy of 70eV and a resolution of 1200 (10% valley). The source temperature was kept at 220°C. The system was used in the multiple ion monitoring mode with a dwell time of about 200 ms per ion recorded and a cycle time of about 2.5 s.

Sterane isomerization values in samples containing low sterane concentrations (intensely pyrolyzed samples, Michigan Basin crude oils) were confirmed by metastable ion monitoring in order to exclude perturbation effects by coeluting compounds. All aromatic hydrocarbon fractions from the pyrolysis experiments were analyzed at higher mass spectrometer resolution (3000, 10% valley) in order to separate key fragments of aromatic steroids (particularly m/z 253) from signals of other aromatic hydrocarbons at the same nominal mass.

RESULTS AND DISCUSSION

Qualitative and Quantitative Evolution of Biological Marker Hydrocarbons

Steranes. Sterane isomerization at C-20 is believed to be an equilibrium reaction involving the epimerization of the biological (20*R*) configuration leading to the (20*S*) isomer which naturally only occurs in the geosphere (Mackenzie et al., 1980). The geochemical "reaction" in sediments starts with a zero concentration of the (20*S*) product and is said to proceed to an equilibrium ratio of $20S/(20S + 20R) = 0.54$ based on geochemical observations (Mackenzie and McKenzie, 1983), and molecular mechanics calculations (van Graas et al., 1982). This is shown in Figure 2.1a (top) as an idealized theoretical increase of the epimer ratio as a function of increasing thermal stress (x-axis) which is induced by the combined effects of geological time and geothermal heat flow in the natural system and by the heating temperature and duration during hydrous pyrolysis.

The laboratory heating experiments on the Toarcian shale sample were able to bring about sterane isomerization which increased from an initial $20S/(20S + 20R)$ ratio of 0.07 in the starting material to a value close to the expected equilibrium value (Fig. 2.1a, bottom). Increasing the hydrous pyrolysis intensity further (e.g., more than 5 days at 340°C or 11 days at 330°C) yielded lower sterane isomerization values again. Considerable scatter in the data was observed, possibly partly because of analytical difficulties in the case of very low absolute concentrations of steranes preserved in the high-temperature pyrolysates, but the majority of the values was close to 0.4 (hatched area in Fig. 2.1a, bottom). The number of experiments performed in the high-temperature range was not sufficient, however, to obtain information on the mode of transition (abrupt or smooth) from the highest values to ratios of about 0.4.

Lewan et al. (1986) already had observed such a reversal of the sterane isomerization trend in their hydrous pyrolysis experiments on Phoshoria Retort Shale. Their isomer ratio in the starting material was 0.42, and after an initial small decline it increased to 0.56 before it fell to about 0.48 at the highest temperature applied (345°C). They observed the maximum to occur earlier in the expelled oil than in the residual bitumen (310°C/3 d and 330°C/3 d, respectively) although the differences in the $20S/(20R + 20R)$ ratios were small for a given temperature ($\leq$

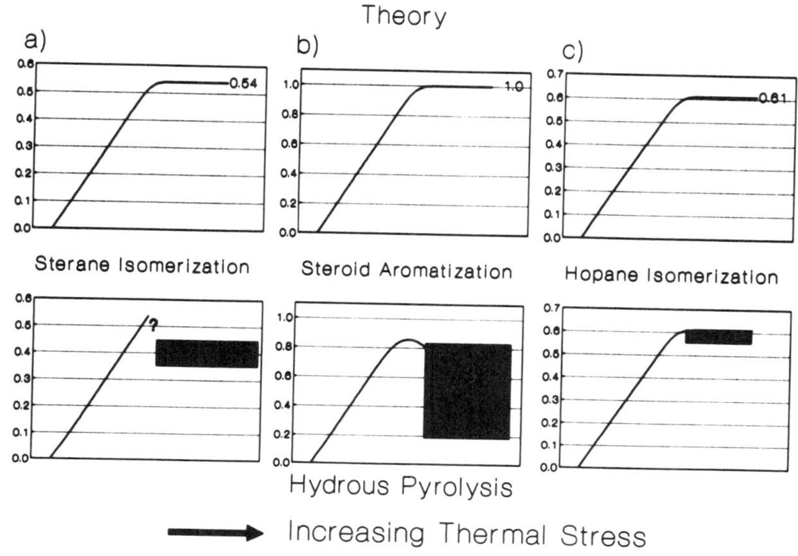

Figure 2.1 Predicted (top) and observed (hydrous pyrolysis, bottom) evolution of biological marker reactions with the x-axis showing the thermal stress in arbitrary units: (a) isomerization of C_{29} steranes at C-20, (b) transformation of C-ring mono- into ABC-ring triaromatic steroid hydrocarbons, and (c) isomerization of 17α(H)-homohopanes at C-22.

0.04). We did not observe a significant difference in sterane isomerization between the expelled material and the bitumen remaining in the rock chips, but we considered differences ≤ 0.03 to be within experimental error limits, particularly when the absolute sterane concentrations were low. The detection of a reversal in the 20*S*/(20*S* + 20*R*) sterane epimer ratios in hydrous pyrolysis experiments is in accordance with the recent observation of relatively low sterane isomerization values in obviously overmature samples (Mackenzie, Rullkötter et al., 1985; Snowdon et al., 1987; see also the following discussion of the Michigan Basin samples). In the case of crude oils, low sterane isomerization values together with low absolute concentrations of steranes may be related to the fact that the oils have taken up less mature organic matter during migration, and occasionally this has been interpreted that way (e.g., Philp and Gilbert, 1982; Rullkötter et al., 1984). The experimental results from hydrous pyrolysis now offer an alternative explanation at least for very mature oils. The progress in isomerization during maturity increase is accompanied by a dramatic decrease of the absolute sterane concentrations both in natural systems (Rullkötter et al., 1984; Mackenzie, Rullkötter, et al., 1985) and in hydrous pyrolysis (Fig. 2.2; see also Eglinton and Douglas, 1988; Abbott et al., 1990). As can be seen from Figure 2.2, the reversed (20*S*/(20*S* + 20*R*) ratios start to appear where the decrease in absolute sterane concentration becomes particularly drastic. As pointed out by Abbott et al. (1990) preferential destruction

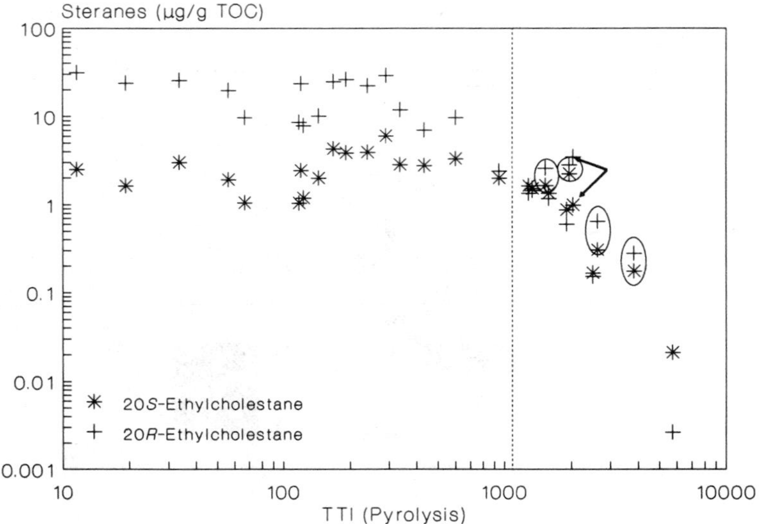

Figure 2.2 Absolute amounts (μg/g C_{org}) of 5α(H),14α(H),17α(H)-24-ethylcholestanes (20*S* and 20*R*) in the free pyrolysates after hydrous pyrolysis of a Lower Toarcian shale sample from northern Germany. X-axis represents pyrolysis intensity expressed as calculated time/temperature index (TTI) values (Waples, 1980) in order to account for variations in both temperature and duration during the experiments. The dotted vertical line indicates where the 20*S*/(20*S* + 20*R*) ratio has passed its maximum. Marked data point (circles, arrows) beyond this indicate 20*S*/(20*S* + 20*R*) ratios 0.5 after the reversal.

of the (20*S*) epimer at high temperatures is the most likely explanation of the decrease in the sterane isomerization ratio.

Aromatic steroid hydrocarbons. Steroid aromatization involves the progressive transformation of C_{27}–C_{29} C-ring monoaromatic steroid hydrocarbons, which start to occur in sediments in an advanced stage of diagenesis (Rullkötter and Welte, 1983), into C_{26}–C_{28} ABC-ring triaromatic steroid hydrocarbons with the loss of the C-10 methyl group (Mackenzie et al., 1981; Moldowan and Fago, 1986; Riolo et al., 1986). The tri-/(tri- + monoaromatic) steroid hydrocarbon ratio as defined by Mackenzie et al. (1981) and Moldowan and Fago (1986) has been said to proceed from zero to unity, that is, to the complete conversion into triaromatic species (Fig. 2.1b, top; Mackenzie and McKenzie, 1983).

In hydrous pyrolysis, steroid aromatization can be simulated, but as in the case of sterane isomerization the aromatization ratio shows a reversal (Fig. 2.1b, bottom). It is fair to assume that the reason for the reaction behavior is a preferred thermal degradation of the triaromatic steroid hydrocarbons by cleavage of the side-chain or through ring D. The decrease of the tri-/(tri- + monoaromatic) steroid hydrocarbon ratio with increasing pyrolysis intensity after the maximum, despite a certain amount of scatter in the data, is more clearly pronounced than in the case of sterane isomerization. The maximum value observed in our hydrous pyrolysis experiments is significantly different from the theoretical maximum 1.0. This may be due to the fact that the thermal destruction of the triaromatic species is kinetically more favored under laboratory conditions than in a sedimentary basin. The observed reversal in the steroid aromatization trend may now explain the presence of small amounts of monoaromatic steroid hydrocarbons in sediments from the North Slope of Alaska where the organic matter according to all other parameters was overmature and where the corresponding triaromatic species could not be detected (Mackenzie, Rullkötter, et al., 1985; see also later discussion of Michigan Basin samples).

Absolute concentrations of the aromatic steroid hydrocarbons follow a trend similar to those of the steranes, that is, there is a decrease of the absolute concentration over about three orders of magnitude within the range of pyrolysis conditions applied (Marzi, 1989), and this is similar to naturally matured sediments (Rullkötter et al., 1984). In addition, there is a significant difference in the extent of steroid aromatization between the free pyrolysate and the residual bitumen in the rock chips (Fig. 2.3) as has been observed before by Lewan et al. (1986). In both cases, the tri-/(tri- + monoaromatic) steroid hydrocarbon ratio is drastically higher in the residual bitumen. As has been pointed out by Lewan et al. (1986), this fractionation probably is a migration (chromatography) effect whereby the more polar triaromatic species are more strongly retained in the rock. This is in accordance with measurements of clay adsorption free energies (Carlson and Cham-

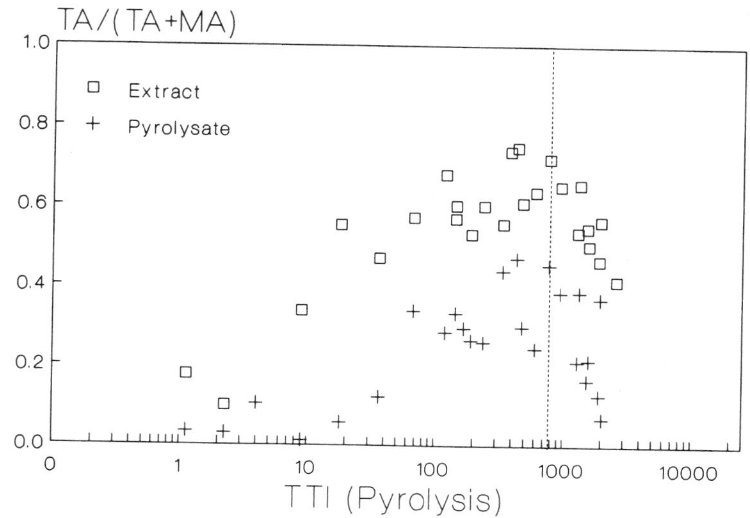

Figure 2.3 Steroid aromatization [TA/(TA + MA)] as a function of pyrolysis intensity (cf. Fig. 2.2 legend) for free pyrolysates and residual bitumens after hydrous pyrolysis.

berlain, 1986) as well as natural observations in natural sediments from the Makaham Delta sediments (Hoffmann et al., 1984).

Hopane isomerization. The isomerization of 17α(H)-homohopanes at C-22, that is, the change from the purely biogenic (22*R*) configuration into an approximately 3:2 mixture of the (22*S*) and (22*R*) diastereomers (Fig. 2.1c, top), is believed to be very similar to that of steranes at C-20 (Ensminger et al., 1977; Mackenzie et al., 1980). Although hopane isomerization could also be advanced in the Toarcian shale from northern Germany by hydrous pyrolysis, there is no reversal in the 22*S*/(22*S* + 22*R*) ratio of the 17α(H)-homohopanes at high pyrolysis intensities (Fig. 2.1c, bottom).

Determination of Kinetic Data

The kinetic constants (activation energy and frequency factor) for the investigated biological marker reactions can be determined as described by Abbott et al. (1985) for laboratory heating experiments and Mackenzie and McKenzie (1983) for naturally matured sample series. Because of the reversal of the sterane isomerization and steroid aromatization ratios observed during hydrous pyrolysis, only the experiments represented by the leading parts of the curves (Fig. 2.1a and 2.1b, bottom) were considered for kinetics evaluation. In addition, we decided to use a combination of our hydrous pyrolysis data and data of natural samples (from Mackenzie and McKenzie, 1983) in order to cover a wide temperature range. For this purpose, data of samples representing overmature organic matter with respect to the maximum of sterane isomerization and steroid aromatization had to be eliminated. This has been achieved by time/temperature index (TTI) calculations (Waples, 1980) which determined the position of the maxima for both natural and laboratory heating data using a comparable, although fairly rough measure. The data selection for sterane isomerization values for North Sea and Pannonian Basin sediments (Mackenzie and McKenzie, 1983) is shown in Figure 2.4.

Data from hydrous pyrolysis experiments and selected natural data were then transferred in an Arrhenius diagram of the natural logarithm of the reaction rate constant as a function of the inverted absolute temperature. Calculation of a combined linear regression then led to an activation energy (E_a) of 169.0 kJ mol^{-1} and a frequency factor (A) of 4.86×10^8 s^{-1} for the isomerization of C_{29} steranes at C-20 (Fig. 2.5). These values are almost identical to those which can be determined from hydrous pyrolysis data alone (E_a = 168.3 kJ mol^{-1}, A = 4.6×10^8 s^{-1}). Unless this is purely coincidental, this may indicate that the reaction in the laboratory experiments does not differ from that in nature. (Note: All sediments included in this study are clay-bearing marine clastics.)

The same data treatment for the steroid aromatization reaction leads to an activation energy of 181.4 kJ mol^{-1} and a frequency factor of 4.85×10^{10} s^{-1} (Fig. 2.6). However, the kinetic constants derived from the combined regression of both natural and hydrous pyrolysis data differ significantly from those which

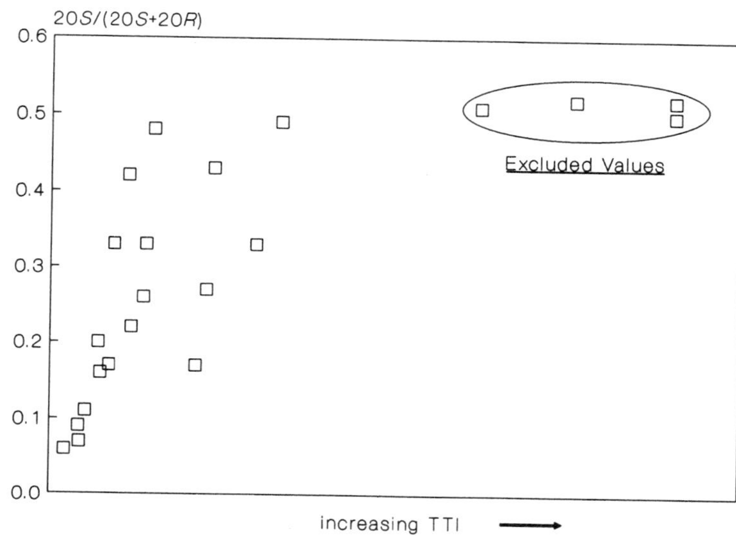

Figure 2.4 Sterane isomerization in North Sea and Pannonian Basin samples (Mackenzie and McKenzie, 1983) as a function of calculated TTI values for the exclusion of data points beyond the reversal of sterane isomerization.

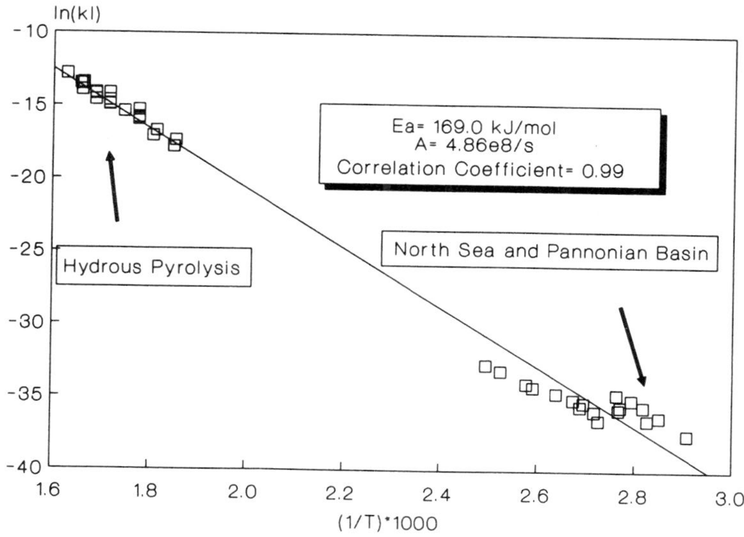

Figure 2.5 Arrhenius diagram for determination of kinetic data of sterane isomerization based on hydrous pyrolysis experiments and selected natural samples from the North Sea and the Pannonian Basin (Mackenzie and McKenzie, 1983).

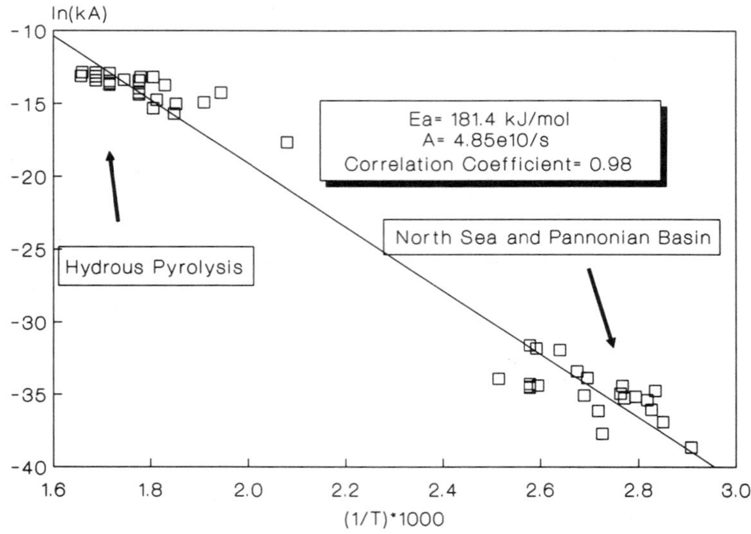

Figure 2.6 Arrhenius diagram for determination of kinetic data for steroid aromatization based on hydrous pyrolysis experiments and selected natural samples from the North Sea and the Pannonian Basin (Mackenzie and McKenzie, 1983).

would be obtained from using either data set alone. As has been pointed out before, largely different steroid aromatization values were measured in free pyrolysates and in the corresponding residual bitumens. Since only the residual bitumen is available in naturally matured sediments, there is no information on the quantity and composition of the material lost by primary migration which could be accounted for in the comparison with the laboratory data regardless if free pyrolyzate, residual bitumen or weighted average values (as we did) were taken to evaluate the laboratory experiments. Alternatively, steroid aromatization during hydrous pyrolysis proceeds differently from that in nature, as has been suggested by Lewan et al. (1986). This then would explain the discrepancy between regression of combined data sets and that of separate treatment of both data sets. This ambiguity could not be resolved by our experiment.

Although isomerization of 17α(H)-homohopanes at C-22 smoothly approached the expected equilibrium value of about 0.6, the evaluation of the kinetics was less satisfactory. From the Arrhenius diagram an activation energy of 168 kJ mol^{-1} and a frequency factor of 8.1×10^8 s^{-1} were determined (Fig. 2.7), but separate treatment of natural and artificial heating data leads to kinetic constants significantly different from those in Figure 2.7. On the other hand, the kinetic data for hopane isomerization are close to those of sterane isomerization, as has been observed before by Mackenzie and McKenzie (1983) based on natural data alone, although the numerical values are quite different when both approaches are compared to each other. The good agreement of the kinetic data for sterane isomer-

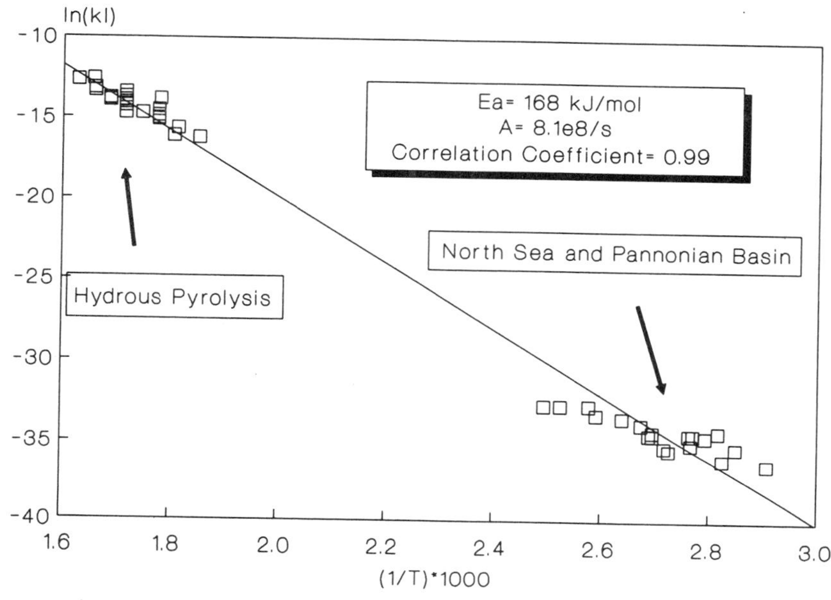

Figure 2.7 Arrhenius diagram for determination of kinetic data for hopane isomerization based on hydrous pyrolysis and selected natural samples from the North Sea and the Pannonian Basin (Mackenzie and McKenzie, 1983).

ization derived both from hydrous pyrolysis alone and from a combination of hydrous pyrolysis and natural data together with the close similarity between the kinetic constants of sterane and hopane isomerization and the alleged similarity of the reactions (Ensminger et al., 1977; Mackenzie and McKenzie, 1983) may provide evidence for the validity of the kinetic constants of both isomerization reactions obtained here.

The differences in the kinetic constants of sterane isomerization and steroid aromatization have been used in the past to distinguish sediments from basins which evolved in different geothermal regimes or to trace back geothermal histories of sedimentary basins (e.g., Mackenzie and McKenzie, 1983; Beaumont et al., 1985; Mackenzie, Beaumont et al., 1985; Rullkötter et al., 1985). Simulation of the progress of sterane isomerization and steroid aromatization using the kinetic data obtained in the course of this work and heating rates of 1°C Ma^{-1} and 10°C Ma^{-1} leads to almost identical results as shown in the aromatization/isomerization (A/I) plot in Figure 2.8. This means that heating rates differing by as much as a factor of ten cannot be distinguished within the common error limits of the determination of biological marker compound ratios. This contrasts, however, the empirical success in applying this approach to several case histories (e.g., Mackenzie and McKenzie, 1983; Beaumont et al., 1985; Mackenzie, Beaumont et al., 1985; Rullkötter et al., 1985) and thus may indicate that the activation energy and fre-

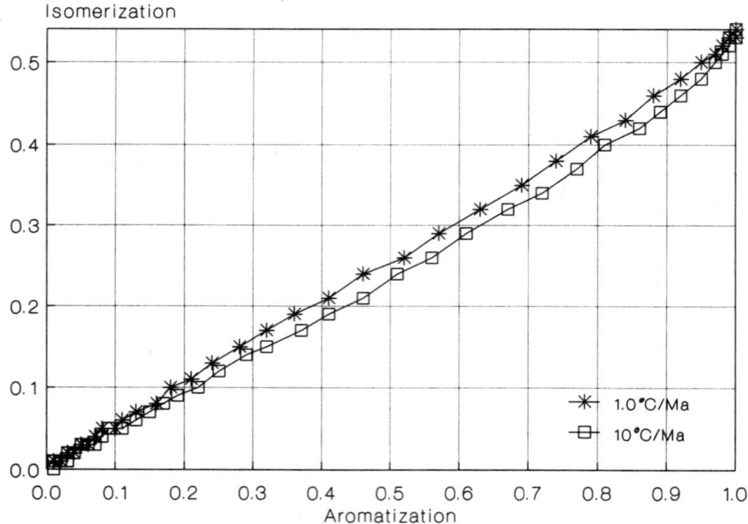

Figure 2.8 Simulated progress of sterane isomerization and steroid aromatization using kinetic data determined by a combination of hydrous pyrolysis and natural samples (Figs. 2.5 and 2.6) and assuming heating rates of 1 and 10°C Ma^{-1}.

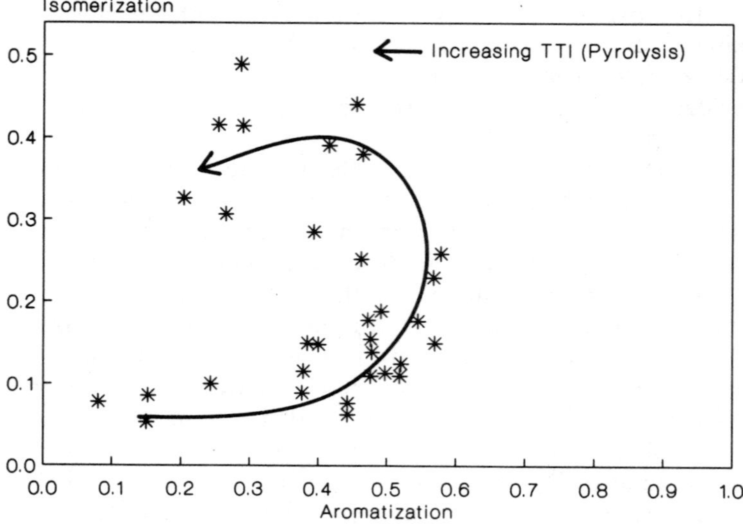

Figure 2.9 Aromatization/isomerization (A/I) representation of data from hydrous pyrolysis showing reversal of the biomarker reactions.

quency factor values of steroid aromatization determined in this study are too low. Higher values would not only separate the two curves further in Figure 2.8, they would also bend them toward the bottom right quadrant which is in accordance with many A/I data sets of sediments and crude oils from natural basins (Hong et al., 1986; J. Rullkötter, unpublished results).

A consequence of the reversal of sterane isomerization and steroid aromatization ratios for their representation in A/I diagrams also is a reversal of the trend as shown in Figure 2.9 (using weighted average values of pyrolysates and residual bitumens). Such a trend has been observed occasionally (J. Rullkötter, unpublished results) where the data points beyond the A/I maximum were represented by mature crude oils and condensates and the ascending part of the steroid aromatization trend were represented by rock samples which in many practical examples of exploration-related geochemical studies contain organic matter less mature than the related oils.

Application of Biological Markers to the Michigan Basin

Geology. The Michigan Basin is composed almost entirely of Paleozoic strata. The basin is situated within the continental craton and is believed to have been formed by gradual subsidence of its central portion throughout Paleozoic times (Dorr and Eschman, 1970; Nunn et al., 1984). The result is a nearly circular structure filled with several kilometers of gently dipping rocks which were deposited as shallow-water sediments. Little deformation is evident in any of the strata, and few rocks of Carboniferous and younger age are found in the Michigan Basin (for a more detailed summary see Rullkötter et al., in press).

Geothermal history and crude oil generation. Organic geochemical studies have shown that crude oils accumulated in Ordovician and Devonian reservoir rocks are mainly of Ordovician origin and that oils in Silurian reservoirs were generated in source rocks of that period (e.g., Vogler et al., 1981; Illich and Grizzle, 1983; Pruitt, 1983; Powell et al., 1984; Rullkötter et al., 1986). In addition, there are geochemical indications that part of the oils in Devonian and younger reservoirs were generated from Devonian rocks (Illich and Grizzle, 1983, 1985; Pruitt, 1983; Rullkötter et al., 1986).

Although major unconformities exist at the base of Middle Ordovician, in the upper Silurian and at the top of Devonian strata (cf. Fig. 2.10), indicating possible uplift and intermittent erosion during Paleozoic time, Nunn et al. (1984) concluded from the scarcity of secondary structural deformation of rock layers that gradual subsidence has dominated in the Michigan Basin throughout the Paleozoic and that the basin has been tectonically stable since the end of the Paleozoic. The almost total lack of rocks younger than Carboniferous obscures the Mesozoic and Cenozoic history of the basin, yet the little-deformed nature of the underlying strata has been taken as evidence that nothing dramatic occurred. The present-day geothermal gradient in the Michigan Basin is an unexceptional 22°C km^{-1} as measured in a deep borehole near the center of the basin (Sleep and Sloss, 1978).

Using this background information and the results from an elastic lithosphere model together with the application of activation energy distributions for kerogen transformation, Nunn et al. (1984) concluded that no significant oil generation has occurred in Silurian and younger sediments in the Michigan Basin although this conflicts with the results of oil/oil and oil/source rock correlation-studies (Vogler et al., 1981; Illich and Grizzle, 1983, 1985; Pruitt, 1983; Rullkötter et al., 1986). In contrast to this, conodont and vitrinite reflectance data, converted to TTI values, were used by Cercone (1984) in order to demonstrate that oil generation was active even in Devonian rocks during the Paleozoic, although for this it had to be assumed that the paleogeothermal gradient once was as high as 45°C km^{-1} and that 1 km of Carboniferous sediments were eroded prior to the Jurassic.

Application of biological markers. In order to get an impression whether one of the conflicting paleogeothermal scenarios of Nunn et al. (1984) and Cercone (1984) is reflected in the molecular composition of the organic matter in Michigan Basin sediments (Marzi, 1989) and crude oils (Rullkötter et al., 1986), the kinetic data of sterane isomerization derived in this study were applied to the geological history of the Michigan Basin. Some qualitative additional evidence is obtained from steroid aromatization data, but we did not use the kinetic data of this reaction due to the ambiguity discussed before and because aromatic steroids, usually present in concentrations lower by a factor of about ten compared to steranes, were not detected in most Michigan Basin crude oils and in several sediments which all contained low biological marker concentrations.

Figure 2.10 shows the subsidence curves (solid lines) for the base of the middle Ordovician, the base of the Silurian and the base of the Devonian for the center of the Michigan Basin based on data from a deep stratigraphic well (Sleep and Sloss, 1978) and the assumption of nondeposition rather than subsidence followed by erosion at the unconformities. A low-temperature scenario, using a geothermal gradient of 31°C km^{-1} for the time prior to the Devonian (Haxby et al., 1976) and the present gradient of 22°C km^{-1} for younger times, was then used to calculate the evolution of sterane isomerization with time as a consequence of increasing subsidence and heating for sediments of the corresponding ages (broken lines; Fig. 2.10). Sterane isomerization in middle Ordovician sediments according to this approach would have reached the equilibrium (maximum) value approximately 200 Ma before present, in the Silurian a value of 0.5 would just be reached at present, whereas sterane isomerization in the Devonian sediments would hardly have advanced (Fig. 2.10).

These results now have to be compared to the extent of sterane isomerization determined on natural samples (cf. bars on the right-hand side of Fig. 2.10). Isomerization values measured on Ordovician rock samples are below those assumed for the equilibrium end value of 0.54. This, together with the fact that higher values are observed in shallower younger sediments (Fig. 2.11a), indicates that the organic

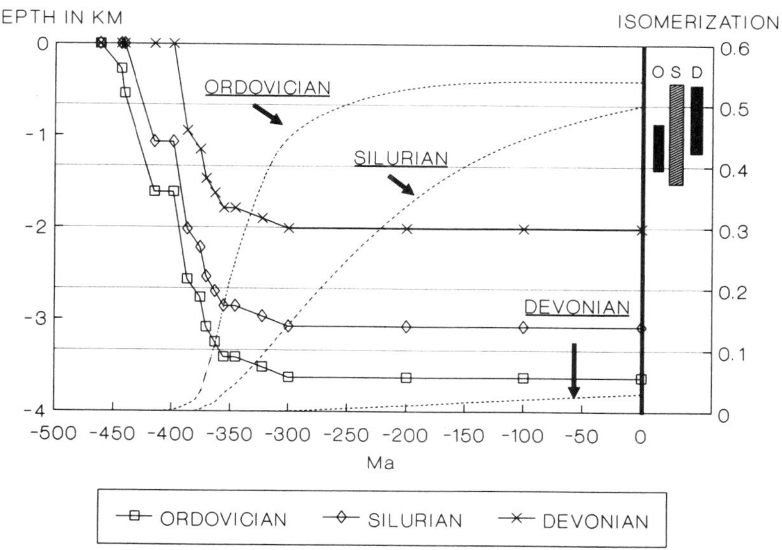

Figure 2.10 Early Paleozoic subsidence (solid lines) and calculated evolution of sterane isomerization based on the geothermal scenario of Nunn et al. (1984) (broken lines) for sediments in the center of the Michigan Basin. Ranges of measurements on natural samples are shown for comparison (O = Ordovician, S = Silurian, D = Devonian).

matter in the Ordovician rocks is very mature and that sterane isomerization has proceeded beyond the maximum (cf. Fig. 2.1). Figure 2.11a shows that sterane isomerization increases from a low value in a Pennsylvanian coal (300 Ma) collected at the surface to values between 0.3 and 0.54 in the Devonian (the data scatter is largely due to the different depth setting of the samples) and decreases again in the Ordovician as would be expected now from the hydrous pyrolysis results for very mature samples. This interpretation is supported by a very similar trend in steroid aromatization (Fig. 2.11b). This also explains sterane isomerization values between 0.4 and 0.5 in many of the Ordovician crude oils accumulated in Ordovician and Devonian reservoirs in the Michigan Basin (Rullkötter et al., 1986).

The calculated sterane isomerization value for Silurian sediments agrees well with those of Silurian oils (no appropriate rock samples were available), although a value of 0.5 may represent a maturity level corresponding to either side of the isomerization maximum. Sterane isomerization in all Devonian samples investigated by far exceeds the calculated value (Fig. 2.10).

Because even in a low-temperature scenario sterane isomerization in Ordovician and Silurian rocks is well advanced and thus has become insensitive to additional heat flow, only Devonian and younger sediments were included in the second calculation using the high-temperature scenario of Cercone (1984), that is, a geothermal gradient of 45°C km^{-1} during the Carboniferous with a linear decline

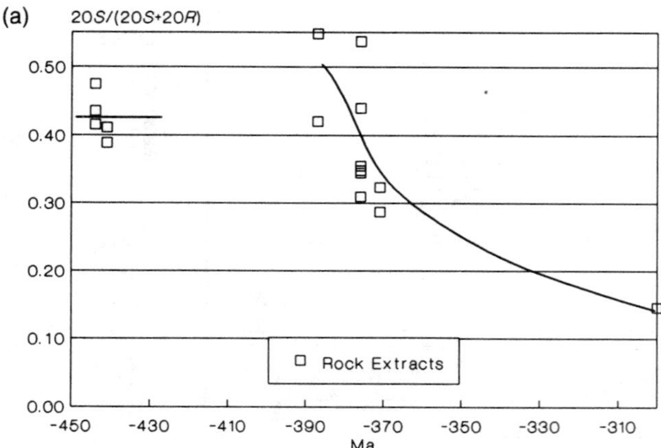

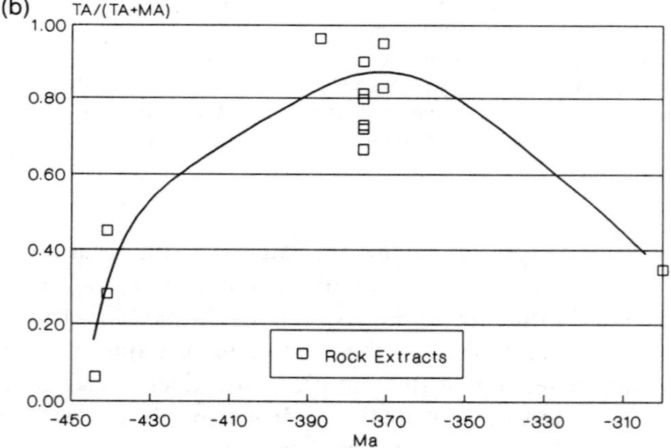

Figure 2.11 Sterane isomerization (a) and steroid aromatization values (b) measured on rock samples from the Michigan Basin. Data scatter particularly for the Devonian samples is due to the fact that these are from a range of depths.

to the present value plus deposition of 1 km of additional early Carboniferous to Pennsylvanian sediments which were eroded until the late Jurassic (Fig. 2.12). The position in the basin was selected to be offset from the center in order to match more closely the actual present depth settings of the sediment samples used for comparison with the results of model calculations. Under these conditions, steranes in early Devonian sediments would have reached a $20S/(20S + 20R)$ isomerization ratio of 0.54 about 310 Ma ago, that is, about 90 Ma after deposition. In contrast to this, sterane isomerization values in the Antrim Shale (middle Devonian) and in the Coldwater Shale (early Carboniferous) would be 0.47 and 0.34, respectively. All three values are in reasonable agreement with the measured data (Fig. 2.12). Based on these measurements and on the model calculations, it is feasible that Devonian source rocks have generated crude oil in the Michigan Basin. Sterane

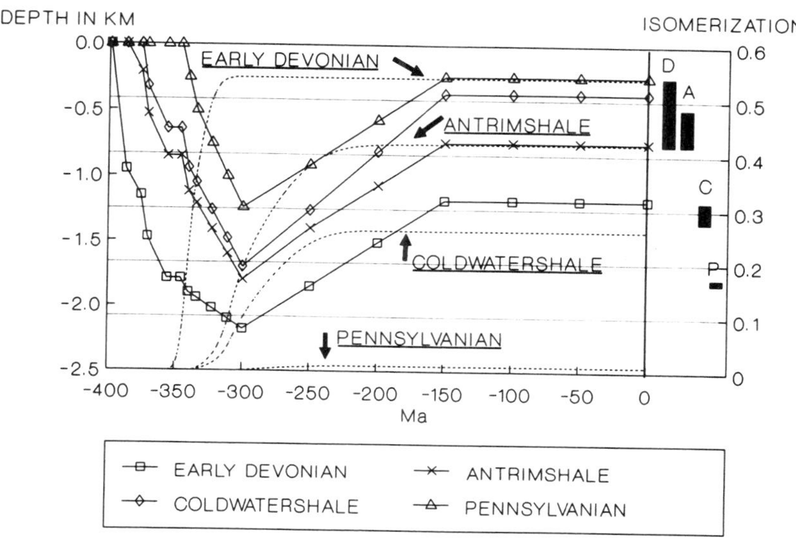

Figure 2.12 Late Paleozoic subsidence off the center of the Michigan Basin (solid lines) and calculated evolution of sterane isomerization based on the scenario of Cercone (1984) (broken lines). Ranges of measurements on natural samples are shown for comparison (D = early Devonian, A = Antrim Shale, C = Coldwater Shale, P = Pennsylvanian coal).

isomerization values of 0.37 and 0.40 measured for oils accumulated in the Berea and Stray Sandstones (Rullkötter et al., 1986) and not correlating well with any other oil from the Michigan Basin, may indicate that oil generation from the nearby Antrim or Coldwater Shales has indeed occurred.

Only the Pennsylvanian coal sample from a surface outcrop shows a significantly higher extent of sterane isomerization than would have been expected from the model calculation (Fig. 2.12). This result is supported by other biomarker data and implies that the coal was buried much more deeply in the past although no direct geological evidence is available for this. In order to account for the higher sterane isomerization value in the Pennsylvanian coal, we have modified the high-temperature scenario of Cercone (1984) somewhat. This also allowed us to refrain from using the unusually and probably unrealistically high geothermal gradient of 45°C km^{-1} in the Carboniferous. Figure 2.13 illustrates these modifications for the final calculation which encompasses the boundary conditions of Nunn et al. (1984) until the end of the lower Carboniferous. Then, the deposition of additional 1800 m of upper Carboniferous sediments is assumed which is accompanied by an increase of the geothermal gradient from 22°C to 35°C km^{-1}. This is followed by uplift and erosion and a decrease of the geothermal gradient to the present value until the beginning of the Jurassic. Under these conditions, all calculated sterane isomerization data match the measured values within the limits of experimental

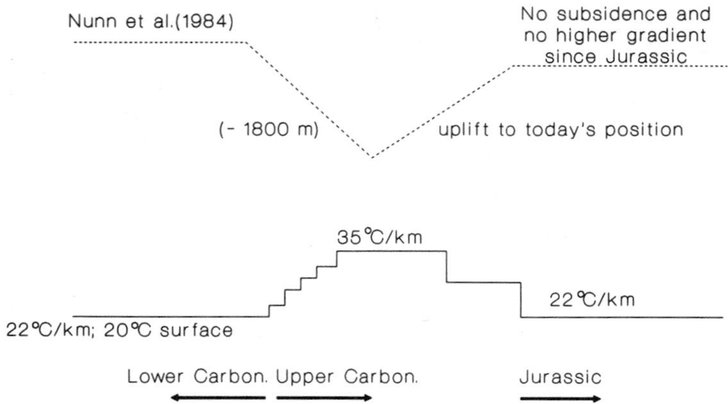

Figure 2.13 Modification of the subsidence and geothermal history of the Michigan Basin for simulation of sterane isomerization progress in Figure 2.14.

error (Fig. 2.14). Both the early Devonian sediments and the Antrim Shale, at the depth setting used for the calculation, may already have exceeded the maximum of sterane isomerization.

The choice of additional subsidence and of the temperature values is somewhat arbitrary. Other combinations are possible, but extremes would be geologically

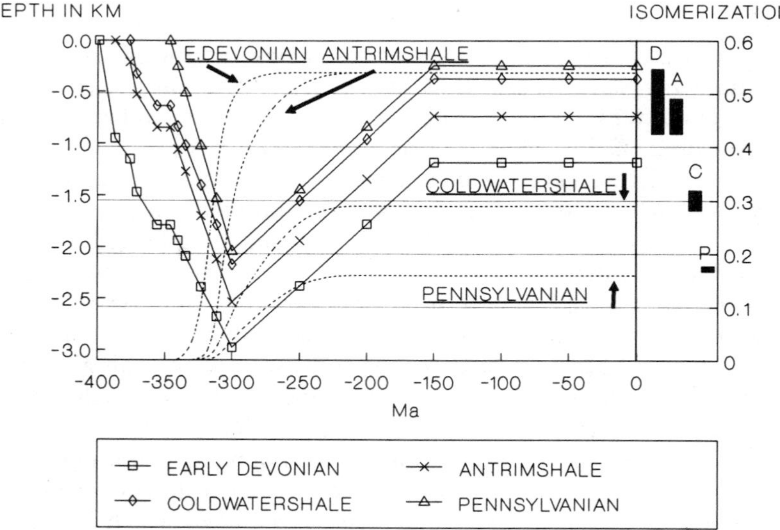

Figure 2.14 Late Paleozoic subsidence off the center of the Michigan Basin (solid lines) and calculated evolution of sterane isomerization based on the scenario in Figure 2.13 (broken lines). Ranges of measurements on natural samples are shown for comparison (cf. Fig. 2.12 legend).

unrealistic. Further constraints can be achieved only by using an additional biological marker reaction with significantly different values of the kinetic parameters.

CONCLUSIONS

Three organic geochemical biological marker reactions, that is, isomerization of C_{29} steranes at C-20, isomerization of 17α(H)-homohopanes at C-22, and transformation of mono- into triaromatic steroid hydrocarbons, could be effectively simulated by hydrous pyrolysis. It was shown that the steroid reactions undergo a reversal instead of leading to a maximum end value, whereas the hopane isomerization behaved as described.

The calculations performed to simulate the progress of sterane isomerization in Michigan Basin sediments are admittedly a rough attempt to define some new constraints on the geological and geothermal history of the basin. Unless a second biological marker reaction with different kinetic constants, in addition to sterane isomerization, is available for the model calculations, the conditions chosen and represented in Figures 2.13 and 2.14 are just one possibility of several others. The reason for this is because with just one reaction the effect of increased temperature (by subsidence or higher heat flow) cannot be satisfactorily separated from the effect of time at a somewhat lower temperature. Despite the limitations of these calculations, the present understanding of the evolution of the Michigan Basin as it is reflected in the literature has to be modified considerably. The organic geochemical data indicate that the Michigan Basin is not as stable an intracratonic basin as has been previously believed. It may rather be a part of or have a connection to the mid-American rift system, which may have caused more tectonic movement in the Michigan Basin than so far understood. Based on an investigation of the regional subsurface fluid movement in the Michigan Basin, Vugrinovitch (1989) came to similar conclusions, particularly with respect to higher heat flow in the past.

Acknowledgments

We thank Prof. Philip A. Meyers, University of Michigan (Ann Arbor, MI), for providing sediment and crude oil samples from the Michigan Basin and for stimulating discussions in the course of a long-term cooperation on the geochemistry of the Michigan Basin. We are grateful to U. Disko and F.J. Keller for their help in GC-MS measurements. The comments and suggestions of two anonymous referees were helpful to improve the quality of this text.

REFERENCES

Abbott, G.D., Lewis, C.A., and Maxwell, J.R. (1985) The kinetics of specific organic reactions in the zone of catagenesis. *Phil. Trans. R. Soc. Lond. A. 315*, 107–122.

Abbott, G.D., Wang, G.Y., Eglinton, T.I., Home, A.K., and Petch, G.S. (1990) The kinetics of biological marker release from vitrinite kerogen during the hydrous pyrolysis of vitrinite kerogen. *Geochim. Cosmochim. Acta 54*, 2451–2461.

Alexander, R., Cumbers, K.M., and Kagi, R.I. (1986) Alkylbiphenyls in ancient sediments and petroleums. In *Advances in Organic Geochemistry 1985* (eds. D. Leythaeuser and J. Rullkötter), pp. 997–1003. Pergamon Press.

Beaumont, C., Boutilier, R., Mackenzie, A.S., and Rullkötter, J. (1985) Isomerization and aromatization of hydrocarbons and the paleothermometry and burial history of Alberta Foreland Basin. *Bull. Amer. Assoc. Petr. Geol. 69*, 546–566.

Carlson, R.M.K. and Chamberlain, D.E. (1986) Steroid biomarker-clay adsorption free energies: Implications to petroleum migration indices. In *Advances in Organic Geochemistry 1985* (eds. D. Leythaeuser and J. Rullkötter), pp. 163–180. Pergamon Press.

Cercone, K.R. (1984) Thermal history of Michigan Basin. *Bull. Amer. Assoc. Petr. Geol. 68*, 130–136.

Dorr, J.A., Jr. and Eschman, D.F. (1970) *Geology of Michigan*. The University of Michigan Press, Ann Arbor.

Ensminger, A., Albrecht, P., Ourisson, G., and Tissot, B. (1977) Evolution of polycyclic hydrocarbons under effect of burial (Early Toarcian shales, Paris Basin). In *Advances in Organic Geochemistry 1975* (eds. R. Campos and J. Goñi), pp. 45–52, Enadimsa.

Eglinton, T.I. and Douglas, A.G. (1988) Quantitative study of biomarker hydrocarbons released from kerogens during hydrous pyrolysis. *Energy & Fuels 2*, 81–88.

Haxby, W.F., Turcotte, D.L., and Bird, J.M. (1976) Thermal and mechanical evolution of the Michigan Basin. In *Sedimentary Basins of Continental Margins and Cratons* (ed. M.H.P. Bott), *Tectonophys. 36*, 57–75.

Hoffmann, C.F., Mackenzie, A.S., Lewis, C.A., Maxwell, J.R., Oudin, J.L., Durand, B., and Vandenbroucke, M. (1984) A biological marker study of coals, shales and oils from the Mahakam Delta, Kalimantan, Indonesia. *Chem. Geol. 42*, 1–23.

Hong, Z.-H., Li, H.-X., Rullkötter, J., and Mackenzie, A.S. (1986) Geochemical application of sterane and triterpane biological marker compounds in the Linyi Basin. In *Advances in Organic Geochemistry 1985* (eds. D. Leythaeuser and J. Rullkötter), pp. 433–439. Pergamon Press.

Illich, H.A. and Grizzle, P.L. (1983) Comment on "Comparison of Michigan Basin crude oils" by Vogler et al. *Geochim. Cosmochim. Acta 47*, 1157–1159.

Illich, H.A. and Grizzle, P.L. (1985) Thermal subsidence and generation of hydrocarbons in Michigan Basin: Discussion. *Bull. Amer. Assoc. Petr. Geol. 69*, 1401–1404.

Lewan, M.D., Dolcater, D.L., and Bjorøy, M. (1986) Effects of thermal maturation on steroid hydrocarbons as determined by hydrous pyrolysis of Phosphoria Retort shale. *Geochim. Cosmochim. Acta 50*, 1977–1987.

Littke, R., Baker, D.R., and Leythaeuser, D. (1988) Microscopic and sedimentologic evidence for the generation and migration of hydrocarbons in Toarcian source rocks of different maturities. In *Advances in Organic Geochemistry 1987* (eds. L. Mattavelli and L. Novelli), pp. 549–559. Pergamon Press.

LITTKE, R., BAKER, D.R., LEYTHAEUSER, D., and RULLKÖTTER, J. (in press) Keys to the depositional history of the Posidonia Shale (Toarcian) in the Hils syncline area, northern Germany. In *Modern and Ancient Continental Shelf Anoxia* (eds. R. Tyson and T. Pearson). Special Publ. Geol. Soc. London, Blackwell.

MACKENZIE, A.S. (1984) Application of biological markers in petroleum geochemistry. In *Advances in Petroleum Geochemistry* (eds. J. Brooks and D.H. Welte), Vol. 1, pp. 115–214. Academic Press.

MACKENZIE, A.S. and MCKENZIE, D.P. (1983) Aromatization and isomerization of hydrocarbons in sedimentary basins formed by extension. *Geol. Magazine 120*, 417–470.

MACKENZIE, A.S., HOFFMANN, C.F., and MAXWELL, J.R. (1981) Molecular parameters of maturation in the Toarcian Shales, Paris Basin, France–III. Changes in aromatic steroid hydrocarbons. *Geochim. Cosmochim. Acta 45*, 1345–1355.

MACKENZIE, A.S., BEAUMONT, C., BOUTILIER, R., and RULLKÖTTER J. (1985) The aromatization and isomerization of hydrocarbons and the thermal and subsidence history of the Nova Scotia margin. *Phil. Trans. R. Soc. London A 315*, 203–232.

MACKENZIE, A.S., RULLKÖTTER, J., WELTE, D.H., and MANKIEWICZ, P. (1985) Reconstruction of oil formation and accumulation in North Slope, Alaska, using quantitative gas chromatography-mass spectrometry. In *Alaska North Slope Oil/Rock Correlation Study* (eds. L.B. Magoon and G.E. Claypool), AAPG Studies in Geology No. 20, pp. 319–377. Amer. Assoc. Petr. Geol., Tulsa.

MACKENZIE, A.S., PATIENCE, R.L., MAXWELL, J.R., VANDENBROUCKE, M., and DURAND, B. (1980) Molecular parameters of maturation in the Toarcian Shales, Paris Basin, France-I. Changes in the configurations of acyclic isoprenoid alkanes, steranes and triterpanes. *Geochim. Cosmochim. Acta 44*, 1709–1721.

MARZI, R. (1989) Kinetik und quantitative Analyse der Isomerisierung und Aromatisierung von fossilen Steroidkohlenwasserstoffen im Experiment und in natürlichen Probensequenzen. *Berichte der Kernforschungsanlage Jülich*, Nr. 2264. KFA Jülich, ISSN 0366-0885.

MOLDOWAN, J.M. and FAGO, F.J. (1986) Structure and significance of a novel rearranged monoaromatic steroid hydrocarbon in petroleum. *Geochim. Cosmochim. Acta 50*, 343–351.

NUNN, J.A., SLEEP, N.H., and MOORE, W.E. (1984) Thermal subsidence and generation of hydrocarbons in Michigan Basin. *Bull. Amer. Assoc. Petr. Geol. 68*, 296–315.

PEAKMAN, T.M. and MAXWELL, J.R. (1988) Early diagenetic pathways of steroid alkenes. In *Advances in Organic Geochemistry 1987* (eds. L. Mattavelli and L. Novelli), pp. 583–592. Pergamon Press.

PHILP, R.P. and GILBERT, T.D. (1982) Unusual distribution of biological markers in an Australian crude oil. *Nature 299*, 245–247.

POWELL, T.G., MACQUEEN, R.W., BARKER, J.F., and BREE, D.G. (1984) Geochemical character and origin of Ontario oils. *Bull. Can. Soc. Petr. Geol. 32*, 289–312.

PRUITT, J.D. (1983) Comment on "Comparison of Michigan Basin crude oils" by Vogler et al. *Geochim. Cosmochim. Acta 47*, 1159–1161.

RADKE, M., WILLSCH, H., and WELTE, D.H. (1980) Preparative hydrocarbon group type determination by automated medium pressure liquid chromatography. *Anal. Chem. 52*, 406–411.

RIOLO, J., HUSSLER, G., ALBRECHT, P., and CONNAN, J. (1986) Distribution of aromatic steroids in geological samples: Their evaluation as geochemical parameters. In *Advances in Organic Geochemistry 1985* (eds. D. Leythaeuser and J. Rullkötter), pp. 981–990. Pergamon Press.

RULLKÖTTER, J., and MARZI, R. (1988) Natural and artificial maturation of biological markers in a Toarcian shale from northern Germany. In *Advances in Organic Geochemistry 1987* (eds. L. Mattavelli and L. Novelli), pp. 639–645. Pergamon Press.

RULLKÖTTER, J. and WELTE, D.H. (1983) Maturation of organic matter in areas of high heat flow: A study of sediments from DSDP Leg 63, offshore California, and Leg 64, Gulf of California. In *Advances in Organic Geochemistry 1981* (eds. M. Bjorøy et al.) pp. 438–448. J. Wiley & Sons.

RULLKÖTTER, J., MACKENZIE, A.S., WELTE, D.H., LEYTHAEUSER, D., and RADKE, M. (1984) Quantitative gas chromatography-mass spectrometry analysis of geological samples. In *Advances in Organic Geochemistry 1983* (eds. P.A. Schenck, J.W. de Leeuw, and G.W.M. Lijmbach), pp. 817–827. Pergamon Press.

RULLKÖTTER, J., SPIRO, B., and NISSENBAUM, A. (1985) Biological marker characteristics of oils and asphalts from carbonate source rocks in a rapidly subsiding graben, Dead Sea, Israel. *Geochim. Cosmochim. Acta 49*, 1357–1370.

RULLKÖTTER, J., MEYERS, P.A., SCHAEFER, R.G., and DUNHAM, K.W. (1986) Oil generation in the Michigan Basin: A biological marker and carbon isotope approach. In *Advances in Organic Geochemistry 1985* (eds. D. Leythaeuser and J. Rullkötter), pp. 359–375. Pergamon Press.

RULLKÖTTER, J., LEYTHAEUSER, D., HORSFIELD, B., LITTKE, R., MANN, U., MÜLLER, P.J., RADKE, M., SCHAEFER, R.G., SCHENK, H.-J., SCHWOCHAU, K., WITTE, E.G., and WELTE, D.H. (1988) Organic matter maturation under the influence of a deep intrusive heat source: A natural experiment for quantitation of hydrocarbon generation and expulsion from a petroleum source rock (Toarcian shale, northern Germany). In *Advances in Organic Geochemistry 1987* (eds. L. Mattavelli and L. Novelli), pp. 847–856. Pergamon Press.

RULLKÖTTER, J., MARZI, R., and MEYERS, P.A. (in press) Biological markers in Paleozoic sedimentary rocks and crude oils from the Michigan Basin: Reassessment of sources and thermal history of organic matter. In *Early Organic Evolution: Implications for Mineral and Energy Resources* (eds. M. Schidlowski, M.M. Kimberley, D.M. McKirdy, P.A. Trudinger, and S. Golubic). Springer-Verlag.

SAJGO, Cs. and LEFLER, J. (1986) A reaction kinetic approach to the temperature-time history of sedimentary basins. In *Lecture Notes in Earth Science* (eds. G. Buntebarth and L. Stegena), Vol. 5, pp. 119–151. Springer-Verlag.

SEIFERT, W.K. and MOLDOWAN, J.M. (1978) Applications of steranes, terpanes and monoaromatics to the maturation, migration and source of crude oils. *Geochim. Cosmochim. Acta 42*, 77–95.

SEIFERT, W.K. and MOLDOWAN, J.M. (1980) The effect of thermal stress on source-rock quality as measured by hopane stereochemistry. In *Advances in Organic Geochemistry 1979* (eds. A.G. Douglas and J.R. Maxwell), pp. 229–237. Pergamon Press.

SEIFERT, W.K. and MOLDOWAN, J.M. (1981) Paleoreconstruction by biological markers. *Geochim. Cosmochim. Acta 45*, 783–794.

SLEEP, N.H. and SLOSS, L.L. (1978) A deep borehole in the Michigan Basin. *J. Geophys. Res. 83*, 5815–5819.

SNOWDON, L.R., BROOKS, P.W., WILLIAMS, G.K., and GOODARZI, F. (1987) Correlation of Canol formation source rock with oil from Norman wells. *Org. Geochem. 11*, 529–548.

STRACHAN, M.G., ALEXANDER, R., VAN BRONSWIJK, W., and KAGI, R.I. (1989) Source and heating rate effects upon maturity parameters based on ratio of 24-ethylcholestane diastereomers. *J. Geochem. Explor. 31*, 285–294.

SUZUKI, N. (1984) Estimation of maximum temperature of mudstone by two kinetic parameters; epimerization of sterane and hopane. *Geochim. Cosmochim. Acta 48*, 2273–2282.

TEN HAVEN, H.L., DE LEEUW, J.W., PEAKMAN, T.M., and MAXWELL, J.R. (1986) Anomalies in steroid and hopanoid maturity indices. *Geochim. Cosmochim. Acta 50*, 853–855.

VAN GRAAS, G., BAAS, J.M.A., VAN DE GRAAF, B., and DE LEEUW, J.W. (1982) Theoretical organic geochemistry I. The thermodynamic stability of several cholestane isomers calculated by molecular mechanics. *Geochim. Cosmochim. Acta 46*, 2399–2402.

VOGLER, E.A., MEYERS, P.A., and MOORE, W.E. (1981) Comparison of Michigan Basin crude oils. *Geochim. Cosmochim. Acta 45*, 2287–2293.

VUGRINOVITCH, R. (1989) Subsurface temperatures and surface heat flow in the Michigan Basin and the relationship to regional subsurface fluid movement. *Mar. Petr. Geol. 6*, 60–70.

WAPLES, D.W. (1980) Time and temperature in petroleum formation: Application of Lopatin's method to petroleum exploration. *Bull. Amer. Assoc. Petr. Geol. 64*, 916–926.

WINTERS, J.C., WILLIAMS, J.A., and LEWAN, M.D. (1983) A laboratory study of petroleum generation by hydrous pyrolysis. In *Advances in Organic Geochemistry 1981* (eds. M. Bjorøy et al.), pp. 524–533. J. Wiley & Sons.

3

A Reinvestigation of Aspects of the Early Diagenetic Pathways of 4-Methylsterenes Based on Molecular Mechanics Calculations and the Acid Catalyzed Isomerization of 4-Methylcholest-4-ene

Josef A. Rechka, Henricus Catharina Cox, Torren M. Peakman, Jan Willem De Leeuw, and James R. Maxwell

Abstract. Molecular mechanics calculations of the relative stabilities of possible products from mild acid treatment of 4-methylcholest-4-ene prior to backbone rearrangement have prompted a reinvestigation of the mixture of components obtained. The calculations indicate that the two products previously believed to be A-ring contracted components are in fact the 5α and 5β isomers of 4-methylcholest-3-ene. This has been verified by reisolation of the components, mass and nuclear magnetic resonance studies of the ruthenium tetroxide oxidation products, and comparison with the 5α isomer prepared independently. These results have been used to reappraise aspects of the sedimentary fate of 4-methyl steroid alcohols.

INTRODUCTION

Studies of the origin and fate of sedimentary steroids remain an active area of research in organic geochemistry (e.g., Summons et al., 1987; Summons and Capon 1988; De Leeuw et al., 1989; Peakman et al., 1989; Goodwin et al., 1988). Evidence is accumulating that the structural variation within the precursor sterols largely determines the variation in steroid hydrocarbon distributions observed in sediments and in crude oils. Recent studies have, therefore, returned to consideration of the sterenes in immature sediments since these are seen as intermediates in the pathway of sterol to sterane conversion during diagenesis (e.g., Wolff et al., 1986a; De Leeuw et al., 1989; Peakman et al., 1989).

A particular class of steroid components, that is, those possessing a methyl group at C-4, have been the subject of recent interest, in view of the potential of certain of them as markers for dinoflagellates (Robinson et al., 1984; Wolff et al., 1986a, 1986b; Summons et al., 1987; Goodwin et al., 1988). A study of the products arising from the acidic treatment of 4-methylcholest-4-ene (prior to backbone rearrangement) accounted for the distribution of C_{28} 4-methylsterenes in immature Messel oil shale (Wolff et al., 1986a). The products were assigned as mainly starting material (I, 75%), accompanied by two components with 5β and 5α stereochemistry and which had apparently undergone A-ring contraction (II, 15% and III, 7%), and low amounts of 4α- and presumed 4β-methylcholest-5-ene (IV and V, both $< 5\%$). Hydrogenation of 4-methylcholest-4-ene gave a mixture of steranes including 4β-methyl-5α-cholestane in higher abundance than its thermodynamically more stable 4α-methyl counterpart, thereby suggesting an explanation for the presence of 4β-methylsteranes in immature sediments through reduction (Wolff et al., 1986a).

Molecular mechanics calculations have been successfully applied to determine the composition of a cholestane isomerate at various temperatures (Van Graas et al., 1982). This approach has been extended to calculating the thermodynamic stabilities of a range of cholestenes (De Leeuw et al., 1989). The calculations reported here were performed using the MM2 force field (Allinger, 1977). To test the force field applied, we decided to calculate the stabilities and equilibrium ratios of the components arising from acid-catalyzed equilibration of 4-methylcholest-4-ene, since the products from mild acid treatment had been examined previously (Wolff et al., 1986a). We reasoned that this could provide a test for the suitability of the force field in dealing with complex polycyclic alkenes such as sterenes.

The results of these calculations indicate that the two components previously assigned as having an A-ring contracted structure (II and III) are unlikely to be present in an equilibrium mixture due to their high $\Delta\Delta G$ values and that the 5β- and 5α- isomers of 4-methylcholest-3-ene (VI and VII) are more likely components. The differing results of the molecular mechanics calculations and those presented by Wolff et al. (1986a) prompted, therefore, a chemical reinvestigation of the acid-catalyzed equilibration of 4-methylcholest-4-ene.

EXPERIMENTAL

Molecular mechanics calculations were performed as described previously (De Leeuw et al., 1989) for steroid alkenes using the MM2 empirical force field (Allinger, 1977). 4-Methylcholest-4-ene was prepared and treated with anhydrous

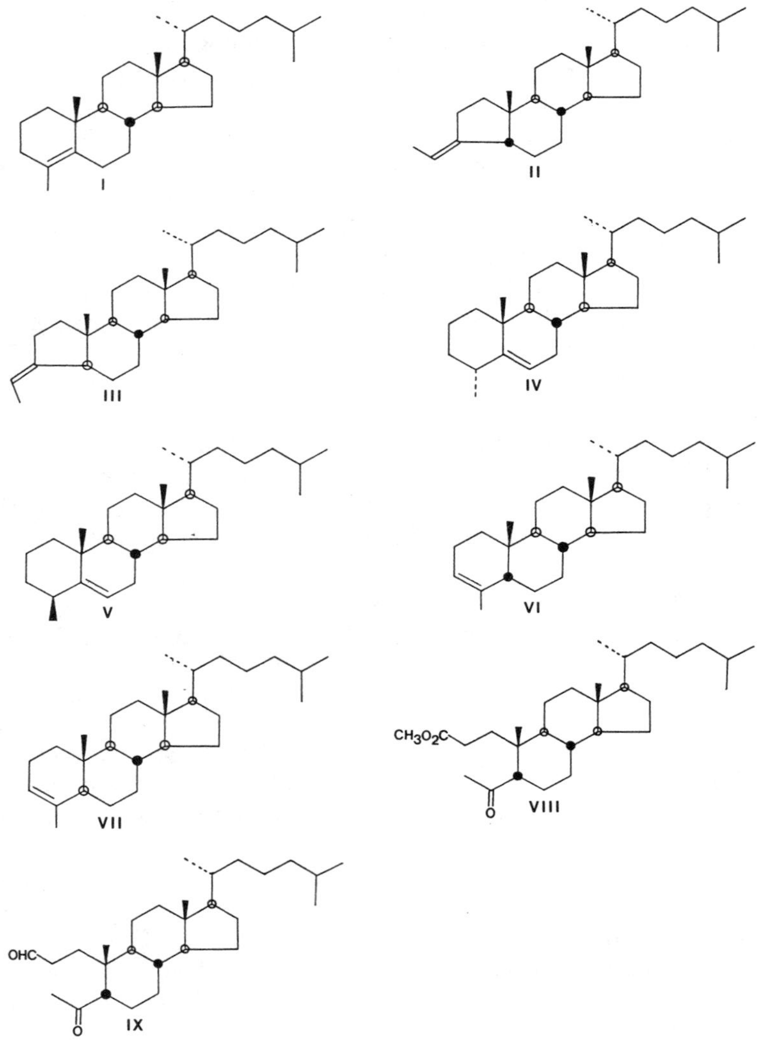

I II III IV V VI VII VIII IX

CH_3O_2C OHC

Figure 3.1

p-toluenesulphonic acid (tosic acid)/acetic acid (AcOH) for various periods of time as described previously (Wolff et al., 1986a and references therein).

Isolation and Oxidation of Acid Treatment Products

The components previously assigned the A-ring contracted structure were obtained by dilution of the product mixture with hexane followed by washing with water, then $NaHCO_3$ solution and finally with brine. After drying with sodium sulphate, filtration and evaporation yielded an oil which was taken up in hexane and passed through a small plug of neutral alumina. Reversed phase high-pressure liquid chromatography (HPLC) with 2 percent aqueous acetone as the mobile phase yielded the individual compounds (cf. Wolff et al., 1986a). The components were oxidized using two methods.

1. A solution of the sterene (2 mg) in 1:1 methanol:dichloromethane (2 ml) was treated with aliquots of a solution of ozone in dichloromethane at −78°C in the presence of $NaHCO_3$ until no further starting material remained, as indicated by thin-layer chromatography. The intermediate alkoxy hydroperoxide was decomposed with dimethyl sulphide and analyzed by GC-MS.
2. A solution of the sterene (2 mg) in a mixture of carbon tetrachloride, acetonitrile and water (1:1:2; 2 ml) was treated with a catalytic amount of ruthenium dioxide and one equivalent of sodium periodate (1.1 mg). The resultant mixture, after the usual workup (Carlsen et al., 1981) and treatment with diazomethane, was chromatographed on silica gel. The major product was isolated using 10 percent ethyl acetate in hexane as the eluent and was analyzed by GC-MS and nuclear magnetic resonance (NMR).

Preparation and Base Treatment of 4α(Me)-5α-Cholestan-3α-(4-toluenesulphonate)

4-(4-Methylbenzenethiomethyl)-cholest-4-ene-3-one (Kirk and Petrow, 1962) was reduced with lithium in liquid ammonia (cf. Sucrow et al., 1977) to give 4α(Me)-5α-cholestan-3-one. Treatment of the ketone (100 mg) with sodium borohydride (100 mg) in ethanol (2 ml) for 4 hours gave a mixture of alcohols isomeric at C-3. The minor 3α-ol isomer was isolated by column chromatography on silica gel eluted with 20 percent ethyl acetate/hexane. This alcohol (6 mg) was then treated with 4-toluenesulphonyl chloride (10 mg) in pyridine (1 ml) for 24 hours at room temperature. Dilution with ether followed by washing with dilute HCl, drying (Na_2SO_4) and evaporation gave the corresponding toluenesulphonyl ester which was used without further purification.

The ester was treated with excess potassium *t*-butoxide in tetrahydrofuran (THF) (2 ml) for 24 hours at room temperature. The products of the elimination reaction were then isolated by partitioning between water and hexane followed by chromatography on silica gel eluted with hexane.

High-Performance Liquid Chromatography (HPLC)

Preparative HPLC was performed using a Waters m6000 pump, with a Rheodyne 7125 injector, and a LDC Refractometer III refractive index detector. The column used was obtained from Phase Separations Ltd. (Spherisorb S5 ODS2 250 × 10 mm i.d.)

Gas Chromatography (GC)

Routine GC analyses were conducted using a Carlo Erba HRGC 5160 (on column injection) fitted with a fused silica column (OV-1, 50 m × 0.32 mm i.d. or DB-1701 60 m × 0.32 mm i.d.). Hydrogen was used as carrier gas; typical temperature program conditions were 50–150°C at 10°C min^{-1} then 150–300°C at 4°C min^{-1}.

Combined Gas Chromatography-Mass Spectrometry (GC-MS)

Electron ionization (EI) spectra were obtained using a Carlo Erba HRGC 5160 (on column injection) interfaced directly with a Finnigan 4500 spectrometer. Data acquisition and processing were performed using an INCOS data system. Temperature programming was as for GC, with helium carrier gas, ion source 170°C, electron energy 45 eV, accelerating voltage 1.2 kV, and filament current 250 μA.

Nuclear Magnetic Resonance Spectroscopy (NMR)

NMR spectra were obtained in deuteriochloroform solution using either Jeol GX270 or GX400 spectrometers.

RESULTS

Table 3.1 shows the $\Delta\Delta G$ values and relative concentrations at the hypothetical equilibrium at 350 K for a series of 4-methylcholestenes, including four A-ring contracted sterenes, calculated by molecular mechanics. The components have been ranked in stability order relative to the most thermodynamically stable structure, 4-methylcholest-4-ene.

Figure 3.2 illustrates the distribution of 4-methylcholestenes obtained from acid-catalyzed treatment (3h) of 4-methylcholest-4-ene at 70°C (343 K). The use of a polar GC column (DB 1701 60 m × 0.32 mm i.d.) gave good resolution among the various sterenes. The abundances after 4h are given in Table 3.2.

The components previously assigned as having an A-ring contracted structure were isolated by reversed phase HPLC. Degradation studies on the compounds indicated that they werė actually 4-methyl-5β- and 4-methyl-5α-cholest-3-ene (VI and VII). Thus, GC-MS analysis of the ozonolysis product of the 5β component showed a mixture of four major components and a number of minor ones. The spectrum of the most abundant (Fig. 3.3a) was consistent with the keto-ester (VIII), methylation having occurred under the reaction conditions. Also present was a component with a molecular ion at $M^{+\cdot}$ 416; this is consistent with the correspond-

Table 3.1 $\Delta\Delta G$ (350 K) VALUES ($kJmol^{-1}$) AND RELATIVE CONCENTRATIONS AT HYPOTHETICAL EQUILIBRIUM FOR A SERIES OF 4-METHYLCHOLESTENES AND A-RING CONTRACTED STERENES AS CALCULATED BY MOLECULAR MECHANICS

Sterene	$\Delta\Delta G$	Relative concentration
Δ^4 (I)	0.00	100
5β-Δ^3 (VI)	5.1	17.4
5α-Δ^3 (VII)	5.4	15.7
4α(Me)5α-Δ^2	6.7	9.9
4α(Me)-Δ^5 (IV)	7.3	8.1
4β(Me)-Δ^5 (V)	14.6	0.7
3(E)-5β (II)	15.3	0.5
4β(Me)5β-Δ^2	17.3	0.3
4β(Me)5α-Δ^2	18.5	0.2
3(E)-5α	21.9	0.1
3(Z)-5β	25.9	0.0
3(Z)-5α (III)	33.0	0.0
4α(Me)5β-Δ^2	41.6	0.0

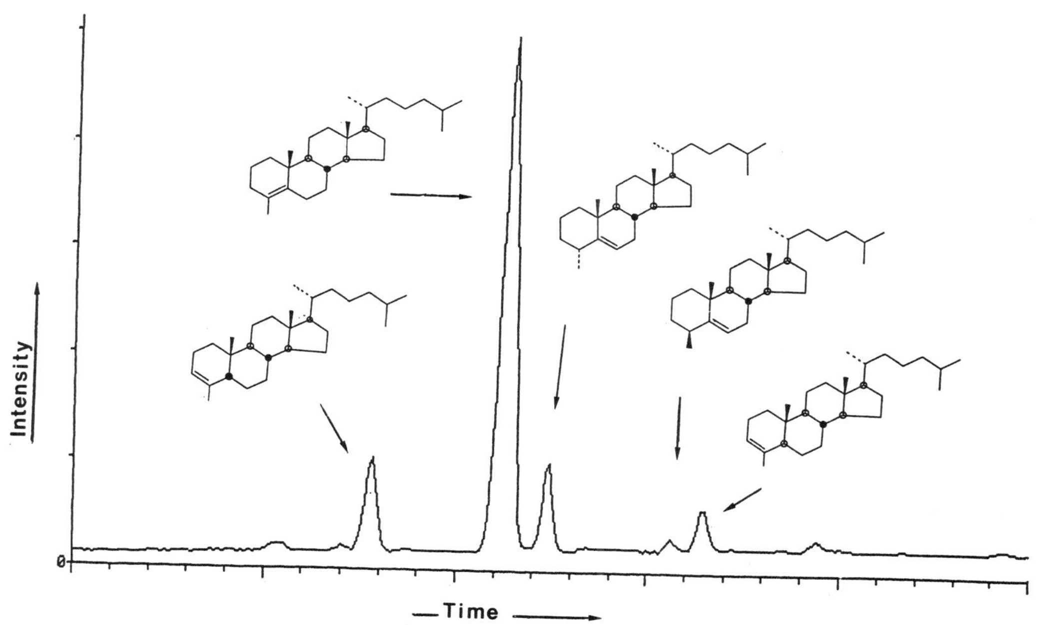

Figure 3.2 Distribution of 4-methylcholestenes from acid (TsOH/AcOH) treatment (3h) of 4-methylcholest-4-ene (60 m × 0.32 mm DB 1701).

Table 3.2 ABUNDANCES OF 4-METHYLCHOLESTENES FROM ACID-CATALYZED TREATMENT (4h) OF 4-METHYLCHOLEST-4-ENE

Sterene	Percent (equil) calculated	Percent experimental
Δ^4 (I)	70.5	60
5β-Δ^3 (VI)	12.3	16
5α-Δ^3 (VII)	11.1	11
4α(Me)-Δ^5 (IV)	5.7	11
4β(Me)-Δ^5 (V)	0.5	3

Comparison of calculated equilibrium values (350 K) and experimental values (343 K).

ing keto-aldehyde (IX). Oxidation with ruthenium tetroxide followed by methylation afforded a cleaner product from which the keto-ester (VIII) could be isolated. The NMR spectrum (Fig. 3.4a) showed clearly the presence of the methyl ketone group as a 3H singlet at δ 2.15 ppm and of the carbomethoxy group as a 3H singlet at δ 3.64 ppm, confirming that the double bond in the original alkene was at C-3. In addition, nuclear Overhauser effect (NOE) difference studies, clearly showed a 3 percent enhancement in the signal from the methyl ketone group and in the signal from the methyl group at C-10 (CH_3-19) when the C-5 hydrogen was irradiated, confirming 5β stereochemistry in the keto-ester and therefore in the original alkene.

The mass spectrum (Fig. 3.3b) of the product from the ruthenium tetroxide oxidation of the other alkene (VII) was similar to that from oxidation of the 5β alkene, again showing that the double bond was at position 3. Thus, signals arising from the methyl ketone group (3H at δ 2.15 ppm) and the methoxyl group (3H at δ 3.67 ppm) were observed in the NMR spectrum (Fig. 3.4b) confirming the double bond position. Irradiation of the C-5 hydrogen during NOE difference studies resulted in a 3 percent enhancement in the signal from the methyl ketone group but, in contrast to the keto-ester (VIII), no enhancement was observed in the signal due to the methyl group at C-10 (CH_3-19), in keeping with the latter having 5α stereochemistry.

GC and GC-MS analysis of the product of base treatment of the toluenesulphonyl ester of 4α-methyl-5α-cholestan-3α-ol showed two major products, as expected from anti elimination (Fig. 3.5). Thus the later eluting product had a mass spectrum very similar to that of 4-methyl-5α-cholest-3-ene (VII), with which it coeluted on OV-1. The mass spectrum of the major product (Fig. 3.3c) was consistent with 4α-methyl-5α-cholest-2-ene, showing an abundant ion at m/z 316 ascribed to a retro Diels Alder fragmentation in ring A (cf. Zaretskii, 1976). Coinjection of the elimination products with the products of the acid catalyzed rearrangement of the 4-ene and examination of the m/z 316 chromatogram of the latter failed to reveal the 2-ene in the rearrangement product mixture.

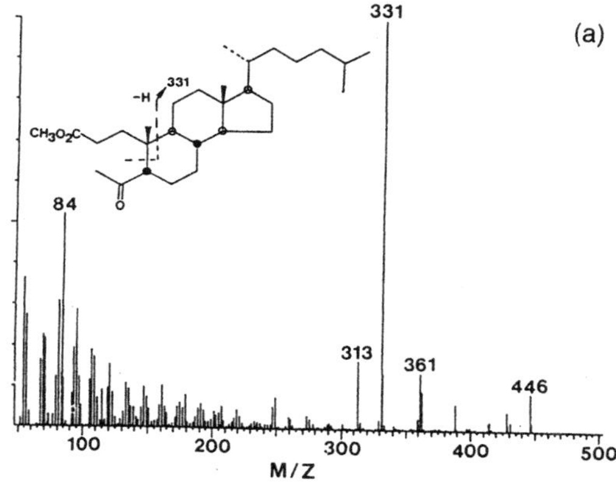

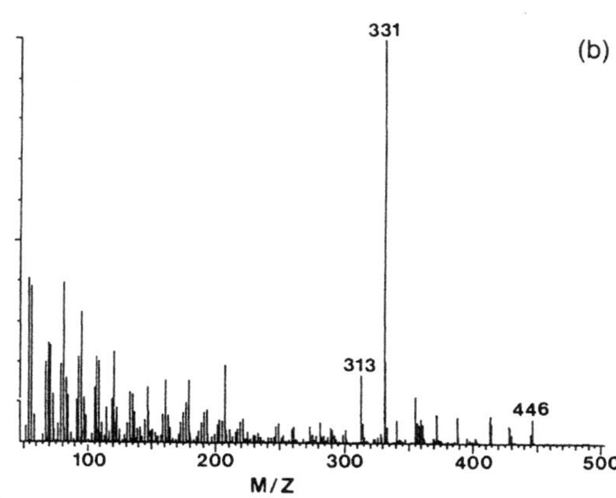

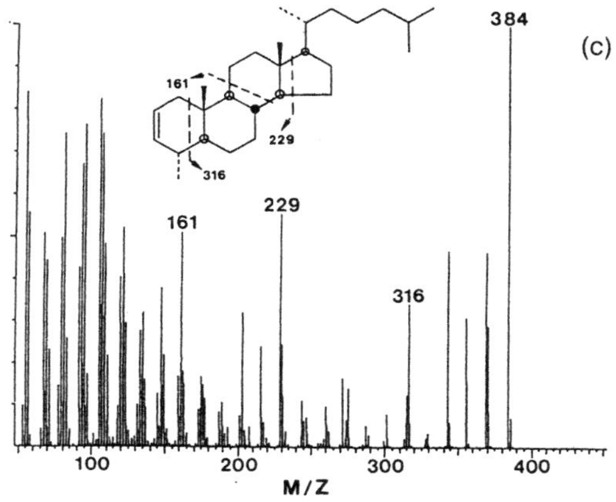

Figure 3.3 Mass spectra (GC-MS, OV-1) of methyl 4-oxo-3,4-seco-A-nor-5β-4-methylcholestan-3-oate (VIII) A; methyl 4-oxo-3,4-seco-A-nor-5α-4-methylcholestan-3-oate B; 4α-methyl-5α-cholest-2-ene C.

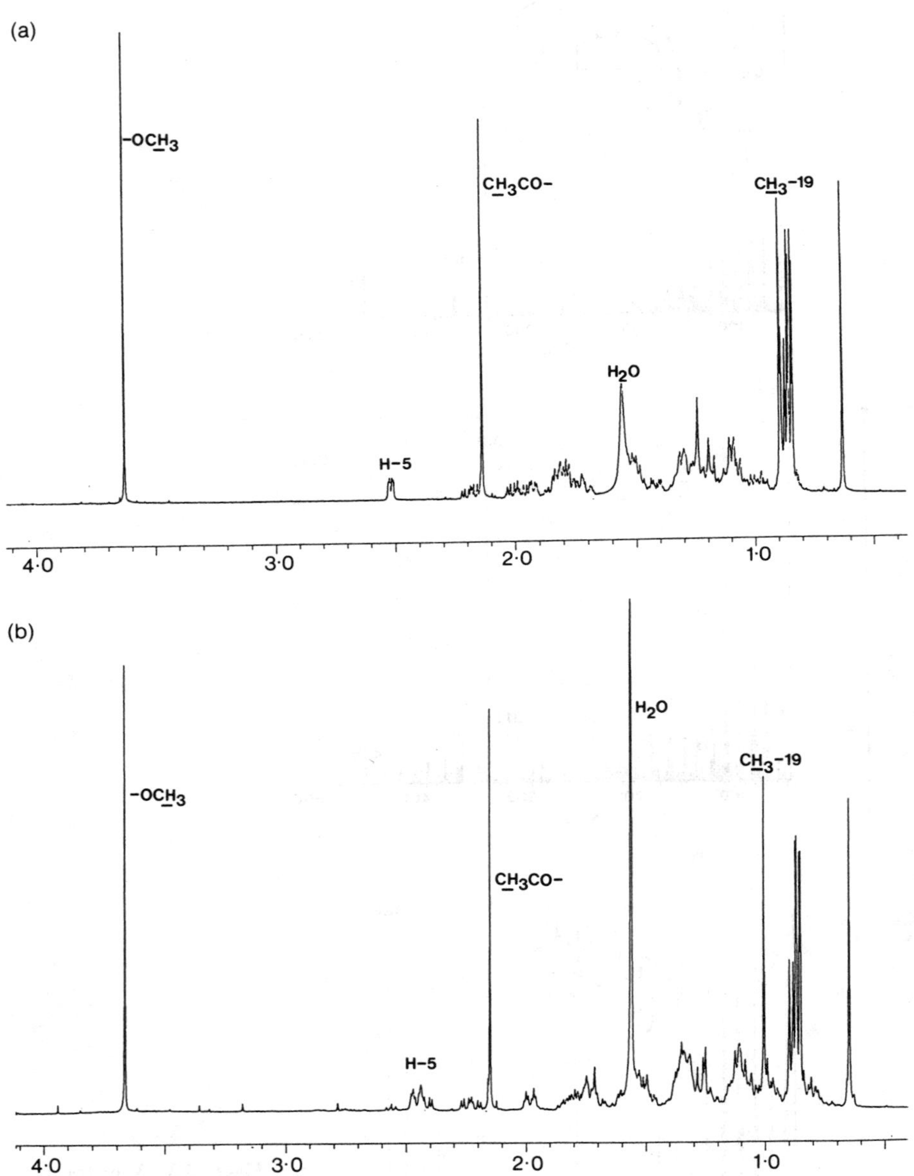

Figure 3.4 NMR spectra ($CDCl_3$, 400 MHz) of methyl 4-oxo-3,4-seco-A-nor-5β-4-methylcholestan-3-oate A; methyl 4-oxo-3,4-seco-A-nor-5α-4-methylcholestan-3-oate B.

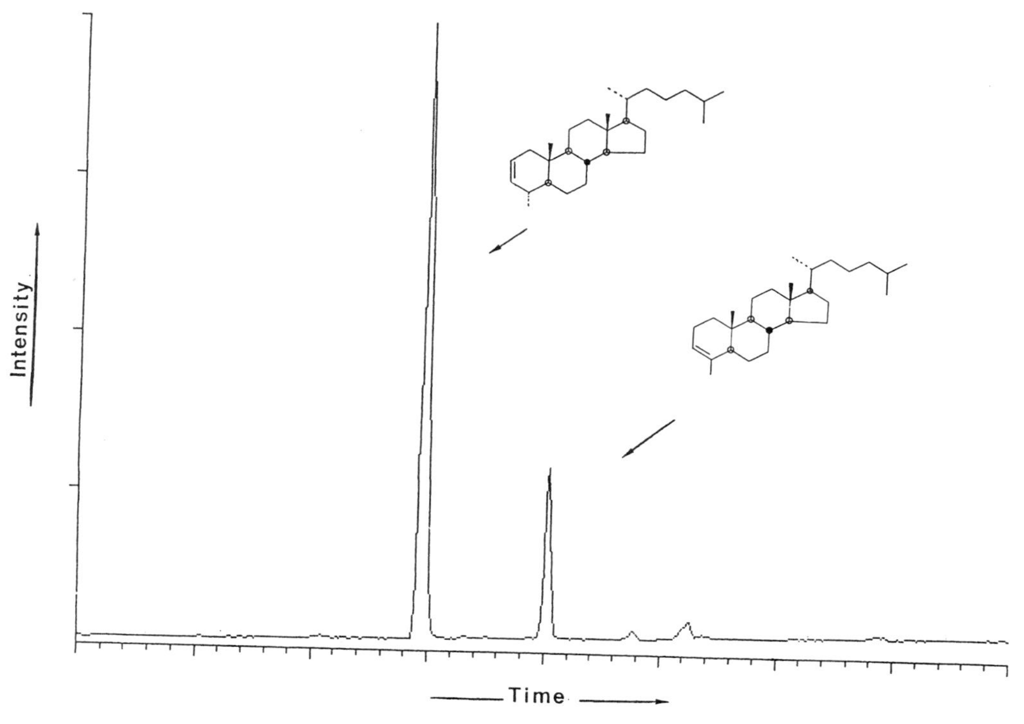

Figure 3.5 Distribution of products from treatment of 4α(Me)-5α-cholestan-3α-(4-toluene-sulphonate).

DISCUSSION

Acidic Treatment of 4-Methylcholest-4-ene

It is now clear that the components occurring in the product mixture from acidic (tosic/acetic acid) treatment of 4-methylcholest-4-ene and in the Messel oil shale, assigned previously as having an A-ring contracted structure (Wolff et al., 1986a), are actually 4-methylcholest-3-enes (VI and VII). The isomerization can be seen as arising essentially only from the tertiary carbocations at C-4 and C-5 (Fig. 3.6). The calculated percentages of the five sterenes I, IV, V, VI, and VII at hypothetical equilibrium are compared with the experimental results (after 4h acid treatment) in Table 3.2. The fit between the calculated and experimental values is reasonably good. The slight discrepancy may result from the experimental data not reflecting pre-equilibrium prior to backbone rearrangement (beyond 4 hours the backbone rearranged sterenes become significant), although such a pre-equilibrium appears possible for cholest-4- and 5-enes (De Leeuw et al., 1989 and references therein). Although the calculations show that 4α-methyl-5α-cholest-2-

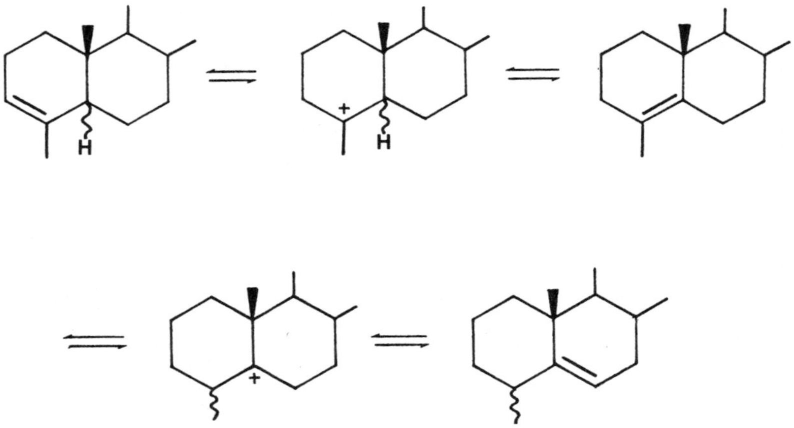

Figure 3.6 Acid-catalyzed isomerization of Δ^3-, Δ^4- and Δ^5-4-methylcholestenes.

ene has a thermodynamic stability comparable to those of the other 4-methylcholestenes in Table 3.1, it was not detected in the product mixture. This would be expected since its formation requires the existence of the secondary C-3 carbocation and such mild acid equilibrations are thought to proceed almost exclusively via tertiary carbocations (De Leeuw et al., 1989).

Low-Temperature Diagenesis of 4-Methylsterols and Stanols

Aspects of the low-temperature diagenesis of 4-methylsteroidal alcohols are now reappraised with reference to the mentioned observations and by comparison with recent conclusions reached for 4-desmethyl steroids (De Leeuw et al., 1989) where selective reduction and isomerization of double bonds via tertiary carbocations are thought to occur.

4α-methylstanols. Dehydration of 4α-methyl-5α-stanols in sediments might be expected to give in the first instance the corresponding Δ^2-4α-methyl- and Δ^3-4α-methyl-5α-sterenes. The pathway of dehydration is not known; a 3β-stanol does not contain the correct *trans*-anti-*trans* arrangement of the alcohol group and neighboring hydrogen unless the ring is twisted into a higher energy conformation. Whether 4α-methyl-5α-stan-3α-ols can be formed in sediments from the corresponding 3β stanols via a ketone intermediate (cf. Mermoud et al., 1984) is not yet known (cf. Wolff et al., 1986b) although such species would be expected to undergo more facile dehydration as they do possess the correct orientation for elimination (cf. Wolff et al., 1986b and references therein). In addition to a mechanism involving dehydration, it is possible that other mechanisms may be involved

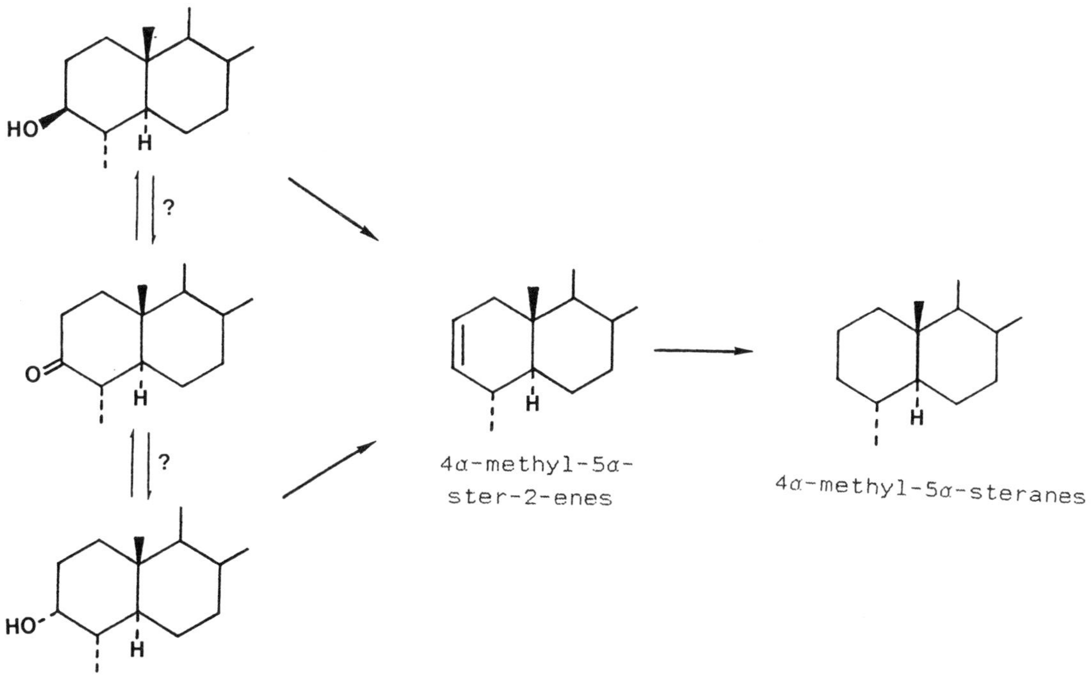

HO
H
?
O
H
?
HO
H
4α-methyl-5α-
ster-2-enes
H
4α-methyl-5α-steranes
H

Figure 3.7 Proposed major low-temperature diagenetic pathways of 4α-methylstanols.

(e.g., SN_2 inversion by phosphate or sulphate under mineral catalysis). Although more studies are clearly required in this area it should be pointed out that Δ^2-4α-methyl-5α-sterenes have been reported in some sediments (De Leeuw and Baas, 1986 and references therein). The fate of these alkenes is probably reduction to

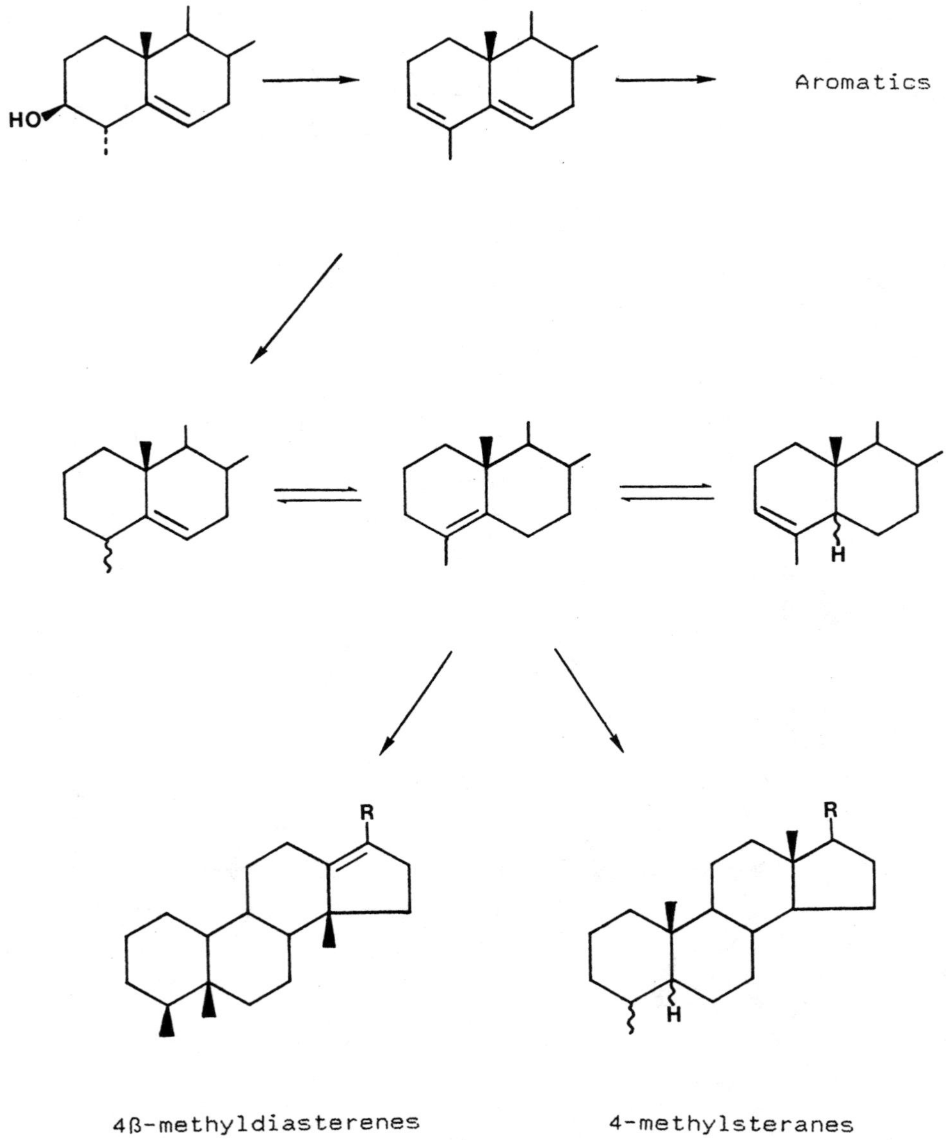

HO

Aromatics

H

R

R

H

4ß-methyldiasterenes

4-methylsteranes

Figure 3.8 Proposed low-temperature diagenetic pathways of Δ^5-4α-methylsterols.

4α-methyl-5α-steranes, a process which is now believed to occur for Δ^2-desmethyl-5α-sterenes (De Leeuw et al., 1989). The suggested early diagenetic pathways of 4α-methyl-5α-stan-3β-ols are summarized in Fig. 3.7.

Δ^5-4α-Methylsterols. In the sedimentary environment, and even in the water column, Δ^5-sterols are believed to undergo facile dehydration to $\Delta^{3,5}$-steradienes (e.g., Wakeham et al., 1984). These dienes appear to undergo a variety of transformations, for example, isomerizations and dehydrogenations to A-ring aromatic steroids (Brassell et al., 1984 and references cited therein) and selective reduction of Δ^5-sterenes (De Leeuw et al., 1989). It is not unreasonable to propose similar diagenetic pathways for Δ^5-4α-methylsterols as illustrated in Figure 3.8. Thus, we can envisage dehydration to $\Delta^{3,5}$-4-methylsteradienes which would undergo aromatization or selective reduction to Δ^5-4-methylsterenes. These Δ^5-enes will rearrange to the corresponding Δ^4-enes and Δ^3-enes which can then undergo either reduction to the isomeric mixtures of 4-methylsteranes observed in immature sediments (e.g., Wolff et al., 1986a) or backbone rearrangement to 4-methyldiasterenes (Rubinstein and Albrecht, 1975; Wolff et al., 1986a; Peakman and Maxwell, 1989; Peakman et al., 1991). It is likely that the reduction of the Δ^3-ene, Δ^4-ene and Δ^5-ene mixture occurs from the less hindered Δ^3-ene and Δ^5-ene isomers since tetrasubstituted double bonds are known to undergo hydrogenation very slowly in the laboratory. Certain aspects of these proposals obviously require confirmation from further chemical studies and investigation of immature sediments.

CONCLUSIONS

Molecular mechanics calculations have indicated that two of the products of the acid catalyzed rearrangement of 4-methylcholest-4-ene which were previously assigned as having an A-ring contracted structure are 4-methylcholest-3-enes. This has been verified by oxidation of the isolated compounds from the mixture.

Aspects of the low-temperature diagenetic pathways of 4-methyl steroid alcohols have been reappraised in view of these findings and by comparison with recent ideas concerning the corresponding 4-desmethyl analogues (De Leeuw et al., 1989).

Acknowledgments

We are grateful to British Petroleum plc. for financial support and for a research fellowship (J.A. Rechka). T.M. Peakman thanks Professor P.A. Schenck (Delft University of Technology) for a research fellowship. We would also like to thank the Natural Environment Research Council for GC-MS facilities (GR3/2951 and GR3/3758) and W.G. Prowse for NMR spectra. J.M.A. Baas and B. Van de Graaf are thanked for making available DELPHI, their computer program for molecular mechanics.

REFERENCES

ALLINGER, N.L. (1977). Conformational analysis. 130. MM2. A hydrocarbon force field utilizing V1 and V2 torsional terms. *J. Amer. Chem. Soc. 99*, 127–134.

BRASSELL, S.C., MCEVOY, J., HOFFMANN, C.F., LAMB, N.A., PEAKMAN, T.M., and MAXWELL, J.R. (1984) Isomerisation, rearrangement and aromatisation of steroids in distinguishing early stages of diagenesis. In *Advances in Organic Geochemistry 1983* (eds. P.A. Schenck, J.W. de Leeuw, and G.M.W. Lijmbach) *Org. Geochem. 6*, 11–23.

CARLSEN, P.H.J., KATSUKI, T., MARTIN, V.S., and SHARPLESS, K.B. (1981). A greatly improved procedure for ruthenium tetroxide catalysed oxidations of organic compounds. *J. Org. Chem. 46*, 3936–3938.

DE LEEUW, J.W. and BAAS, M. (1986) Early stage diagenesis of steroids. In *Methods in Geochemistry and Geophysics 24. Biological markers in the sedimentary record.* (ed. R.B. Johns), pp. 101–123. Elsevier, Amsterdam.

DE LEEUW, J.W., COX, H.C., VAN GRAAS, G., VAN DE MEER, F.W., PEAKMAN, T.M., BAAS, J.M.A., and VAN DE GRAAF, B. (1989) Limited isomerisation and selected hydrogenation of sterenes during early diagenesis. *Geochim. Cosmochim. Acta. 53*, 903–909.

GOODWIN, N.S., MANN, A.L., and PATIENCE, R.L. (1988) Structure and significance of C_{30} 4-methyl steranes in lacustrine shales and oils. *Org. Geochem. 12*, 495–506.

KIRK, D.N. and PETROW, V. (1962) Modified steroid hormones Part XXVIII. A new route to 4-methyl-3-oxo-Δ^4-steroids. *J. Chem. Soc.* 1091–1096.

MERMOUD, F., WÜNSCHE, L., CLERC, O., GÜLAÇAR, F.O., and BUCHS, A. (1984). Steroidal ketones in the early diagenetic transformations of Δ^5-sterols in different types of sediments. In *Advances in Organic Geochemistry 1983.* (eds. P.A. Schenck, J.W. de Leeuw, and G.M.W. Lijmbach) *Org. Geochem. 6*, 25–29.

PEAKMAN, T.M., TEN HAVEN, H.L., RECHKA, J.A., de LEEUW, J.W. and MAXWELL, J.R. (1989) Occurrence of (20*R*)- and (20*S*)-$\Delta^{8,14}$- and Δ^{14}-5α-sterenes and the origin of 5α,14β,17β-steranes in immature sediments. *Geochim. Cosmochim. Acta. 53*, 2001–2009.

PEAKMAN, T.M. and MAXWELL, J.R. (1989) Early diagenetic pathways of steroid alkenes. In *Advances in Organic Geochemistry 1987.* (eds. L. Mattavelli, and L. Novelli). *Org. Geochem. 13*, 583–592.

PEAKMAN, T.M., JERVOISE, A., WOLFF, G.A., and MAXWELL, J.R. (1991) Acid-catalyzed rearrangements of steroid alkenes. Part 4. An initial reinvestigation of the backbone rearrangement of 4-methylcholest-4-ene. Chapter 4 in this text.

ROBINSON, N., EGLINTON, G., BRASSELL, S.C., and CRANWELL, P.A. (1984). Dinoflagellate origin for sedimentary 4-methyl steroids and 5α-stanols. *Nature. 308*, 439–441.

RUBINSTEIN, I. and ALBRECHT, P. (1975) The occurrence of nuclear methylated sterenes in a shale. *J. Chem. Soc., Chem. Commun.*, 957–958.

SUCROW, W., LITTMANN, W., and RADÜCHEL (1977). Synthese von cistrostadienol, lophenol, und 24-methylenlophenol. *Chem. Ber. 110*, 1523–1531.

SUMMONS, R.E., VOLKMAN, J.K., and BOREHAM, C.J. (1987) Dinosterane and other steroidal hydrocarbons of dinoflagellate origin in sediments and petroleum. *Geochim. Cosmochim. Acta. 51*, 3075–3082.

SUMMONS, R.E. and CAPON, R.J. (1988) Fossil steranes with unprecedented methylation in ring-A. *Geochim. Cosmochim. Acta. 52*, 2733–2736.

VAN GRAAS, G., BAAS, J.M.A., VAN DE GRAAF, B., and DE LEEUW, J.W. (1982) Theoretical organic geochemistry I. The thermodynamic stability of several cholestane isomers calculated by molecular mechanics. *Geochim. Cosmochim. Acta. 46*, 2399–2402.

WAKEHAM, S.G., GAGOSIAN, R.B., FARRINGTON, J.W., and CANUEL, E.A. (1984). Sterenes in suspended particulate matter in the eastern tropical north Pacific. *Nature. 308*, 840–843.

WOLFF, G.A., LAMB, N.A., and MAXWELL, J.R. (1986a). The origin and fate of 4-methyl-steroid hydrocarbons I. 4-methylsterenes. *Geochim. Cosmochim. Acta. 50*, 335–342.

WOLFF, G.A., LAMB, N.A., and MAXWELL, J.R. (1986b). The origin and fate of 4-methyl-steroid hydrocarbons II. Dehydration of stanols and occurrence of C_{30} 4-methylsteranes. In *Advances in Organic Geochemistry 1985*. (eds. D. Leythaeuser and J. Rullkötter). *Org. Geochem. 10*, 965–974.

ZARETSKII (1976). *Mass Spectroscopy of Steroids.* Halsted Press.

4

Acid-Catalyzed Rearrangements of Steroid Alkenes. Part 4. An Initial Reinvestigation of the Backbone Rearrangement of 4-Methylcholest-4-ene.[1]

Torren M. Peakman, Anthony Jervoise, George A. Wolff, and James R. Maxwell

Abstract. An initial reinvestigation into the acid-catalyzed rearrangement of 4-methylcholest-4-ene (**1**) indicates that the major products of backbone rearrangement are (20*R*)- and (20*S*)- 4β-methyl-10α-diacholest-13(17)-enes (**4a** and **4b**). A minor series of 4-methyldiacholest-13(17)-enes, possibly having the corresponding 4α-methyl stereochemistry (**6a** and **6b**), and a series of ring A/B rearranged 4-methylspirocholest-13(17)-enes, possibly having 4α-methyl-10β stereochemistry (**5a** and **5b**), were also encountered. We would predict that the 4-methyldiaster-13(17)-enes and rearranged 4-methylspiroster-13(17)-enes observed in immature sediments have these general structures.

INTRODUCTION

Chapter 3 (Rechka et al.) describes a reinvestigation of the products arising from acidic treatment of 4-methylcholest-4-ene (**1**) prior to backbone rearrangement. This initial distribution was shown to contain starting material (**1**), 4-methyl-5α-

[1]Part 3: Peakman and Maxwell (1988a).

and 4-methyl-5β-cholest-3-ene (**2a** and **2b**) and 4α- and 4β-methylcholest-5-ene (**3a** and **3b**), and a similar distribution of 4-methylcholestenes has been shown to occur in the immature Eocene Messel oil shale (Wolff et al., 1986a). Upon continued acid treatment, backbone rearrangement occurs to give 4-methyldiacholest-13(17)-enes (Rubinstein and Albrecht, 1975; Ensminger et al., 1978; Wolff et al., 1986a), which have also been reported in sediments (e.g., Rubinstein and Albrecht, 1975; Rubinstein et al., 1975; Barnes et al., 1979; Brassel et al., 1980; Brassel, 1984; Macquaker et al., 1986; Peakman and Maxwell, 1988b). In none of these studies, however, has the stereochemistry at C-4 been determined. In addition, there are components of the rearranged spiroster-13(17)-ene type in the backbone rearrangement mixture which also occur in sediments (Wolff et al., 1986b; Peakman and Maxwell, 1988b).

In this chapter we report our results from initial studies regarding the structures of the backbone rearrangement products of 4-methylcholest-4-ene (**1**). In addition to our interest in the sedimentary occurrence of these compounds, which is discussed briefly, it was considered worthwhile to investigate the effect of the 4-methyl group on the outcome of the rearrangement when compared with the 4-desmethyl situation. For reference purposes refer to Peakman et al. (1988) for the products arising from the backbone rearrangement of cholest-4-ene and 5-ene (**7**) and to Peakman and Maxwell (1988b) for their occurrences in immature sediments.

EXPERIMENTAL

High-Performance Liquid Chromatography (HPLC)

Semipreparative HPLC (cf. Wolff et al., 1985) was performed under reversed phase conditions using a Waters 6000A pump fitted with a Rheodyne 7125 injector (20 μl loop), and an LDC Refractometer III was used as the detector. The column (Spherisorb 5W ODS2; 250 mm × 10 mm i.d.) was obtained from Phase Separations Ltd. The mobile phase was 2 percent water in acetone, degassed by continuous helium purge, at a flow rate of 4 ml min^{-1}.

Gas Chromatography (GC)

GC analyses were performed on a Carlo Erba Mega 5160 chromatograph (on-column injection) fitted with a 25 m OV-1 fused silica capillary column obtained from Hewlett-Packard. Hydrogen was the carrier gas. Samples were injected in hexane at 40°C and programmed to 300°C at 4°C min^{-1}.

Gas Chromatography-Mass Spectrometry (GC-MS)

GC-MS was performed using a similar chromatograph and column as described. The chromatograph was interfaced to a Finningan 4000 mass spectrometer (ionizing temperature ca. 250°C, electron energy 35 eV, emission current 350 μA, accelerating voltage ca. 2 kV). The scan range was m/z 50–450 with a cycle time of 1 s.

Nuclear Magnetic Resonance Spectroscopy (NMR)

NMR spectra were recorded in deuterochloroform using tetramethylsilane as internal standard on either a Jeol FX200 spectrometer (operating at 200 MHz for ^{1}H and 50 MHz for ^{13}C) or a Jeol GX400 spectrometer (operating at 400 MHz for ^{1}H).

Anhydrous *p*-Toluenesulphonic Acid/Acetic Acid

p-Toluenesulphonic acid monohydrate (7.5 g), cyclohexane (75 ml) and glacial acetic acid (260 ml) were heated in a distillation apparatus until the distillation temperature reached 117°C, leaving undistilled anhydrous *p*-toluenesulphonic acid/ acetic acid.

4-Methylcholest-4-ene (**1**)

4-Methylcholest-4-ene (**1**) was prepared by standard methods (Wolff et al., 1986a and references cited therein).

Rearrangement Studies of 4-Methylcholest-4-ene (**1**)

A series of 1-ml reactivials (Pierce) containing **1** (5 mg) and anhydrous *p*-toluenesulphonic acid/acetic acid (0.5 ml) were heated at 70°C for various times (Table 4.1). Each vial was worked up by the addition of water (10 ml) followed by extraction with dichloromethane (3 × 10 ml). The combined organic fractions were neutralized with aq. sodium bicarbonate and dried (magnesium sulphate). After filtering, the solvent was removed under reduced pressure. The residue was chromatographed on a small alumina column. Elution with hexane afforded the steroid hydrocarbon products.

On a larger scale, a sealed conical flask containing **1** (700 mg) in anhydrous *p*-toluenesulphonic acid/acetic acid (70 ml) was heated in an oven (48 h). The mixture was worked up in a similar manner to that described.

Initial separation of the reaction mixture was achieved on a 50 g silica column impregnated with silver nitrate (5 g). The steroid hydrocarbons were eluted with hexane. A total of 55 fractions (of ca. 25 ml each) were collected. Similar fractions, as judged by GC analyses, were combined to give five overall fractions (A to E). The relative elution order of the various steroid hydrocarbons on the silver nitrate-impregnated column was 4-methyl-5β-cholest-3-ene (**2b**) > 4-methylcholest-4-ene (**1**) > 4-methyl-5α-cholest-3-ene (**2a**) > (20*R*)- and (20*S*)-rearranged 4α-methyl-10β-spirocholest-13(17)-enes (**5a** and **5b**) and (20*S*)-4β-methyl-10α-diacholest-13(17)-ene (**4b**) > 4α-methylcholest-5-ene (**3a**) > (20*R*)-4β-methyl-10α-diacholest-13(17)-ene (**4a**) and (20*R*)-4α-methyl-10α-diacholest-13(17)-ene (**6a**).

Fractions B, C, and E were further fractionated by reversed phase HPLC. Fraction B yielded (20*R*)-rearranged 4α-methyl-10β-spirocholest-13(17)-ene (**5a**), t_R = 13 min; (20*S*)-4β-methyl-10α-diacholest-13(17)-ene (**4b**), t_R = 16 min and 4-methylcholest-4-ene (**1**), t_R = 20 min. Fraction C yielded (20*S*)-4β-methyl-10α-diacholest-13(17)-ene (**4b**), t_R = 16 min and 4α-methylcholest-5-ene (**3a**), t_R = 21 min. Fraction E yielded (20*R*)-4α-methyl-10α-diacholest-13(17)-ene (**6a**), t_R = 13 min; (20*R*)-4β-methyl-10α-diacholest-13(17)-ene (**4a**), t_R = 14 min and 4α-methyl-cholest-5-ene (**3a**), t_R = 21 min.

RESULTS AND DISCUSSION

Time Course Study

Initially, small aliquots of 4-methylcholest-4-ene (**1**) were heated in anhydrous *p*-toluenesulphonic acid/acetic acid at 70°C for various periods of time to monitor the course of the rearrangement. Table 4.1 indicates the typical product distributions of 4-methylsterenes as determined from peak areas in GC chromatograms.

Isolation of Individual Components

From the initial time course study (Table 4.1) a large scale rearrangement reaction was carried out over 2 days (Fig. 4.1). The products were initially fractionated by column chromatography over silica gel impregnated with 10 percent silver nitrate into five fractions (A to E). Individual components were obtained from these fractions by reversed-phase HPLC (cf. Wolff et al., 1985) in sufficient purity for structural studies by NMR. Details are covered in the experimental section.

Structural Studies of Isolated Components

(20*R*)-4β,5β,14β-trimethyl-18,19-dinor-8α,9β,10α-cholest-13(17)-ene

[(20*R*)-4β-methyl-10α-diacholest-13(17)-ene], **4a**

Table 4.1 PRODUCT DISTRIBUTIONS (%) FROM ISOMERIZATION OF 4-METHYLCHOLEST-4-ENE (1)

Reaction time	Component							
	1 + 3a	**2a**	**2b**	**3b**	**4a + 6a**	**4b (+ 6b?)**	**5a**	**5b**
1 h	78	7	13	1				
4 h	75	7	14	1	2			
8 h	74	7	13	1	4			
13 h	64	5	11	1	13	1	3	
24 h	50	4	9		24	3	4	
72 h	19	2	5		43	15	5	2
168 h					50	36	3	3
480 h					51	47		

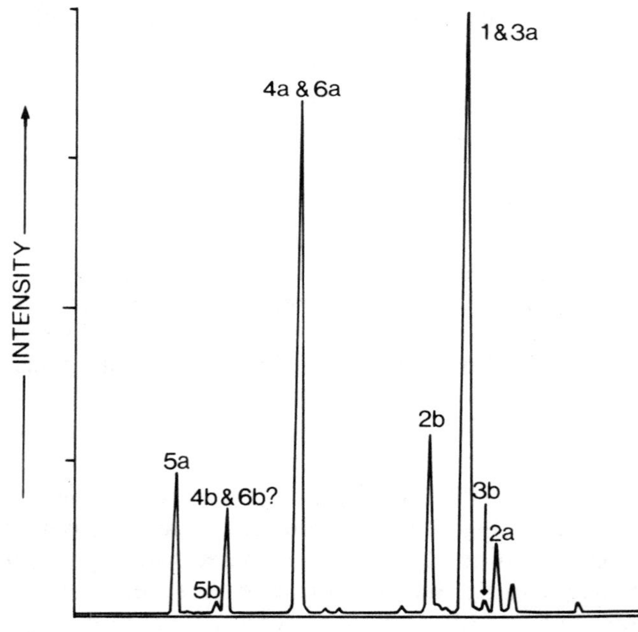

Figure 4.1 Reconstructed total ion chromatogram (GC-MS) of products from isomerization of 4-methylcholest-4-ene (**1**) with anhydrous *p*-toluenesulphonic acid/acetic acid (70°C, 2 days).

This compound showed a typical diaster-13(17)-ene mass spectrum with a base peak at m/z 271 corresponding to loss of the side chain and a weak molecular ion accompanied by a somewhat more intense loss of 15 amu (Fig. 4.2a) (cf. Kirk and Shaw, 1975; Rubinstein et al., 1975; Peakman et al., 1988). The 400 MHz ^{1}H NMR spectrum also showed features characteristic of a diaster-13(17)-ene skeleton (cf. Kirk and Shaw, 1975; Rubinstein et al., 1975; Peakman et al., 1988; Sieskind et

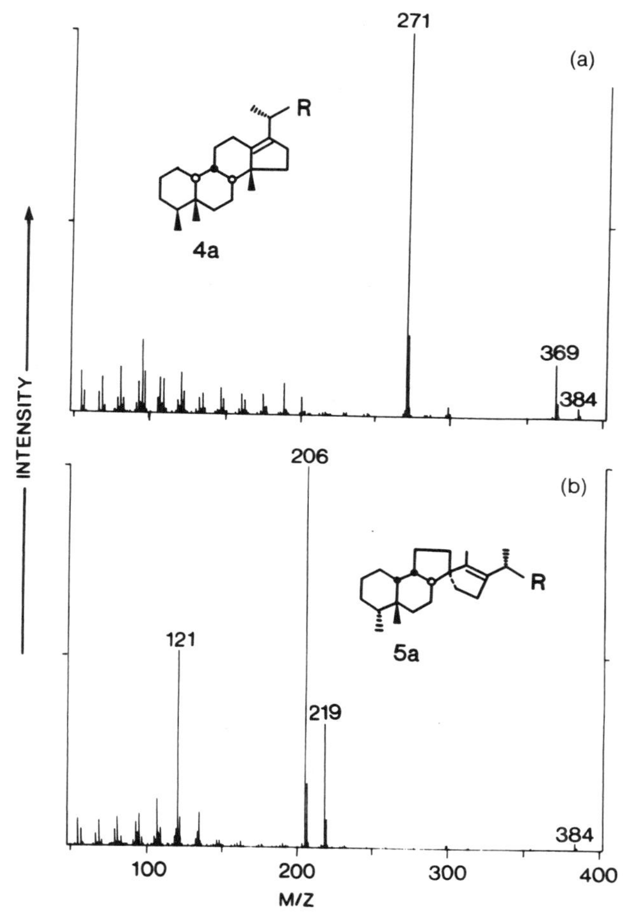

Figure 4.2 Mass spectra (GC-MS) of (a) (20*R*)-4β-methyl-10α-diacholest-13(17)-ene (**4a**) and (b) (20*R*)-rearranged 4α-methyl-10β-spirocholest-13(17)-ene (**5a**).

al., 1989). The chemical shifts of the methyl resonances were compared with those for (20*R*)-10α-diacholest-13(17)-ene (**8a**) (cf. Kirk and Shaw, 1975; Rubinstein et al., 1975; Peakman et al., 1988) and are summarised in Table 4.2. The chemical shifts for the C-21 H_3 doublets were almost identical indicating (20*R*) stereochemistry. A substantial difference was noted, however, for the C-5 β-CH_3 resonance which was shielded by 0.115 ppm in the 4-methyl component (**4a**), compared with the 4-desmethyl counterpart (**8a**), obviously due to the presence of the 4-methyl group. Initial evidence for 4β-methyl stereochemistry came from the ^{1}H-^{1}H correlated spectrum (COSY) which indicated that the C-4 hydrogen was in the axial position (i.e., α; Fig. 4.3a) since it resonated over a range of ca. 38 Hz. This is consistent with ^{3}J coupling (i.e., coupling through three bonds, e.g., <u>H</u>—C—C—<u>H</u>) to the C-4 methyl group (J = 6.4 Hz), ^{3}J axial-axial coupling to the C-3 β-hydrogen (J ca. 12 Hz) and ^{3}J axial-equatorial coupling to the C-3 α-hydrogen (J ca. 4 Hz).

Table 4.2 ^{1}H NMR CHEMICAL SHIFT ASSIGNMENTS OF REARRANGED STERENES

	Component						
	4a[1]	**4b**[2]	**8a**[2]	**8b**[2]	**5a**[2]	**9a**[2]	**6a**[2]
C-4 CH_3	0.758, d	0.761, d			0.721, d		0.912, d
	J = 6.4 Hz	J = 6.3 Hz			J = 6.8 Hz		J = 6.6 Hz
C-5β-CH_3	0.710, s	0.713, s	0.825, s	0.823, s	0.818, s	0.962, s	0.937, s
C-11 ax-H	0.6, m	0.6, m	0.6, m	0.6, m	n.o.	n.o.	0.6, m
C-12 eq-H	2.3, m	2.3, m	2.3, m	2.3, m	n.o.	n.o.	2.3, m
C-14β-CH_3	0.877, s	0.873, s	0.884, s	0.882, s			0.885, s
C-16 H_2	2.1, m	2.1, m	2.1, m	2.1, m	2.1, m	2.0, m	2.1, m
C-18 H_3					1.460, dd	1.468, dd	
					J = 2 Hz	J = 2 Hz	
C-20 H_3	2.4, m	2.4, m	2.4, m	2.4, m	2.5, m	2.5, m	2.4, m
C-21 H_3	0.945, d	0.900, d	0.945, d	0.888, d	0.928, d	0.938, d	0.945, d
	J = 6.8 Hz	J = 6.8 Hz	J = 6.8 Hz	J = 6.8 Hz	J = 6.8 Hz	J = 6.8 Hz	J = 6.6 Hz
C-26 H_3	0.835, d	0.857, d	0.835, d	0.857, d			0.835, d
	J = 6.6 Hz	J = 6.5 Hz	J = 6.4 Hz	J = 6.8 Hz	0.834, d	0.835, d	J = 6.6 Hz
C-27 H_3	0.830, d	0.851, d	0.830, d	0.850, d	J = 6.6 Hz	J = 6.6 Hz	0.830, d
	J = 6.6 Hz	J = 6.5 Hz	J = 6.6 Hz	J = 6.7 Hz			J = 6.6 Hz

[1]400MHz

[2]200MHz

n.o. = not observed

(a)

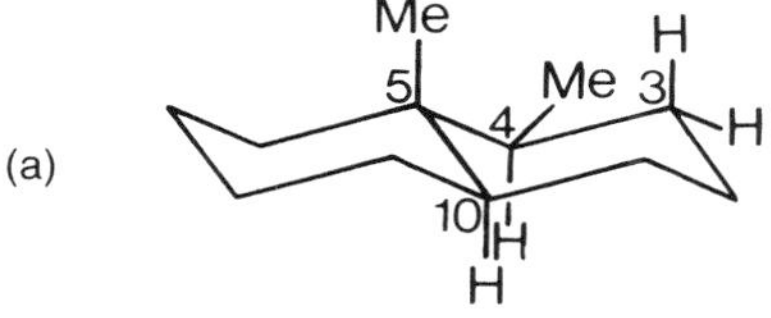

(b)

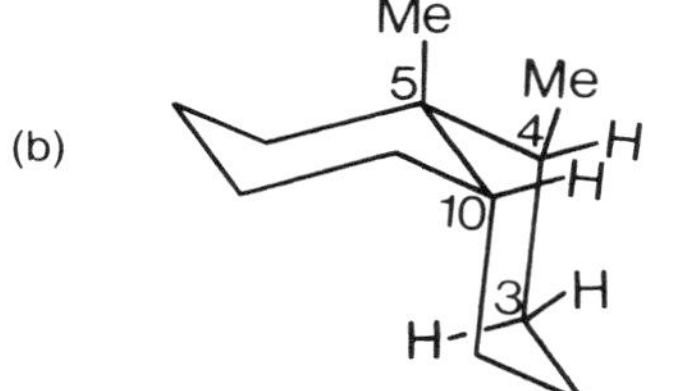

Figure 4.3 Three dimensional perspectives of rings A and B for (a) 4β-methyl-10α-diacholest-13(17)-enes and (b) rearranged 4α-methyl-10β-spirocholest-13(17)-enes.

If the C-4 hydrogen was equatorial (i.e., β), the resonating range would be expected to be about 29 Hz (i.e., $^3J4,4\text{-}CH_3$ = 6.4 Hz; $^3J4\text{-}3\alpha$ ca. 4 Hz and $^3J4\text{-}3\beta$ ca. 4 Hz) (e.g., Kemp, 1986). Unfortunately, we could not locate in the $^1H\text{-}^1H$ NMR spectrum the cross peaks for the C-3 hydrogens, whose patterns would have been expected to give confirmatory evidence for 4β-methyl stereochemistry. Analysis of the ^{13}C NMR spectrum, using chemical shift comparisons with those given for selected carbons in (20*R*)-10α-diacholest-13(17)-ene (**8a**) (i.e., C-7 to C-9, C-11 to C-27 and C-14 β-CH_3; Table 4.3) and by calculations, using the method of Beierbeck et al. (1977), for ring A and B carbons (taking into account different stereochemistry at C-4 and C-10; Table 4.4), clearly indicated the (20*R*)-4β-methyl-10α-diacholest-13(17)-ene structure (**4a**). The chemical shifts for the carbons in ring A were also in accord with the corresponding carbons in 1β, 10β-dimethyldecalin (Ayer et al., 1978).

(20*S*)-4β,5β,14β-trimethyl-18,19-dinor-8α-9β,10α-cholest-13(17)-ene

[(20*S*)-4β-methyl-10α-diacholest-13(17)-ene], **4b**

As expected, this compound had an almost identical mass spectrum to that of (20*R*)-4β-methyl-10α-diacholest-13(17)-ene (**4a**, Fig. 4.2a). The 200 MHz ^1H NMR spectrum (Table 4.2) clearly indicated the full structure by comparison of the chemical shifts of the C-4 β-CH_3 and C-5 β-CH_3 resonances with those obtained for (20*R*)-4β-methyl-10α-diacholest-13(17)-ene (**4a**) and of the C-21 H_3 resonance with that obtained for (20*S*)-10α-diacholest-13(17)-ene (**8b**) (cf. Kirk and Shaw, 1975; Rubinstein et al., 1975; Peakman et al., 1988; Table 4.2).

(20*R*)-4α, 5β-dimethyl-19-nor-12(13→14)-*abeo*-8α,9β,10β-cholest-13(17)-ene

[(20*R*)-rearranged 4α-methyl-10β-spirocholest-13(17)-ene], **5a**

The mass spectrum of this compound (Fig. 4.2b) showed a molecular ion at *m*/*z* 384, two major fragment ions at *m*/*z* 206 and 121, and an additional ion at *m*/*z*

Table 4.3 ^{13}C **NMR ASSIGNMENTS OF REARRANGED STERENES**

Carbon number	Component			
	4a	**8a**	**5a**	**9a**
C-1	24.4	24.4	24.4	24.4
C-2	26.9	27.3	28.5	28.5
C-3	30.6	21.6	31.2	21.5
C-4	43.5	42.2	30.3	42.5
C-4 CH_3	15.5		15.7	
C-5	36.7	33.8	36.5	29.7
C-5β-CH_3	11.8	16.9	21.9	27.4
C-6	38.7	42.4	37.9	31.2
C-7	22.2	22.1	21.3	22.3
C-8	54.9	55.4	52.5	52.1
C-9	36.3	36.5	39.6	40.1
C-10	51.9	50.7	50.3	48.1
C-11	31.7	31.3	(a)	(c)
C-12	23.0	23.0	(a)	(c)
C-13	141.4	141.5	(b)	(d)
C-14	50.2	50.2	60.4	60.5
C-14β-CH_3	18.2	18.3		
C-15	37.9	37.9	(a)	(c)
C-16	27.8	27.8	(a)	(c)
C-17	133.9	133.9	(b)	(d)
C-18			9.5	9.5
C-20	31.4	31.4	32.6	32.6
C-21	20.3	20.3	19.7	19.8
C-22	35.8	35.8	35.8	35.8
C-23	25.5	25.5	25.6	25.6
C-24	39.0	39.0	39.1	39.1
C-25	28.0	28.0	28.0	28.0
C-26	22.7	22.7	22.6	22.6
C-27	22.6	22.6	22.6	22.6

(a) From 21.0, 28.5, 34.0, and 35.1
(b) From 134.1 and 139.7
(c) From 20.5, 28.5, 33.7, and 35.0
(d) From 134.1 and 139.7

219. Such a mass spectrum is characteristic of ster-13(17)-enes possessing a spiro C-D ring junction (Anastasia et al., 1978; Peakman and Maxwell, 1988c; Peakman et al., 1988). The exact structure of this component has, to date, proved somewhat difficult to define precisely. The 200 MHz ^1H NMR spectrum clearly indicated a spiroster-13(17)-ene structure (Table 4.2; Anastasia et al., 1978; Peakman and Maxwell, 1988c). Comparison of the chemical shifts of the C-21 H_3, C-26 and C-27 H_3 doublets with those for (20*R*)-rearranged-10β-spirocholest-13(17)-ene **(9a)** (Peakman et al., 1988) indicated (20*R*) stereochemistry (Table 4.2). Since this compound has arisen by a backbone rearrangement, we can assume 8α, 9β, and

Table 4.4 CALCULATED ^{13}C NMR CHEMICAL SHIFTS OF CARBON ATOMS IN RINGS A (AND B) HAVING DIFFERENT STEREOCHEMISTRIES AT C-4 AND C-10

	4α-methyl-10α(H)	4β-methyl-10α(H)	4α-methyl-10β(H)	4β-methyl-10β(H)
C-1	26.7	26.7	22.2	22.2
C-2	22.2	26.7	26.7	22.2
C-3	28.6	31.3	31.3	28.6
C-4	40.7	43.4	43.4	40.7
C-4 CH_3	12.0	16.5	16.5	12.0
C-5	37.6	37.6	37.6	37.6
C-5 CH_3	16.5	12.0	21.1	21.1
C-6	33.1	37.7	28.6	33.1
C-10	47.9	52.5	49.8	45.2

5β-methyl stereochemistry (cf. Kirk and Shaw, 1975; Peakman et al., 1988) but what of the stereochemistry at C-4 and C-10? The chemical shifts of the C-4 and C-5 methyls are completely different from those of (20*R*)-4β-methyl-10α-diacholest-13(17)-ene (**4a**), therefore, we can eliminate 4β-methyl-10α stereochemistry. 4α-Methyl-10α stereochemistry was also eliminated on the basis of the ^{13}C chemical shift of the 5β-CH_3 carbon which indicated 10β stereochemistry. This rationale follows from detailed ^{13}C NMR studies on normal steroids which indicate that the ring A/B methyl group (C-19) is deshielded by 11-12 ppm on changing the ring stereochemistry from *trans* to *cis* (e.g., Gough et al., 1972). This feature is also observed in going from (20*R*)-10α-diacholest-13(17)-ene (**8a**) to (20*R*)-10β-diacholest-13(17)-ene (**10a**) (Peakman et al., 1988). 10β Stereochemistry is also in accord with the GC retention time of this component as it is the first eluting compound observed in the reaction mixture product (steroid hydrocarbons with a *cis* A/B ring junction elute earlier than those with a *trans*). This early GC retention time is also in accord with (20*R*) stereochemistry (spiroster-13(17)-enes with (20*R*) stereochemistry elute earlier than those possessing (20*S*) stereochemistry; Peakman and Maxwell, 1988c). With regard to the stereochemistry at C-4, we favor the 4α-methyl structure on the basis of the ^1H-^1H correlated spectrum (COSY) which suggests that the C-4 hydrogen is axial (i.e., β; Fig. 4.3b) since it resonates over a range of about 39 Hz. This is in accord with ^3J coupling to the C-4 methyl group (J = 6.8 Hz), ^3J axial-axial coupling to the C-3α-hydrogen (J ca. 12 Hz) and ^3J axial-equatorial coupling to the C-3 β-hydrogen (J ca. 4 Hz). If the C-4 hydrogen was equatorial (i.e., α) its resonating range would be expected to be ca. 29 Hz (i.e., $^3J4,4\text{-}CH_3$ = 6.8 Hz; $^3J4\text{-}3\alpha$ ca. 4 Hz and $^3J4\text{-}3\beta$ ca. 4 Hz). Unfortunately, again, we could not locate the crosspeaks in the ^1H-^1H correlated spectrum for the C-3 hydrogens. Their cross-peak patterns would be expected to distinguish between 4α-and 4β-methyl stereochemistry.

We also encountered some difficulties with the complete assignment of the ^{13}C NMR spectrum (Table 4.3), especially in rationalising a chemical shift of 30.3 ppm for the C-4 carbon and a shift of 37.9 ppm for C-6 when compared with the calculated values (Table 4.4). Despite this, many features of the ^{13}C NMR were in accord with the structure, especially when (20*R*)-rearranged-10β-spirocholest-13(17)-ene (**9a**) was used as a model compound (Peakman et al., 1988; Table 4.3).

(20*S*)-4α,5β-dimethyl-19-nor-12(13→14)-*abeo*-8α,9β,10β-cholest-13(17)-ene

[(20*S*)-rearranged 4α-methyl-10β-spirocholest-13(17)-ene], **5b**

This compound was not isolated from the reaction mixture but evidence for its presence came from GC-MS data (Fig. 4.1), the known isomerization at C-20 in rearranged ster-13(17)-enes (Kirk and Shaw, 1975) and the known GC elution order of ster-13(17)-enes possessing a spiro C-D ring junction [i.e., (20*R*) before (20*S*)] (Peakman and Maxwell, 1988c).

(20*R*)-4α,5β,14β-trimethyl-18,19-dinor-8α,9β,10α-cholest-13(17)-ene

[(20*R*)-4β-methyl-10β-diacholest-13(17)-ene], **6a**

In the HPLC fractionation of fraction E, we obtained another 4-methyldiacholest-13(17)-ene, albeit in minor amount, in addition to (20*R*)-4β-methyl-10α-diacholest-13(17)-ene (**4a**). This initially surprising result was due to coelution of these two compounds on the apolar GC column employed (Fig. 4.1). Comparison of the chemical shift of the C-21 H_3 doublet with that of (20*R*)-4β-methyl-10α-diacholest-13(17)-ene (**4a**) indicated (20*R*) stereochemistry (Table 4.2). The chemical shift for C-4 CH_3 was completely different from those of (20*R*)-4β-methyl-10α-diacholest-13(17)-ene (**4a**) and (20*R*)-rearranged-4α-methyl-10β-spirocholest-13(17)-ene (**5a**) indicating a third combination of the four C-4, C-10 stereochemical possibilities.

At present, we believe that we are dealing with 4α-methyl-10α stereochemistry on the basis of a characteristic feature, seen in the ^{1}H NMR spectrum, which distinguishes 10α- from 10β-diacholest-13(17)-enes. This is the chemical shift of the strongly shielded C-11 axial hydrogen at ca. 0.6 ppm in the 10α-series (Table 4.2) which is easily observed as a multiplet in the ^{1}H NMR spectrum of (20*R*)-10α-diacholest-13(17)-ene (**8a**), (20*R*)-4β-methyl-10α-diacholest-13(17)-ene (**4a**) but not in (20*R*)-10β-diacholest-13(17)-ene (**10a**) in which it is deshielded and now lies under the methyl resonances (Peakman, unpublished observations; cf. Sieskind et al., 1989). The presence of the characteristic multiplet at 0.6 ppm in the spectrum of this additional 4-methyldiacholest-13(17)-ene indicates, therefore, 10α stereochemistry. Hence, since we have already characterized (20*R*)-4β-methyl-10α-diacholest-13(17)-ene (**4a**), then this compound appears to be (20*R*)-4α-methyl-10α-diacholest-13(17)-ene (**6a**). The corresponding (20*S*) isomer (**6b**) presumably coelutes with (20*S*)-4β-methyl-10α-diacholest-13(17)-ene (**4b**) since GC-MS analyses (cf. Fig. 4.1) do not indicate additional peaks possessing a mass spectrum with the features of a 4-methyldiacholest-13(17)-ene.

The Backbone Rearrangement of 4-Methylcholest-4-ene (1) and Comparison with the 4-Desmethyl Situation

The backbone rearrangement of cholest-4-ene and 5-ene (**7**) gives rise to two major products (20*R*)- and (20*S*)-10α-diacholest-13(17)-ene (**8a** and **8b**) (Kirk and Shaw, 1975). Minor products are the corresponding 10β-diacholest-13(17)-enes (**10a** and **10b**) and rearranged 10α-and 10β-spirocholest-13(17)-enes (**9a**, **9b** and **11a**, **11b**) (Peakman et al., 1988). In the 4-methyl situation, we also obtain two major compounds [(20*R*)-and (20*S*)-4β-methyl-10α-diacholest-13(17)-ene (**4a** and **4b**)] but interestingly, instead of the three additional minor series possible (i.e., 4α-methyl-10α, 4α-methyl-10β, 4β-methyl-10β) we only see the 4α-methyl-10α components above the detection level. Within the rearranged spiroster-13(17)-ene components, we observe both possible series in the desmethyl case but only one, having 4α-methyl-10β-stereochemistries (**5a**, **5b**), above the detection level, of the possible four in the 4-methyl situation. Noteworthy is that this series possesses 10β stereochemistry, a feature not observed for the 4-methyldiacholest-13(17)-enes. Interestingly, this feature of 10β stereochemistry in the rearranged 4-methylspirocholest-13(17)-enes, occurring at an early point within the backbone rearrangement (i.e., before equilibrium), is also an important feature in the desmethyl situation (Peakman et al., 1988). Taking these observations for the des-methyl and the 4-methyl series together it is tempting to suggest some form of kinetic control governing the fate of the C-14 carbonium ion, during the course of the backbone rearrangement, which depends on the stereochemistry at C-10. Thus, 10α stereochemistry kinetically favors methyl migration from C-13 to C-14 giving a diaster-13(17)-ene, whereas 10β stereochemistry favors contraction of ring C giving a rearranged spiroster-13(17)-ene. The idea that the stereochemistry at C-10 has a kinetic effect on the distribution of rearranged products is borne out when one considers the composition of the final thermodynamic equilibrium mixtures. In the des-methyl situation, the two possible series of rearranged spiroster-13(17)-enes become very low in abundance (both ca. 1%; Peakman et al., 1988) and in the 4-methyl case they were not even detected (Table 4.1). Since the backbone rearrangement of cholest-4-ene and 5-ene (**7**) has been shown to be at least partly reversible (Peakman et al., 1988), then it is possible that the rearranged 4α-methyl-10β-spirocholest-13(17)-enes (**5a** and **5b**) become transformed into the 4α-methyl-10α-diacholest-13(17)-enes (**6a** and **6b**) via the intermediates indicated in Fig. 4.4. The abundance of the minor 4α-methyl-10α-diacholest-13(17)-ene (**6a**) and its (20*S*) counterpart (**6b**) in the final equilibrium (i.e., 480 h reaction; Table 4.1), however, is not yet known because of GC coelution problems (Fig. 4.1).

In the backbone rearrangement of cholest-4-ene and 5-ene (**7**) components of partial backbone rearrangement (**12**; up to ca. 7%) are observed (Peakman et al., 1988) although their 4-methyl counterparts account for no more than 1 percent of the products in the backbone rearrangement of 4-methylcholest-4-ene (**1**).

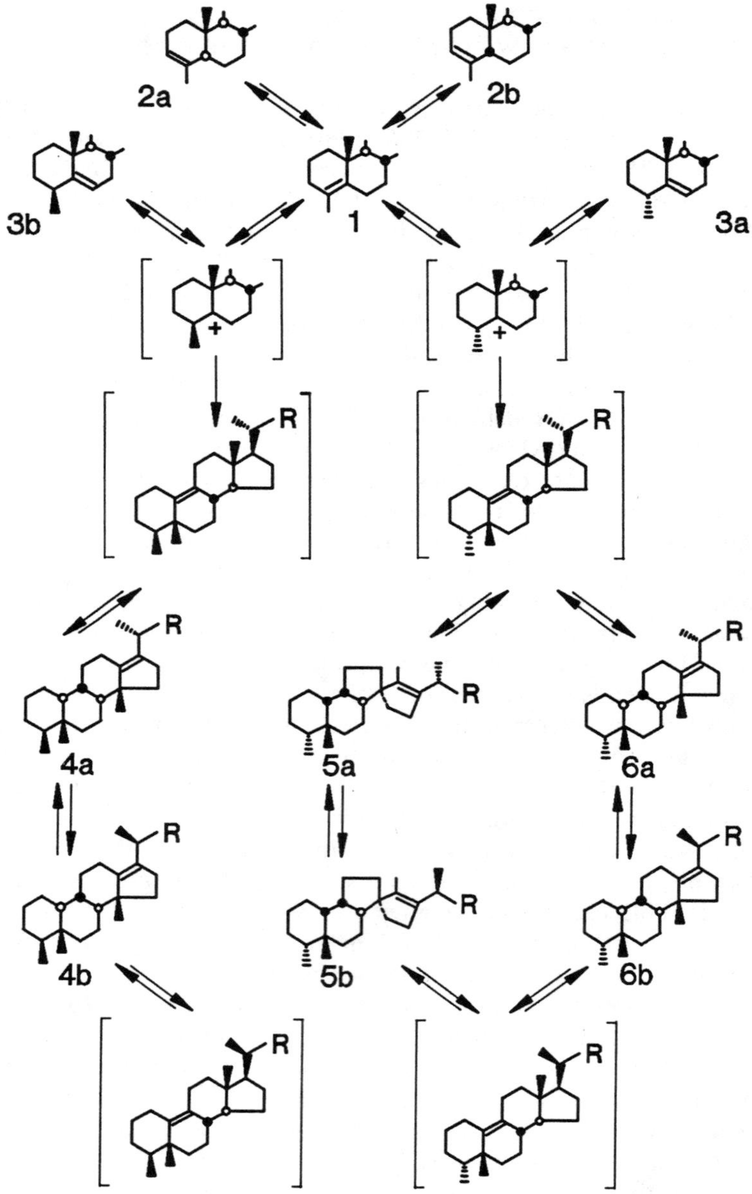

Figure 4.4 Scheme showing the acid catalyzed rearrangement of 4-methylcholest-4-ene (**1**). Square brackets indicate putative transient alkenes and/or selected carbocation species.

A postulated scheme for the course of the backbone rearrangement of 4-methylcholest-4-ene (**1**) is given in Fig. 4.4.

We can conclude from these comparisons that the 4-methyl group does have some subtle control over the product distributions obtained during backbone rearrangement.

Sedimentary Occurrences and Distributions of 4-Methyldiaster-13(17)-enes and Rearranged 4-Methylspiroster-13(17)-enes

There are few literature reports of 4-methyldiaster-13(17)-enes in sediments although a study of numerous black shale deposits (Farrimond, 1987) indicates their more widespread occurrence. From the initial results given herein, we would predict that the major sedimentary 4-methyldiaster-13(17)-enes have the (20*R*)- and (20*S*)-4β-methyl-10α-diaster-13(17)-ene structure (e.g., **4a** and **4b**). A minor series would also be expected having the corresponding 4α-methyl structure (e.g., **6a** and **6b**) although further studies of sediment extracts employing GC columns which can separate these series are required. The occurrence of rearranged 4-methylspirosterenes (probably with the 4α-methyl-10β structure; e.g., **5a** and **5b**) in a black shale from Corfu (Greece) has previously been demonstrated (Wolff et al., 1986b; Peakman and Maxwell, 1988b).

The sedimentary transformation sequence of 4-methylsterenes to 4-methyldiaster-13(17)-enes and rearranged 4-methylspiroster-13(17)-enes would be expected to parallel the results of the laboratory time course study as indicated previously for the cholest-4-ene and 5-ene (**7**) rearrangement (Peakman and Maxwell, 1988b).

With increasing diagenesis, we can expect formation of 4-methyldiasteranes with structures matching those of the 4-methyldiaster-13(17)-enes. Thus, we would predict that the sedimentary reduction products of the major 4β-methyl-10α-diaster-13(17)-enes would be 4β-methyl-10α,13β,17α(H)-diasteranes (major) and 4β-methyl-10α,13α,17β(H)-diasteranes (minor; cf. Ensminger et al., 1978). We might also expect minor amounts of the corresponding 4α-methyldiasteranes which would result from reduction of the minor 4α-methyl-10α-diaster-13(17)-enes.

Since spirosteranes have not been identified in sediments and, as indicated earlier, they become very low in abundance as the backbone rearrangement approaches equilibrium, then their sedimentary fate is possibly rearrangement to 4α-methyl-10α-diaster-13(17)-enes (cf. Fig. 4.4) which then undergo reduction.

From a survey of published 4-methyldiaster-13(17)-ene distributions, an interesting observation can be made concerning the C_{30} components. In some samples there appear to be significant amounts of two C_{30} (20*R*)-4β-methyl-10α-diaster-13(17)-enes (Barnes et al., 1979; Brassell, 1984; Brassell et al., 1980) while in others the first eluting component dominates (Peakman and Maxwell, 1988b; Macquaker et al., 1986). In view of the recent interest in 4-methylsteranes possessing either the 23,24-dimethyl (dinosterane) or 24-ethyl side-chain (Summons et al., 1987; Goodwin et al., 1988) these observations could be of interest.

CONCLUSIONS

1. The major products from the acid-catalyzed backbone rearrangement of 4-methylcholest-4-ene (**1**) are (20*R*)-and (20*S*)-4β-methyl-10α-diacholest-13 (17)-enes (**4a** and **4b**).
2. Minor products may be (20*R*)- and (20*S*)-4α-methyl-10α-diacholest-13(17)-enes (**6a** and **6b**) and (20*R*)- and (20*S*)-rearranged 4α-methyl-10β-spiro-cholest-13(17)-enes (**5a** and **5b**).
3. Only a limited number of the possible rearrangement products are formed above the detection level.
4. We predict that the corresponding 4-methyl components reported in sediments have the same structures as those observed herein and that a sequence of transformations occurs in sediments which parallels those seen in the laboratory time course study.

Acknowledgments

We are grateful to British Petroleum plc and the donors of the Petroleum Research Fund, administered by the American Chemical Society, for financial support (Bristol) and to the Alexander von Humboldt-Stiftung for a fellowship to T.M. Peakman (KFA, Jülich). GC-MS facilities (Bristol) were made available by grants from the NERC (GR3/2951 and GR3/3758).

APPENDIX

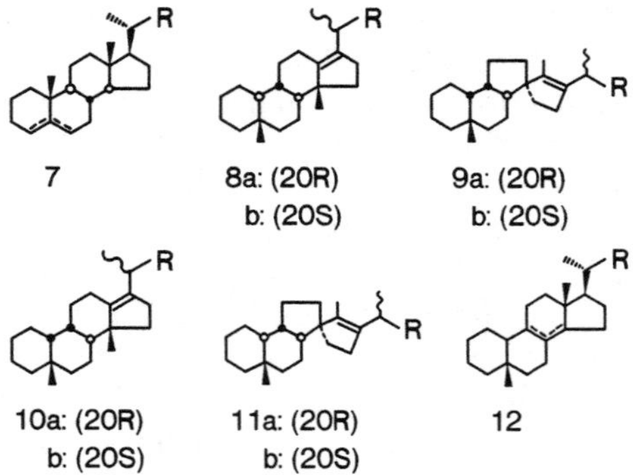

Figure 4.5

REFERENCES

ANASTASIA, M., SOAVE, A.M., and SCALA, A. (1978) Acid catalysed backbone rearrangement involving the C-D ring junction in normal steroid series. *J. Chem. Soc. Perkin Trans. I*, 1131–1132.

AYER, W.A., BROWNE, L.M., FUNG, S., and STOTHERS, J.B. (1978) Carbon-13 NMR studies. 73. Carbon-13 spectra of several 10-methyl-trans-decalins. Further definition of the deshielding anti-periplanar γ effect. *Org. Mag. Res. 11*, 73–80.

BARNES, P.J., BRASSELL, S.C., COMET, P.A., EGLINTON, G., MCEVOY, J., MAXWELL, J.R., WARDROPER, A.M.K., and VOLKMAN, J.K. (1979) Preliminary lipid analyses of core sections 18, 24 and 30 from Hole 402A. In *Initial Reports of the Deep Sea Drilling Project* (eds. L. Monterdat, D.G. Roberts et al.), Vol. 48, pp. 965–976. U.S. Government Printing Office, Wash.

BEIERBECK, H., SAUNDERS, J.K., and APSIMON, J.W. (1977) The semiempirical derivation of ^{13}C nuclear magnetic resonance chemical shifts. Hydrocarbons, alcohols, amines, ketones and olefins. *Can. J. Chem. 55*, 2813–2828.

BRASSELL, S.C. (1984) Alphatic hydrocarbons of a Cenomian black shale and its adjacent green claystone from the southern Angola Basin, Deep Sea Drilling Project Leg 75. In *Initial Reports of the Deep Sea Drilling Project* (eds. W.W. Hays, J-C. Sibuct et al.), Vol. 75, pp. 965–976. U.S. Government Printing Office, Wash.

BRASSELL, S.C., COMET, P.A., EGLINTON, G., MCEVOY, J., MAXWELL, J.R., QUIRKE, J.M.E., and VOLKMAN, J.K. (1980) Preliminary lipid analyses of cores 14, 18 and 28 from Deep Sea Drilling Project Hole 416A. In *Initial Reports of the Deep Sea Drilling Project* (eds. Y. Lancelot, E.L. Winterer et al.), Vol. 50, 647–664. U.S. Government Printing Office, Wash.

ENSMINGER, A., JOLY, G., and ALBRECHT, P. (1978) Rearranged steranes in sediments and crude oils. *Tetrahedron Lett.*, 1575–1578.

FARRIMOND, P. (1987) The Toarcian and Cenomanian/Turonian oceanic anoxic events. Ph.D Thesis, University of Bristol, U.K.

GOODWIN, N.S., MANN, A.L., and PATIENCE, R.L. (1988) Structure and significance of C_{30} 4-methyl steranes in lacustrine shales and oils. *Org. Geochem. 12*, 495–506.

GOUGH, J.L., GUTHRIE, J.P., and STOTHERS, J.B. (1972) Stereochemical assignments in steroids by ^{13}C nuclear resonance spectroscopy: Configuration of the A/B ring junction. *J. Chem. Soc., Chem. Commun.*, 979–980.

KEMP, W. (1986) *NMR in Chemistry—A Multinuclear Introduction.* MacMillan, Basingstoke, U.K.

KIRK, D.N. and SHAW, P.M. (1975) Backbone rearrangement of steroidal 5-enes. *J. Chem. Soc., Perkin Trans. I*, 2284–2294.

MACQUAKER, J.H.S., FARRIMOND, P., and BRASSELL, S.C. (1986) Biological markers in the Rhaetian black shales of south west Britain. In *Advances in Organic Geochemistry 1985* (eds. D. Leythaeuser and J. Rullkötter); *Org. Geochem. 10*, 93–100.

PEAKMAN, T.M. and MAXWELL, J.R. (1988a) Acid-catalysed rearrangements of steroid alkenes-III. Backbone rearrangement of de-A-steroid alkenes and preparation of an aromatic de-A-steroid. *Tetrahedron 44*, 1559–1565.

PEAKMAN, T.M. and MAXWELL, J.R. (1988b) Early diagenetic pathways of steroid alkenes.

In *Advances in Organic Geochemistry 1987* (eds. L. Mattavelli and L. Novelli); *Org. Geochem. 13*, 583–592.

PEAKMAN, T.M. and MAXWELL, J.R. (1988c) Acid-catalysed rearrangements of steroid alkenes-Part 1. Rearrangement of 5α-cholest-7-ene. *J. Chem. Soc., Perkin Trans. I*, 1065–1070.

PEAKMAN, T.M., ELLIS, K., and MAXWELL, J.R. (1988) Acid-catalysed rearrangements of steroid alkenes-Part 2. A re-investigation of the backbone rearrangement of cholest-5-ene. *J. Chem. Soc., Perkin Trans. I*, 1071–1075.

RECHKA, J.A., COX, H.C., PEAKMAN, T.M., DE LEEUW, J.W., and MAXWELL, J.R. (1990) A reinvestigation of aspects of the early diagenetic pathways of 4-methylsterenes based on molecular mechanics calculations and the acid catalysed rearrangement of 4-methyl-cholest-4-ene (Chapter 3 of this text.)

RUBINSTEIN, I. and ALBRECHT, P. (1975) The occurrence of nuclear methylated steranes in a shale. *J. Chem. Soc., Chem. Commun.*, 957–958.

RUBINSTEIN, I., SIESKIND, O., and ALBRECHT, P. (1975) Rearranged sterenes in a shale: Occurrence and simulated formation. *J. Chem. Soc., Perkin Trans. I*, 1833–1836.

SIESKIND, O., KINTZINGER, J-P., and ALBRECHT, P. (1989) Novel backbone rearrangement of steroids: Formation of (20*R* and 20*S*) 1β,14β-dimethyl-18,19-dinor-5β,8α,9β,10α-cholest-13(17)-enes. *J. Chem. Soc., Chem. Commun.*, 133–135.

SUMMONS, R.E., VOLKMAN, J.K., and BOREHAM, C.J. (1987) Dinosterane and other steroidal hydrocarbons of dinoflagellate origin in sediments and petroleum. *Geochim. Cosmochim. Acta 51*, 3075–3082.

WOLFF, G.A., PEAKMAN, T.M., and MAXWELL, J.R. (1985) Semi-preparative HPLC of some geologically interesting steroid hydrocarbons. *J. High Res. Chrom. Chrom. Commun. 8*, 695–696.

WOLFF, G.A., LAMB, N.A., and MAXWELL, J.R. (1986a) The origin and fate of 4-methylsteroid hydrocarbons I. Diagenesis of 4-methylsterenes. *Geochim. Cosmochim. Acta 50*, 335–342.

WOLFF, G.A., LAMB, N.A., and MAXWELL, J.R. (1986b) The origin and fate of 4-methylsteroid hydrocarbons II. Dehydration of stanols and the occurrence of C_{30} 4-methylsteranes. In *Advances in Organic Geochemistry 1987* (eds. D. Leythaeuser and J. Rullkötter); *Org. Geochem. 10*, 965–974.

5

NMR Structural Studies of C_{21} and C_{27} Monoaromatic Steroid Hydrocarbon Biological Markers

Robert M. K. Carlson, William R. Croasmun, Daniel E. Chamberlain, and J. Michael Moldowan

Abstract. Ring-C monoaromatic steroidal hydrocarbons (MA-steroids) identical to those of the corresponding chemical fossils with various stereochemical configurations were prepared by a one-step catalytic dehydrogenation-isomerization (CDI) of 5α-cholestane. Each of the four major MA-steroids was isolated in good purity from the complex product mixture by repeated nonaqueous reversed phase (NARP) and adsorptive HPLC. Structures were investigated using a combination of NMR techniques, including difference nuclear Overhauser enhancement (NOE) and decoupling, spin-lattice relaxation measurements using partial relaxation methods, solvent-induced shifts, binuclear shift reagents, two-dimensional *J* spectroscopy (2D *J*), ^1H-^1H correlated spectroscopy (COSY), and dynamic NMR at 400 or 500 MHz. Various ^{13}C spectral measurements were acquired for the MA-steroid (**2**) of highest final yield. Structural information provided by these various techniques is appraised; comparisons are drawn which should be of general interest to those doing structural elucidations on very small geochemical samples. Solvent-induced shifts provided useful resolution improvements at 500 MHz. Dynamic NMR studies (in the 200–300 K range) show **3** and **4** to be conformationally rigid (5α), but show **1** and **2** to have temperature-dependent conformational equilibria (5β). Even at 500 MHz and employing 2D *J* spectroscopy, the A- and B-ring resonances of C_{27} MA-steroids **3** and **4** remained unresolved. The C_{21} homologs (also present in petroleum) were prepared by CDI of 5α-pregnane. The 5α-pregnane (which lacks the C-20 asymmetric center) produces only two products, supporting a single fixed configuration (17β-methyl) for MA-steroids generated by the CDI method and found in petroleum. The 2D *J* and COSY spectral results for the C_{21} homologs provide clear ^{1}H spectral assignments. Comparisons of C_{21} and C_{27} MA-steroid ^{1}H spectral data help clarify the assignments for the higher homologs. On the basis of NMR spectral arguments, the structures of the four major C_{27} MA-steroids of petroleum could be assigned as **1** (5β, 20*S*), **2** (5β, 20*R*), **3** (5α, 20*S*), and **4** (5α, 20*R*) isomers of 18-nor-17β-methylcholestra-8,11,13-triene (numbers **1** to **4** correspond to gas chromatographic elution orders), assignments which correlate with those made through rigorous syntheses (Riolo et al., 1985). The CDI-HPLC-NMR strategy is proposed as a general approach to the elucidation of chemical fossil structures. It can also provide partial structural information to help in the choice of candidate chemical fossil structures for synthesis, even where none may be apparent, and provide novel compounds for use as references.

INTRODUCTION

The ring-C monoaromatic steroid chemical fossils (MA-steroids), which have a widespread occurrence among petroleums (Seifert and Moldowan, 1978, 1986; Riolo et al., 1986) are important intermediates in the geologic dehydrogenation processes leading to the fully aromatized triaromatic geosteroids (Mackenzie, 1984; Mackenzie et al., 1981; Mackenzie et al., *Nature*, 1982). Upon burial and exposure to diagenetic and catagenetic processes, various precursor sterols (Carlson et al., 1980) are converted through a number of sterol, sterene, and rearranged intermediates to the steranes, rearranged steranes, and aromatic steroid hydrocarbons (Mackenzie et al., 1982) found in petroleum.

MA-steroids occur predominantly as ring-C components in mature petroleum (Mackenzie et al., 1982; Mackenzie et al., *Nature*, 1982; Seifert et al., 1983; Riolo et al., 1986), although ring-A monoaromatics have been found in shales (Hussler et al., 1981). The precise precursors of the ring-C MA-steroids and the mechanism of their formation are unknown, but it has been suggested that their formation is initiated by migration of a double bond in the side chain of certain sterols (e.g., Δ^{22} sterols) (Moldowan and Fago, 1986; Peakman, 1986). They appear to carry the signature of their precursors into petroleum and have been useful in addressing petroleum sourcing questions (Moldowan et al., 1985).

Monoaromatic steroids are formed early in diagenesis and are thought to give rise to fully aromatized triaromatic (TA) steroid hydrocarbons during petroleum source rock maturation (Mackenzie et al., *Nature*, 1982). The kinetics of this proposed conversion have been studied and applied to basin analysis (Mackenzie and McKenzie, 1983). However, the conversion of MA-steroids to TA-steroids may not be as simple as originally postulated due to common occurrences of several series of rearranged MA-steroids, which may also be TA-steroid progenitors (Moldowan and Fago, 1986; Riolo and Albrecht, 1985). Nevertheless, MA-steroid analyses provide information regarding petroleum catagenic processes, as well as being useful in oil correlation work.

Since the presence of MA-steroids in petroleum was first recognized over 30 years ago (O'Neal and Hood, 1956), a number of studies have largely determined the structures of various series primarily through contributions from Albrecht and co-workers (Schaefle et al., 1978; Ludwig et al., 1981; Riolo et al., 1985; Riolo and Albrecht, 1985). The methods employed in many of these efforts involved lengthy syntheses based on tentative proposed structures to obtain standards for high resolution capillary GC-MS coinjection studies with petroleum.

Noble metal catalyzed dehydrogenation-isomerization (CDI) of biologically related steranes was first recognized by Petrov and co-workers to give rise to mixtures containing the complete series of saturated and monoaromatic sterane isomers found in petroleum (Petrov et al., 1976; Zubenko et al., 1980, 1981). These products have capillary GC-MS properties identical with those of the petroleum natural products (Seifert and Moldowan, 1979; Seifert et al., 1983). In addition, triaromatic steroids and other steroidal chemical fossils and intermediates are also formed by the CDI method. Only steranes with GC-MS properties identical to petroleum steranes are generated in significant amounts; other possible isomers are absent or are present at low concentrations. Furthermore, the concentration ratios of the various saturated and aromatic products reflect the ratios found in petroleum. These similarities with the corresponding steroids in petroleum argue strongly that the two compound groups are identical (Seifert et al., 1983).

Using HPLC employing alternate nonaqueous reversed phase and adsorptive techniques, individual MA-steroid isomers can be isolated from cholestane CDI reaction mixtures. Even at high purity, as judged by capillary GC, GC-MS, HPLC, and NMR, the individual isomers do not crystallize. Thus, structural elucidation by X-ray diffraction methods is difficult or impossible.

The mass spectra of all ring-C MA-steroids are virtually identical (Seifert et al., 1983; Riolo et al., 1985). The high intensity m/z 253 ion [cleavage of the C-17(20) bond with charge retention at the quaternary-benzylic center, C-17] and a less intense m/z 143 ion (ring-B cleavage) are indicative of a ring-C aromatic structure, but further structural details cannot be inferred from the mass spectral data alone.

Since other physical instrumental methods provide inadequate data, NMR spectroscopy became the key method for obtaining detailed structural information for these MA-steroids.

The MA-steroids are complex tetracyclic molecules containing over 40 hydrogens and over 26 carbons which, to a large extent, are in similar chemical and magnetic environments. Thus, high-field NMR measurements are essential for useful structural studies (Croasmun and Carlson, 1987). Even at 500 MHz, the ^1H spectra of some of the compounds are so intractable that a further strategy of preparing lower molecular weight homologs by the CDI-HPLC sequence was developed to aid in structural studies. Moreover, to generate structural information for the lower molecular weight homologs, data from a variety of NMR techniques must be combined, including two-dimensional methods (Croasmun and Carlson, 1987). We have extended our previous communication (Seifert et al., 1983) to apply NMR in a detailed stereochemical structural study of the four major C_{27} chemical fossil compounds prepared by the general CDI technique and isolated by HPLC.

EXPERIMENTAL SECTION

Preparation of ring-C monoaromatic steroid hydrocarbons. Dehydrogenation-isomerization of cholestane, similar to the method devised by Zubenko et al. (1980, 1981) was used to prepare the ring-C, C_{27}, MA-steroids (Seifert et al., 1983). Cholestane ("gold label," Aldrich Chemical Company), 400 mg, and the catalyst (10% Pd on carbon, Aldrich Chemical Company), 300 mg, were sealed under vacuum (< 30 millitorr) in thick-walled, 50-mL Pyrex flasks (the reaction generates gases that can rupture the reaction flask if too large a proportion of reactants is used). Flasks were maintained at 300°C (in a Sola Basic Lindberg Model 51422 muffle furnace equipped with a Model 919 Eurotherm temperature controller) for 254 hours. Studies on MA-product variations observed for various reaction times, temperatures, catalysts, and catalyst-cholestane proportions are reported elsewhere (Seifert et al., 1983).

After cooling, the flasks (which contained slight positive pressure) were cautiously opened and the products (a clear semiviscous oil in the catalyst) were taken up in dichloromethane (Burdick and Jackson Laboratories, Inc.), then combined and filtered (0.5 μm "Zefluor" membranes, Rainin Scientific, Inc.). The catalyst (*pyrophoric*) was washed three times with dichloromethane and all filtrates were combined and evaporated under a stream of dry N_2. The total yield for the cholestane CDI product mixture was 91.3 percent, of which the ring-C MA-steroids accounted for 16.4 wt percent (an overall yield of MA-steroids of 15.0 wt %). The MA-steroids **1** through **4** accounted for 31, 30, 21, and 17 wt percent, respectively, of the MA-steroid mixture. Ring-C monoaromatic pregnanes were prepared by the same CDI method used to prepare the C_{27} MA-steroids. The C_{21} products were prepared from 5α-pregnane (Steraloids, Inc.), 201 mg, and 10 percent Pd on carbon, 204 mg, sealed under vacuum in thick-walled 7 cm × 10 mm diameter quartz tubes, and heated to 275°C for 258 hours. The clear, semiviscous product (163 mg) was subjected to preparative thin-layer chromatography (2 mm silica gel, 60 layers on 20 cm × 20 cm plate, E. Merck, and hexane developing solvent) by which the monoaromatic pregnanes were isolated (15 mg) from the 0.30 Rf band. The fraction contained two major compounds, **5** and **6** at 52 wt percent and 29 wt percent, respectively, with 20 wt percent of five minor compounds, as determined by GC (OV-17).

Gas chromatographic analyses. Routine GC analyses of products and HPLC fractions (in cyclohexane) were performed using a Hewlett-Packard Model 5710A chromatograph (with FID detection) equipped with a Model 7671A autosampler fitted with a Hamilton 1.0-μL syringe. Carrier gas (He, 2 mL/min) inlet pressure was controlled with a 0-30 psi Brooks regulator (Emerson Electric Company) in line with a Nupro metering valve (S-series, fine) to allow use of support-coated open tubular (SCOT) columns: Dexsil 300 (stainless steel 100 ft × 0.02-in. I.D., Perkin-Elmer), OV-101, and OV-17 (glass 50 m × 0.5 mm I.D., Alltech Associates). All C_{27} MA-steroid analyses were done isothermally at 250°C (Dexsil 300), 220°C (OV-101), and 240°C (OV-17). Kovats Retention Indices were measured relative to *n*-tricosane and *n*-octacosane (OV-101), *n*-pentacosane and *n*-triacontane (OV-17), *n*-pentacosane, and *n*-nonacosane (Dexsil 300) internal reference standards, with data reduction performed by a Hewlett-Packard Model 3388A reporting integrator. In the GC systems (Dexsil 300, OV-101, OV-17), the MA-steroids exhibit the following Kovats Retention Indices, respectively: **1** (2636, 2555, 2762), **2** (2696, 2615, 2819), **3** (2708, 2620, 2844), and **4** (2771, 2684, 2904). Retention indices are reported as the mean of three separate measurements for each phase. The C_{21} MA-pregnanes were analyzed using the OV-17 system at 220°C, where **5** and **6** eluted at 17.5 min and 22.5 min, respectively.

HPLC methods. Isolation and purification of individual MA-steroids were achieved by nonaqueous reversed-phase HPLC (NARP-HPLC) followed by adsorptive HPLC. Reversed-phase equipment consisted of an Altex Model 100 pump (with preparative pumping head), Rheodyne Model 7125 injection valve (with 2.0 mL loop), Whatman M20 10/50 ODS-II (octadecylsilane-bonded phase, high loading, 10 μm diameter particles), and a Waters Associates Model 403 refractive index detector. Crude cholestane CDI product mixtures (1.5 g) were split into fractions via NARP-HPLC employing an acetonitrile-dichloromethane (all HPLC solvents were Burdick and Jackson Laboratories, Inc.) (50:50, v/v) mobile phase at a 10 mL/min flow rate. MA-steroids eluted as a single fraction at 20 min which (150 mg loading for semipreparative runs) was further separated by repeated NARP-HPLC using a pure acetonitrile mobile phase at 10 mL/min, **3** elutes at 114 min, **1** and **4** (unresolved) at 120 min, and **2** at 126 min. Further purification required adsorptive HPLC: Constametric-II pump (Laboratory Data Control, Inc.), Altex Model 210 injector (with 0.1 mL loop), a Zorbax, 5 μm particle size silica gel, 4.6 mm I.D. × 250 mm column (Du Pont) and Ultrasphere (Altex) 5 μm particle size silica gel, 10 mm I.D. × 250 mm column connected in series, and an Altex Model 153 fixed wavelength (254 nm) UV detector. The NARP-HPLC fractions (typically 3 mg loading) were separated and purified using a pure hexane mobile phase at 2.0 mL/min: **1** and **2** (unresolved) eluting at 11.9 min and **3** and **4** (unresolved) eluting at 12.3 min. Repeated chromatography using these two systems provided samples of MA-steroids **1** through **4** with purities between 90 and 98 wt percent, as determined by GC and NMR. The monoaromatized pregnanes were isolated using the same NARP-HPLC method, except that a Whatman M9 10/50 ODS-II column was employed and a 3.00 mL/min flow rate of acetonitrile. Under those conditions, **5** elutes in 28.8 min and **6** in 30.7 min. The adsorption HPLC system for the monoaromatized pregnanes was identical to that used for the MA-steroids (except that an additional Ultrasphere silica gel column was connected in series and a 2.50 mL/min flow rate was used), and under those conditions **5** elutes in 17.3 min and **6** in 18.1 min.

NMR measurements. The 500 MHz ^{1}H NMR spectra were recorded on a Brüker WM-500 spectrometer operating in the Fourier transform mode. Each spectrum consists of 16,384 or 32,768 points generated from 32 to 1000 free induction decays. Except as noted, measurements were made at the ambient probe temperature, 298 K. MA-steroid solutions (10 to 100 mM) were prepared in high purity chloroform-d, cyclohexane-d_{12}, and benzene-d_6 (Norell, Inc.), and dichloromethane-d_2 and pyridine-d_5 (Merck and Company, Inc.). Chemical shifts were measured relative to the solvent signals, except for the shift reagent and variable temperature studies which employed a tetramethylsilane internal reference. Variable temperature studies were performed in dichloromethane-d_2 (203 K to ambient) and benzene-d_6 (ambient to 343 K). For variable temperature work, the probe temperature was calibrated using the temperature dependence of the chemical shift difference between the two resonances from a neat sample of methanol (low temperature) or ethylene glycol (high temperature). Decoupled spectra were obtained using pulsed, single frequency irradiation of 1 to 10 milliwatts RF power.

Spin-lattice relaxation rates were obtained by the inversion-recovery method (Hall and Sukumar, 1980). Samples were not degassed. NOE difference spectra were obtained using inverse-gated single frequency irradiation at low RF power (0.02–0.2 milliwatts). Every 16 scans, the contents of computer memory were negated and the irradiation frequency toggled between an on-resonance and an off-resonance value so that a difference spectrum was accumulated directly.

The 125 MHz ^{13}C NMR spectra required free induction decays of 32,768 points per spectrum. Broad band or single frequency ^{1}H irradiation of 3 watt power was used during each 0.57 seconds accumulation. Between scans decoupler power was reduced to 0.5 watt for 6 seconds to allow the sample to cool but maintain nuclear Overhauser enhancement.

The 500 MHz 2D *J* spectra were obtained using a (90°-t_1-180°-t_1-acquisition) sequence and processed with standard Brüker software. 128 Spectra of 8192 data points each were acquired, at 16 scans per spectrum. The data were weighted with a sine bell in the f_2 dimension and a Lorentzian to Gaussian lineshape transformation was used in the f_1 dimension. Zero filling in the f_1 dimension gave a digital resolution of 0.244 Hz in f_1 (spectral width 62.5 Hz) and 0.488 Hz in f_2 (spectral width 2000 Hz). Data are displayed in absolute value mode. After transformation, tilting, and projection, the data were correlated to artificially enforce symmetry with respect to the $f_1 = 0$ axis, thereby improving signal-to-noise.

The 400 MHz double quantum filtered COSY spectrum was obtained in pure adsorption mode via the TPPI technique. The data were processed using the FTNMR software package (D. Hare, Woodinville, Washington). At 64 scans per spectrum, 520 spectra of 2048 data points each were acquired. Spectral width was reduced to 1500 Hz, excluding the two aromatic proton signals and solvent peak, yielding a digital resolution of 1.45 Hz per point in each dimension after zero filling in f_1. The data were weighted with a 30-degree shifted sine bell in each dimension. No symmetrization was performed.

Binuclear shift reagent. Tris(6,6,7,7,8,8,8-heptafluoro-2, 2-dimethyl-3, 5-octanedionato)-ytterbium [$Yb(fod)_3$] was purchased from Stohler Isotope Chemicals. The Ag(fod) was prepared by the method of Wenzel et al. (1980). An aqueous silver nitrate (Alpha Products) solution (5.5 g/75 mL) was added to the methanolic 6,6,7,7,8,8,8-heptafluoro-2, 2-dimethyl-3,5-octanedione (ICN-K&K Laboratories) solution (9.6 g/5 mL) which had been neutralized by additions of 4M aqueous NaOH. Our Ag(fod) precipitate was gray rather than white after drying under vacuum over P_2O_5. The Ag(fod) was purified by dissolving it in cold acetone, filtering the solution using a millipore 0.5 μM Teflon membrane filter and assembly, evaporating the filtrate at room temperature and reduced pressure, and drying the white powder under vacuum over P_2O_5. The binuclear shift reagent [equal molar quantities of Ag(fod) and $Yb(fod)_3$] was used in chloroform-d solution according to the methods of Wenzel and Sievers (1981). Concentrations of Ag(fod)-$Yb(fod)_3$ employed must be considered as approximate due to a slight precipitation of the reagent during the NMR measurement (Wenzel and Sievers, 1981; Smith, 1981).

RESULTS AND DISCUSSION

The GC elution orders of the four major C_{27} petroleum MA-steroids formed by the CDI procedure are the same on each of the three separate SCOT liquid phases (Dexsil 300, OV-17, and OV-101), although resolution factors vary among the phases (OV-17 providing the best resolution). Elution orders are also identical on Dexsil 400 (Seifert et al., 1983). Accordingly, the MA-steroids are numbered (**1** through **4**) according to their elution order. The HPLC elution orders do vary, however, providing the selectivity necessary for separation. The elution orders: **3**, **1** plus **4** (unresolved) and **2** on NARP-HPLC and **1** plus **2** (unresolved), and **3** plus **4** (unresolved) on adsorptive HPLC provide a method for the isolation and purification of each of the four MA-steroids. Both **2** and **3** are isolated in reasonable purity by the higher loading NARP-HPLC technique alone. Since **2** is a major MA-steroid formed by the CDI method, it is used in most preliminary NMR experiments and in experiments requiring larger amounts.

Maximizing ^{1}H NMR spectral resolution. Each of the C_{27} MA-steroids (Fig.

5.1) contains 42 hydrogens as shown by mass spectral data (Seifert et al., 1983). Although their aromatic nature results in sizable chemical shifts for some positions, the large number of protons results in substantial methylene resonance peak overlap even at 500 MHz. Important information is obtained from the clear methyl and aromatic resonances (discussed later), but firm, unambiguous, and detailed stereochemical structural assignments required a careful study of the numerous A- and B-ring resonances between 1 and 3 ppm. The spectrum of **2** in $CDCl_3$ showed extensive resonance overlap, which obscured coupling patterns and made decoupling and NOE experiments ambiguous or impossible (Croasmun and Carlson, 1987).

Solvent-induced shifts were used in an attempt to improve resolution for double irradiation and 2D NMR experiments and proved very effective at 500 MHz. An 8-line benzylic resonance and an 11-line methine resonance (subsequently assigned to C-7α and C-20 protons, respectively) are cleanly resolved in spectra recorded using benzene-d_6 as solvent but completely obscured in spectra recorded using cyclohexane-d_{12} as solvent. In the solvent sequence cyclohexane-d_{12}, chloroform-d_1, dichloromethane-d_2, benzene-d_6, pyridine-d_5, the low-field C-16 proton resonance shifts steadily downfield while the low-field C-6 proton resonance shifts steadily upfield, with a concomitant improvement of resolution of C-1 and C-6 resonances. This allows selective irradiation of C-1 (but not C-6) with cyclohexane as solvent and selective irradiation of C-6 (but not C-1) with benzene or pyridine as solvent (Croasmun and Carlson, 1987). Benzene-d_6 was the best solvent for selective irradiations involving benzylic protons, primarily due to the anomalous collapse (into a "deceptively simple" narrow triplet) of the C-15 benzylic protons, which obscures one of the C-7 benzylic resonances in other solvents. Benzene-d_6 also provided the best spectral resolution in the 1.5 to 1.7 ppm range.

Another powerful means of resolving overlapping signals and obtaining *J* coupling information is two-dimensional NMR spectroscopy (Croasmun and Carlson, 1987). However, by performing the 2D *J* experiments on **2**, we found that the important ring-A methylene protons have multiple strong couplings and that they overlap with side chain methylenes. This resulted in 2D *J* spectra of very low intensity with many cross-peaks from the strong couplings. The 2D data are already complicated by the large number (16) of resonances in a small region of the spec-

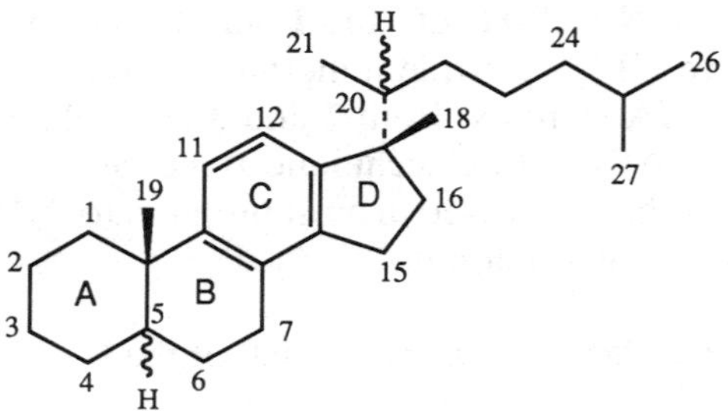

Figure 5.1 Carbon skeleton form of the major C_{27} petroleum monoaromatic steroid hydrocarbons isolated from the CDI product mixture from cholestane.

trum. Consequently, the 2D *J* technique did not provide us with useful data for the C_{27} compounds. Furthermore, COSY analyses of the C_{27} MA-steroids would suffer from the same problems and were not attempted. However, these 2D NMR techniques did provide useful data (discussed later) for the C_{21} monoaromatic steroids, which lack side-chain methylenes.

NMR shift reagents also were used in an attempt to improve spectral resolution further. Shift reagent studies of unfunctionalized aromatics such as the MA-steroids are possible through binuclear shift reagents (mixtures of Ag and Yb β-diketonates) (Wenzel et al., 1980; Wenzel and Sievers, 1981). Spectra of **2** were measured with varying equimolar amounts of Ag(fod)-Yb$(fod)_3$ in $CDCl_3$ (Table 5.1). Although substantial resonance shifts were observed, the shifts of overlapping resonances in the 1 to 3 ppm range were all of comparable magnitude toward low field (resulting in unimproved resolution), were obscured by the fod methyl resonances at ~2 ppm, and were broadened so as to obscure *J*-splitting. The binuclear shift reagent experiments provided relative internuclear distance data for **2**. As expected, the two aromatic protons exhibit the largest paramagnetic shifts with increasing shift reagent concentrations (Table 5.1). However, reasons for differential shifting rates for the two aromatic protons and for the two singlet methyls are unclear, although this effect is most reasonably attributed to stereochemical hindrance affecting the geometry of the Ag(fod)-ring-C aromatic center association. The benzylic protons shift at comparable rates and with greater magnitudes than resonances (subsequently assigned to C-1, 6, and 16) of protons more distant from the C-ring. The C-21 methyl resonance exhibits a low but measurable (Table 5.1) shift rate, but the side-chain methyls (C-26 and C-27), distant from the C-ring,

Table 5.1 BINUCLEAR SHIFT REAGENT INDUCED RESONANCE SHIFTS FOR 2

Proton(s)	Slope[a] Δδppm/molal reagent
12	15.0
11	13.4
19(Me)	13.1
Me at 17	9.6
7α + (15α + β)	9.0
7β	8.9
1α	5.4
16α	4.5
6β	4.4
21(Me)	2.9
26(Me) + 27(Me)	0

[a]Slopes (linear coefficient of determination values between 0.95 and 1.00) for the change in chemical shift (relative to the unshifted resonance) of various **2** protons (**2** at 9.1 millimolal) as a function of binuclear shift reagent concentrations between 3.3 and 33 millimolal) in $CDCl_3$.

exhibit no measurable shift within the shift reagent concentration range studied. Shift reagent experiments did not provide the specific increases in spectral resolution that we needed to perform critical selective double resonance experiments for **2**.

Thus, solvent-induced shifts appear to be the method of choice for improving spectral resolution for the C_{27} MA-steroids, and benzene-d_6 was used as the solvent for subsequent 500 MHz NMR measurements (except for low temperature studies).

Five hundred MHz spectra. Complete ^{1}H spectra (with C_6D_6 as solvent) of **1** through **4** are presented in Figure 5.2, along with assignments developed in the following discussion. The spectra demonstrate the purity with which MA-steroids can be isolated by HPLC from the cholestane CDI product mixture. Spectra of all four compounds have similar features in the methyl (0.7–1.3 ppm) region; that is, each has two methyl singlets, one isolated methyl doublet, and two closely spaced methyl doublets. Also, the spectrum of each compound exhibits two aromatic protons between 6.9 and 7.3 ppm in a loose AB pattern with ortho coupling constants, 7.9 to 8.2 Hz. These methyl and aromatic spectral features shared by **1** through **4** demonstrate that each MA-steroid has: (1) an intact C_8 "sterol-type" side chain with a terminal isopropyl group (Fig. 5.1) and a C-21 methyl group, (2) two nuclear methyls attached at quaternary carbons, and (3) an aromatic C-ring with no substituents at C-11 or C-12 (Seifert et al., 1983). The features are consistent with a common skeletal form (Fig. 5.1) for the C/D-ring and side chain.

Both the methyl resonances and the benzylic and other resonances in the 1.8 to 2.8 ppm region have very similar patterns for alternate pairs (Fig. 5.2). MA-steroids **1** and **2** form one pair having closely spaced methyl singlets with similar chemical shifts (1.240 and 1.235 ppm in **1** and 1.267 and 1.251 ppm in **2**), whereas **3** and **4** form another pair with better resolved methyl singlets (1.249 and 1.074 ppm in **3** and 1.279 and 1.085 in **4**). The **1**, **2** pair also exhibit spectra with nearly identical patterns in the 1.8 to 2.8 ppm region (as do the **3**, **4** pair). Both the singlet methyl and 1.8 to 2.8 ppm region exhibit resonances associated with nuclear features which strongly suggest common nuclear stereochemical features in the **1**, **2** pair and different common nuclear stereochemical features in the **3**, **4** pair.

In contrast, the isolated methyl doublet (C-21) chemical shifts of **1** and **3** (0.956 and 0.965 ppm, respectively, Fig. 5.2) also form a pair. Those of **2** and **4** (0.812 and 0.818 ppm, respectively, Fig. 5.2) form another pair. The same **1**, **3** and **2**, **4** pairing is seen for the closely spaced methyl doublet (C-26 and 27) chemical shifts (0.821, 0.812, and 0.822–0.808 ppm for the **1**, **3** pair, respectively; 0.891, 0.885, and 0.894–0.886 ppm for the **2**, **4** pair, respectively). Also, a one-proton multiplet (subsequently assigned to C-20) is present as a well-resolved resonance at 1.73 ppm in the spectrum of **2** and **4**, but is lost under overlapping resonances at ~1.66 ppm in the spectrum of **1** and **3**. These features arise from side-chain structures, thus suggesting that the side-chain stereochemistry of **1** and **3** is identical and distinct from that of **2** and **4** (which also form a pair with apparently identical side-chain stereochemistry).

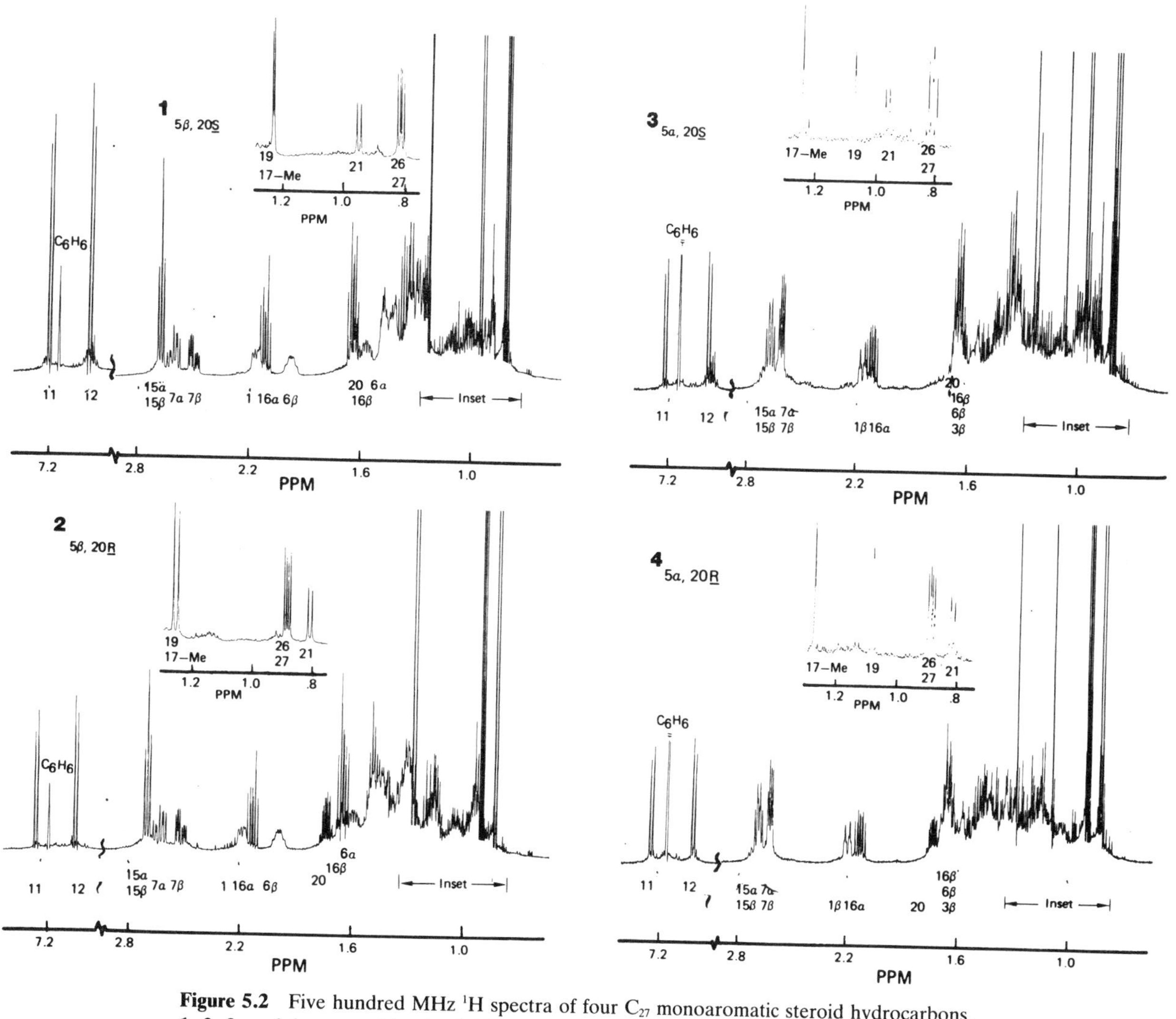

Figure 5.2 Five hundred MHz ^1H spectra of four C_{27} monoaromatic steroid hydrocarbons, **1**, **2**, **3**, and **4** prepared by the CDI method. Sample concentrations in benzene-d_6 were ~2 mg/mL for **3** and **4** and ~6 mg/mL for **1** and **2**. Insets are attenuated traces of the methyl regions (0.7–1.3 ppm).

Double resonance studies involving 2. Results of difference NOE and spin decoupling experiments for **2** are presented in Table 5.2. Irradiation of the C-21 methyl doublet decouples the 1.726 ppm multiplet, allowing its assignment to the C-20 methine shown in Figure 5.3. The difference NOEs induced by C-21 methyl irradiation locate protons in the D-ring and side chain vicinity. The NOE of the 7.005 ppm resonance induced by C-21 irradiation supports its assignment to the C-12 aromatic proton; NOEs of the deceptively simple 2.704 ppm two-proton benzylic triplet support its assignment to the geminal protons at C-15 in ring-D, and NOE of the 2.130 ppm eight-line, one-proton multiplet permits it to be assigned to one of the protons at C-16 (corresponding to the C-2 protons of indane) in the D-ring. Those NOEs are illustrated in Figure 5.4 (where *R* is the side chain and could be 20*R* or *S* although only the 20*S* configuration is shown), which represents the C- and D-rings. Rotation around the 17(20) bond is apparent from the C-21 irradiation-induced NOEs of both the 12 and 16α protons. These ring-D assignments were supported by spin decoupling of the 2.130 ppm (C-16α) resonance which resulted in changes in the 2.704 ppm resonance (C-15 protons) but also collapse (loss of a 13 Hz splitting) of the 1.644 ppm eight-line, one-proton multiplet (allowing its assignment to the remaining C-16 proton). Subsequent decoupling

Table 5.2 CHEMICAL SHIFT AND CONNECTIVITY DATA FOR 2 DETERMINED BY DECOUPLING AND DIFFERENCE NOE

Proton(s) irradiated	σppm	Decoupling	NOE
1α	2.187	1β, [2α + β][a]	
[1β][a]	~1.33	1α, [2α + β]	
[2α + β]	~1.45	1α, [1β][b]	
[5]	~1.48	6β[b,c]	
6α	1.58	6β, 7α, 7β[c]	
6β	1.98	6α, 7α, 7β, [5]	7
7α	2.635	7β, 6α, 6β	
7β	2.518	7α, 6α, 6β	
11	7.220	12	1α, 12
12	7.005	11	11
15α + β	2.704	16α, 16β	
16α	2.130	16β, 15α + β	15, 16β
[16β]	1.644	16α, 15α + β	
Me at 17	1.251	S	
19Me	1.267	S	1α
20	1.726	21[b]	
21Me	0.812	20	12, 15, 16α, 20

[a]Bracket indicates signals with overlapping resonances, reducing irradiation specificity (see text).

[b]Other unassigned resonances decoupled, consistent with assigned structure (see text).

[c]Decouplings obscured by proximity of irradiation frequency.

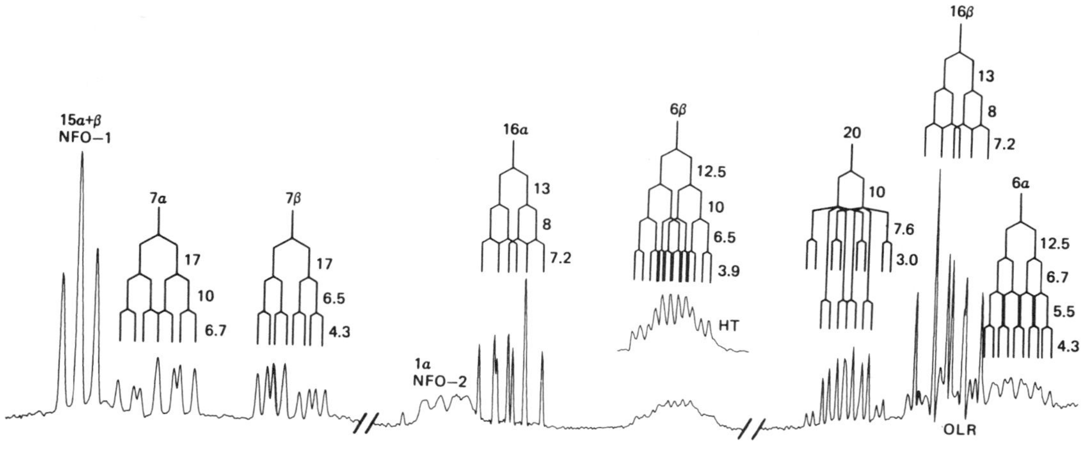

Figure 5.3 Coupling patterns and constants for resonances in the 1.5–2.7 ppm region of the 500 MHz spectrum of **2**. NFO-1 is a nonfirst order, deceptively simple triplet. NFO-2 is nonfirst-order pattern presumably due to virtual coupling, see text. HT indicates higher temperature (343 K) spectral result. OLR refers to an overlapping resonance at ~1.65 ppm.

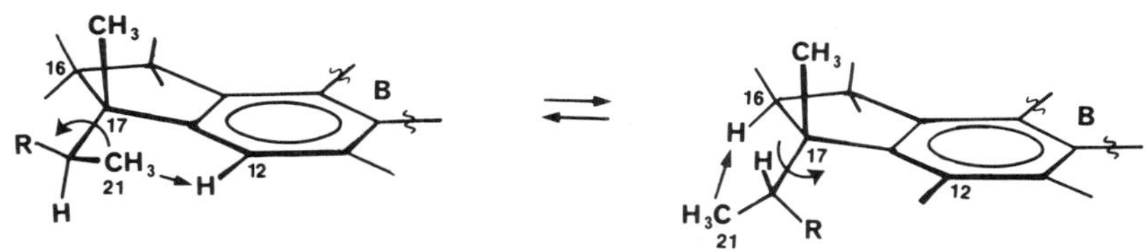

Figure 5.4 Stereochemical representation of monoaromatic steroid C- and D-rings showing rotation around the C-17 (C-20) bond and the close approach of hydrogens on the C-21 methyl to those attached at C-12 and C-16.

(Table 5.2) of the 2.704 ppm (C-15) and 1.644 ppm (C-16) resonances supported these assignments, with the 7.220 ppm resonance assigned to the C-11 proton. The eight-line coupling patterns of the two C-16 protons in Fig. 5.3 show that they have vicinal couplings only to the C-15 protons (8 and 7.2 Hz); thus there is no C-17 methine (also supported by the C-20 coupling pattern, which only shows J coupling to C-21 and C-22 protons, see Fig. 5.3). This means that the C-18 angular methyl has migrated to C-17 (C-17 becoming quaternary) during aromatization in the CDI reaction. These results define the basic C- and D-ring and side-chain structure of **2** as shown in Fig. 5.1, with details of side-chain stereochemistry considered in a later section.

Resonances associated with the A- and B-rings of **2** were considered next. The remaining benzylic resonances (both eight-line multiplets) at 2.635 ppm and 2.518 ppm (which are coupled, see Table 5.2, and show a slight second-order skewing) are assigned to the C-7 geminal pair, with their clear spin coupling pattern indicating two (C-6 α and β) vicinal couplings apparent in Figure 5.3. Irradiation of either of the C-7 resonances results in decoupling (Table 5.2) of resonances at 1.98 ppm (broad, featureless at ambient temperatures) and 1.58 ppm (10-line multiplet). The broad 1.98 ppm resonance sharpens into an 11-line multiplet at 343°C seen in Figure 5.3. The C-6 geminal protons are each coupled to the two C-7 vicinal benzylic protons and one additional proton at ~1.48 ppm (Table 5.2), presumably the C-5 methine.

Decoupling experiments (Table 5.2) showed that the broadened 2.187 ppm signal in Figure 5.3 is not coupled to either of the C-6 protons. Irradiation of the 2.187 ppm resonance indicated couplings to signals at 1.33 and 1.45 ppm. A large, geminal coupling to the 1.33 ppm resonance was observed with smaller vicinal coupling to the 1.45 ppm signal. The assignment of this isolated resonance was clarified by further NOE studies. Irradiation of the 7.220 ppm (C-11) signal induces an NOE (Table 5.2) in the 2.187 signal, strongly supporting its assignment to the C-1 α(H) on stereochemical grounds as illustrated in Figure 5.5.

Irradiation of the singlet methyls (which are not sufficiently resolved for selective irradiation) also induces an NOE in the 2.187 ppm resonance (presumably from C-19). Decoupling (Table 5.2) allowed the assignment of the C-1 geminal partner to 1.33 ppm with the two strongly coupled C-2 protons coresonating at 1.45 ppm, giving rise to some non-first-order character in the 2.19 ppm resonances, presumably as a result of virtual coupling. Many of the assignments made through the selective decoupling experiments are now more readily achieved using COSY data discussed later. These results, along with mass spectral data (Seifert et al., 1983), define the complete skeleton of **2** as shown in Figure 5.1 with stereochemistry being considered in a later section.

^{13}C NMR data. The noise decoupled spectrum of **2** exhibits signals for all 27 carbons (Table 5.3). Clear multiplicity data were obtained in 1981 from off-resonance decoupled spectra for 19 of the signals, with uncertainties in the other multiplicities arising from signal overlap and signal-to-noise problems (Table 5.3). The Distortionless Enhancement by Polarization Transfer (DEPT) technique (Doddrell et al., 1982) has greatly improved carbon multiplicity assignments and is now the method of choice.

Guided by the ^{1}H NMR results, we have made tentative assignments with the aid of the ring-C monoaromatic C_{21} substituted steroid ^{13}C NMR data (Cheung

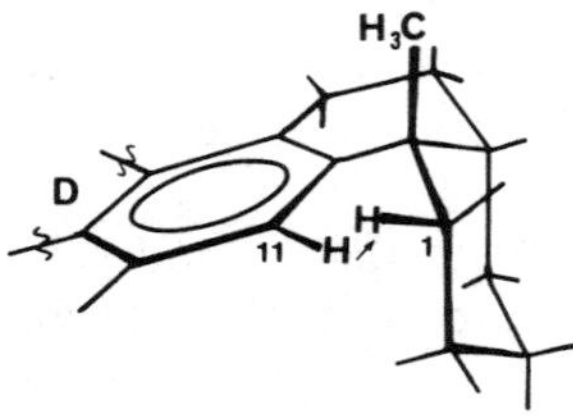

Figure 5.5 Stereochemical representation of a C-ring monoaromatic steroid possessing a *cis* A/B-ring juncture showing the close approach of the 1α hydrogen to the aromatic hydrogen attached at C-11.

Table 5.3 ^{13}C CHEMICAL SHIFT AND MULTIPLICITY DATA FOR 2

Carbon number	Calculated or referenced ^{13}C chemical shift, ppm	Measured ^{13}C chemical shift, ppm	Multiplicity
1	38.4[b], 38.5[d]	38.7	T
2	22.7[b], 24.8[d]	23.0	[T][a]
3	23.1[d]	[25.1]	[T]
4	28.4[b], 27.3[d]	28.2	T
5	41.9[b], 41.7[d]	41.2	D
6	25.8[b], 25.5[d]	26.0	T
7	23.6[b], 28.5[d]	24.2	[T]
8	135.5[d]	131.4	S
9	144.1[d]	141.7	S
10	37.3[b], 37.6[d]	37.2	S
11	125.3[d]	123.8	D
12		120.4	D
13	152.5[e]	147.7	S
14	142.8[e]	141.5	S
15	30.2[e]	[29.4]	T
16	35.6[e]	[34.3]	T
17	48.3[e]	[51.1]	S
Me at 17		[26.3]	[Q]
19Me	32.2[b], 31.9[d]	31.9	[Q]
20		[41.1]	D
21Me		[15.2]	Q
22		[31.9]	T
23	23.9[c]	24.7	[T]
24	39.5[c]	39.3	T
25	28.0[c]	28.0	D
26 Pro-*R*	22.5[c]	22.5	[Q]
27 Pro-*S*	22.8[c]	22.8	[Q]

[a]Bracket indicates uncertainty in assignment (see text).

[b]Calculated using data from Blunt and Stothers (1977), Cheung et al. (1979), Bridgewater et al. (1980).

[c]Data for the cholesterol side chain (Joseph-Nathan et al., 1979).

[d]Data for 4α-methyl-*cis*-1,2,3,4,4a,9,10,10a-octahydrophenanthrene (Zubenko et al., 1981).

[e]Data for spiro [indan-1,1′-cyclohexane] (Zubenko et al., 1981).

et al., 1979; Bridgewater et al., 1980) 4a-methyl-*cis*-1,2,3,4,4a,9,10,10a-octahydrophenanthrene and spiro [indan-1,1′-cyclohexane] data reported by Zubenko et al. (1981), steroid ^{13}C NMR spectral compilations by Blunt and Stothers (1977), Smith (1978), and Eggert and Djerassi (1981), and C_{27} sterol side chain ^{13}C NMR data (Joseph-Nathan et al., 1979), as well as the observed multiplicities and relative signal intensities.

The six aromatic carbon resonances are readily assigned by making use of the off-resonance decoupled spectrum, the reduced NOE of the isolated C-9 carbon, the shielding of C-12 relative to C-11 by C-17 substituents, and the indane and C-ring aromatic steroid assignments of Cheung et al. (1979).

The remoteness of the C_{27} terminal side-chain carbons (C-24 to C-27) allows direct resonance assignments (Table 5.3) for **2** by correlation with the reported side-chain assignments. Multiplicity data, although somewhat obscured for C-26 and C-27 (Table 5.3), are consistent with the side-chain assignments.

The A- and B-ring assignments were developed by generating correction factors for the ^{13}C chemical shifts of each A- and B-ring carbon (except C-3) of 5β-cholestane relative to 5β-cholestan-3-one from data tabulated by Blunt and Stothers (1977). The factors were tested by calculating the ^{13}C chemical shifts of 5β-androstane and 5β-pregnane from the 5β-androstane-3,17-dione, and 5β pregnan-3,20-dione spectral data. The calculated shifts agreed with the reported shifts for each A- and B-ring carbon and C-19 carbon within 0.3 ppm in each case. The correction factors were then applied to data reported by Cheung et al. (1979) for 17β-methyl-18-nor-5β-pregna-8,11,13-trien-20,21-diol-3-one in order to calculate the A- and B-ring chemical shifts for a 5β-ring-C monoaromatic steroid hydrocarbon. The calculated shifts (Table 5.3) correlated with observed resonances to within 0.3 ppm for C-1-2-4-6-10 and -19 and within 0.7 ppm for resonances we assign to C-5 and -7, allowing the tentative assignments for these signals in Table 5.3. These assignments are consistent with observed multiplicities and correlate well with those for corresponding carbons in 4α-methyl-*cis*-1,2,3,4,4a,9,10,10a-octahydrophenanthrene reported by Zubenko et al. (1981).

The remaining singlet is assigned to C-17, the remaining doublet to C-20, and the remaining quartets to the C-21 (15.2 ppm) and C-17 (26.3 ppm) methyls. C-16 and C-15 are assigned by correlation with the spectrum of 17β-methyl-18-nor-5β-pregna-8,11,13-trien-20,21-diol-3-one and spiro [indan-1, 1′-cyclohexane] (Table 5.3). The remaining assignments are based on multiplicity data and suffer from a lack of model compound data for the proposed ring-C-side chain structure of **2**.

Double resonance studies involving 3. Results of NOE, spin decoupling, and spin-lattice relaxation experiments for **3** are shown in Table 5.4. Irradiation of the C-21 methyl located the C-20 multiplet, partially resolved from overlapping signals at 1.67 ppm. (See Fig. 5.6.) Conversely, irradiation at 1.67 ppm collapses the C-21 doublet, as expected.

Decoupling identifies a coupled four-spin system consisting of a geminal pair at 2.125 and 1.655 ppm and a second geminal pair with both protons at 2.684 ppm. NOE from the C-21 methyl to the 2.125 ppm signal locates the first geminal pair on C-16; the second is therefore on C-15.

The 6.992 ppm signal exhibits NOE from the C-21 or C-17 methyls and may be assigned to C-12. Irradiation of the C-17 methyl also induces an NOE (Table 5.4) in one of the C-16 proton resonances. These results show that the basic structure

Table 5.4 CHEMICAL SHIFT, SPIN-LATTICE RELAXATION RATES AND CONNECTIVITY DATA FOR 3 DETERMINED BY DECOUPLING AND DIFFERENCE NOE

Proton(s) irradiated	σppm	Decoupling	NOE
[1α][a]	1.37	1β, [2]	
1β	2.177	[1α, 2, 3]	
[2]	1.51	1β[b,c]	
[3]	1.63	1β[b]	
[6]	1.44	7α + β[c]	
[6]	1.64	7α + β[c]	
7α + β	2.622	[6α, 6β]	
11	7.230	12	1β, 12
12	6.992	11	11
15α + β	2.684	16α, [16β]	
16α	2.125	[16β], 15α + β	
[16β]	1.655	16α, 15α + β	
Me at 17	1.249	S	12, 16β
19Me	1.074	S	11
[20]	1.67	21[b]	
21Me	0.965	[20]	12, 16α, 20
26Me	0.822		
27Me	0.808		

[a]Bracket indicates signals with overlapping resonances, reducing irradiation specificity (see text).
[b]Other unassigned resonances decoupled, consistent with assigned structure (see text).
[c]Decouplings obscured by proximity of irradiation frequency.

of the C-and D-ring and side chain of **3** is identical to that of **2** (except for stereochemistry, as mentioned).

The 2.177 ppm resonance of **3** (C-1β, see Fig. 5.7) exhibits NOE properties similar to those of the 2.187 ppm C-1α signal of **2** (see Fig. 5.5), that is, showing NOE from the C-11 proton. This 2.177 ppm signal also shows a strong geminal coupling (13 Hz to the 1.37 ppm resonance), two vicinal couplings (4 Hz each, both at ~1.5 ppm), and a long-range coupling (~1 Hz).

The C-6 protons are decoupled by irradiation of the 2.622 ppm (C-7 geminal pair) multiplet (Fig. 5.6) and are found (at 1.64 and 1.44 ppm) amid overlapping resonances so that their coupling patterns cannot be observed. Again, assignments made through multiple selective decoupling experiments are now best done using COSY data discussed later.

Neither difference decoupling, with irradiation of the C-7 benzylic protons, nor the method of nulling overlapping resonances by obtaining partially relaxed spectra succeeded in resolving the C-6 resonance pattern or finding the C-5 methine if present in **3**.

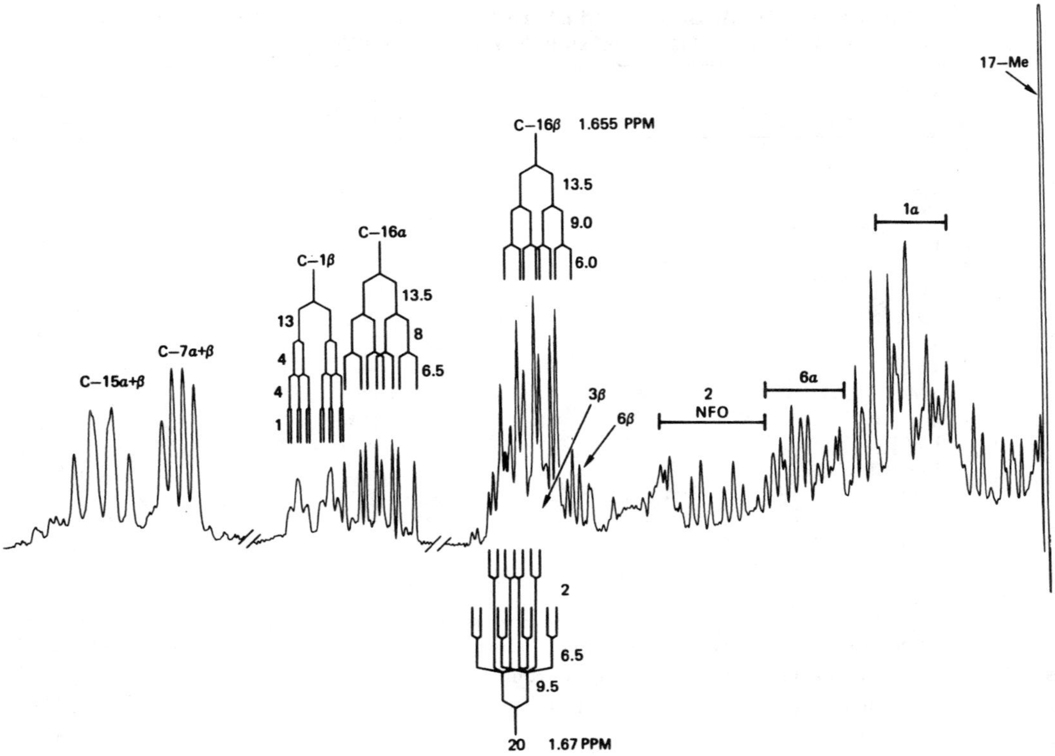

Figure 5.6 Coupling patterns and constants for resonances in the 1.2–2.7 ppm region of the 500 MHz ^{1}H spectrum of **3**. NFO refers to nonfirst order pattern.

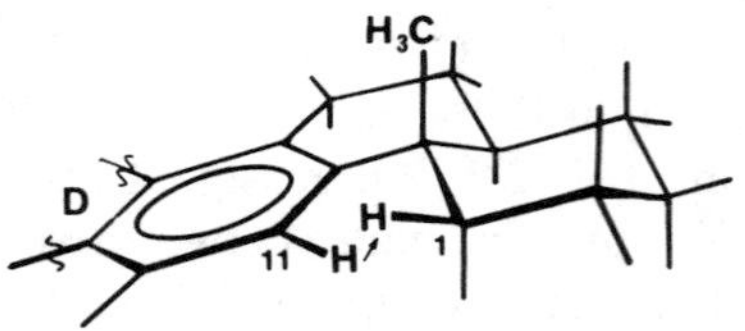

Figure 5.7 Stereochemical representation of a C-ring monoaromatic steroid possessing a *trans* A/B-ring juncture showing the close approach of the 1β hydrogen to the aromatic hydrogen attached at C-11.

C_{21} Ring-C monoaromatic steroids 5 and 6. The A- and B-ring ^{1}H resonance assignments for the C_{27} MA-steroids could not be made by 2D *J* spectroscopic analyses because of the severe ^{1}H spectral overlap problems already discussed. Therefore, we prepared a C_{21} ring-C MA-steroid in which the numerous interfering side-chain methylene resonances are not present. The C_{21} ring-C MA-steroids (which also occur in petroleum) were prepared by CDI treatment of 5α-pregnane. This eliminated 12 side-chain proton resonances present in the C_{27} MA-steroids. The ^{1}H NMR spectra shown in Figure 5.8 of **5** and **6** are very similar to the spectra

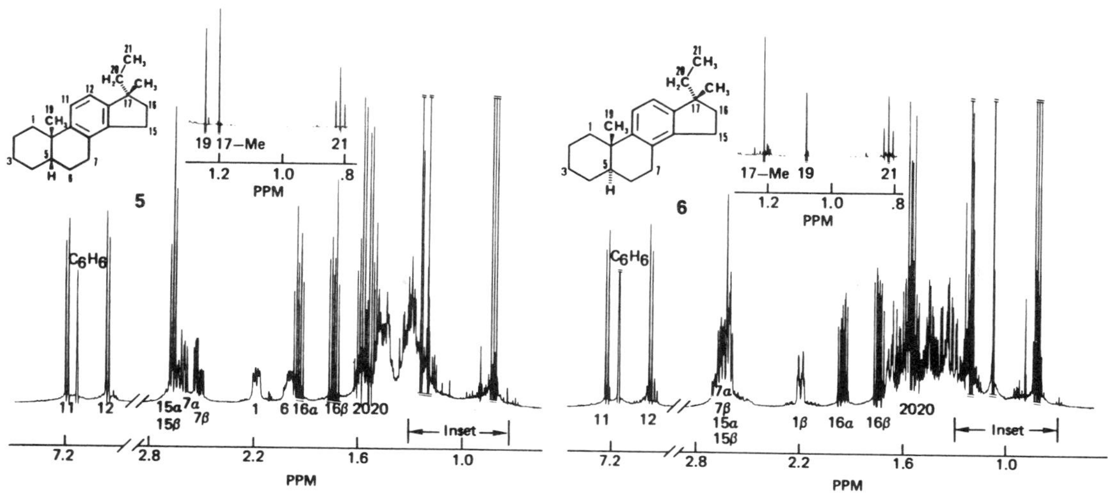

Figure 5.8 Five hundred MHz ^{1}H spectra (in benzene-d_6) of ring-C monoaromatic pregnanes **5** and **6** prepared by the CDI method.

of the C_{27} MA-steroids shown in Figure 5.2. The spectra of **5** and **6** exhibit two A/B coupled (ortho) aromatic protons. Since **5** and **6** exhibit the m/z 253 mass spectral ion, it is clear that the aromatic protons correspond to C-11 and C-12 as for **1** through **4**. Both **5** and **6** exhibit two methyl singlets (C-17 methyl and C-19) and one methyl triplet (C-21). The two C-20 protons are identified by decoupling the C-21 triplet (Table 5.5). The benzylic resonances of C_{21} MA-steroids are very similar to the corresponding patterns of C_{27} MA-steroids **1** and **2** in Figure 5.3 with C-15α and β present as a deceptively simple triplet slightly downfield from the 7α and 7β signals which are both eight-line patterns showing slight second-order character. Furthermore, C-1, C-6α, C-16α, and C-16β are resolved in spectra of **5** (Fig. 5.8) and resemble patterns for the corresponding protons of **1** and **2** in Figure 5.2. The spectrum of **6** (Fig. 5.8) resembles that of **3** and **4** (Fig. 5.2), although the C-7 and C-15 benzylic protons coresonate in **6** but are resolved in **3** and **4**.

2D *J* spectrum of **6**. The projection and cross-sections from the 2D *J* resolved spectrum of **6** shown in Figure 5.9 were accumulated in order to observe A- and B-ring ^{1}H resonance patterns and thus complete the stereochemical structural assignments for **6**, and **3** and **4** (by analogy). The 2D *J* spectrum of **6** provides resonance patterns for 9 of the 11 nonbenzylic A- and B-ring protons. The chemical shifts of the two missing signals, 2β and 3α (Fig. 5.9), are indicated by decoupling and NOE data (discussed later).

One of the C-1 proton resonances of **6** is significantly deshielded relative to the other nonbenzylic A- and B-ring protons. Evidence presented below supports a rigid 5α(H)-A/B-*trans* "steroid" structure for **3** and **4**, the C_{27} homologs of **6**. If

Table 5.5 CHEMICAL SHIFT AND CONNECTIVITY DATA FOR 6 DETERMINED BY DECOUPLING, DIFFERENCE NOE, AND 2-D NMR

Proton(s) irradiated	σppm	Decoupling
[1α][a]	1.376	1β[c]
1β	2.195	[1α, 2α + β, 3β]
[2α]	1.44	1β[c]
[2β]	1.53	1β[c]
[3α]	~1.21	[3β][c]
[3β]	1.678	1β, 2α, 2β, 4β, 4α
[4α]	1.357	[3β][c]
[4β]	1.301	[3β][c]
[5]	1.321	[6β][c]
[6α]	1.449	[6β], 7α + β
[6β]	1.639	[6α], 7α + β
7α + β	2.60–2.71	[6α, 6β]
11	7.221	12
12	6.966	11
[15α + β]	2.60–2.71	16α, 16β
16α	1.952	16β, [15α + β]
16β	1.743	16α, [15α + β]
Me at 17	1.216	S
19Me[e]	1.080	S
[20 Pro-*R*][d]	1.531	21
[20 Pro-*S*][d]	1.581	21
21Me	0.082	[20,20]

[a]Bracket indicates signals with overlapping resonances, reducing irradiation specificity (see text).

[c]Decouplings obscured by proximity of irradiation frequency.

[d]Interchangeable-assignment uncertain.

[e]Irradiation of 19Me induces difference NOEs in 1β, 2β, 4β, and 6β.

a 5α(H)-A/B-*trans* form is assumed, then the C-1 signal at 2.195 ppm must be the C-1β proton. Beginning from the resolved 1β resonance, the remaining A-ring protons on C-1 through C-4 may be assigned in a straightforward fashion, using decoupling, the resolved resonance patterns from the 2D *J* spectrum, and the results of an NOE difference experiment in which the 19-methyl was irradiated. Note that:

1. There is a long-range "W" coupling between 1β and 3β (dotted line in Fig. 5.10).
2. Irradiation of the 19-methyl yields NOEs of axial protons on the β face (2β, 4β, and 6β) and the adjacent 1β (Fig. 5.10). This result indicates a 5α structure.

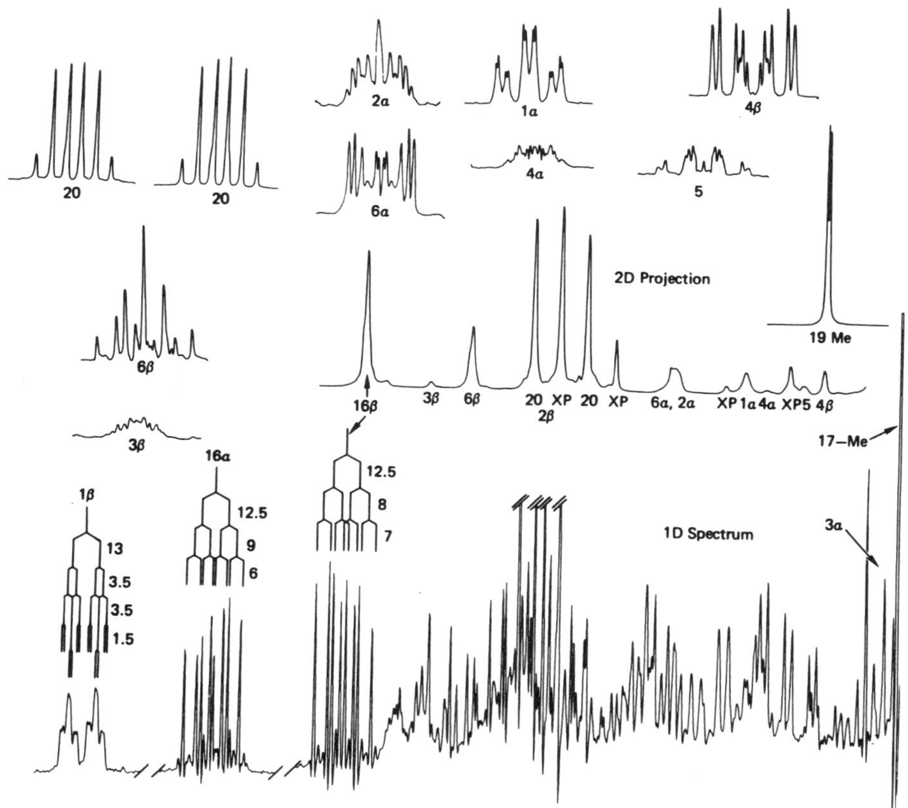

Figure 5.9 Two-dimensional *J* resolved 500 MHz spectroscopic results and assignments for C_{21} monoaromatic steroid **6**. The 2D projection is surrounded by cross-sections representing the *J* coupling patterns of individual resonances.

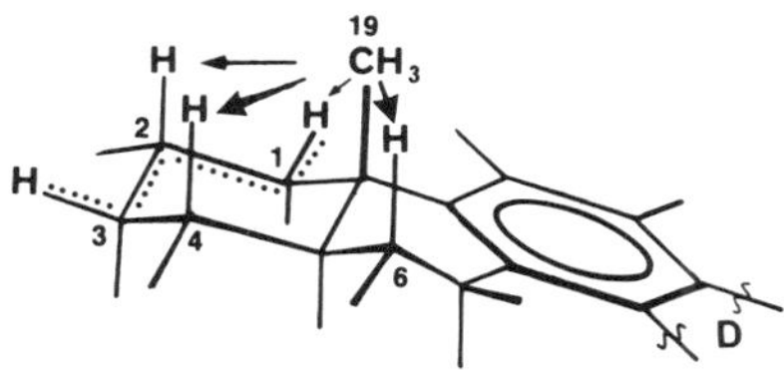

Figure 5.10 Stereochemical representation of a C-ring monoaromatic steroid possessing a *trans* A/B-ring juncture showing the close approach of the 1β, 2β, 4β, and 6β hydrogens to the hydrogens on the C-19 methyl group.

3. The 2D *J* patterns are consistent with the expected number of large (geminal, axial-axial) and small (axial-equatorial, equatorial-equatorial) couplings for each resonance, again assuming a 5α structure. (For a detailed discussion of 5α steroid coupling patterns see Croasmun and Carlson, 1987.)

4. There is a small coupling between 1α and the 19-methyl, which appears as a resolved doublet in the 2D *J* spectrum—a phenomenon typical of 5α steroids (Croasmun and Carlson, 1987).
5. The downfield shifts of 1β and 2β are consistent with ring current effects resulting from the proximity and orientation of the C-ring relative to these protons in a 5α configuration.

Similarly, the C-5 and C-6 protons may be assigned by beginning from the C-7 protons; 6β is distinguished from 6α by its NOE from the 19-methyl. Contour plots of the 2D *J* data were required to disentangle the 6α pattern from the nearly isochronous 2α.

The decoupling and difference NOE data (Table 5.5) and 2D *J* resolved spectrum allow a complete ^{1}H NMR spectral assignment for **6** consistent with the 17β-methyl-18-nor-5α-pregna-8,11,13-triene structure. The homologous nature of **6** with **3** and **4** allows the structural assignments for **3** and **4** by direct spectral correlation.

COSY spectrum of 5. The ^{1}H-^{1}H COSY 2-D NMR spectrum of the C_{21} monoaromatic steroid **5** was run at 400 MHz and is shown in Figure 5.11. The cross-peaks associated with coupled spins in the B- and D-rings, and part of the A-ring, and with the C_2 side chain are marked; the aromatic region has been excluded. The spectrum shows that the side chain and D-ring are each isolated spin networks with all cross-peaks clearly visible. The A-ring C-1 and C-2 hydrogens show cross-peaks as indicated in Figure 5.11, but hydrogens on C-3 and C-4 have resonances which are packed into an extremely narrow spectral region ~1.3–1.4 ppm. All cross-peaks associated with the spin coupling between hydrogens on C-6 and the benzylic C-7 are clearly seen. However, the hydrogen on C-5 (which connects the A-ring and B-ring spin systems) is obscured at 1.5–1.6 ppm. Consequently, the existence of the C-5 hydrogen is not revealed by COSY cross-peaks, but is apparent from the coupling patterns of the C-6 hydrogens, each showing couplings to a minimum of four other hydrogen atoms.

Stereochemical assignments. The C_{27} MA-steroids have four asymmetric centers (C-5, 10, 17, and 20); the C_{21} monoaromatics have three (C-5, 10, and 17). Since epimerization at C-10 would require breaking a carbon-carbon bond, isomerism at that site is unlikely. Isomerism at C-17 must be considered because the C-18 methyl migrates to C-17 from C-13 during MA-steroid formation. Thus, eight C_{27} stereoisomers and four C_{21} stereoisomers are possible. The fact that only four and two are obtained, respectively, by CDI indicates that the C-20 site *can* generate C_{27} epimers but one of the sites C-5 or C-17 does not. Since the C-18 methyl is β, it is reasonable that the C-13 to C-17 methyl migration occurs on the β face. This would result in a single 17β methyl configuration for all of the MA-steroids **1** through **4** (17*S*) and the two C_{21} monoaromatics **5** and **6** (17*R*). Indeed, the NMR

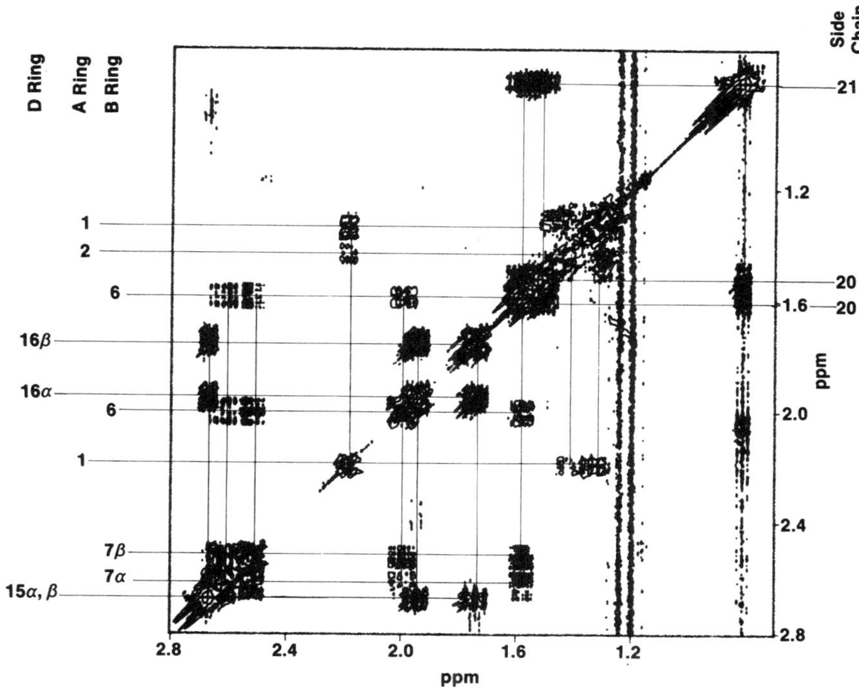

Figure 5.11 ^1H-^1H COSY 2D NMR 400 MHz spectrum (with aromatic spectral region excluded) of C_{21} monoaromatic steroid **5** with cross-peaks associated with A-, B-, and D-rings and side chain indicated.

spectra of **5** and **6**, exhibit spectral differences arising from A- and B-ring resonances alone; the spectral patterns of their D-ring and side chain resonances are virtually identical as shown in Figure 5.8. Using the 17*S* stereochemical assignment for compounds **2** and **3** and difference NOE data generated by consecutively irradiating the C-21 side chain methyl (to locate 16α) and C-17β methyl (to locate 16β), we could make specific stereochemical assignments for D-ring proton resonances listed in Tables 5.2 and 5.4.

The analysis of the 2D *J* spectrum strongly supports the assignment of **6** (and by comparison **3** and **4**) as a 5α, A/B-*trans* steroid. The MA-steroids **1** and **2**, which also have C-5 methines, are assigned as 5β, A/B-*cis* steroids. The 5α/5β assignment for the MA-steroids is supported by other data as follows.

Spectra were recorded for each MA-steroid, **1** through **4**, over the 203 K to ambient temperature range. The C-1 and C-6, A- and B-ring signals, of **1** and **2** showed a distinct temperature dependence, although those of **3** and **4** did not, as shown in Figure 5.12. The temperature dependence of the A- and B-ring resonances of **1** and **2** is consistent with conformational flexibility (Riddell and Robinson, 1965;

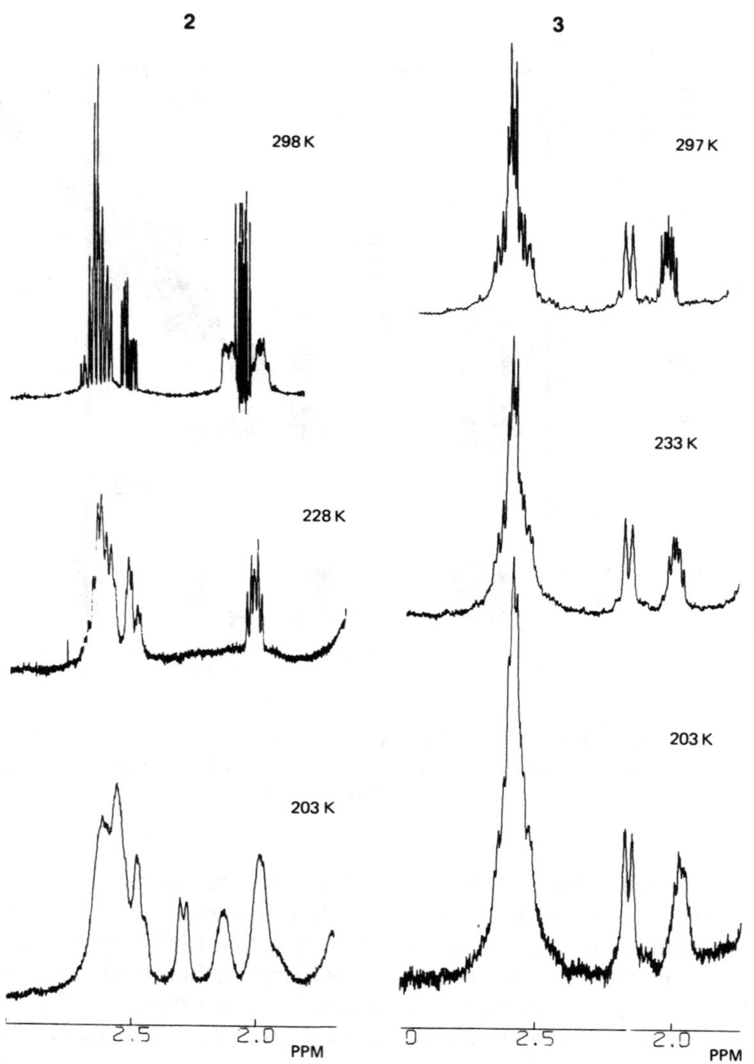

Figure 5.12 Variable temperature 500 MHz ^{1}H spectra of **2** and **3** (~2 mg/mL in CD_2Cl_2).

Gerig and Roberts, 1966; Dalling et al., 1971) illustrated by Figure 5.13 in 5β, A/B-*cis* steroids. Conversely, the lack of temperature dependence in the spectra of **3** and **4** is consistent with the rigid 5α, A/B-*trans* structure.

The singlet methyl chemical shifts observed for the **1**, **2** and **3**, **4** pairs correlate with the chemical shifts reported for the 5β and 5α epimers of the ring-C aromatic androstane, 17,17-dimethyl-18-nor-androsta-8,11,13-triene prepared by Turner

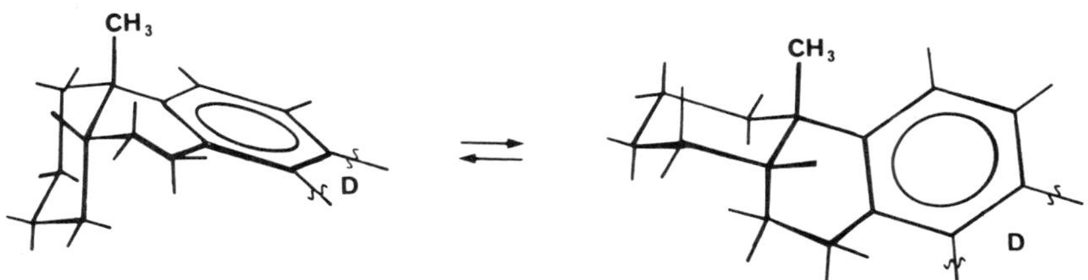

Figure 5.13 Stereochemical representation of a C-ring monoaromatic steroid possessing a *cis* A/B-ring juncture showing conformational flexibility.

(1972). Furthermore, singlet methyl chemical shifts of 5β and 5α epimers of a series of 12-methyl-ring-C monoaromatic cholestanes prepared by Schaefle et al. (1978), as noted earlier (Seifert et al., 1983), and 4a-methyloctahydrophenanthrenes prepared by Wenkert et al. (1965) and Campbell et al. (1979), also correlate with the singlet methyl chemical shifts of the **1**, **2** and **3**, **4** pairs. The greater deshielding of the C-19 methyl of the ring-C monoaromatic A/B *cis* steroids relative to the A/B *trans* steroids is consistent with the conformational flexibility in the former, which allows 5β conformers in which the C-19 methyl approaches the plane of the aromatic C-ring more closely than in the rigid A/B *trans* steroids. Thus low-temperature spectral data, ring-C monoaromatic pregnane spectra, 2D *J* spectral analysis of **6** and methyl shift correlations between **1** through **4** and related synthetic steroids combine to provide strong evidence for the C_{27} MA-steroid A/B-ring stereochemistry. Compounds **1** and **2** have 5β, A/B-*cis* stereochemistry and compounds **3** and **4** have 5α, A/B-*trans* stereochemistry.

Tentative stereochemical assignments can be made for the C-1, C-6, and C-7 proton resonances from NOE data, *J* couplings, and ring current shielding contributions. The low field C-7 ^{1}H signal of **2** has a *J* coupling of 10 Hz to the low field C-6 signal and a *J* coupling of 6.7 Hz to the high field C-6 signal in Figure 5.3. The low field C-7 signal is coupled to both C-6 protons with constants significantly lower than 10 Hz. The conformation of **2** (or **1**) depicted above (Figure 5.5) brings the C-1α proton to within 0.3 Å of the C-11 aromatic proton, which (from the Haigh-Mallion ring current tables (Haigh and Mallion, 1971, 1980) account for the large deshielding of the C-1α proton relative to C-1β. Also, in the above conformation, the C-1α proton is within 1.8 Å distance from the C-11 proton, accounting for the C-1α NOE induced by C-11 ^{1}H irradiation noted in Table 5.2. The same **2** conformation makes the C-6β and C-7α protons pseudo axial-axial, which accounts for their large 10 Hz *J* coupling seen in Figure 5.3.

Only the assignment of the C-20 stereochemistry of **1** through **4** remains. The C-21 methyl doublet resonances in the spectra of **1** and **3** have significantly different chemical shifts from those observed in **2** and **4**. A similar situation occurs in cholestane, where the chemical nonequivalence of the C-21, 20*R* and 20*S*, methyls is thought to arise from their diastereotopic nature rather than restricted rotation around the 17(20) bond (Osawa et al., 1979). In **2** and **3**, nuclear Overhauser enhancements at both the 12 and 16α protons upon irradiation of the 21 methyl noted in Tables 5.2 and 5.4 demonstrate rotation about the C-17(20) bond at ambient temperatures. Lowering the NMR probe temperature to 203 K has no effect on the chemical shifts of the C-21 methyl or other side chain proton resonances of **1** through **4**. There is thus no evidence for temperature-dependent hindered rotation. The nonequivalence of the two C-20 hydrogens of the ring-C monoaromatic pregnane two-carbon side chain in Figure 5.8 underscores the diastereotopic nature of C-20 spectral effects in the C_{27} higher homologs.

Thus, the C-20 configurations of **1** through **4** cannot be deduced from the individual NMR spectra alone and have been rigorously assigned through synthetic studies by Riolo et al. (1985). Before these synthetic results were reported, we made tentative assignments by correlation of the spectra of **1** through **4** with data reported by Ludwig et al. (1981) for a C_{26} triaromatic steroidal hydrocarbon of known configuration, 17β-methyl-18,19-dinor-17*S*,20*R*-cholesta-1,3,5(10),6,8(9), 11,13(14)-heptaene, reasoning as follows.

Since the first order structures for the C- and D-rings and side chain of the Ludwig et al. (1981) C_{26} triaromatic steroid are identical to those of the C_{27} MA-steroids **1** through **4**, it is reasonable that the MA-steroids should exhibit D-ring and side chain NMR spectral features similar to those of the C_{26} triaromatic steroid. Ring current shifts for side-chain protons arising from the aromatic A- and B-rings must contribute to the larger ring-C ring current shifts in an additive manner in the C_{26} triaromatic steroid. Since the C-21 methyl doublet resonates at higher field than the C-26 and 27 methyls in that 18,19-dinor-17β methyl-20*R*-C_{26} triaromatic steroid, the same relative chemical shifts would be expected for the C_{27} MA-steroids of identical C-17 and 20 stereochemistry. Therefore, it is reasonable to assert that the C_{26} triaromatic steroid and MA-steroids **2** and **4** have an identical 20*R* configuration. The MA-steroids **1** through **3** must both then be 20*S*, which correlates with synthetic work (Ludwig et al., 1981; Mackenzie et al., *Nature*, 1982). These correlations are verified by the rigorous synthesis of Riolo et al. (1985) and Riolo and Albrecht (1985).

SUMMARY AND CONCLUSIONS

The major C_{27} ring-C MA-steroid chemical fossils of petroleum, **1** through **4** were prepared by CDI treatment of 5α-cholestane and isolated from the complex reaction mixture by NARP and adsorptive HPLC. The purified MA-steroid biological markers, which are oils and could not be studied by X-ray crystallographic methods, were each studied using various high-field NMR techniques in an effort to establish their exact structures. Even at 500 MHz, spectral resolution problems were experienced, but marked resolution improvement could be achieved using solvent-induced shifts. Comparisons of the spectra suggested the same basic structural form for **1** through **4**, with **1** and **2** having one nuclear configuration and **3** and **4** having another. Also, spectral patterns showed that **1** and **3** have one side-chain configuration and **2** and **4** another. Difference NOEs and decouplings allowed the basic assignment of **2** (and **1** by direct correlation). The C- and D-ring resonances of **3** (and **4**) could also be assigned in this way, but the A- and B-ring resonances of these compounds were insufficiently resolved to allow unambiguous nuclear assignments.

Neither two-dimensional *J* resolved spectra nor partial relaxation methods could resolve overlapping C_{27} A- and B-ring resonances. The C_{21} ring-C MA-steroids (**5** and **6**), homologous with **1** through **4**, were prepared by the CDI treatment of 5α-pregnane. Complete assignment of the 2D *J* spectrum of **6** as a 5α, A/B-*trans* steroid was achieved with the aid of decoupling and difference NOE data. Correlation of the 2D *J* results for **6** with the ^{1}H spectral data for **3** and **4** allowed their assignment as 5α, A/B-*trans* steroids. COSY data were used to assign the ^{1}H spectrum of **5**. Support for the 5α, A/B-*trans* assignments for **3** and **4** and for the 5β, A/B-*cis* assignment for **1** and **2** came from *J* coupling patterns and low temperature ^{1}H NMR data which showed **3** and **4** to be conformationally rigid (5α), but **1** and **2** to have temperature-dependent conformational equilibria (5β). The formation of only two C_{21} MA-steroids by CDI methods supports a single configuration at C-17 (17β-methyl) for both the C_{21} and C_{27} steroids (as well as the corresponding higher homologs in petroleum). Correlation of the C-21, C-26, and C-27 methyl resonance of **1** through **4** with a synthetic triaromatic steroid of known C-20 configuration (Ludwig et al., 1981) allowed tentative C-20 configurational assignments for **1** through **4** which were verified by rigorous synthetic studies by Riolo et al. (1985) and Riolo and Albrecht (1985). The combined NMR evidence strongly supports 17β-methyl-18-nor-[5α/5β,17*S*,20*R*/20*S*]-cholesta-8,11,13-triene structures for **1** through **4** prepared by the CDI-HPLC method.

Some comment on the relative utility of the various NMR methods employed here for steroid structure elucidation is perhaps appropriate.

Binuclear shift reagents, selective signal nulling by partial relaxation methods, and dynamic NMR offered limited help in our studies for reasons discussed earlier.

Among the proliferation of ^{1}H-^{1}H correlation methods (COSY and related sequences), the power and efficiency of 2D *J*-resolved NMR for steroids should not be overlooked (Croasmun and Carlson, 1987). Two-dimensional *J*-resolved spectral studies (with double resonance studies) provided data for a complete assignment of the 500 MHz ^{1}H spectrum of our C_{21} compound. The individual 2D *J*-resolved ^{1}H coupling patterns (and even signal widths where patterns were unclear) and long-range couplings (some of which are not resolved in the one-dimensional spectra) provided us with detailed structural and configurational information. Our ^{13}C NMR spectral study was hampered by sample size and resolution limitations. Both the more modern INEPT (Burum and Ernst, 1980) and DEPT (Doddrell et al., 1982) techniques for determining multiplicities and the recently developed ^{1}H-detected 1H-^{13}C correlation spectroscopy (Bax and Subramanian, 1986) are readily applicable to steroids. These methods have relatively good sensitivity, but there will always be samples small enough whereby ^{1}H NMR is possible but ^{13}C is not.

In summary, in this study careful solvent selection and variation with successive double resonance experiments followed by 2D *J* spectral studies provided the most structural information with the smallest samples and least amount of spectrometer time. Obviously, as an alternative to decoupling measurements, a COSY spectrum may be acquired; time may or may not be saved in this way, depending on the details of the spectrum and level of the analysis.

The methods and strategies that we used in the study of the structures of these ring-C MA-steroid biological markers should be directly applicable to the resolution of structures of other chemical fossils (e.g., di- and triterpanes). Even partial structural information derived from CDI-HPLC-NMR studies could help direct the synthetic efforts of natural products geochemists to the best chemical fossil candidate structures. Furthermore, our approach could be used to prepare previously unstudied CDI reference compounds for use in a GC-MS search for new biological markers in geologic samples.

Acknowledgments

The 500 MHz spectra were obtained at the Southern California Regional NMR Facility (California Institute of Technology, Pasadena), supported by NSF Grant No. CHE 79-16324.

REFERENCES

Bax, A. and Subramanian, S. (1986). Sensitivity-enhanced two-dimensional heteronuclear shift correlation NMR spectroscopy. *J. Magn. Reson. 67*, 565–569.

Blunt, J.W. and Stothers, J.B. (1977). Carbon-13 NMR studies. Part 69. Carbon-13 NMR spectra of steroids—a survey and commentary. *Org. Magn. Reson. 9*, 439–464.

Bridgewater, A.J., Cheung, H.T.A., Vadasz, A., and Watson, T.R. (1980). Ring C aromatic steroids. Part 2. Rearrangement of 16-alpha, 17-alpha-epoxy- and 17-alpha-hydroxy-5,7-dienes. *J. Chem. Soc. PI*, 556–562.

Burum, D. P. and Ernst, R.R. (1980). Net polarization transfer via a J-ordered state for signal enhancement of low sensitivity nuclei. *J. Magn. Reson. 39*, 163–168.

Campbell, A.L., Leader, H.N., Sierra, M.G., Spencer, C.L., and McChesney, J.D. (1979). Substituted 4-alpha-methyloctahydrophenanthrenes: Conformation and proton magnetic resonance characteristics. *J. Org. Chem. 44*, 2755–2757.

Carlson, R.M.K., Tarchini, C., and Djerassi, C. (1980). Implications of recent advancements in the marine sterol field. In *Frontiers of Bioorganic Chemistry and Molecular Biology* (ed. S.N. Ananchenke), pp. 211–224, Pergamon, New York.

Cheung, H.T.A., McQueen, R.G., Vadasz, A., and Watson, T.R. (1979). Ring-C aromatic steroids. 17-Beta-methyl-18-norpregna-8,11,13-trienes. *J. Chem. Soc. PI*, 1048–1055.

Croasmun, W.R. and Carlson, R.M.K. (1987). Steroid structural analysis by two-dimensional NMR. In *Two-Dimensional NMR Spectroscopy-Applications for Chemists and Biochemists* (eds. W.R. Croasman and R.M.K. Carlson), Methods in Stereochemical Analysis, Vol. 9, pp. 387–424, VCH Publishers.

Dalling, D.R., Grant, D.M., and Johnson, L.F. (1971). Conformational inversion rates in the dimethylcyclohexanes and in some *cis*-decalins. *J. Am. Chem. Soc. 93*, 3678–3682.

Doddrell, D.M., Pegg, D.T., and Bendall, M.R. (1982). Distortionless enhancement of NMR signals by polarization transfer. *J. Magn. Reson. 48*, 323–327.

Eggert, H. and Djerassi, C. (1981). Carbon-13 nuclear magnetic resonance spectra of monounsaturated steroids. Evaluation of rules for predicting their chemical shifts. *J. Org. Chem. 46*, 5399–5401.

Gerig, J.T. and Roberts, J.D. (1966). Nuclear magnetic resonance spectroscopy. Conformational equilibration of *cis*-decalins. *J. Am. Chem. Soc. 88*:12, 2791.

Haigh, C.W. and Mallion, R.B. (1971). New tables of ring current shielding in proton magnetic resonance. *Org. Magn. Reson. 4*, 203–228.

Haigh, C.W. and Mallion, R.B. (1980). Ring current theories in nuclear magnetic resonance. *Prog. Nucl. Magn. Reson. Spectrosc. 13*, 303–344.

Hall, L.D. and Sukumar, S. (1980). Phase-sensitive displays for proton 2D-J spectra. *J. Magn. Reson. 38*, 555–558.

Hussler, G., Chappe, B., Wehrung, P., and Albrecht, P. (1981). C_{27}-C_{29} ring A monoaromatic steroids in cretaceous black shales. *Nature 294*, 556–558.

Joseph-Nathan, P., Mejia, G., and Abramo-Bruno, D. (1979). ^{13}C NMR assignment of the side-chain methyls of C_{27} steroids. *J. Am. Chem. Soc. 101*, 1289–1291.

Ludwig, B., Hussler, G., Wehrung, P., and Albrecht, P. (1981). C_{26}-C_{29} triaromatic steroid derivatives in sediments and petroleums. *Tetrahedron Lett. 22*, 3313–3316.

Mackenzie, A.S., Hoffman, C.F., and Maxwell, J.R. (1981). Molecular parameters of maturation in the Toarcian shales, Paris Basin, France-III. Changes in aromatic steroid hydrocarbons. *Geochim. Cosmochim. Acta 45*, 1345–1355.

Mackenzie, A.S., Brassell, S.C., Eglinton, G., and Maxwell, J.R. (1982). Chemical fossils: The geological fate of steroids. *Science 217*, 491–504.

Mackenzie, A.S., Lamb, N.A., and Maxwell, J.R. (1982). Steroid hydrocarbons and the thermal history of sediments. *Nature 295*, 223–226.

Mackenzie, A.S. and McKenzie, D. (1983). Isomerization and aromatization of hydrocarbons in sedimentary basins formed by extension. *Geol. Mag. 120*, 417–470.

Mackenzie, A.S. (1984). Application of biological markers in petroleum geochemistry. In *Advances in Petroleum Geochemistry, Vol. 1* (eds. J. Brooks and D.H. Welte), Academic Press, pp. 115–214.

Moldowan, J.M., Seifert, W.K., and Gallegos, E.J. (1985). Relationship between petroleum composition and depositional environment of petroleum source rocks. *Amer. Assoc. Petrol. Geol. Bull. 69*, 1255–1268.

Moldowan, J.M. and Fago, F.J. (1986). Structure and significance of a novel rearranged monoaromatic steroid hydrocarbon in petroleum. *Geochim. Cosmochim. Acta 50*, 343–351.

O'Neal, M.J. and Hood, A. (1956). Mass spectrometric analysis of polycyclic hydrocarbons. *Am. Chem. Soc. Div. Pet. Chem.*, Atlantic City, Preprints *1*(*4*), 127–135.

OSAWA, E., SHIRAHAMA, H., and MATSUMOTO, T. (1979). Application of force field calculations to organic chemistry. 8. Internal rotation in simple to congested hydrocarbons including 2,3-dimethylbutane, 1,1,2,2-tetra-tert-butylethane, 2,2,4,4,5,5,7,7-octamethyloctane, and cholestane. *J. Am. Chem. Soc. 101*, 4824–4832.

PEAKMAN, T.M. (1986). Synthesis, occurrence and low temperature diagenesis of steroid hydrocarbons. Ph.D. Dissertation, University of Bristol, U.K.

PETROV, A.A., PUSTIL'NIKOVA, S.D., ABRYUTINA, N.N., and KAGRAMONOVA, G.R. (1976). Petroleum steranes and triterpanes. *Neftekhimiya 16*, 411–427.

RIDDELL, F.G. and ROBINSON, J.T. (1965). The rate of interconversion of the conformations of a derivative of *cis*-decalin. *J. Chem. Soc. Chem. Commun.*, 227–228.

RIOLO, J., LUDWIG, B., and ALBRECHT, P. (1985). Synthesis of ring C monoaromatic steroid hydrocarbons occurring in geological samples. *Tetrahedron Lett. 26*, 2697–2700.

RIOLO, J. and ALBRECHT, P. (1985). Novel rearranged ring C monoaromatic steroid hydrocarbons in sediments and petroleum. *Tetrahedron Lett. 26*, 2701–2704.

RIOLO, J., HUSSLER, G., ALBRECHT, P., and CONNAN, J. (1986). Distribution of aromatic steroids in geological samples: Their evaluation as geochemical parameters. In: *Advances in Organic Geochemistry 1985* (eds. D. Leythaeuser and J. Rullkötter), Pergamon Journals, 981–990.

SCHAEFLE, J., LUDWIG, B., ALBRECHT, P., and OURISSON, G. (1978). Aromatic hydrocarbons from geological sources. VI. New aromatic steroid derivatives in sediments and crude oils. *Tetrahedron Lett. 43*, 4163–4166.

SEIFERT, W.K. and MOLDOWAN, J.M. (1978). Applications of steranes, terpanes, and monoaromatics to the maturation, migration, and source of crude oils. *Geochim. Cosmochim. Acta 42*, 77–95.

SEIFERT, W.K. and MOLDOWAN, J.M. (1979). The effect of biodegradation on steranes and terpanes in crude oils. *Geochim. Cosmochim. Acta 43*, 111–126.

SEIFERT, W.K. and MOLDOWAN, J.M. (1986). Use of biological markers in petroleum exploration. In: *Methods in Geochemistry and Geophysics* (ed. R.B. Johns), *24*, 261–290.

SEIFERT, W.K., CARLSON, R.M.K., and MOLDOWAN, J.M. (1983). Geometric synthesis, structure assignment, and geochemical correlation application of monoaromatized petroleum steranes. In: *Advances in Organic Geochemistry 1981* (eds. M. Bjoroy et al.), pp. 710–724, J. Wiley and Sons, New York.

SMITH, W.B. (1978). Carbon-13 NMR spectroscopy of steroids. *Annu. Rep. NMR Spectros. 8*, 199–226.

SMITH, W.B. (1981). The carbon-13 NMR of olefins complexed with a binuclear silver (I)-ytterbium (III) chelate. *Org. Magn. Reson. 17*, 124–126.

TURNER, A.B. (1972). Selective ring C aromatization of steroids. *Chem. and Ind. 13*, 932–933.

WENKERT, E., AFONSO, A., BEAK, P., CARNEY, R.W.J., JEFFS, P.W., and MCCHESNEY, J.D. (1965). The proton magnetic resonance spectral characteristics of tricyclic diterpenic substances. *J. Org. Chem. 30*, 713–722.

WENZEL, T.J., BETTES, T.C., SADLOWSKI, J.E., and SIEVERS, R.E. (1980). New binuclear lanthanide NMR shift reagents effective for aromatic compounds. *J. Am. Chem. Soc. 102*, 5903–5904.

WENZEL, T.J. and SIEVERS, R.E. (1981). Binuclear complexes of lanthanide (III) and silver (I) and their function as shift reagents for olefins, aromatics, and halogenated compounds. *Anal. Chem. 53*, 393–399.

ZUBENKO, V.G., PUSTIL'NIKOVA, S.D., ABRYUTINA, N.W., and PETROV, AL.A. (1980). Petroleum monoaromatic hydrocarbons of the steroid type. *Neftekhimiya 20*, 490–497.

ZUBENKO, V.G., VOROBYOVA, N.S., ZEMSKOVA, Z.K., PEKHK, T.I., and PETROV, A.A. (1981) On the equilibrium of *cis* and *trans*-isomers in octahydrophenanthrenes-structural fragments of monoaromatic steranes. *Neftekhimiya 21*, 323–328.

6

Biomarker Distributions in Crude Oils as Determined by Tandem Mass Spectrometry

R. Paul Philp and Jung-N. Oung

Abstract. The rapid development of biomarker geochemistry and its application to petroleum exploration has been due in no small part to the development of gas chromatography-mass spectrometry (GC-MS). The recent availability of hybrid and triple stage quadrupole mass spectrometers has introduced novel ways for the detection and identification of biomarkers in complex mixtures of organic compounds.

A triple stage quadrupole mass spectrometer has been used, both with a gas chromatograph and direct insertion probe, to develop novel methods for biomarker analyses. Selected ions formed in the ion source of the mass spectrometer are permitted to enter the collision cell where, following collision with an inert gas, the daughter ions are separated using the third quadrupole. In this manner specific parent/daughter ion relationships can be established and utilized to monitor and resolve classes of biomarkers or individual components from within a complex mixture. Furthermore, deuterated analogues of naturally occurring biomarkers can be utilized for quantitation since the parent/daughter relationship for the standard will differ, depending on the number of deuterium atoms present, even though the relative retention times are identical.

The ability of MS/MS to resolve components spectrometrically as well as chromatographically greatly enhances the analytical capabilities of this type of system compared to GC-MS. This paper discusses the use of GC-MS/MS and direct insertion probe MS/MS to analyze complex mixtures both chromatographically and spectrometrically. In addition, the use of deuterated cholestane to quantify biomarker distributions obtained by direct insertion probe MS/MS analysis from oils derived from different source materials will be discussed.

INTRODUCTION

One of the major reasons for the rapid development of petroleum geochemistry and its application to petroleum exploration in the past decade has been the significant advances made in the analytical equipment available for characterization of geochemical samples. Extracts from geological samples such as rocks, oils, sediments, or shales are complex mixtures of organic compounds. Many of these compounds are referred to as biomarkers, and they are ideally suited for analysis by gas chromatography-mass spectrometry (Philp and Lewis, 1987). Despite major advances in capillary column chromatography in the past few years, many components in these geological samples still coelute, even on ultrahigh resolution capillary columns. This can make it extremely difficult to obtain unique distributions for selected classes of compounds by single ion monitoring or to identify unknown components that may coelute with other components.

Biomarkers are commonly used to obtain information on the origin and history of crude oils and the fate of organic matter in the geological record. Particular emphasis has been placed on biomarker evaluation to obtain information on source, maturity, depositional environment, biodegradation, and migration (Philp and Lewis, 1987). In addition, biomarker "fingerprints" have been used extensively for oil/oil and oil/source rock correlations, for example Seifert and Moldowan (1979) and Shi Ji-Yang et al. (1982).

Until recently the method of choice for biomarker determination has been gas chromatography-mass spectrometry (GC-MS). Operation of the mass spectrometer in the single ion monitoring (SIM), or multiple ion detection (MID), mode permits the fingerprints of different classes of biomarkers to be readily obtained. A great deal of geochemical information can be obtained from just the sterane (*m/z* 217) and terpane (*m/z* 191) chromatograms.

The development of hybrid mass spectrometers, such as the triple stage quadrupole (TSQ) has provided an additional dimension for biomarker separations. Using this method, it is possible to separate compounds spectrometrically as well as chromatographically (Philp and Oung, 1988). The TSQ mass spectrometer consists of three quadrupoles in tandem. The middle quadrupole is a collision cell that operates in the R_F mode only. The TSQ is typically operated in the so-called "parent" or "daughter" modes for analysis of geochemical samples. In these modes, selected ions formed in the ion source are permitted to enter the collision cell where they collide with an inert gas, commonly argon, and form daughter ions.

The ions produced in the collision cell are then analyzed by the third quadrupole. The use of established parent/daughter ion relationships permits selective separation of compound classes. It is important to note that spectrometric, rather than chromatographic separation of biomarkers was the subject of some pioneering work by Gallegos (1976).

This chapter illustrates some applications of MS/MS to various geochemical problems and to describe the data obtainable from such applications. It describes how fingerprints of different classes of biomarkers can be obtained and quantified using the direct insertion probe, without any GC separation. The resulting biomarker distributions are comparable to those obtained from GC-MS except that total separation of isomers is not possible. However, for the purposes of rapid screening of samples this is not always essential and is certainly not detrimental to the approach described herein.

For additional information, refer to several recent papers which further describe this type of application (Philp et al., 1988; Philp and Oung, 1988; Summons, 1987; Summons et al., 1988).

EXPERIMENTAL

Crude oil samples from various basins in Taiwan, New Zealand, China, and the Middle East, were utilized for this study. Samples were analyzed as either whole oils or fractionated into saturates, aromatics, and NSO fractions by thin-layer chromatography. A deuterated cholestane standard was added to every sample prior to fractionation or analysis. The fractions, or whole oils, were subsequently analyzed by either direct insertion probe mass spectrometry or GC-MS/MS using a Finnigan MAT TSQ 70[1] mass spectrometer. Chromatographic separations were completed using a 25 m × 0.25 m aluminum coated fused silica capillary column (Scientific Glass Engineering) coated with a 0.1 μm film of the HT-1 phase. The column was temperature programmed from 40° to 330° at 2° per minute with an injector temperature of 280°C. The transfer line temperature was set at 280° and the ion source temperature was 200°C. The ion source was operated in the electron impact (EI) mode at an electron energy of 70 eV and collision activated decomposition (CAD) spectra were obtained with argon as collision gas at 1 millitorr and collision energy was generally −10 eV. For the most part, the parent/daughter ion combinations used consisted of the molecular ions (parents) and characteristic base peak fragments (daughters) commonly observed in the EI spectra. Exceptions to these parent/daughter combinations are noted in the text. Similar conditions were used when the samples were analyzed by the direct insertion probe.

[1]TSQ 70 is a registered trademark of Finnigan MAT.

DISCUSSION

The geochemical literature demonstrates an increasing trend toward the use of hybrid mass spectrometers for the analysis of oils and source rock extracts (i.e., Philp and Oung, 1988; Summons et al., 1988; Gallegos and Moldowan, 1989). As biomarker geochemistry evolves further, MS/MS will continue to play an important role in identification of unknown compounds particularly those in the molecular weight region above C_{40} or more polar compounds from fractions previously investigated in great detail.

A convenient way to introduce the GC-MS/MS concept is to use sterane and diasterane distributions in crude oils as an example. It is well documented that there are several coeluting isomers and homologues in the C_{27}–C_{30} sterane region of a crude oil chromatogram (Mackenzie et al., 1982). To accurately assess the relative proportions of the various homologues, it is desirable to completely resolve them. For most geochemical samples, such separations are not possible with even the best capillary columns. However, mass spectrometric separation of the various C_{27}–C_{30} sterane isomers and homologues can be accomplished by monitoring specific parent/daughter ion pairs. Figure 6.1 shows the resulting chromatogram obtained by monitoring the molecular ions for each homologue producing a daughter ion at m/z 217. This GC-MS/MS chromatogram has a similar appearance to the single ion chromatogram obtained by monitoring the ion at m/z 217. However, it is possible to deconvolute the data and separate individual homologues from within this chromatogram. This is illustrated in the top chromatogram (Fig. 6.1) where the C_{28} steranes have been resolved from the other homologues. Diasteranes and regular steranes can be further separated by monitoring the parent ions of both m/z 189 and 217, respectively (Fig. 6.2). This combination of GC and MS/MS permits complete separation of nearly all diasteranes and steranes homologues and isomers although Gallegos and Moldowan (1989 and 1991) have recently proposed that even more isomers can be separated by using very long GC temperature program rates. Another useful application of MS/MS to sterane analyses is the ability to distinguish C_{30}-regular and 4-methylsteranes. The former have an intense daughter ion at m/z 217 and the latter at m/z 231. Hence monitoring m/z 414 $\rightarrow$ 217 and m/z 414 $\rightarrow$ 231 parent/daughter pairs provides a rapid means of distinguishing between the two types of steranes.

Polycyclic terpanes of the type found in geochemical samples, produce an intense base peak at m/z 191. When samples are analyzed in the MID mode, complex mass fragmentograms make it difficult to obtain information on the distribution of classes of terpanes within each sample. Molecular weights or parent ions of the homologous series will differ by multiples of 2 mass units depending on the molecular formula of the particular series of terpanes.

There are many examples that can be used to illustrate the application of MS/MS to separate various classes of terpanes not previously correctly identified.

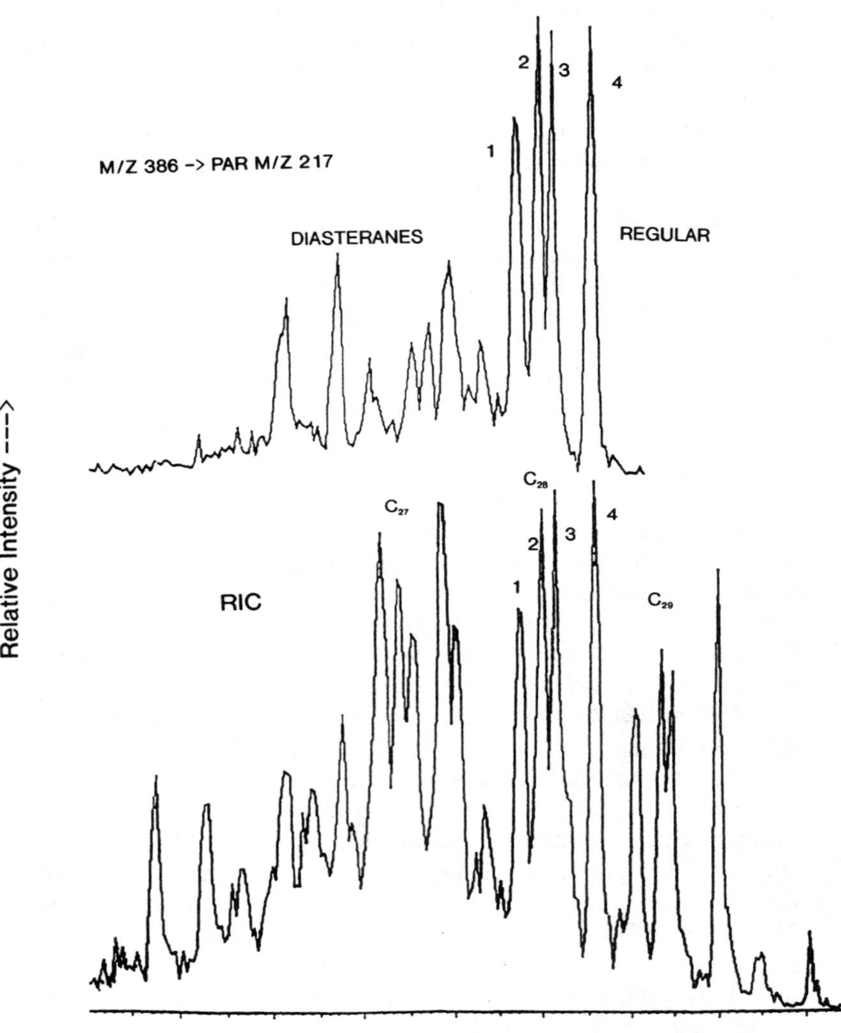

Figure 6.1 Separation of the C_{28}-steranes and diasteranes from the total sterane distribution by monitoring the parent/daughter relationship of *m/z* 386 → 217. (Peaks labeled 1 and 4 are 14α,17α-20*S* and 20*R*-C_{28} steranes, respectively and peaks 2 and 3 are the 14β,17β-20*R* and 20*S*-C_{28} steranes, respectively.)

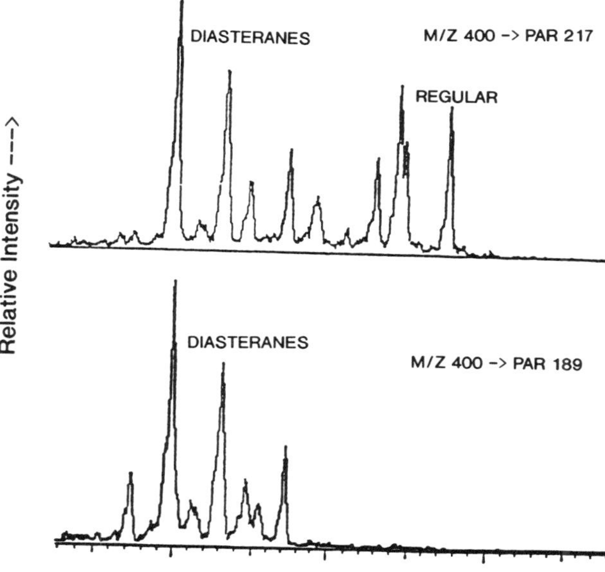

Figure 6.2 Diasteranes can be further resolved from the sterane/diasterane mixture by monitoring the *m/z* 400 → 189 relationship, as well as the *m/z* 400 → 217 relationship for total C_{29} steranes.

One example is a C_{30}-terpane which elutes between the two C_{27} trisnorhopanes, T_s and T_m (Fig. 6.3). Although this C_{30}-pentacyclic terpane has yet to be unequivocally identified, its relative concentration appears to be significantly higher in samples of terrigenous origin. In addition to this C_{30} terpane, there is a component which frequently elutes after T_m in the region of the 17β(H)-trisnorhopane. Analyses of a variety of crude oil samples has lead us to suggest that in many samples this component is not necessarily 17β(H)-trisnorhopane but rather a C_{28} pentacyclic terpane with a parent ion at *m/z* 384 (Fig. 6.4). This component shows an intense *m/z* 384 → 191 parent/daughter relationship and not the intense *m/z* 370 → 191 parent/daughter relationship shown by the T_s and T_m components which elute in the same region. Seifert and Moldowan (1978) noted the presence of a $C_{28}H_{48}$ pentacyclic triterpane eluting immediately after T_m on the basis of monitoring the ions at *m/z* 191 and 384. Both their data and the data presented herein illustrate a possible pitfall of only using the *m/z* 191 fingerprints where that component could have been incorrectly assigned as 17β(H)-trisnorhopane.

Additional terpanes, albeit in trace amounts, in various regions of the chromatogram can be observed by use of the TSQ mass spectrometer operating in the parent mode. For example Fig. 6.5 shows results for the analysis of a Chinese crude oil which illustrate the distribution of C_{30} terpanes as determined from the *m/z* 412 → 191 parent/daughter relationship. The presence of several C_{30}-terpanes can be observed in the chromatogram and, although their identities have still to

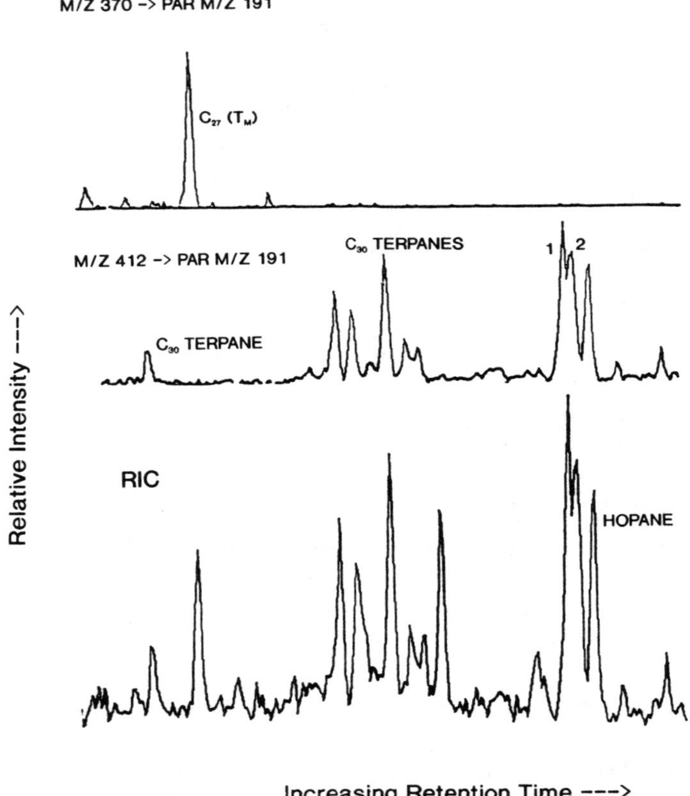

Figure 6.3 Monitoring the m/z 412 → 191 parent/daughter ion relationship for an Indonesian oil sample revealed the presence of a number of C_{30} terpanes in these chromatograms. One of these C_{30}-terpanes was found to elute between the C_{27}-trisnorhopanes, T_s and T_m. Note also two isomers of oleanane in this chromatogram, namely, 18α(H)- and 18β(H)-oleanane (peaks 1 and 2, respectively).

be unambiguously established, their presence may permit their use as additional correlation parameters.

Another use of the MS/MS approach is to separate coeluting components of different molecular weights but having the same daughter ion. Rinaldi et al. (1988) commented on the presence of hexacyclic-C_{31}-hopanoids in certain oils that have a tendency to coelute with the C_{31}-22R-homohopanes. Fortunately, these hexacyclic compounds have a parent ion at m/z 424 compared to m/z 426 for the pentacyclic terpanes. Hence, simultaneous monitoring of the m/z 426/191 and m/z 424/191 parent/daughter relationships can be used to demonstrate the presence of pentacyclic and hexacyclic terpanes in many crude oils (Fig. 6.6).

From the preceding examples it can be seen that a large amount of additional

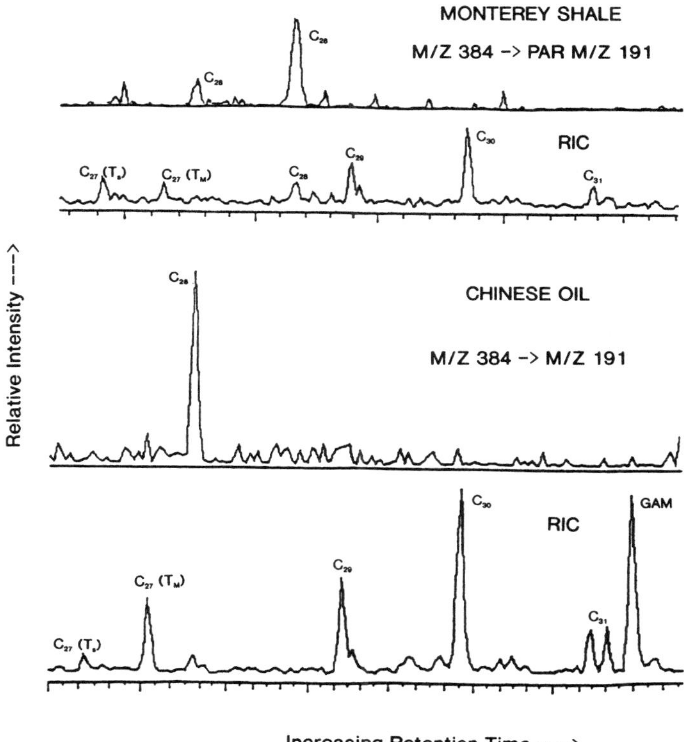

Figure 6.4 The upper two chromatograms in this figure are from an oil derived from the Monterey Formation, Calif. Note the characteristically high concentrations of the C_{28}-bisnorhopane with the longer retention time than the other C_{28} terpane. The bottom chromatograms are derived from a Chinese oil and unlike the Monterey sample, clearly show the predominance of the early eluting C_{28} terpane which elutes much closer to the C_{27}-trisnorhopanes. Other hopanes are identified by carbon number on the chromatograms along with gammacerane which is labeled GAM.

geochemical information can be obtained by using MS/MS. Many of these additional components remain unidentified. However, this is not necessarily a major problem, since these components can still be incorporated into maturity, source, and depositional environment parameters and their precise identities established at a later date.

A partial approach to establishing the identity of these unknown components is to alternately operate the MS/MS system in the parent and daughter modes. One complete sequence of parent experiments is performed and then a daughter experiment is performed and this process is repeated throughout the analysis. The daughter experiments are performed on parent ions of components known to elute in various regions of the chromatogram. The oil used to obtain the data in

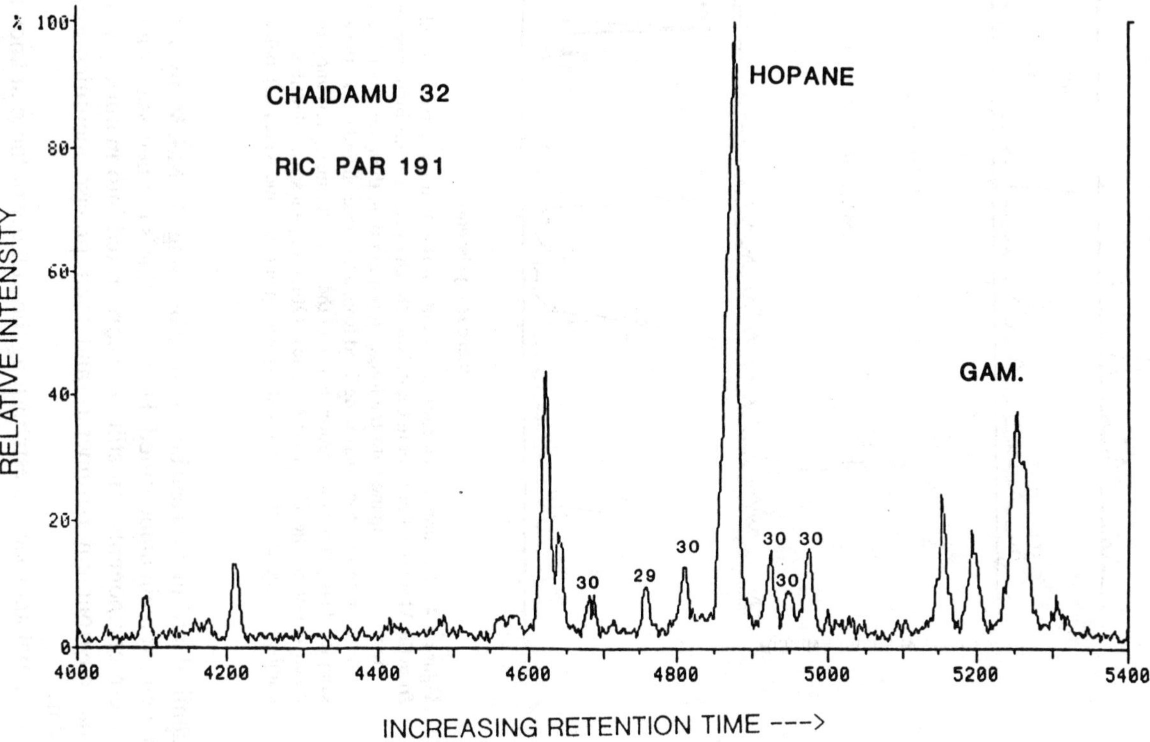

Figure 6.5 Monitoring of the m/z 412 → 191 parent/daughter relationship for this Chinese oil sample illustrates the presence of several C_{30}-terpanes in addition to hopane.

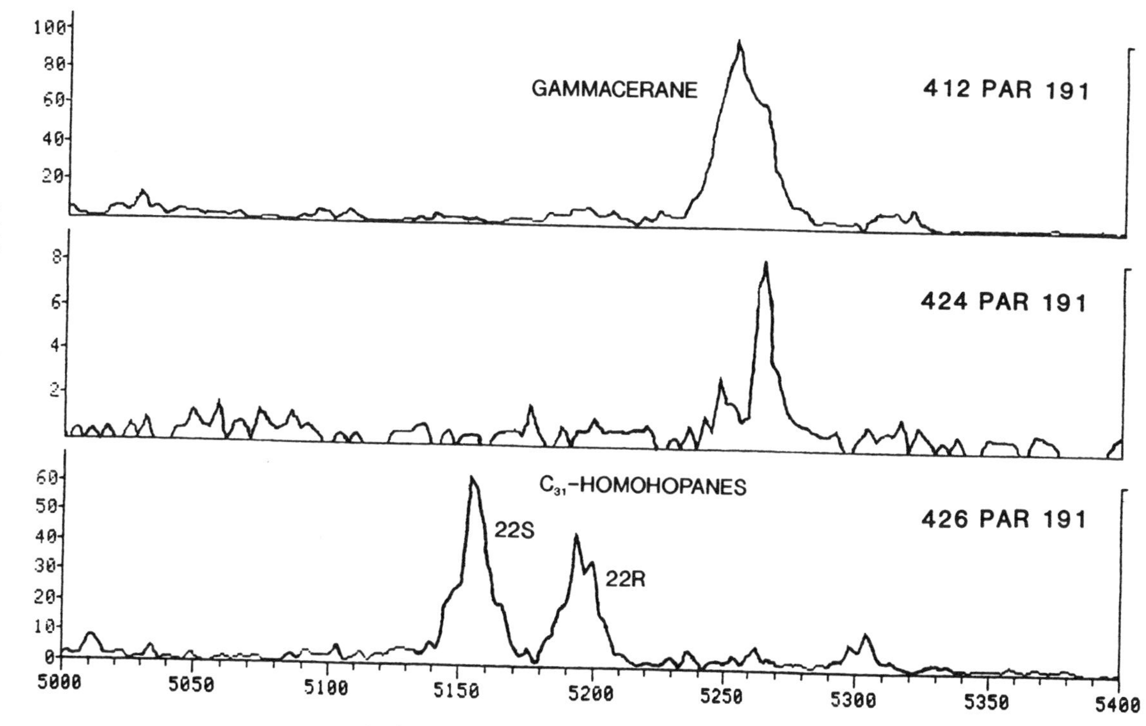

Figure 6.6 Coeluting components are readily resolved using the MS/MS approach. For example, it was anticipated that this Chinese oil sample contained a C_{31}-hexacyclic hopane which coeluted with gammacerane. Monitoring of the m/z 412 → 191 and m/z 424 → 191 parent/daughter relationship completely resolved these two components.

Figure 6.7 was known to contain a number of previously unidentified C_{30} pentacyclic terpanes. Daughter spectra of the parent ion at *m/z* 412 were collected and at the end of the experiment, the parent and daughter data were deconvoluted and complete daughter spectra for each component obtained. Daughter spectra for two of the major peaks in the daughter ion chromatogram, namely, 18α(H)-oleanane and hopane are shown in Figure 6.8a and 6.8b, respectively.

A major advantage of obtaining the daughter spectra is that the need to undertake background subtractions is eliminated. Because only the parent ion is permitted to enter the collision cell all background or interfering components are eliminated, and the daughter spectra subsequently obtained will be only of that one parent ion. This is a major improvement over the regular EI spectra obtained with a single stage analyzer system (GC-MS) where it is necessary to remove interfering ions by subtracting background spectra.

A C_{24}-tetracyclic terpane in many oils, particularly those of terrestrial origin have been discussed in previous papers. It is commonly assumed that this compound

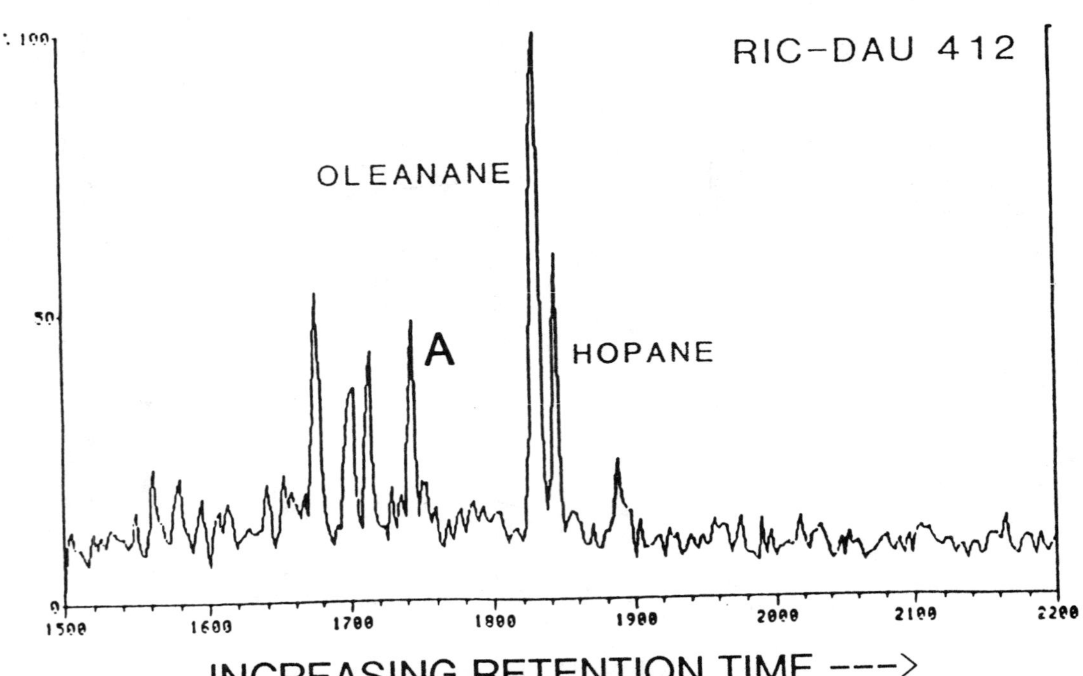

Figure 6.7 In addition to obtaining parent data using the MS/MS approach it is possible to obtain daughter data on various components simultaneously. In this diagram all of the daughter spectra for the ion at *m/z* 412 have been deconvoluted from the parent data and summed to produce this total ion current chromatogram.

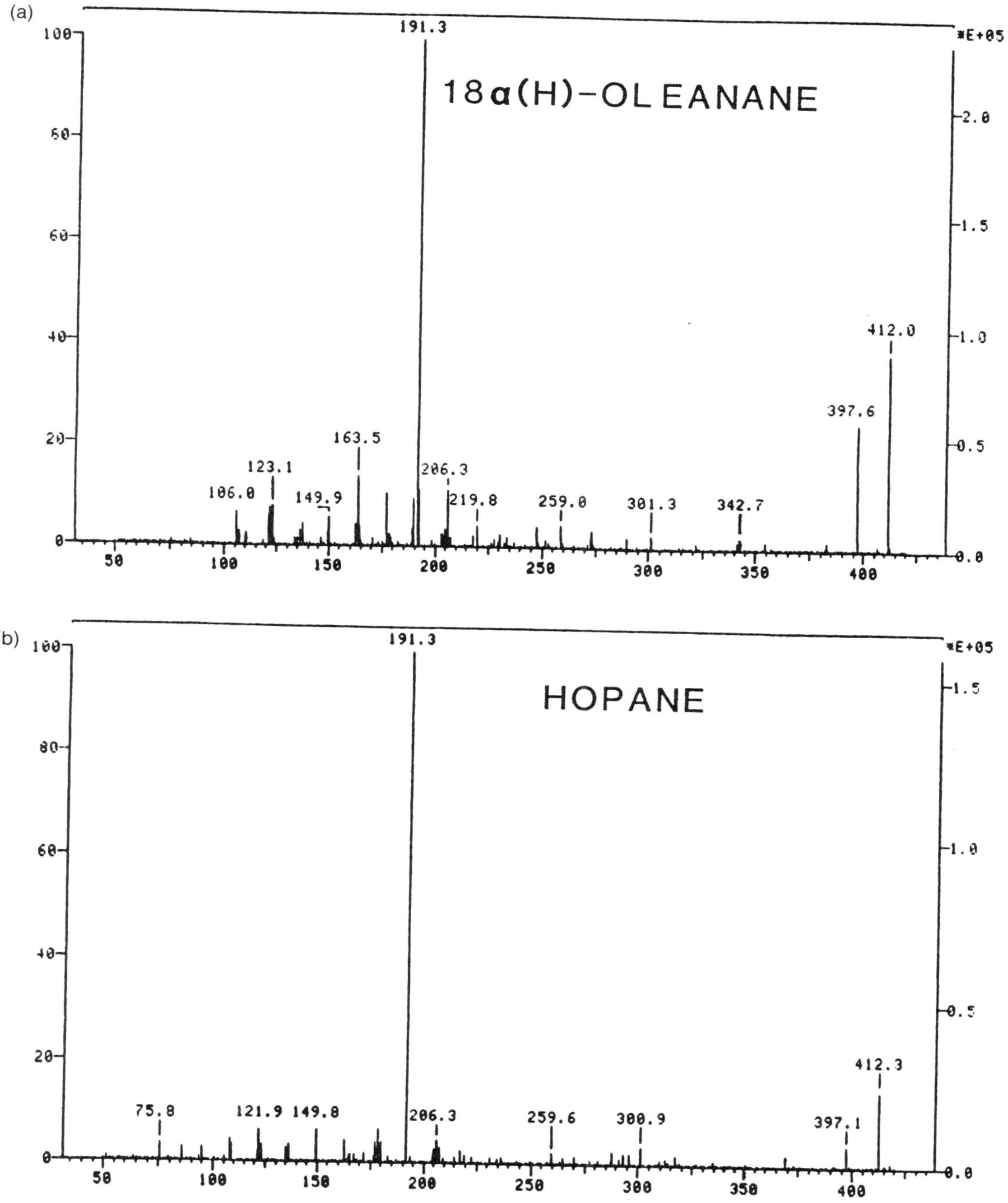

Figure 6.8 Collision activated decomposition spectra (CAD) of the m/z 412 parent ion for (a) 18α(H)-oleanane and (b) hopane.

is derived from the degradation of the hopane E ring and is a 17,21-secohopane. However, our analyses of oils for tetracyclic terpanes showed that in some oils there were at least four, possibly five, tetracyclic (C_{24}) terpanes and several other C_{25}, C_{26}, and C_{27} terpanes (Fig. 6.9). From the chromatogram two series of tetracyclic terpanes can be observed along with the three additional C_{24} tetracyclic terpanes. Operation of the system in the simultaneous parent-daughter mode permits one to obtain complete daughter spectra for each of the C_{24} tetracyclic compounds and one has been subsequently identified as the 17,21-secohopane (labeled S) and the other as a de-A-lupane derivative (labeled L). The latter has been shown previously to be prevalent in samples containing terrestrial source materials (Corbet et al., 1980).

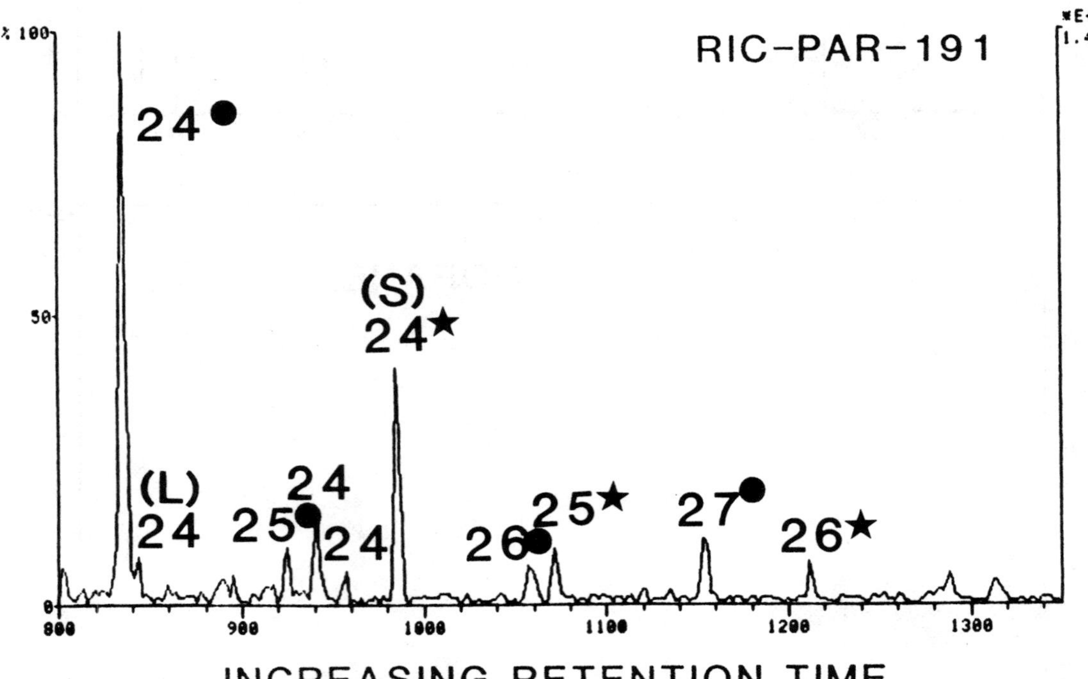

Figure 6.9 Determination of tetracyclic terpanes by monitoring parent/daughter relationships for the C_{24}–C_{27} compounds revealed several C_{24} components and two homologous series of the terpanes from C_{24}–C_{27}. One series is indicated by peaks labeled 24☆ and the other by 24•. The peak labeled L corresponds to the de-A-lupane and S to the 17,21-secohopane.

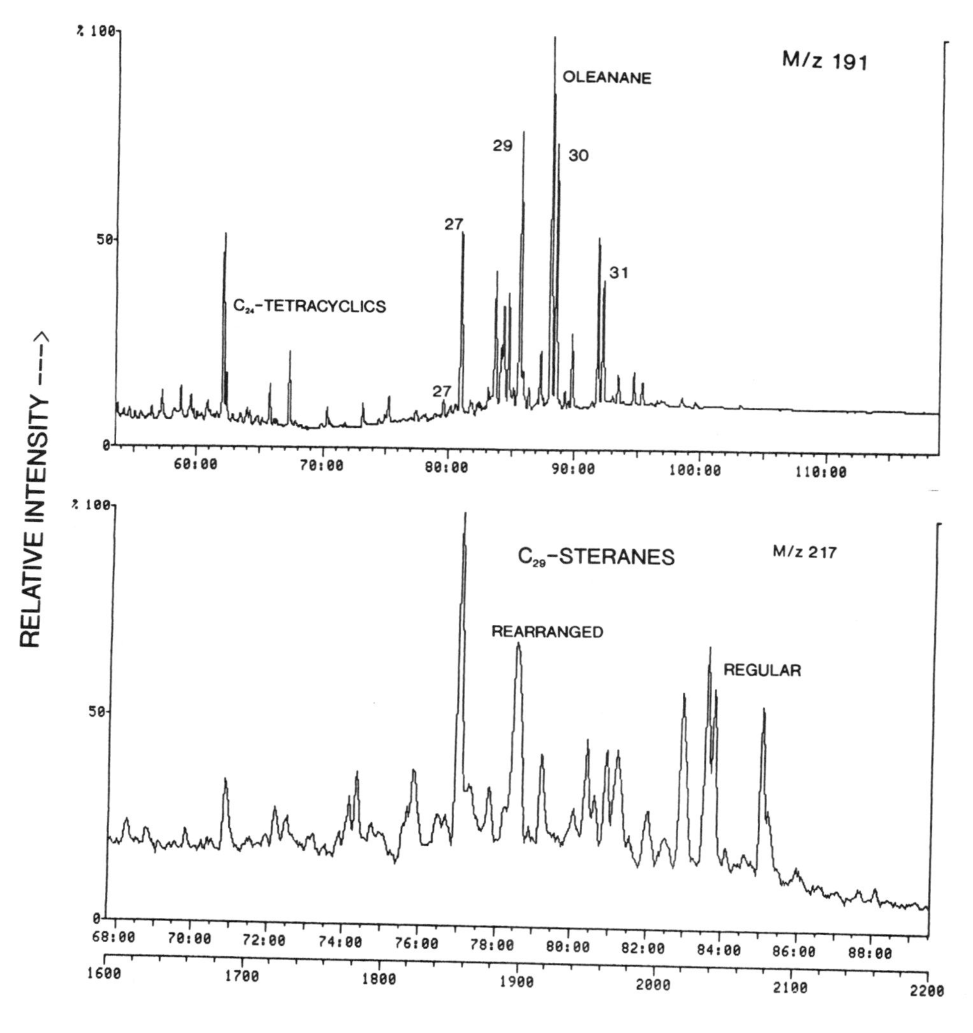

Figure 6.10 (a) *m/z* chromatogram for a Taiwan oil derived from terrestrial source material which is characterized by the presence of the C_{24}-tetracyclic terpanes and the regular hopane series (peaks labeled 27–31) plus 18α(H)-oleanane. (b) *m/z* 217 chromatogram for the Taiwan oil showing the predominance of the C_{29} rearranged and regular steranes.

Quantitation and Comparison between Direct Insertion Probe and GC/MS/MS

In order to demonstrate quantitative biomarker differences between oils derived from terrestrial versus marine sources, four samples of each oil type were selected. Typical distributions of steranes and terpanes obtained by the conventional single ion monitoring of the ions at *m/z* 191 and 217 for one of the oils derived from terrestrial source materials from offshore Taiwan are shown in Figure 6.10a and 6.10b, respectively. The biomarker distribution resulting from analysis of the same sample using the direct insertion probe with the mass spectrometer operating in the MS/MS parent mode is shown in Figure 6.11. d_2-Cholestane which has a major daughter ion at *m/z* 219 rather than *m/z* 217 was used as an internal standard.

The saturated hydrocarbon fractions, plus the deuterated standard, were distilled off the probe into the ion source. After all the spectra were acquired, they were summed to produce a composite spectrum. For example, Figure 6.11 shows parent ions for the various classes of compounds along with the internal standard. Based on the amount of d_2-cholestane added to the original sample, absolute amounts of steranes and amounts of terpanes relative to the internal standard can also be determined (Table 6.1).

It is not possible to directly compare the biomarker distributions obtained by MS/MS probe versus MID data. The MS/MS probe data reflect the intensity of various parent ions in the original sample and each parent ion will contain a contribution from several components. For example, the parent peak at *m/z* 330 does not represent one component but a composite of all the compounds having a parent ion at *m/z* 330. Likewise, the *m/z* 372 parent corresponds to all the C_{27} regular and rearranged steranes. By comparison, MID data only reflect variations

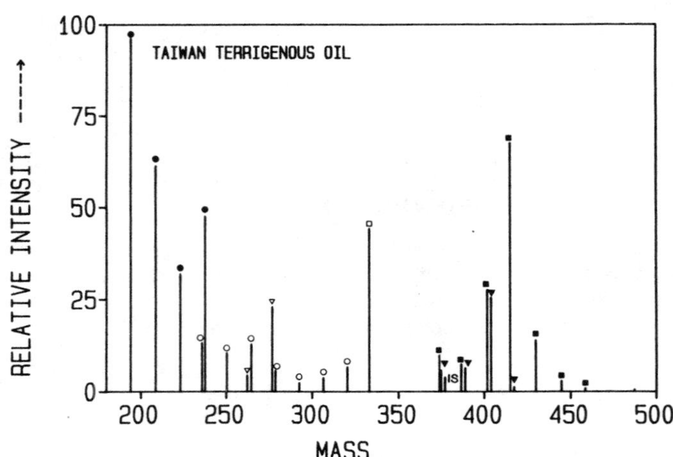

Figure 6.11 MS/MS of the Taiwan sample using the direct insertion probe. Peaks labeled with the ● are bicyclic sesquiterpanes, ▽ diterpanes, ○ tricyclic terpanes, □ tetracyclic terpanes, ■ hopanes, ▼ steranes, and the ion at *m/z* 374 labeled IS is from the C_{27} deuterated standard.

Table 6.1 QUANTITATION DATA FOR STERANES AND TERPANES IN MARINE AND TERRESTRIAL OILS

Sample Location	372[a]	386[a]	400[a]	414[a]	398[a]	412[a]	318[a]	330[a]	372/400	372 + 386 + 400[a]	398/412	217/191	318/330
Offshore Taiwan (Terrestrial)	8.87	8.21	26.28	1.60	29.20	47.30	17.20	51.40	0.34	43.46	76.50	0.57	0.33
	12.56	8.76	37.17	3.49	24.05	47.55	11.20	103.90	0.34	58.48	71.60	0.82	0.11
	8.09	8.71	33.81	1.46	36.65	89.65	8.80	59.05	0.24	50.61	126.30	0.40	0.15
	9.27	8.35	22.00	1.43	9.25	22.05	8.60	18.85	0.42	39.62	31.30	1.27	0.46
Average	8.09	7.83	25.33	1.89	19.93	36.00	14.31	43.95	0.33	41.25	55.93	0.80	0.36
Marine sourced samples													
Williston Basin	31.28	22.03	29.51	11.38	22.29	17.26	37.89	18.49	1.06	82.82	39.55	2.09	2.05
New Mexico	66.40	31.40	42.74	8.67	11.27	19.55	29.57	25.60	1.55	140.54	30.82	4.56	1.16
Gulf of Suez	34.98	29.65	30.32	12.11	19.04	15.75	17.65	8.34	1.15	94.95	34.79	2.73	2.12
Dubai	33.13	27.67	29.16	15.18	20.13	17.28	18.96	7.56	1.14	89.96	37.41	2.50	2.51
Norway	51.35	43.00	42.00	19.48	12.38	18.38	19.45	12.53	1.22	136.35	31.23	4.37	1.55
Average	43.43	30.75	34.75	13.36	17.02	17.74	24.70	14.50	1.23	108.92	34.76	3.23	1.88

[a]Expressed in ng/ul using d_2-cholestane as internal standard.

in intensities of major fragments of different compound classes and for the most part one peak in the chromatogram represents one component. Despite the inability to resolve individual isomers, the direct insertion MS/MS probe approach is valuable for rapid and quantitative screening of oils and source rock extracts.

Table 6.1 summarizes quantitative data from the analysis of the four terrestrial and four marine oils. The results are based on the d_2-cholestane internal standard and hence represent absolute values for sterane concentrations and relative values for the terpanes and other compounds in the samples. The data in Table 6.1 show differences between oils derived from marine and terrestrial source materials on the basis of the biomarker quantitation. The steranes in oils derived from terrestrial sources are dominated by the C_{29} compounds with relatively low C_{30} sterane concentrations. The marine oils have nearly equal concentrations of C_{27}, C_{28}, and C_{29} sterane concentrations and high C_{30} sterane concentrations relative to the C_{30} sterane concentrations in the oils from terrestrial sources. In addition, the terrestrial oils have high concentrations of the C_{30} terpanes and C_{24} tetracyclic terpanes. The high m/z 412 terpane content is due to the presence of 18α(H)-oleanane and hopane, both of which are C_{30} terpanes. The marine oils generally have a higher concentration of the C_{23} tricyclic terpane than is observed in the terrestrial oils. Various ratios (Table 6.1) obtained from analysis of these oils via the direct insertion probe also permit the oils to be distinguished.

SUMMARY

This chapter has provided several examples in which the MS/MS approach has been used to analyze complex geochemical samples. The MS/MS system can be operated in the parent or daughter mode to provide valuable geochemical data. In many cases these data may be in the form of tentative identification of previously unidentified components. Alternatively, MS/MS may be used to provide information on the distribution of novel biomarkers. It is also a useful technique for separating components which may have different molecular weights but coelute with each other. Finally, operation in the daughter mode provides a method for getting collision spectra of specific daughter ions which may be used to identify unknown components.

The data presented have also demonstrated that by use of tandem mass spectrometry and the direct insertion probe it is possible to differentiate oils derived from different sources by a quantitative analysis of the biomarkers. This is achieved by use of one or more deuterated internal standards. Despite the fact that not all isomers are resolved when using the direct insertion probe, the method provides a fast reproducible method for oil/oil correlations, source evaluation, and quantitative determination of various classes of biomarkers. It is also important to note that since the gas chromatography step is eliminated, the analysis time for each sample using the direct insertion probe can be reduced from approximately two hours to a few minutes.

Acknowledgments

Funds to purchase the Finnigan Triple Stage quadrupole mass spectrometer were obtained from the National Science Foundation Instrumentation Program (grant no. EAR 8517312); the Sarkeys Foundation and the ARCO Foundation.

REFERENCES

CORBET, B., ALBRECHT, P., and OURISSON, G. (1980). Photochemical or photomimetic fossil triterpenoids in sediments and petroleum. *J. Am. Chem. Soc.*, *101*, 1171–1173.

GALLEGOS, E.J. (1976). Analysis of organic mixtures using metastable transition spectra. *Anal. Chem.*, *48*(9), 1348–1351.

GALLEGOS, E.J. and MOLDOWAN, J.M. (1989). The effect of injection hold time on the GC resolution and the effect of collision gas on the mass spectra in geochemical biomarker research. In: *Div. of Petroleum Chem. Preprints*, *34*(1), 159–169, 1989.

GALLEGOS, E.J. and MOLDOWAN, J.M. (1991). The effect of injection hold time on GC resolution and the effect of collision gas on the mass spectra in geochemical biomarker research. Chapter 9 in this text.

MACKENZIE, A.S., BRASSELL, S.C., EGLINTON, G., and MAXWELL, J.R. (1982). Chemical fossils: The geochemical fate of steroids. *Science* 217, 491–504.

PHILP, R.P. and LEWIS, C.A. (1987). Organic geochemistry of biomarkers. *Ann Rev. Earth Planet. Sci.*, *15*, 363–395.

PHILP, R.P. and OUNG, J.N. (1988). Biomarkers, occurrence, formation and detection. *Anal. Chem.*, *60*(10), 887A–896A.

PHILP, R.P., OUNG, J., and LEWIS, C.A. (1988). Biomarker determinations in crude oils using a triple stage quadrupole mass spectrometer. *J. Chromatog.*, *446*, 3–16.

RINALDI, G.G.L., LEOPOLD, V.M., and KOONS, C.B. (1988). Presence of benzohopanes, monoaromatic secohopanes, and saturated hexacyclic hydrocarbons in petroleum from carbonate environments. In: *Geochemical Markers* (eds. T.F. Yen and J.M. Moldowan). Harwood Academic Publishers, New York, p. 331–355.

SEIFERT, W.K. and MOLDOWAN, J.M. (1978). Applications of steranes, terpanes and monoaromatics to the maturation, migration and source of crude oils. *Geochim. Cosmochim. Acta*, *42*, 77–95.

SEIFERT, W.K. and MOLDOWAN, J.M. (1979). The effect of biodegradation on steranes and terpanes in crude oils. *Geochim. Cosmochim. Acta*, *43*, 111–126.

SHI JI-YANG, MACKENZIE, A.S., ALEXANDER, R., EGLINTON, G., GOWAR, A.P., WOLFF, G.A., and MAXWELL, J.R. (1982). A biological marker investigation of petroleums and shales from the Shengli Oilfield, The People's Republic of China. *Chem. Geol.*, *35*, 1–31.

SUMMONS, R.E. (1987). Branched alkanes from ancient and modern sediments: Isomer discrimination by GC/MS with multiple reaction monitoring. *Org. Geochem.*, *11*, 281–289.

SUMMONS, R.E., POWELL, T.G., and BOREHAM, C.J. (1988). Petroleum geology and geochemistry of the Middle Proterozoic McArthur Basin, Northern Australia: III. Composition of extractable hydrocarbons. *Geochim. Cosmochim. Acta* *52*, 1747–1763.

7

Biomarker Maturation in Contemporary Hydrothermal Systems, Alteration of Immature Organic Matter in Zero Geological Time

Bernd R. T. Simoneit, Orest E. Kawka, and Gong-Ming Wang

Abstract. Assessments of the thermal maturation of oils and source rocks by molecular analyses are based on the correlation between biomarker transformations and the thermal history of the organic matter in sedimentary basins. Thermal maturation and the evolution of petroleum normally occur over extended periods of geological time (> 1 My) under low geothermal gradients (maximum T ≤ 150°C). The kinetic parameters of the isomerization and aromatization reactions of steranes and hopanes under those conditions have been well documented. Comparison of the extent of each reaction can be used to assess the type of basin and the geothermal gradient. The instantaneous generation of petroleum at higher temperatures (> 150°C) can occur in virtually zero geological time as is evidenced by the oils found in the hydrothermal system of Guaymas Basin, Gulf of California. This instantaneous maturation in less than 100 ky is also reflected in the associated biomarker distributions in the seabed oils. Under hydrothermal conditions, the aromatization of the monoaromatic to the triaromatic steroid hydrocarbons is usually enhanced relative to the sterane isomerizations. This result as expected for the former reaction is more temperature-dependent. The isomerization of the hopanes is also much more rapid than that of the steranes and is near equilibrium in most of the oils. This aspect is more comparable with the hopane isomerization observed during simulations than with those in basin models. This supports previous suggestions that the mechanism of hopane isomerization may differ depending on the thermal regimes. An unusual appearance of a C_{18} compound (Diels' hydrocarbon) in the short-alkyl-chain series of the triaromatic steroid hydrocarbons in the mature oils may be a reflection of such rapid thermal maturation. The rapid biomarker evolution, with characteristically enhanced hopane isomerizations and steroid hydrocarbon aromatizations make the hydrothermal process most similar to laboratory simulations of petroleum maturation.

INTRODUCTION

Deeply buried organic matter of sedimentary basins can be transformed to petroleum by extended exposure to elevated temperatures. This thermal maturation and generation of petroleum normally occurs over periods of millions of years at relatively low temperatures, 60–150°C (Hunt, 1979; Tissot and Welte, 1984). Hydrothermal systems can also generate petroleum as is evidenced by oils found at the seabed of Guaymas Basin, Gulf of California (Simoneit and Lonsdale, 1982; Simoneit, 1985; Kawka and Simoneit, 1987). Crustal accretion and hydrothermal activity at this young spreading center generate the petroleum at depth in the sediments through a rapid pyrolysis of immature organic matter (Simoneit et al., 1984). The interpreted geological history of Guaymas Basin suggests that the oldest sediments in the rift areas are < 200 ky old (Curray et al., 1982). The occurrence of hydrothermal oils, sourced from these immature sediments, indicates that the instantaneous generation of petroleum at high temperatures (> 150°C) can occur in virtually zero geological time. The hydrothermal petroleums at the seafloor in the south rift have ^{14}C ages of < 5000 years (Peter et al., 1991).

Biomarker analysis provides a means to assess the thermal maturation of sediments and oils. It is based on the observed correlation between the biomarker transformations and the thermal history of the organic matter in well-studied sedimentary basins and on similar pyrolytic conversions observed during laboratory simulations. The extent of and the kinetic constraints on the isomerization and aromatization reactions of the steranes and hopanes under both in situ conditions (Ensminger et al., 1977; Mackenzie and McKenzie, 1983; Mackenzie et al., 1980, 1981a, *Nature*, 1982, 1984; McKenzie et al., 1983; Sajgó and Lefler, 1986; Seifert and Moldowan, 1980; Suzuki, 1984) and during laboratory pyrolysis (Abbott et al., 1984, 1985; Lewan et al., 1986; Mackenzie et al., 1981b; Suzuki, 1984; van Dorsselaer et al., 1977; Zumberge et al., 1984) have been extensively reported in the literature. Due to the temperature dependence of these reactions, differences would be expected in the relative extents of the biomarker transformations in high-temperature hydrothermal systems and basins with low-geothermal gradients.

We have analyzed the sterane, hopane, and triaromatic steroid hydrocarbon contents of the hydrothermal petroleums from Guaymas Basin and herein interpret their distribution in terms of the effects of rapid high-temperature maturation. Of special interest is the distribution of triaromatic steroid hydrocarbons in these oils. Comparison with laboratory simulations suggests a possible biomarker series characteristic of high-temperature processes of organic matter maturation.

SAMPLING AND ANALYSIS

Hydrothermal Oils

The locations, collection, isolation, and bulk chromatographic fractionation of the hydrothermal oils from the Guaymas Basin seabed have been described previously (Simoneit and Lonsdale, 1982; Simoneit, 1983a, 1983b; Kawka and Simoneit, 1987). Initial screening of the saturate and aromatic hydrocarbon fractions of the oils by high-resolution gas chromatography (HRGC) and high-resolution gas chromatography-mass spectrometry (HRGC-MS) was performed as described in mentioned references.

More detailed quantitative assessment of the sterane, hopane, and aromatic steroid hydrocarbon distributions in the oils was obtained using similar HRGC-MS conditions but with operation in the multiple ion detection (MID) mode. The relative amounts of the various components described here were obtained by integration and selective summation (described in the figure captions) of the ions characteristic of each of the biomarker series. MS dwell times of either 105 or 210 msec per mass (1 amu wide) was used, with the ions monitored in descriptors of 6 to 10 masses each. The composition of these descriptors (total scan times of 1.5 to 2.0 sec each) was set to facilitate computer-controlled time-programming of the masses scanned so as to have them coincide with the expected HRGC elution times of the compounds of interest. Monitoring of the biomarker distributions in the aromatic fractions was concurrent with the quantitation of various polycyclic aromatic hydrocarbons.

Assessing the extent of aromatization of the monoaromatic (base peak m/z 253) to the triaromatic steroid (base peak m/z 231) hydrocarbons was complicated by their separate elution in the saturated and aromatic hydrocarbon fractions, respectively, during liquid-solid chromatography. The proportions of the two series in each oil was obtained by quantitation of the integrated areas of their base peaks relative to that of an internal standard, perdeuterated anthracene (base peak m/z 188) added to each fraction before HRGC-MS.

Laboratory Simulations

Hydrous pyrolysis of immature sediments. A sample of thermally unaltered mud from the seabed of Guaymas Basin (sample 1176-PC2, 10–20 cm) was chosen for a thermal alteration experiment. The hydrous pyrolysis was carried out in a chromium-lined, stainless steel vessel with 100 g mud (with 34% pore water) plus 500 g simulated seawater. The temperature program was 20–330°C over 5 hours, at 330°C for 1 hour and cooling to 30°C over 2 hours. The pyrolysis products were extracted from both the aqueous and solid phases with chloroform.

Generation of Diels' hydrocarbon from cholesterol. Diels' hydrocarbon [1,2-(3′-methylcyclopenteno)phenanthrene] was prepared from cholesteryl chloride by heating with selenium in a manner analogous to the original procedures (Diels and Rickert, 1935; Diels and Karstens, 1930; Diels et al., 1927; Harper et al., 1934).

The isolation, separation, and analysis by HRGC-MS of the pyrolysates were similar, unless otherwise noted, to the procedures described earlier for the hydrothermal oils.

RESULTS AND DISCUSSION

Thermal Maturation

Hydrothermal oils. The rapid heating of the sediments in Guaymas Basin by hydrothermal processes results in an instantaneous thermal maturation of biomarkers. The triterpenoid content of the thermally unaltered sediments is normally characterized by significant concentrations of triterpenes and hopanes with their biologically derived 17β(H),21β(H)-configurations, as exemplified by the distribution presented in Figure 7.1a. The oils generated by the hydrothermal pyrolysis, on the other hand, contain primarily hopanes in their thermally most stable 17α(H),21β(H)-configuration (Ensminger et al., 1974, 1977; Seifert and Moldowan, 1978) and lesser but variable amounts of the intermediate 17β(H),21α(H)-hopanes (moretanes) and the biologically derived ββ-series (see Figs. 7.1b and c). Epimerization in the extended series of 17α(H),21β(H)-hopanes ($\geq C_{31}$) has also occurred with the relative amounts of the 22*S* and 22*R* (biologically derived) diastereomers dependent on the degree of maturation. The epimer ratios (22*S*/22*R*) for the C_{31} and C_{32} hopanes in oil sample 1173-3 (Fig. 7.1b) are 0.87 and 0.68, respectively; which are less than that reported for diastereomers at thermal equilibrium, that is, a ratio of ca. 1.5 (Ensminger et al., 1977; Seifert and Moldowan, 1980). The ratios for the extended hopane series in another oil, 1172-4, are 1.1 and 1.2, respectively; this increased epimerization is consistent with the absence of the 17β(H),21β(H)-hopanes and the relative decrease in the amount of 17β(H),21α(H)-hopanes (cf. Figs. 7.1b and c).

The steroid hydrocarbons in the sediments are also transformed rapidly to their thermally more stable configurations during the hydrothermal generation of petroleum. Thermally-unaltered surface sediments of Guaymas Basin contain only sterenes and diasterenes but no steranes in any configuration (Simoneit et al., 1979, 1984; Simoneit, 1983b). Hydrothermal pyrolysis of the sediments at depth results in the generation of steranes and a relative decrease in the unsaturated analogs (Simoneit et al., 1984). The sterane distributions in the pyrolysates appear to reflect

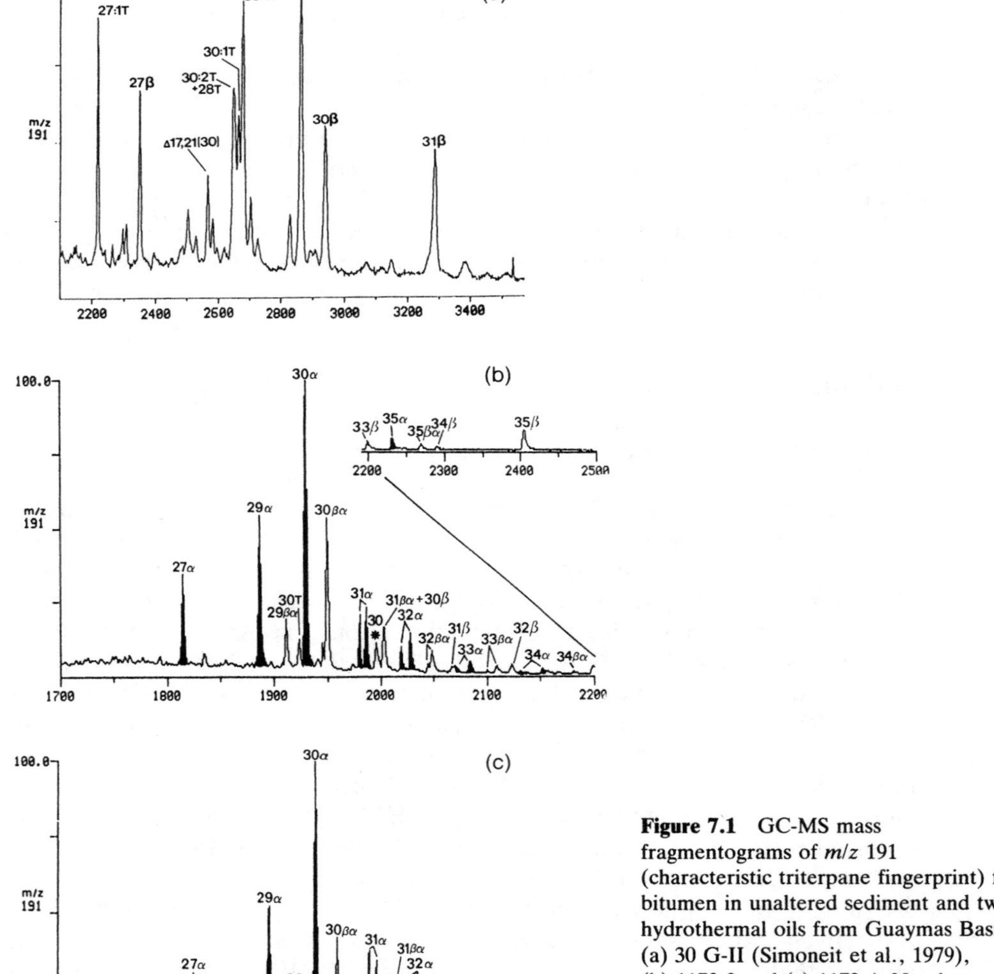

Figure 7.1 GC-MS mass fragmentograms of m/z 191 (characteristic triterpane fingerprint) for bitumen in unaltered sediment and two hydrothermal oils from Guaymas Basin: (a) 30 G-II (Simoneit et al., 1979), (b) 1173-3 and (c) 1172-4. Numbers refer to carbon skeleton size while suffixes are their configuration: α = 17α(H),21β(H)- (shaded), β = 17β(H),21β(H)-, βα = 17β(H),21α(H)-hopane series and: i = hopenes; * = suspected gammacerane.

variable degrees of thermal maturation, as typified by the patterns in the two seabed oils (Fig. 7.2). The primary components of 1173-3 (Fig. 7.2a) are the C_{27}–C_{29} steranes with the 5α(H),14α(H),17α(H)-20*R* configuration, while the thermally less stable 5β(H)-steranes (Philp, 1985) are found in lower amounts. The thermally

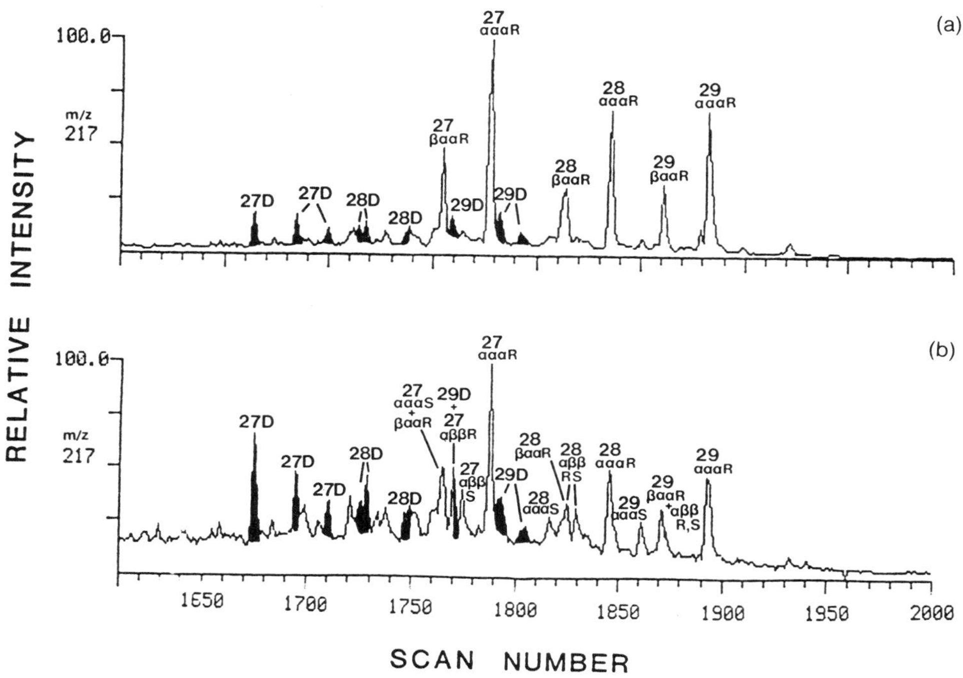

Figure 7.2 GC-MS mass fragmentograms of *m/z* 217 (characteristic sterane fingerprint) for two hydrothermal oils from Guaymas Basin: (a) 1173-3 and (b) 1172-4. Numbers refer to carbon skeleton size while letters represent the structure: α or β = hydrogen configuration at C-5, C-14, C-17, respectively; *R* or *S* = epimer at C-20; D = rearranged sterane/diasterane (shaded).

induced epimerization at C-20, from the biologically derived *R* to the *S* configuration (Seifert and Moldowan, 1978; Mackenzie et al., 1980), is minor; and the 14β(H),17β(H) series is absent in this oil. The greater degree of thermal maturation inferred for oil 1172-4, based on its hopane distribution (see Fig. 7.1c), is also indicated by its sterane distribution (Fig. 7.2b). The primary components range from C_{27} to C_{29} and the epimerization at C-20 is extensive, resulting in a 20*S*/*R* ratio of 0.35. This degree of epimerization is, however, still far from complete since the theoretically calculated 20*S*/20*R* ratio is 1.17 (van Graas et al., 1982). Generation of the 14β(H),17β(H)-(20*R*,*S*) sterane series has also occurred and is consistent with the increased maturation. The amount of diasteranes, relative to steranes, is also greater in this sample (cf. Fig. 7.2a, b). An increase in this proportion with increasing maturation is found for all the hydrothermal oils.

Hydrous pyrolysis (simulation). In general, the hydrocarbon content of the simulation product is very unlike that of the unaltered substrate but is similar to that of the hydrothermal oils (Simoneit, 1989). A comparison of the pyrolysate, hydrothermal oils, and bitumen from unaltered sediment is presented in Table 7.1. The *n*-alkanes generated by the simulation range from C_{14}–C_{36} with a minor odd carbon number predominance (CPI = 1.09) and C_{max} at C_{23} (Simoneit, 1989) which compares well with that of the hydrothermal petroleums (Kawka and Simoneit, 1987). Pristane and phytane are generated with a Pr/Ph of 0.56, which is lower than that for many of the hydrothermal petroleums, where this ratio is about 1.0. The naphthenic hump or UCM of the hydrothermal petroleums is generally broad extending over the full GC retention range, whereas the lipids of the unaltered sediments exhibit a narrow UCM centered at about *n*-C_{19} elution time. The UCM of the simulation bitumen is essentially the same as observed from the hydrothermal petroleums (Simoneit, 1989).

The polycyclic aromatic hydrocarbons (PAH) of the pyrolysate are significantly different from the unaltered sediments but similar to the hydrothermal petroleums of Guaymas Basin, although not identical (see Table 7.1). The hydrothermal petroleums contain PAH consisting primarily of the unsubstituted parent compounds (Simoneit, 1984), whereas the unaltered sediments (simulation substrate) have only perylene from diagenetic sources (Simoneit et al., 1979). Phenanthrene and anthracene as well as the alkylphenanthrenes and methylene-phenanthrene are generated in this pyrolysis with essentially the same composition as in typical Guaymas Basin hydrothermal petroleum (Simoneit, 1989). Triaromatic steroid hydrocarbons were generated in the hydrous pyrolysis. However, the higher molecular weight PAH (e.g., benzopyrenes, benzo(g,h,i)perylene, coronene), excluding perylene, were not generated in the simulation. This may indicate that

Table 7.1 SUMMARY OF COMPOUND CONVERSIONS IN LABORATORY HYDROTHERMAL ALTERATION OF GUAYMAS BASIN SEDIMENT

Compounds	Laboratory simulation (330°C)	Unaltered sediment	Hydrothermal petroleums
n-Alkanes and UCM	+ mature	+ immature	+ mature
n-Alkanoic acids	+ altered	+ immature	trace
Alkylcyclohexanes	+	+	+
Biomarkers			
17β(H)-hopanes and hopenes	trace	+	− (tr.)
17α(H)-hopanes	+	trace	+
steranes and diasteranes	trace	−	+
triaromatic steroid HC (Diels HC)	+	−	+
Aromatic hydrocarbons			
perylene	+	+	trace
benzopyrenes	trace	−	+
phenanthrene series	+	−	+
fluoranthene/pyrene	+	−	+
high-molecular weight PAH[a]	−	−	+

[a]For example, benzoperylene, coronene, (exclusive of perylene).

aromatization of steroids proceeds more easily than the formation of the heavy PAH during pyrolysis at such temperatures (~330°C) over short time periods (see Sampling and Analysis section).

The sterane and hopane distributions in the hydrous pyrolysate indicate that thermal maturation has occurred, but the extent of it is less than that of most of the hydrothermal oils (Simoneit, 1989). The triterpanes generated during the simulation experiment are of an intermediate maturity. The αβ-hopanes are predominant, but a significant amount of the ββ-series remains and a major amount of the intermediate 17β(H),21α(H)-hopane series (βα, moretanes) is also present, not unlike that for the 1173-3 oil. It should be noted that αβ-norhopane is the dominant triterpane in the simulation sample ($C_{29}/C_{30} = 1.7$) as compared to αβ-hopane in the hydrothermal petroleums ($C_{29}/C_{30} = 0.45$), and the C_{29} and C_{30} moretanes are also abundant in the simulation sample. The extended αβ-hopane epimer ratios for C_{31} and C_{32} 22*S*/*R* are 0.3 and 0.5, respectively, much less than those for the oils; this confirms the immature nature of the triterpanes from the simulation. The steranes generated during the hydrous pyrolysis consist mainly of a relatively immature assemblage of the 5α(H),14α(H),17α(H)-20*R* and the lesser 5β(H)-series (Simoneit, 1989). Neither epimerization from *R* to *S* at C-20 nor the generation of the 14β(H),17β(H) series has occurred to any measurable extent in the pyrolysate.

Isomerization/Aromatization Trends

The relationships between the extents of sterane and hopane isomerization (I) and steroid hydrocarbon aromatization (A) in sediments and oils have been used to differentiate heating rate effects (Mackenzie and Maxwell, 1981; Mackenzie and McKenzie, 1983; Mackenzie et al., *Nature*, 1982, 1984; McKenzie et al., 1983). An AI plot using the isomerization at C-20 of the C_{29} steranes and the aromatization of the C_{29} monoaromatic to the C_{28} triaromatic steroid hydrocarbons is presented in Figure 7.3a. The extent of reaction in a number of the Guaymas Basin hydrothermal oils is plotted. For most of these petroleums, the steroid hydrocarbon aromatization has been considerable, with the parameter above 50 percent for all the oils. The sterane isomerization, in general, is not as extensive with the reaction less than 50 percent complete [$S/(S + R) = 0.54$ at equilibrium] for more than half of the oils. Such a pattern is consistent with the inferred high-heating rates in Guaymas Basin. A high-heating rate would tend to enhance the aromatization rate, relative to sterane isomerization, since it is the more temperature-dependent reaction according to the estimates of the kinetic parameters, activation energies, and frequency factors (Mackenzie and McKenzie, 1983; Mackenzie et al., *Nature*, 1982, 1984).

Although general comments regarding the thermal genesis of the oils can be made, specific inferences regarding time-temperature history are difficult. Since these oils have been moved to the seabed by both bulk transport and in solution with the hydrothermal fluids (Kawka and Simoneit, 1987), migrational effects on

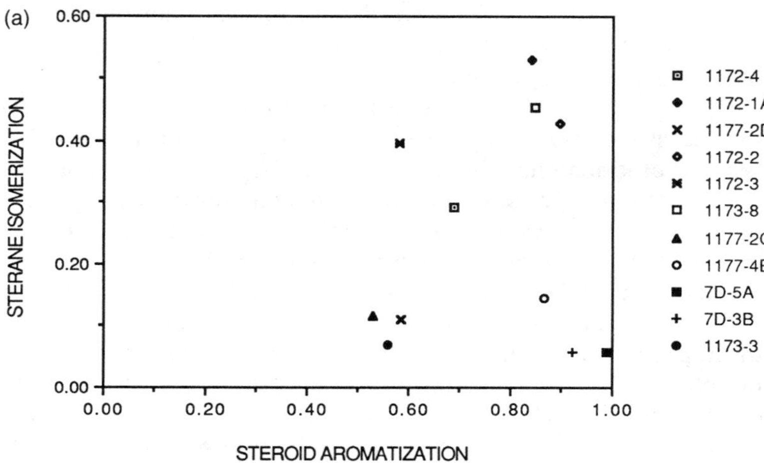

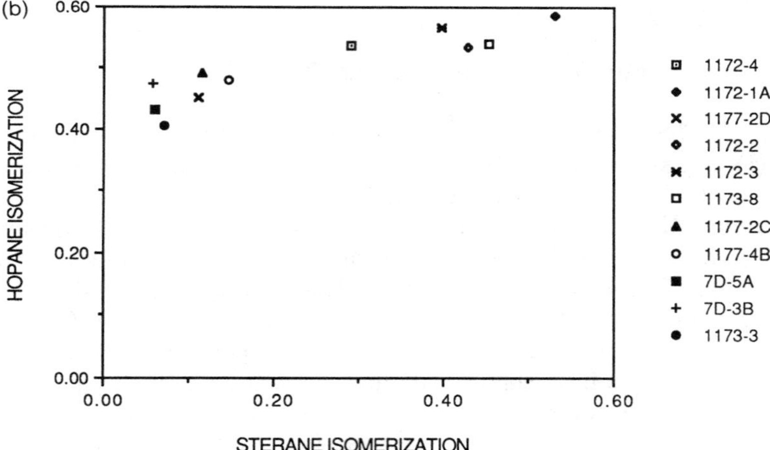

Figure 7.3 Relationship between sterane isomerization and aromatization and hopane isomerization trends in hydrothermal oils from the Guaymas Basin seabed: (a) AI plot of C_{29} sterane isomerization [$S/(S + R)$] at C-20 versus degree of aromatization [TA/(TA + MA)] of 5β and 5α C_{29}-20R monoaromatic (MA) to the C_{28}-20R triaromatic (TA), measured as described in Mackenzie et al. (1981a); (b) II plot of C_{32} hopane isomerization [$S/(S + R)$] at C-22 versus sterane isomerization. Abundances calculated using integrated areas or heights of the base peak (mass spectral) response for each compound type: sterane (m/z 217), hopane (m/z 191), MA (m/z 253), TA (m/z 231). Numbers in legend refer to sample identifications as described in Kawka and Simoneit (1987).

the aromatization parameter may have been intensified. Lewan et al. (1986) found that the aromatization ratio was different in the expelled oils and bitumen extracts of a hydrous pyrolysate of Phosphoria Retort Shale. Such migrational effects may explain the high scatter in the data of Figure 7.3a.

Another comparison is made in Figure 7.3b, which contains a II plot of the extents of epimerization at C-22 in the C_{32} hopanes and at C-20 in the C_{29} steranes. For more than half of the hydrothermal oils, the C_{29} sterane isomerization is low, while the extent for the C_{32} hopane is consistently high. Such an enhancement of the hopane relative to the sterane isomerization is consistent with an early suggestion by Mackenzie and McKenzie (1983) that the actual mechanism of the former may vary depending on the thermal regime. The kinetic parameters for the two reactions derived from pyrolysis studies (Suzuki, 1984; Zumberge et al., 1984) also predict that the hopane isomerization would be much more enhanced relative to the sterane isomerization at high temperatures. This is especially relevant in the case of hydrothermal maturation in Guaymas Basin where high-pyrolytic temperatures exist.

The extensive sterane isomerization for a few of the samples may indicate unusually high temperatures of genesis. Although migration should not affect the patterns in this type of plot (Fig. 7.3b) as it may have in the previous diagram (Fig. 7.3a), recent reports on maturation of steranes appear to question the validity of using sterane isomerizations as maturity indicators in such systems. Recent hydrous pyrolysis studies have shown a reversal of the sterane isomerization trends at high levels of maturity (Lewan et al., 1986; Rullkötter and Marzi, 1989). Therefore, the inherent assumption in maturity analyses by isomerization ratios, namely irreversibility, may be violated at high temperatures and/or maturities.

Triaromatic Steroid Hydrocarbons

The triaromatic steroid hydrocarbons found in petroleums and source rocks occur in a number of structural forms that differ in the number of nuclear methyls they contain. In a study of the Toarcian shales in Paris Basin, France, Mackenzie et al., (1981a) observed that the triaromatic steroid hydrocarbons with one (*m/z* 231 base peak) or two (*m/z* 245 base peak) nuclear methyls were the dominant series, while the other types with either no or three nuclear methyls (*m/z* 217 and *m/z* 259 base peaks, respectively) were in relatively lower concentration. This usual dominance of the *m/z* 231 series in sediments and oils (Mackenzie et al., 1982) is exemplified by the two crude oils, Jiang Han and North Sea (Fig. 7.4). The triaromatic steroid hydrocarbons (C_{26}–C_{28}) with long alkyl side chains are the major components in both of the oils and occur in diastereomer pairs (C-20*S* and *R*). The short-alkyl-chain triaromatic steroid hydrocarbons (C_{20} and C_{21}) also occur but their abundance relative to the long-alkyl-chain counterparts differs in the two oils. A catagenetically induced cleavage of the alkyl side chain of the steroid hydrocarbons, first suggested by Seifert and Moldowan (1978), has been offered as an explanation

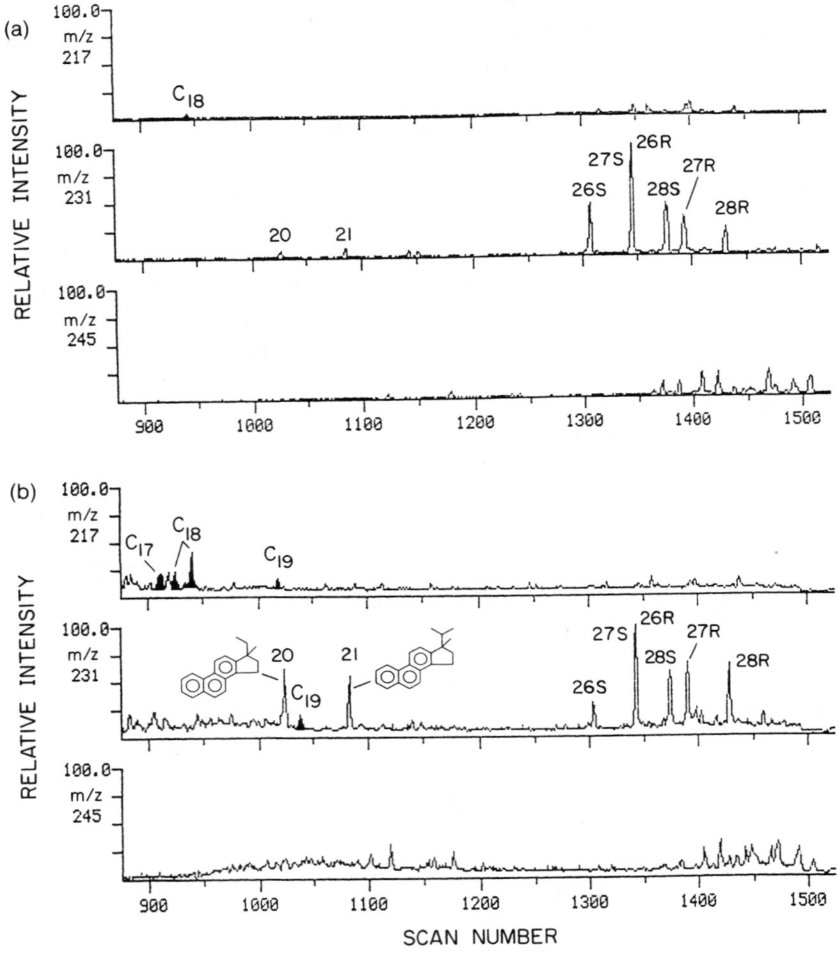

Figure 7.4 Relative distributions of the three series (normalized) of triaromatic steroid hydrocarbons in two crude oils: (a) Jiang Han and (b) North Sea. GC-MS mass fragmentograms for *m/z* 217 (no nuclear methyls), *m/z* 231 (one nuclear methyl), and *m/z* 245 (two nuclear methyls). Arabic numerals refer to total number of carbons, while *S* and *R* denote configuration at C-20, where applicable. Peaks shaded and identified by C_{17}, C_{18}, or C_{19} refer to compounds described in Figure 7.5.

for the relative increase in the short-alkyl-chain triaromatic steroid hydrocarbons at high levels of thermal maturation (Mackenzie et al., 1981a).

The relative distributions of the same three series of triaromatic steroid hydrocarbons in two of the hydrothermal oils from Guaymas Basin, 1173-3 and 1172-4, are shown in Figure 7.5. Unlike the crude oils in Figure 7.4, the hydro-

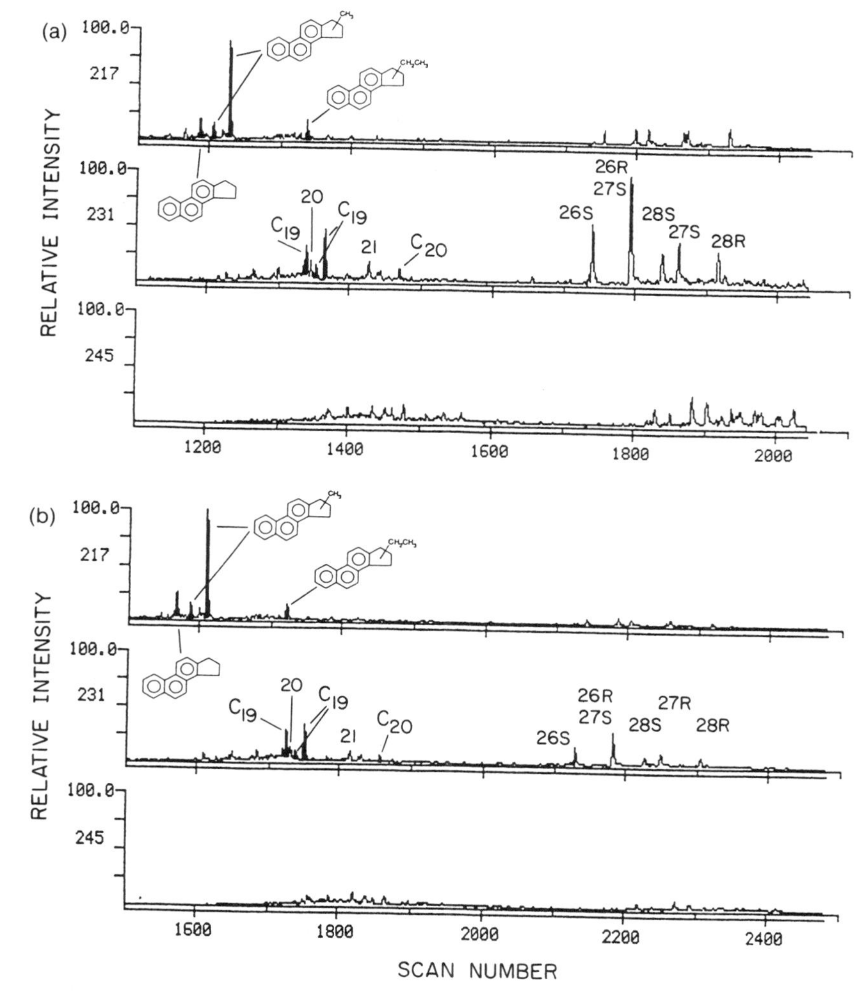

Figure 7.5 Relative distributions of the three series (normalized) of triaromatic steroid hydrocarbons in two hydrothermal oils from Guaymas Basin: (a) 1173-3 and (b) 1172-4. Peak identities as in Figure 7.4. Compounds characteristic of the hydrothermal oils are shaded. In addition to the structures provided for C_{17}, C_{18} (Diels HC and isomer), and C_{19} in the *m/z* 217 series, these compounds also include the peaks in the *m/z* 231 series, identified by C_{19} and C_{20}, whose structures are not known (see text). Note that the 20- and 21-carbon compounds (not subscripted) normally found in oils (see Fig. 7.4) are present in low proportions relative to the unknown series.

thermal oils contains major amounts of triaromatic steroid hydrocarbons with no nuclear methyls (*m/z* 217). In addition, the short-alkyl-chain components are significantly more abundant than the long-alkyl-chain components of this series. The

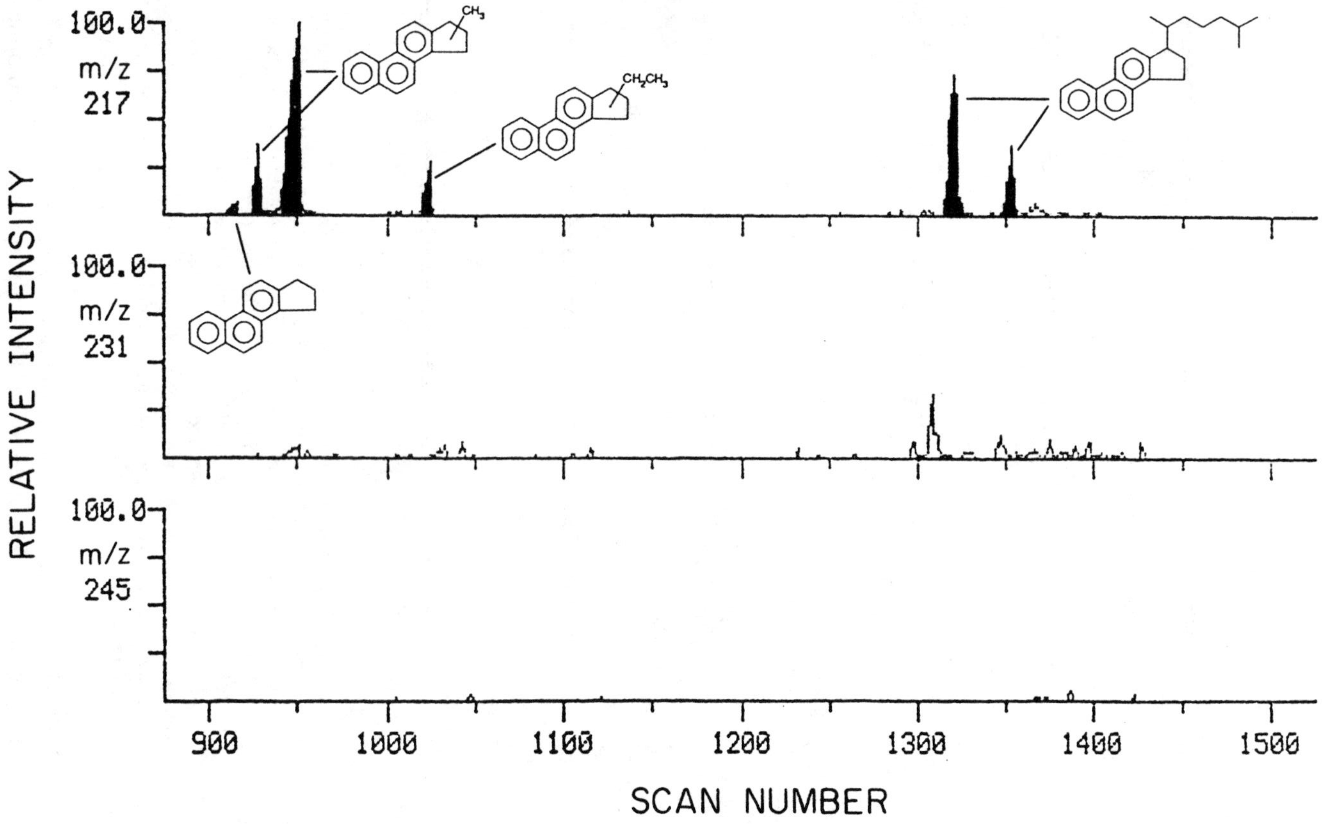

RELATIVE INTENSITY

100.0
m/z
217

CH_3

CH_2CH_3

100.0
m/z
231

100.0
m/z
245

900 1000 1100 1200 1300 1400 1500

SCAN NUMBER

Figure 7.6 Relative distributions (normalized) of triaromatic steroid hydrocarbons in pyrolysate of cholesteryl chloride and selenium. Peak identities as in Figure 7.5.

most prominent compound in the *m/z* 217 series of the 1173-3 oil (Fig. 7.5a) is identified as a $C_{18}H_{16}$ compound, Diels' hydrocarbon [1,2-(3′-methylcyclopenteno) phenanthrene] by comparison of its mass spectrum and retention time with that of the major product derived from laboratory pyrolysis of cholesteryl chloride and selenium (Fig. 7.6). Other minor components in this *m/z* 217 series of oil 1173-3 are a second isomer of the C_{18} compound (methyl position on cyclopenteno-ring unknown), a homologous $C_{19}H_{18}$ compound, and 1,2-cyclopentenophenanthrene, $C_{17}H_{14}$. These components are also evident in the products of the laboratory synthesis (see Fig. 7.6). The proportion of Diels' and these aforementioned hydrocarbons relative to all the triaromatic steroid hydrocarbons is greater in oil 1172-4 than in oil 1173-3 (cf. Figs. 7.5a and b). This former oil is more mature in terms of its sterane and hopane distributions, which suggests that thermal maturation may lead to an increased generation of Diels' hydrocarbon. The formation of the triaromatic steroid hydrocarbons at the high temperature encountered during hydrothermal pyrolysis appears to proceed with a preferential loss of all nuclear methyls. Such increased demethylation during aromatization at high temperatures was previously suggested as an explanation for changes in the relative abundances of the *m/z* 245 and *m/z* 231 series with increasing burial depth in the Paris Basin (Mackenzie et al., 1981a).

The generation of Diels' hydrocarbon in large quantities during hydrous pyrolysis has been reported (Rowland et al., 1986), wherein its proportion increased with increasing pyrolysis temperature at the expense of the other triaromatic steroid hydrocarbons. Our own hydrous pyrolysis of immature seabed sediments from Guaymas Basin (see previous discussion, Simoneit, 1989) also generated Diels' hydrocarbon in an amount similar in quantity to the other triaromatic steroid hydrocarbons (*m/z* 231), even though the hopane and sterane maturities of the pyrolysate were less than most of the hydrothermal oils.

Another significant different between the crude and hydrothermal oils (cf. Figs. 7.4 and 7.5) is the presence in the latter of an unusual series of short-alkyl-chain triaromatic steroid hydrocarbons of the *m/z* 231 type, which predominate over the 20- and 21-carbon (marked as 20 and 21 in Figs. 7.4 and 7.5) compounds normally found in oils. These unusual compounds, indicated as C_{19} and C_{20} in Figs. 7.4 and 7.5, have mass spectra which are similar to the other compounds with base peak *m/z* 231. Most likely, they differ structurally in the position of the nuclear methyl group, perhaps on ring-A or ring-C rather than at position C-3′ on the cyclopenteno-ring (C-17 on the steroid skeleton). The GC elution pattern of the multiple isomers of the unknown C_{19} compound and one isomer of the unknown C_{20} compound in the *m/z* 231 series is strikingly similar to that of the short-alkyl-chain compounds in the *m/z* 217 series (see Fig. 7.5). This suggests that the C_{17}, C_{18} (Diels' hydrocarbon), and C_{19} in the *m/z* 217 series and the unusual group of C_{19} and C_{20} compounds in the *m/z* 231 series may be related structurally and were possibly generated in a similar manner.

The high proportion of Diels' hydrocarbon in hydrothermal oils generated at high temperatures and in laboratory hydrous pyrolysates leads us to suggest that it may be a good indicator of high-temperature processes. On the other hand, the low concentration of Diels' hydrocarbon in hydrothermal petroleum from Escanaba Trough (Kvenvolden and Simoneit, 1989) may indicate that the organic character of the source material is also important. This C_{18} compound and the other apparently related C_{19} and C_{20} compounds of the *m/z* 217 and *m/z* 231 series should be monitored in oils and sediments to assess their significance and possible utility as indicators of high-temperature catagenesis.

CONCLUSION

The biomarker distribution in the hydrothermal oils of Guaymas Basin reflect the high temperatures required for generation of petroleum in zero geological time. The initially immature biomarker signatures (primarily triterpenes, triterpanes, and sterenes in their biologically derived configurations) are rapidly converted to distributions found in mature sediments and oils.

Certain trends in the hydrothermal maturation of biomarkers, though, are more similar to those observed during laboratory simulations and may, therefore, be characteristic of high-temperature processes. Under hydrothermal conditions, the aromatization of the monoaromatic to triaromatic steroids is enhanced relative to the sterane isomerization. This result is as expected, for kinetic models of the two reactions have indicated that the aromatization reaction is more temperature-dependent. The hopane isomerization is also much more rapid (near equilibrium in most of the oils) than that of the steranes. This trend is most comparable to laboratory simulations of maturation and supports the suggestion that the actual mechanism of the hopane isomerization may depend on the thermal regime.

A preliminary laboratory simulation (330°C hydrous pyrolysis) of Guaymas Basin sediment alteration indicates that the products are intermediate in terms of maturity between source sediment and hydrothermal petroleum. The duration of heating was too brief to convert all precursors to their mature end products. The major simulation products were the same as in the hydrothermal oils. However, the biomarkers were not as mature and heavy PAH had not yet formed.

Comparison of the triaromatic steroid hydrocarbon distributions in the hydrothermal oils, crude oils, and laboratory pyrolysates results in the identification of a number of compounds, including Diels' hydrocarbon ($C_{18}H_{16}$), which appear to be characteristic of rapid thermal maturation. If their occurrence is independent of source, these compounds may be useful tracers of such high-temperature processes.

Acknowledgments

We thank the crews and pilots of the D.S.V. *Alvin*, R.V. *Lulu*, and R.V. *Atlantis II* for their skillful recoveries of hydrothermal samples. Support for this research from the National Science Foundation, Division of Ocean Sciences (Grants

OCE-8312036, OCE-8512832, and OCE-8601316), and partial support by the Donors of the Petroleum Research Fund', administered by the American Chemical Society, are gratefully acknowledged.

REFERENCES

ABBOTT, G.D., LEWIS, C.A., and MAXWELL, J.R. (1984) Laboratory simulation studies of steroid aromatisation and alkane isomerisation. *Org. Geochem. 6*, 31–38.

ABBOTT, G.D., LEWIS, C.A., and MAXWELL, J.R. (1985) Laboratory models for aromatization and isomerization in sedimentary basins. *Nature 318*, 651–653.

CURRAY, J.R., MOORE, D.G. et al. (1982) *Initial Reports of the Deep Sea Drilling Project, 64, Parts I and II*, U.S. Government Printing Office, Washington, D.C., 1314 p.

DIELS, O. and KARSTENS, A. (1930) Über Dehydrierungen mit Selen. *Liebigs Ann. Chem. 478*, 129–137.

DIELS, O. and RICKERT, H.F. (1935) Über den Identitäts-Nachweis des Dehydrierungs-Kohlenwasserstoffes $C_{18}H_{16}$ aus Sterinen und Geninen mit γ-Methyl-cyclopentenophenanthren. *Ber. Chem. Ges. 68*, 267–272.

DIELS, O., GÄDKE, W., and KÖRDING, P. (1927) Über die Dehydrierung des Cholesterins. *Liebigs Ann. Chem. 459*, 1–26.

ENSMINGER, A., VAN DORSSELAER, A., SPYCKERELLE, C., ALBRECHT, P., and OURISSON, G. (1974) Pentacyclic triterpanes of the hopane type as ubiquitous geochemical markers: Origin and significance. In: *Advances in Organic Geochemistry 1973* (eds. B. Tissot and F. Bienner), Editions Technip, Paris. pp. 245–260.

ENSMINGER, A., ALBRECHT, P., OURISSON, G., and TISSOT, B. (1977) Evolution of polycyclic alkanes under the effect of burial (early Toarcian shales, Paris Basin). In *Advances in Organic Geochemistry 1975* (eds. R. Campos and J. Goni), Enadimsa, Madrid. pp. 45–52.

HARPER, S.H., KON, G.A.R., and RUZICKA, F.C.J. (1934) Synthesis of polycyclic compounds related to the sterols. Part II. Diels's hydrocarbon $C_{18}H_{16}$. *J. Chem. Soc. (Lond.) 1934*, 124–128.

HUNT, J.M. (1979) *Petroleum Geochemistry and Geology*, W.H. Freeman and Company, San Francisco, 617p.

KAWKA, O.E. and SIMONEIT, B.R.T. (1987) Survey of hydrothermally generated petroleums from the Guaymas Basin spreading center. *Org. Geochem. 11*, 311–328.

KVENVOLDEN, K.A. and SIMONEIT, B.R.T. (1989) Hydrothermally derived petroleum: Examples from Guaymas Basin, Gulf of California and Escanaba Trough, Northeast Pacific Ocean. *Amer. Assoc. Petrol. Geol. Bull.*, in press.

LEWAN, M.D., BJORØY, M., and DOLCATER, D.L. (1986) Effects of thermal maturation on steroid hydrocarbons as determined by hydrous pyrolysis of Phosphoria Retort Shale. *Geochim. Cosmochim. Acta 50*, 1977–1987.

MACKENZIE, A.S. and MAXWELL, J.R. (1981) Assessment of thermal maturation in sedimentary rocks by molecular measurements. In *Organic Maturation Studies and Fossil Fuel Exploration* (ed. J. Brooks), Academic Press, London. pp. 239–254.

MACKENZIE, A.S. and MCKENZIE, D. (1983) Isomerisation and aromatisation of hydrocarbons in sedimentary basins formed by extension. *Geological Magazine 120*, 417–470.

MACKENZIE, A.S., PATIENCE, R.L., MAXWELL, J.R., VANDENBROUCKE, M., and DURAND, B. (1980) Molecular parameters of maturation in the Toarcian shales, Paris Basin, France I. Changes in the configurations of acyclic isoprenoid alkanes, steranes and triterpanes. *Geochim. Cosmochim. Acta 44*, 1709–1721.

MACKENZIE, A.S., HOFFMANN, C.F., and MAXWELL, J.R. (1981a) Molecular parameters of maturation in the Toarcian shales, Paris Basin, France III. Changes in aromatic steroid hydrocarbons. *Geochim. Cosmochim. Acta 45*, 1345–1355.

MACKENZIE, A.S., LEWIS, C.A., and MAXWELL, J.R. (1981b) Molecular parameters of maturation in the Toarcian shales, Paris Basin, France IV. Laboratory thermal alteration studies. *Geochim. Cosmochim. Acta 45*, 2369–2376.

MACKENZIE, A.S., LAMB, N.A., and MAXWELL, J.R. (1982) Steroid hydrocarbons and the thermal history of sediments. *Nature 295*, 223–226.

MACKENZIE, A.S., BRASSELL, S.C., EGLINTON, G., and MAXWELL, J.R. (1982) Chemical fossils: The geological fate of steroids. *Science 217*, 491–504.

MACKENZIE, A.S., BEAUMONT, C., and MCKENZIE, D.P. (1984) Estimation of the kinetics of geochemical reactions with geophysical models of sedimentary basins and applications. In *Advances in Organic Geochemistry 1983* (eds. P.A. Schenck, J.W. de Leeuw, and G.W.M. Lijmbach). *Org. Geochem. 6*, 875–884.

MCKENZIE, D., MACKENZIE, A.S., MAXWELL, J.R., and SAJGÓ, CS. (1983) Isomerisation and aromatisation of hydrocarbons in stretched sedimentary basins. *Nature 301*, 504–506.

PETER, J.M., PELTONEN, P., SCOTT, S.D., SIMONEIT, B.R.T., and KAWKA, O.E. (1991) ^{14}C Ages of hydrothermal petroleum and carbonate in Guaymas Basin, Gulf of California: implications for oil generation, expulsion and migration. *Geology 19*, 253–256.

ROWLAND, S.J., AARESKJOLD, K., XUEMIN, G., and DOUGLAS, A.G. (1986) Hydrous pyrolysis of sediments: Composition and proportions of aromatic hydrocarbons in pyrolysates. *Org. Geochem. 10*, 1033–1040.

RULLKÖTTER, J. and MARZI, R. (1989) New aspects of the application of sterane isomerization and steroid aromatization to petroleum exploration and the reconstruction of geothermal histories of sedimentary basins. In *Prepr., Div. Petr. Chem., Amer. Chem. Soc. 34*, 126–131.

SAJGÓ, CS. and LEFLER, J. (1986) A reaction kinetic approach to the temperature-time history of sedimentary basins. In *Paleogeothermics. Lecture Notes in Earth Sciences Vol. 5* (eds. G. Buntebarth and C. Stegena), Springer-Verlag, Berlin. pp. 120–151.

SEIFERT, W.K. and MOLDOWAN, J.M. (1978) Applications of steranes, terpanes and monoaromatics to the maturation, migration and source of crude oils. *Geochim. Cosmochim. Acta 42*, 77–95.

SEIFERT, W.K. and MOLDOWAN, J.M. (1980) The effect of thermal stress on source rock quality as measured by hopane stereochemistry. In *Advances in Organic Geochemistry 1979* (eds. A.G. Douglas and J.R. Maxwell), Pergamon Press, Oxford. pp. 229–237.

SIMONEIT, B.R.T. (1983a) Organic matter maturation and petroleum genesis: Geothermal versus hydrothermal. In *The Role of Heat in the Development of Energy and Mineral Resources in the Northern Basin and Range Province.* Geotherm. Res. Council, Special Report No. 13, Davis, California, pp. 215–241.

SIMONEIT, B.R.T. (1983b) Effects of hydrothermal activity on sedimentary organic matter: Guaymas Basin, Gulf of California—Petroleum genesis and protokerogen degradation. In *Hydrothermal Processes at Seafloor Spreading Centers* (eds. P.A. Rona, K. Boström, L. Laubier, and K.L. Smith, Jr.), NATO-ARI Series, Plenum Press, New York. pp. 453–474.

SIMONEIT, B.R.T. (1984) Hydrothermal effects on organic matter—high versus low temperature components. In *Advances in Organic Geochemistry 1983* (eds. P.A. Schenck, J.W. de Leeuw, and G.W.M. Lijmbach). *Org. Geochem. 6*, 857–864.

SIMONEIT, B.R.T. (1985) Hydrothermal petroleum: Genesis, migration and deposition in Guaymas Basin, Gulf of California. *Can. J. Earth Sci. 22*, 1919–1929.

SIMONEIT, B.R.T. (1989) Natural hydrous pyrolysis–Petroleum generation in submarine hydrothermal systems. In *Productivity, Accumulation and Preservation of Organic Matter in Recent and Ancient Sediments* (eds. J.K. Whelan and J.W. Farrington). Columbia University Press, submitted.

SIMONEIT, B.R.T. and LONSDALE, P.F. (1982) Hydrothermal petroleum in mineralized mounds at the seabed of Guaymas Basin. *Nature 295*, 198–202.

SIMONEIT, B.R.T., MAZUREK, M.A., BRENNER, S., CRISP, P.T., and KAPLAN, I.R. (1979) Organic geochemistry of recent sediments from Guaymas Basin, Gulf of California. *Deep-Sea Res. 26A*, 879–891.

SIMONEIT, B.R.T., PHILP, R.P., JENDEN, P.D., and GALIMOV, E.M. (1984) Organic geochemistry of Deep Sea Drilling Project sediments from the Gulf of California—Hydrothermal effects on unconsolidated diatom ooze. *Org. Geochem. 7*, 173–205.

SUZUKI, N. (1984) Estimation of maximum temperature of mudstone by two kinetic parameters; epimerization of sterane and hopane. *Geochim. Cosmochim. Acta 48*, 2273–2282.

TISSOT, B.P. and WELTE, D.H. (1984) *Petroleum Formation and Occurrence: A New Approach to Oil and Gas Exploration*, p. 699. Springer Verlag, Berlin.

VAN DORSSELAER, A., ALBRECHT, P., and OURISSON, G. (1977) Identification of novel 17α(H)-hopanes in shales, coals, lignites, sediments and petroleum. *Bull. Soc. Chim. France*, 165–170.

VAN GRAAS, G., BAAS, J.M.A., VAN DER GRAAF, B., and DE LEEUW, J.W. (1982) Theoretical organic geochemistry. I. The thermodynamic stability of several cholestane isomers calculated from molecular dynamics. *Geochim. Cosmochim. Acta 46*, 2399–2402.

ZUMBERGE, J.E., PALMER, S.E., and SCHIEFELBEIN, C.J. (1984) Kinetics of sterane and hopane epimerization: Laboratory simulation by hydrous pyrolysis. In *Abstracts with Programs, GAS 97th Annual Meeting*, Vol. 16, No. 6, p. 706.

8

Hydrocarbon Biological Markers in Carboniferous Coals of Different Maturities

Hans Lodewijk ten Haven, Ralf Littke, and Jürgen Rullkötter

Abstract. A great variety of biological markers have been found in Carboniferous coal samples covering a vitrinite reflectance range from 0.65 to 0.92 percent R_o. Changes in the palaeodepositional environment are reflected by the distribution of bacterial hydrocarbons (hopanoids and C_{32}–C_{40} head-to-head- and head-to-tail linked isoprenoids). These biological markers make a significant contribution (5 to 10%) to the aliphatic hydrocarbon fraction of the low-maturity coal samples. With increasing maturity, the concentration of hopanoids decreases, a decrease that is most pronounced for the 17β(H),21α(H)-hopanes. These latter compounds do not seem to be converted to their 17α(H),21β(H)-counterparts. Preliminary results indicate that the concentration of 22,29,30-trisnor-17α(H)-hopane decreases more rapidly than that of 17α(H),21β(H)-hopane.

Biological markers for gymnosperms (C_{20} tetracyclic diterpenoids) were observed in all samples, but are most pronounced in the low-maturity range. The occurrence of these compounds is in accordance with the phylogenetic evolution of the plant kingdom during the Carboniferous and make these samples amongst the oldest so far in which these biological markers have been detected.

INTRODUCTION

The elucidation of structures of biological markers and the investigation of their distribution patterns in geological samples is a basic goal of organic geochemistry. Numerous organic geochemical studies have been dedicated to this objective, however, only a few have dealt with coal samples. Coal deposits and their precursors (peat, lignite) occur over a wide range of geological time, and are, therefore, suitable for monitoring, by specific biological markers, the phylogenetic evolution of the plant kingdom.

Terrigenous triterpenoids (amyrins and related compounds) have previously been observed in high abundance in coal samples (Hollerbach, 1980; Hoffmann et al., 1984), but their occurrence is related, and therefore restricted, to the development of angiosperms in the early Cretaceous. Recently, it has been proposed by Strachan et al. (1988) that 1,2,7-trimethylnaphthalene is a specific degradation product of amyrin-type triterpenoids, and, hence its occurrence in geological time is also restricted. The tentative identification of oleanane in Jurassic samples from the Beacon Supergroup, Antarctica (Matsumoto et al., 1987), points to a pre-Cretaceous development of angiosperms, as recently suggested by Martin et al. (1989).

Potential biological markers for gymnosperms include tetracyclic diterpenoids, such as ent-beyerane, phyllocladanes and entkauranes (Noble et al., 1985). It has been noted that with increasing maturation, the 16α(H)-isomer of phyllocladane is apparently converted to the thermodynamically more stable 16β(H)-isomer (Noble et al., 1985; Alexander et al., 1987). The first occurrence of gymnosperms in the geological record is known to be during the end of the Carboniferous and, more specifically, there is an increase in their abundance during Westphalian time (Phillips and Peppers, 1984). Recently, it has been suggested that primitive gymnosperms might have already developed in the early Carboniferous (Galtier and Rowe, 1989).

Triterpenoids with a hopanoid skeleton, which have no age relationship, occur ubiquitously in the geosphere (Ourisson et al., 1979), and have also been found in coals of various ranks and ages (e.g., Ourisson et al., 1979; Hollerbach, 1980; Hoffmann et al., 1984; Hazai et al., 1988). They can represent a major portion of the total organic extract (Ourisson et al., 1987).

Carboniferous coal-bearing strata in northwest Germany first occur in the upper Namurian (≈315 ma) and extend to the Stephanian (≈295 ma). Here we report on the identification and distribution of hydrocarbon biological markers from four coal seams of Westphalian B age (≈305 ma), which bear molecular evidence for the presence of gymnosperms. For comparative purposes, we have also analyzed one coal sample of Pennsylvanian age (equivalent to Westphalian; Phillips and Peppers, 1984) from the Midcontinent, in the United States. Both sampling areas were located just north of the equator at the time of coal deposition according to continental reconstruction maps of the late Carboniferous.

EXPERIMENTAL

Ten samples from four coal seams from borehole Nesberg 1, located in the Ruhr area (FRG), and one sample from borehole Kelly No. 1, located in northeast Oklahoma,, Midcontinent (USA) were selected for this study. Geological background information can be found elsewhere (Littke and ten Haven, 1989; Wenger and Baker, 1986). Sample preparation and procedures for organic petrography, solvent extraction, and compound class separation of the extract, as well as instrumental conditions for microscopic and molecular analyses, have been described previously (Littke and ten Haven, 1989; ten Haven and Rullkötter, 1988). Prior to extraction, a known amount of an internal standard (squalane) was added to the samples in order to calculate the absolute concentration of selected compounds. Because of coelution of 30-nor-17α(H),21β(H)-hopane with tricosane, the quantification of hopanoids was carried out after removal of the *n*-alkanes by 5Å molecular sieves. Concentrations of pristane, phytane and 22,29,30-trisnor-17α(H)-hopane calculated before and after molecular sieve treatment revealed that the results obtained from the nonadducted hydrocarbon fraction are consistent with those from the total hydrocarbon fraction. Metastable ion monitoring ($M^{+\cdot} \rightarrow 217$) was applied to confirm the presence of steranes, and, because of their very low concentration, to calculate maturity parameters based on these compounds. Selected ions (*m*/*z* 191, 205, 274) were recorded to obtain information about distribution patterns of hopanoids, methylated hopanoids, and tetracyclic diterpenoids. Coinjection of 16α(H)-phyllocladane was performed to confirm its presence in our samples.

RESULTS AND DISCUSSION

Background information on the samples, including vitrinite reflectance and maceral composition, is given in Table 8.1. The ten samples from the Ruhr area, which are representative of a large sample suite (Littke and ten Haven, 1989), can be subdivided into four groups on the basis of maceral composition, viz. "average coal" (22360, 22791), sporinite-rich coal (22358, 22383, 22391), vitrinite-rich coal (22795, 22796, 22797), and mudstone partings (22359, 22387). The differences in maceral composition are related to differences in the palaeodepositional environment (Littke and ten Haven, 1989); vitrinite-rich coals are representative of depositional conditions under which organic-matter preservation was very good, whereas sporinite-rich coals (with high contents of inertinites) were deposited under more oxic conditions. Mudstone partings are fluvial deposits covering former peat swamps.

All Ruhr area samples contain the same classes of aliphatic hydrocarbons, of

Table 8.1 DEPTH, TOTAL ORGANIC CARBON CONTENT, MEAN RANDOM VITRINITE REFLECTANCE (R_o) AND MACERAL COMPOSITION OF COAL SEAM SAMPLES

Sample	Type[a]	Depth (m)	TOC (%)	R_o (%)	Maceral composition			
					Vitr.	Iner.	Lipt.[b]	Min.
Kelly No. 1, Midcontinent, U.S.A.								
21229	B	67.50[c]	78.9[c]	0.65	88	6	3	3
Nesberg 1, Ruhr area, FRG								
22358	C	1198.32	76.4	0.70	33	40	27	0
22795	B	1198.35	76.0	0.70	82	17	1	0
22359	D	1198.50	23.1	0.70	21	9	10	60
22360	A	1198.72	73.1	0.70	39	48	8	5
22796	B	1382.08	75.1	0.78	88	11	1	0
22383	C	1429.04	81.4	0.81	63	18	19	0
22797	B	1429.14	70.4	0.81	100	0	0	0
22791	A	1429.33	77.1	0.81	75	14	8	3
22387	D	1429.50	20.3	0.81	17	6	7	70
22391	C	1529.55	86.2	0.92	56	23	21	0

[a]A = average coal, B = vitrinite-rich, C = sporinite-rich, and D = mudstone parting.

[b]The most important maceral of the liptinites is sporinite.

[c]Data from Wenger (1987).

which the major ones, listed in order of decreasing abundance, are: *n*-alkanes ≥ C_{16}, C_{18}–C_{20} isoprenoids > C_{27}–C_{35} hopanes > C_{32}–C_{40} head-to-head linked isoprenoids > C_{32}–C_{40} head-to-tail linked isoprenoids. However, variations in the relative abundances of these compounds were observed. Present in minor quantities and in more or less equal amounts, are the following classes of compounds: isoprenoids (excluding those listed), C_{24}–C_{27} 17,21-secohopanes, C_{28}–C_{35} A-ring methylated hopanes, C_{27}–C_{29} steranes (including diasteranes), C_{15}–C_{17} bicyclic sesquiterpanes including 4β(H)-eudesmane and 8β(H)-drimane, C_{19}–C_{20} tricyclic terpanes, C_{19}–C_{20} tetracyclic diterpenoids, C_{22}–C_{26} midchain methylalkanes and C_{14}–C_{31} alkylcyclohexanes. The 20*S*/(20*R* + 20*S*) isomerisation ratio of C_{29} steranes, a thermal maturity parameter, varies between 0.46 and 0.50 for all samples, whereas the ratio of ββ/(αα + ββ) C_{29} steranes has a wider range, between 0.5 for the least mature Ruhr basin samples and 0.7 for the most mature samples.

The sample from the U.S. Midcontinent also contains most of these compound classes, but the short-chain isoprenoids (C_{15}, C_{16}, C_{18}–C_{20}) are by far the most abundant compounds (cf. Wenger and Baker, 1986; Wenger, 1987). The significance of a few selected classes of compounds will be discussed hereafter.

Hopane Stereochemistry

Examples of gas chromatograms of two "average coal" (cf. Table 8.1) extracts at maturity levels of 0.70 and 0.81 percent R_o are shown in Figure 8.1. *n*-Alkanes are very abundant in both samples, but the concentration of the C_{15}–C_{29} alkanes is much higher in the more mature coal (≈900 μg/g C_{org}) than in the less mature coal (≈300 μg/g C_{org}). More details concerning the absolute concentration of alkanes and short chain isoprenoids are published elsewhere (Littke et al., 1990). The relative abundance of hopanes decreases with increasing maturity. The concentration of selected triterpenoid hydrocarbons is given in Table 8.2. From these data it can be seen that the absolute concentration of all hopanoids decreases with increasing maturity (cf. Rullkötter et al., 1984). The concentrations of the 17β(H),21α(H)-hopanes (moretanes) are reduced preferentially to an extent such that they could not be quantified reliably in the more mature samples; this indicates that these compounds are less stable than their 17α(H),21β(H)-counterparts. Although this observation is well known (e.g., Seifert and Moldowan, 1980), it is difficult to judge from these data whether the moretanes are actually converted to 17α(H),21β(H)-hopanes, or are thermally destroyed by cleavage of the ring system, or other processes. Average absolute concentrations of individual hopanoids for the two coal seams with vitrinite reflectance values of 0.70 and 0.81 percent, respectively, are presented in Figure 8.2 (the same trends occur when individual coal facies, e.g., vitrinite-rich coals, are compared with each other). These concentrations are normalized to the most abundant hopanoid (30-nor-17α(H),21β(H)-hopane). It is obvious again that the decrease of 17β(H),21α(H)-hopane is the strongest. Among the other compounds the decrease in abundance is more pro-

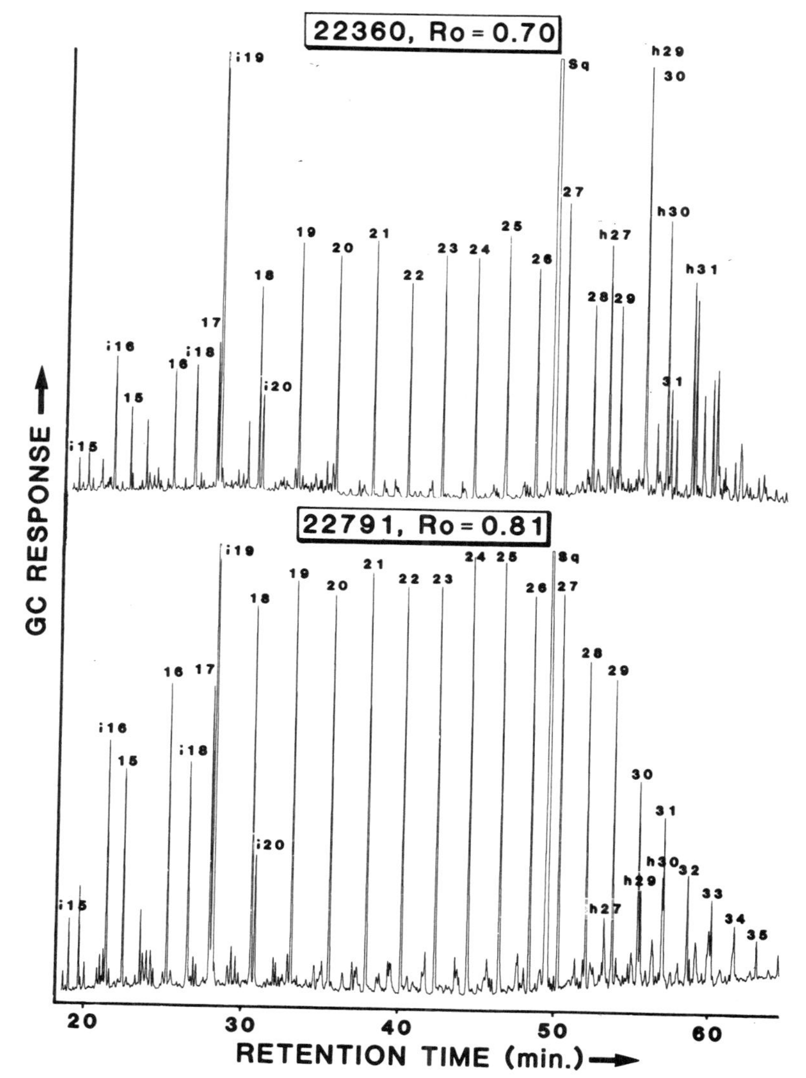

Figure 8.1 Gas chromatograms of the aliphatic hydrocarbon fraction of "average coal" samples at maturity levels 0.70 and 0.81 R_o. *n*-Alkanes are indicated by their carbon number, isoprenoids with the letter *i* followed by their carbon number, and hopanes by the letter *h* followed by their carbon number. Sq = squalane (internal standard).

Table 8.2 EXTRACT YIELD AND CONCENTRATION OF SELECTED HYDROCARBONS[a] IN COAL SAMPLES FROM THE RUHR AREA

		Extract	C15+ St.hyd.	17α(H)-22,29,30-trisnorhop.			17α(H),21β(H)-30-norhopane			17α(H),21β(H)-hopane			17β(H),21α(H)-hopane		
Sample	Type	ppm	ppm	ppm	% of C15+	μg/gCorg	ppm	% of C15+	μg/gCorg	ppm	% of C15+	μg/gCorg	ppm	% of C15+	μg/gCorg
22358	C	47620	3238	20.1	0.62	26	24.0	0.74	32	20.0	0.62	26	6.8	0.21	9
22795	B	59160	591	7.6	1.29	10	7.8	1.32	10	5.9	1.00	8	2.0	0.34	3
22359	D	21700	369	5.8	1.57	25	8.3	2.25	36	5.2	1.38	22	2.0	0.54	9
22360	A	55080	1269	18.6	1.47	25	23.6	1.86	32	18.8	1.47	26	5.7	0.45	8
22796	B	81026	2997	11.2	0.37	15	18.8	0.62	25	25.2	0.84	34		n.d.	
22383	C	42600	4430	8.5	0.19	10	11.2	0.25	14	16.2	0.37	20		n.d.	
22797	B	40385	2302	6.0	0.26	8	7.1	0.31	10	8.6	0.37	12		n.d.	
22791	A	37800	2457	8.5	0.34	11	12.8	0.52	17	12.8	0.52	17		n.d.	
22387	D	8600	481	0.7	0.14	3	1.1	0.22	5	1.0	0.22	5		n.d.	
22391	C	34200	2701	5.6	0.21	7	6.5	0.24	8	9.6	0.35	11		n.d.	

[a]Calculated after removal of *n*-alkanes.

[b]A = average coal, B = vitrinite-rich, C = sporinite-rich, and D = mudstone parting.

n.d., not determined due to very low concentrations.

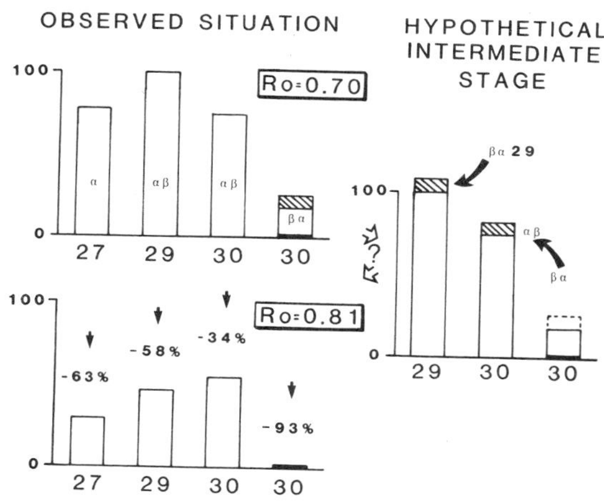

Figure 8.2 Bar graphs showing the distribution of selected hopanoids in two coal seams at the maturity levels 0.70 and 0.81 R_o (left). Concentrations were calculated by averaging the four samples of each seam (as μg/g C_{org}; Table 8.2) and then normalized to the highest component. On the right a hypothetical intermediate stage is presented assuming a conversion of C_{29} and C_{30} moretanes to hopanes.

nounced for 22,29,30-trisnor-17α(H)-hopane (63%) than for 17α(H),21β(H)-hopane (34%). The 30-nor-17α(H),21β(H)-hopane shows an intermediate decrease (58%). In the same figure, a hypothetical situation is given indicating how absolute concentrations should change if 17β(H),21α(H)-hopanes are transformed to 17α(H),21β(H)-hopanes. In this case, one would expect that the absolute concentrations of 17α(H),21β(H)-hopanes increase during a certain maturation stage, a situation which is not observed in our sample suite.

The reaction mechanism (if indeed this reaction proceeds at all) for the (stepwise?) interconversion of 17β(H),21α(H)-hopanes to 17α(H),21β(H)-hopanes (changes required at two asymmetric centers) should be similar to that of the interconversion of 17β(H),21β(H)-hopanes to their 17α(H), 21β(H) and/or 17β(H), 21α(H) counterparts. On the other hand it has recently been suggested that 17β(H),21β(H)-hopanes, based on a simulation experiment, do not convert to 17α(H),21β(H)- and/or 17β(H),21α(H)-hopanes, as previously thought, but that 17β(H),21β(H)-hopanes are quantitatively lost (Larcher et al., 1988). This experiment involved heating of Rundle oil shale alkanes, containing 17β(H),21β(H)-, 17α(H),21β(H)-, and 17β(H),21α(H)-hopanes, in the presence of aluminium montmorillonite. In contrast to the loss of the 17β(H),21β(H)-hopanes, the relative concentrations of the other hopanes remained unchanged. If this experiment is representative of diagenetic/catagenetic processes taking place in the geosphere, then it is unlikely that a transformation of 17β(H),21α(H)- to 17α(H),21β(H)-hopanes has occurred in the Carboniferous coals studied here. In fact, this may indicate that this reaction does not happen at all in sediments and that changes in relative abundance are a consequence of a progressive loss of the more labile isomer(s).

The differences in the experimentally determined decomposition rates of the 17β(H)-C_{27}, relative to the 17β(H),21β(H)-C_{29} or C_{30} hopanes (Larcher et al., 1988) showed a trend opposite to what we have observed for the 17α(H)-members, that is, the decrease in abundance was most pronounced for the C_{30} compound. A similar observation has been made by Monthioux and Landais (1989) for an artificially matured coal. Larcher et al. (1988) attributed the lack of reactivity of the 22,29,30-trisnor-17β(H)-hopane to the absence of a side chain at C-21, making C-21 secondary instead of tertiary, and to hindered E-ring opening reactions via β-cleavages because carbocation formation at C-22 is not possible. These arguments should be independent of the configuration at C-17, and would, therefore, also hold for 22,29,30-trisnor-17α(H)-hopane. Hence, our data are different from the experimentally derived data of Larcher et al. (1988). The change in configuration at C-17 may indeed be an important factor, although it is unclear at the moment why this should be.

Palaeoenvironmental Assessment

The effect of different palaeodepositional environments on the distribution of biological markers from samples of the same maturity is most striking for a sporinite-rich and a vitrinite-rich coal, both with R_o = 0.70 percent. Partial reconstructed total ion chromatograms of these two samples (after removal of the *n*-alkanes by 5Å molecular sieves) are given in Figure 8.3. The distribution of the bacterial hopanes (Ourisson et al., 1979) is almost identical in both samples, but the A-ring methylated hopanes (2- or 3-methyl) are enriched in relative concentration in the sporinite-rich sample, whereas the head-to-head linked isoprenoids are very abundant in the vitrinite-rich sample. Both classes of compounds are thought to be of bacterial origin and are found in sediments of varying maturities and crude oils (Ourisson et al., 1979; Moldowan and Seifert, 1979). The head-to-head linked isoprenoids have been proposed to be biological markers for strictly anaerobic methanogens (Ourisson et al., 1982). We relate the observed differences in the two coal samples to different bacterial communities, which thrived in the respective palaeodepositional environments, that is, anoxic conditions during the deposition of the precursor material of the vitrinite-rich coals and more oxic conditions during other stages of peak formation. As stated, the petrographic results indicate a very good preservation during deposition of "vitrinite-rich coals" (Littke and ten Haven, 1989) and, based on the relative high abundance of head-to-head linked isoprenoids in these samples, the good preservation can be related to anoxic conditions.

A comparison of an "average coal" sample of the Ruhr basin with the coal sample from the Midcontinent shows differences in the relative abundances of the short-chain isoprenoids and the hopanoids, but, apart from these concentration differences, the samples are much alike (Fig. 8.4; Table 8.3).

Molecular Stratigraphy

All samples contain a series of tetracyclic diterpenoids in minor amounts (see inset, Fig. 8.4). Several of these compounds have mass spectral fragmentation patterns similar to those of the recently discovered C_{19} tetracyclic diterpenoids

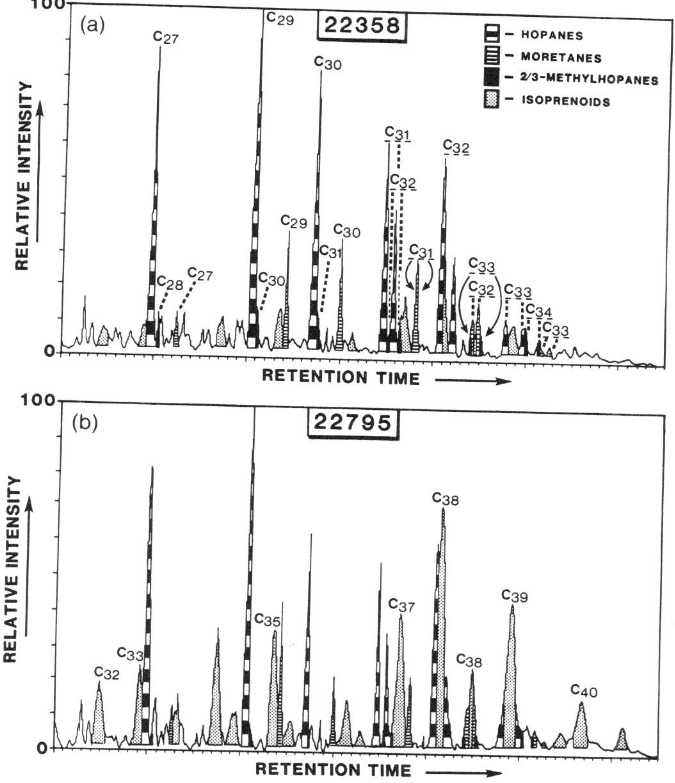

Figure 8.3 Partial reconstructed ion chromatograms of nonadducted hydrocarbon fractions of (a) a sporinite-rich sample (22358) and (b) a vitrinite-rich sample (22795). 17α(H),21β(H)-hopanes, 17β(H),21α(H)-hopanes (= moretanes), and A-ring methylated hopanes are indicated by the number of carbon atoms in the skeleton in (a), while the numbers in (b) denote the carbon skeleton number of the head-to-head linked isoprenoids. The nonlabeled isoprenoids are regular (i.e., head-to-tail).

Table 8.3 COMPOUND CLASSES IN CARBONIFEROUS COALS (Fig. 8.4)

a. isoprenoids (regular)	f. secohopanes
b. bicyclic sesquiterpanes	g. steranes
c. alkylcyclohexanes	h. hopanes
d. tricyclic terpanes	i. moretanes
e. C_{20}-tetracyclic diterpenoids	j. isoprenoids (head-to-head)

(Petrov et al., 1988). The distribution pattern of the C_{20} compounds, displayed by the *m/z* 274 mass chromatogram is shown in Figure 8.5 for a sporinite-rich coal (R_o = 0.70%). The following diterpenoids were identified: ent-beyerane, 16β(H)-phyllocladane, 16α(H)-ent-kaurane, 16α(H)-phyllocladane (confirmed by coinjection with an authentic standard) and 16β(H)-ent-kaurane. The ratio of the 16β/(16β + 16α) phyllocladanes in the least mature sample from the Ruhr area has already reached its endpoint of the presumed equilibrium value (0.75–0.80; Alexander et al., 1987). These compounds are thought to be biological markers for

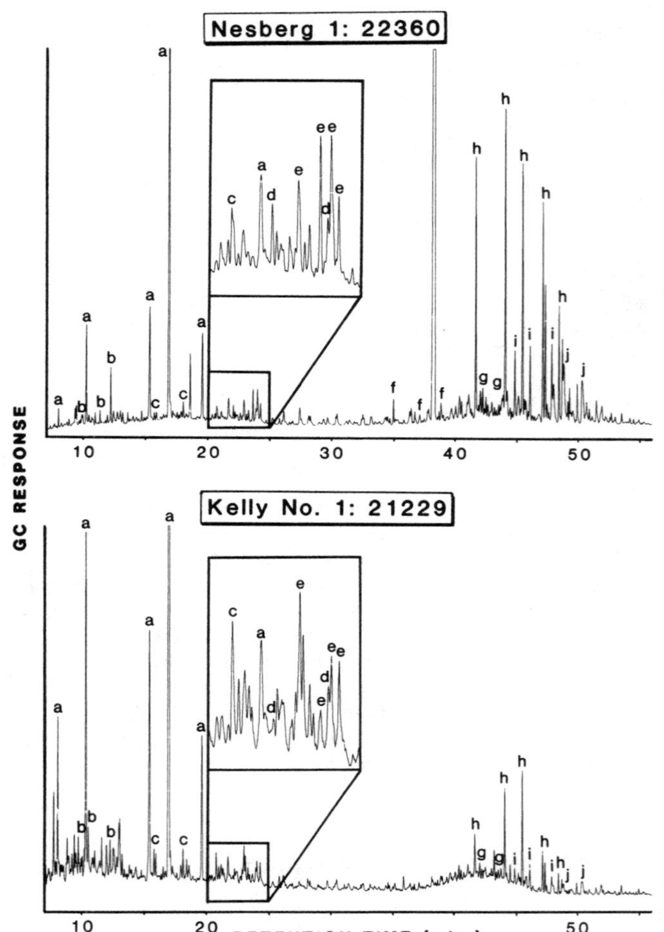

Figure 8.4 Gas chromatograms of the nonadducted aliphatic hydrocarbon fraction of an "average coal" sample from the Ruhr area and of the coal sample from the Midcontinent. Identification of letter labeled peaks is given in Table 8.3.

certain gymnosperms (Noble et al., 1985). The advent of gymnosperms near the end of the Carboniferous makes these samples among the oldest in which these compounds would be expected to occur. Only a single occurrence of phyllocladanes in a Carboniferous wood sample from Scotland has been reported before (Raymond et al., 1989).

CONCLUSIONS

The present study illustrates two aspects of biological markers in coals which will also be relevant to other organic matter types in sediments. The measured decrease in relative and absolute concentrations of hopanes with increasing maturity provides

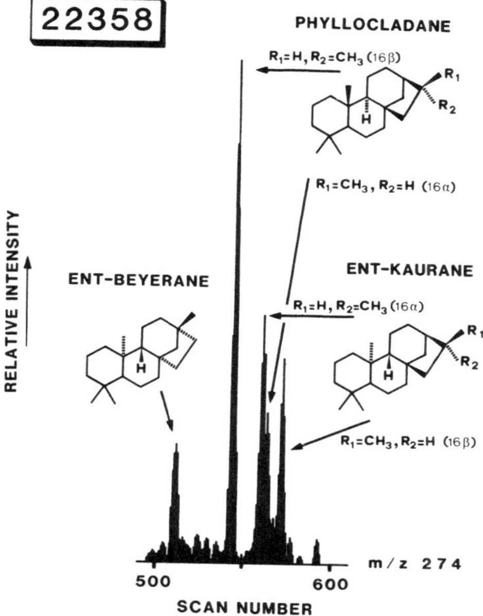

Figure 8.5 Mass chromatogram of *m/z* 274, showing the distribution of tetracyclic diterpenoids in a sporinite-rich sample (22358).

supporting evidence to the recent observations of several authors that the change in saturated hydrocarbon epimer ratios as a function of thermal stress is not due to isomerization but rather to the different kinetic constants of the thermal destruction reactions of the epimers involved.

In the palaeoenvironmental assessment, it was found that the relative concentration of head-to-head linked isoprenoids believed to be biosynthesized by strictly anaerobic microorganisms may be an indicator of the redox conditions during deposition. Phyllocladane may be used as molecular stratigraphic tool related to the advent of early higher land plants (gymnosperms).

Acknowledgments

We would like to thank Dr. D. Baker for providing us the Pennsylvanian coal sample and for useful comments, Dr. J. Grimalt for the authentic standard of 16α(H)-phyllocladane (courtesy of Dr. R. Alexander), Drs.. M. Radke and R. Schaefer for supervision of the analytical procedures, and Dr. T.M. Peakman and two anonymous reviewers for critically reading the manuscript. This research was supported by the Deutsche Forschungsgemeinschaft (grant no. We 346/27).

REFERENCES

ALEXANDER, G., HAZAI, I., GRIMALT, J., and ALBAIGES, J. (1987) Occurrence and transformation of phyllocladanes in brown coals from Nograd basin, Hungary. *Geochim. Cosmochim. Acta 51*, 2065–2073.

GALTIER, J. and ROWE, N.P. (1989) A primitive seed-like structure and its implications for early gymnosperm evolution. *Nature 340*, 225–227.

HAZAI, I., ALEXANDER, G., ESSIGER, B., and SZEKELY, T. (1988) Identification of aliphatic biological markers in brown coals. *Fuel 67*, 973–982.

HOFFMANN, C.F., MACKENZIE, A.S., LEWIS, C.A., MAXWELL, J.R., OUDIN, J.L., DURAND, B., and VANDENBROUCKE, M. (1984) A biological marker study of coals, shales and oils from the Mahakam delta, Kalimantan, Indonesia. *Chem. Geol. 42*, 1–23.

HOLLERBACH, A. (1980) Organische Substanzen biologischen Ursprungs in Erdölen und Kohlen. *Erdöl Kohle Erdgas Petrochem. 33*, 362–367.

LARCHER, A.V., ALEXANDER, R., and KAGI, R.I. (1988) Differences in reactivities of sedimentary hopane diastereomers when heated in the presence of clay. In *Advances in Organic Geochemistry 1987* (eds. L. Mattavelli and L. Novelli), *Org. Geochem. 13*, 665–669.

LITTKE, R. and TEN HAVEN, H.L. (1989) Palaeoecologic trends and petroleum potential of upper Carboniferous coal seams of western Germany as revealed by their petrographic and organic geochemical characteristics. *Int. J. Coal Geol. 13*, 529–574.

LITTKE, R., LEYTHAEUSER, D., RADKE, M., and SCHAEFER, R.G. (1990) Petroleum generation and migration in coal seams of the Carboniferous Ruhr Basin, northwest Germany. In *Advances in Organic Geochemistry 1989* (eds. B. Durand and F. Behar), *Org. Geochem. 16*, 247–258.

MARTIN, W., GIERL, A., and SAEDLER, H. (1989) Molecular evidence for pre-Cretaceous angiosperm origins. *Nature 339*, 46–48.

MATSUMOTO, G.I., MACHIHARA, T., SUZUKI, N., FUNAKI, M., and WATANKI, K. (1987) Steranes and triterpanes in the Beacon Supergroup samples from southern Victoria Land in Antarctica. *Geochim. Cosmochim. Acta 51*, 2668–2671.

MOLDOWAN, J.M. and SEIFERT, W.K. (1979) Head-to-head linked isoprenoid hydrocarbons in petroleum. *Science, 204*, 169–171.

MONTHIOUX, M. and LANDAIS, P. (1989) Natural and artificial maturation of coal: Hopanoid stereochemistry. *Chem Geol. 75*, 209–220.

NOBLE, R.A., ALEXANDER, R., KAGI, R.I., and KNOX, J. (1985) Tetracyclic diterpenoid hydrocarbons in some Australian coals, sediments and crude oils. *Geochim. Cosmochim. Acta 49*, 2141–2147 (see also Erratum, 1986, *Geochim. Cosmochim. Acta 50*, 489).

OURISSON, G., ALBRECHT, P., and ROHMER, M. (1979) The hopanoids: Palaeochemistry and biochemistry of a group of natural products. *Pure Appl. Chem. 51*, 709–729.

OURISSON, G., ALBRECHT, P., and ROHMER, M. (1982) Predictive microbial biochemistry—from molecular fossils to procaryotic membranes. *Trends Bioch. Sci. 7*, 236–239.

OURISSON, G., ROHMER, M., and PORALLA, K. (1987) Prokaryotic hopanoids and other polyterpenoid steroid surrogates. *Ann Rev. Microbiol. 41*, 301–333.

PETROV, A.A., PEHK, T.Y., VOROBIEVA, N.S., and ZEMSKOVA, Z.K. (1988) Identification of some novel tetracyclic diterpene hydrocarbons in petroleum. *Org. Geochem. 12*, 151–156.

Phillips, T.L. and Peppers, R.A. (1984) Changing patterns of Pennsylvanian coal-swamp vegetation and implications of climatic control on coal occurrence. *Int. J. Coal Geol. 3*, 205–255.

Raymond, A.C., Liu, S.Y., Murchinson, D.G., and Taylor, G.H. (1989) The influence of microbial degradation and volcanic activity on a Carboniferous wood. *Fuel 68*, 66–73.

Rullkötter, J., Mackenzie, A.S., Welte, D.H., Leythauser, D., and Radke, M. (1984) Quantitative gas chromatography-mass spectrometry analysis of geological samples. In *Advances in Organic Geochemistry 1983* (eds. P.A. Schenck, J.W. de Leeuw, G.W.M. Lijmbach), *Org. Geochem. 6*, 817–827.

Strachan, M.G., Alexander, R., and Kagi, R.I. (1988) Trimethyl naphthalenes in crude oils and sediments: effects of source and maturity. *Geochim. Cosmochim. Acta 52*, 1255–1264.

Seifert, W.K. and Moldowan, J.M. (1980) The effect of thermal stress on source-rock quality as measured by hopane stereochemistry. In *Advances in Organic Geochemistry 1979* (eds. A.G. Douglas and J.R. Maxwell), Pergamon, Oxford. pp. 229–237.

ten Haven, H.L. and Rullkötter, J. (1988) The diagenetic fate of taraxer-14-ene and oleanene isomers. *Geochim. Cosmochim. Acta 52*, 2543–2548.

Wenger, L.M. (1987) Variations in organic geochemistry of anoxic-oxic black shale-carbonate sequences in the Pennsylvanian of the Midcontinent, U.S.A. Ph.D. dissertation, Rice University, Houston.

Wenger, L.M. and Baker, D.R. (1986) Variations in organic geochemistry of anoxic-oxic black shale-carbonate sequences in the Pennsylvanian of the Midcontinent, U.S.A. In *Advances in Organic Geochemistry 1985* (eds. D. Leythaeuser and J. Rullkötter), *Org. Geochem. 10*, 85–92.

9

The Effect of Injection Hold Time on GC Resolution and the Effect of Collision Gas on Mass Spectra in Geochemical "Biomarker" Research

Emilio J. Gallegos and J. Michael Moldowan

Abstract. The peak resolution, Rs, of a capillary column can be increased by a factor of 2 or greater if the hold time, t_1, after injection is increased from a few minutes to several hours at the initial temperature (T_1) prior to programming the column to a final temperature (T_2).

$$Rs = \frac{t(a) - t(b)}{W(0.5a) + W(0.5b)}$$

where

t = retention time of components a or b
W(0.5) = peak width at half peak height

A demonstration using a sterane epimer mixture showed a near linear increase in resolution with increased hold times of 0, 16, and 48 hours. In these experiments $T_1 = 150°C$ and $T_2 = 300°C$, with a temperature programming rate of 2°C per minute. H_2 was used as the carrier gas in all the experiments. Up to 10 new sterane epimers were revealed using this technique.

Using a synthetic stigmastane isomerizate, a natural petroleum saturate fraction, and greatly extended GC hold times, it was shown that the peak for the 5α,14α,17α(H),20*S* C_{29}-sterane isomer is significantly contaminated with two coeluting C_{29}-sterane epimers under normal GC-MS operating conditions. Thus, all previously reported C_{29}-sterane 20*S*/20*R* ratios are probably too high.

The extended hold time GC separation technique was also demonstrated to improve resolution of mono- and triaromatic steroids. This technique combined with collision-activated daughter ion analysis on a triple quadrupole mass spectrometer provide a new dimension in aromatic steroid biomarker analysis.

A triple quadrupole mass spectrometer is an excellent device for analysis of "biomarker" samples. Several MS/MS experiments along with a full scan experiment may be done in a single GC-MS run. The switching time between experiments usually requires tens to hundreds of milliseconds, which precludes removal of collision gas prior to the full scan experiment. A comparison of normal full scan mass spectra and that of mass spectra obtained in the presence of collision gas are made for a number of C_{29}-sterane epimers. Generally, "collision gas" full scan mass spectra show more intense lower mass fragments compared to those of normal mass spectra. However, the diagnostic integrity of the mass spectra is not compromised.

INTRODUCTION

A triple quadrupole mass spectrometer coupled with a gas chromatograph is capable of doing several sequential MS/MS and full scan experiments. Applied to biomarker research, for example, the system can monitor all parent ions of *m/z* 217 and 191; look at daughters of *m/z* 372, *m/z* 386, and *m/z* 400; look for a neutral loss of an ethyl group; and record full mass spectra by scanning the first quadrupole (Q1MS) or the third quadrupole (Q3MS) (Philp et al., 1988). If all the experiments were run within a nominal time of 2 seconds, each experiment would be accomplished in slightly under 0.3 second. MS/MS experiments require that the collision gas be present. Hence, Q1MS and Q3MS full scan mass spectra must be obtained with the collision gas present since this gas cannot be pumped out fast enough between experiments. A comparison of these mass spectra and those obtained under normal conditions show that the diagnostic integrity of the mass spectra is retained.

Sterane mixtures in petroleums and rock extracts have been widely studicd because of their application to geochemical correlations in petroleum exploration. They are complex mixtures which cannot be completely resolved by GC/MS (Seifert and Moldowan, 1979). GC-MS/MS improves their analysis but additional unresolved coeluting compounds remain (Warburton and Zumberge, 1983). We discovered that increasing the hold time in a temperature-programmed GC run enhances the separation of certain steranes. In this chapter we describe the application of these new separation and analysis techniques to sterane biomarker analysis.

INSTRUMENTATION

Data were obtained on a Finnigan MAT TSQ 70 triple sector quadrupole mass spectrometer GC-MS system and a 7070H VG Micromass double focusing GC-MS system. The TSQ 70 is equipped with a Varian Model 3400 gas chromatograph and the 7070H with a Hewlett-Packard Model 5790A gas chromatograph. Both chromatographs are interfaced to the mass spectral system and use a 60-meter, 0.25-mm ID, DB-1 fused silica capillary column from J and W Scientific. In the analyses of steranes, the gas chromatographs were programmed from 150°C to 300°C at a rate of 2°C per minute. Different temperature programs were used for aromatic steroid analyses (cf. Figures 9.11 and 9.12).

The experiments described in Figures 9.1 to 9.10 and 9.13 to 9.17 were made on a mixture of C_{27}, C_{28}, and C_{29}-sterane epimers prepared by pyrolysis of cholestane, ergostane, and stigmastane over Pd/C in an evacuated, sealed glass tube at 260°C for 68 hours as described previously (Seifert et al., 1983).

The experimental setup for the TSQ 70 experiments is the following:

Parent scan time	0.2 s
Collision offset voltage	−10.0
Collision gas	Ar
Set mass	217.2
Parents	217.2, 358.3, 372.4, 386.4, 400.4, 414.4
Ionization mode	Electron impact 70 eV
Q1MS full scan	
Scan range	40,430
Collision offset voltage	−40
Multiplier voltage	1800
Electrometer gain	8
MSMSC	40

The VG experimental setup is as follows:

Scan range	20–600
Quadratic up	1.4 s
Quadratic down	0
Top	0
Bottom	0.1
Ionization mode	Electron impact 70 eV
Multiplier	2000 v
Electrometer gain	5

RESULTS AND DISCUSSION

Hold Time Experiments on Steranes

The hold time phenomenon experiments on steranes will focus only on selected C_{29} epimers. Figure 9.1 shows the GC-MS/MS trace of the *m/z* 400 parents

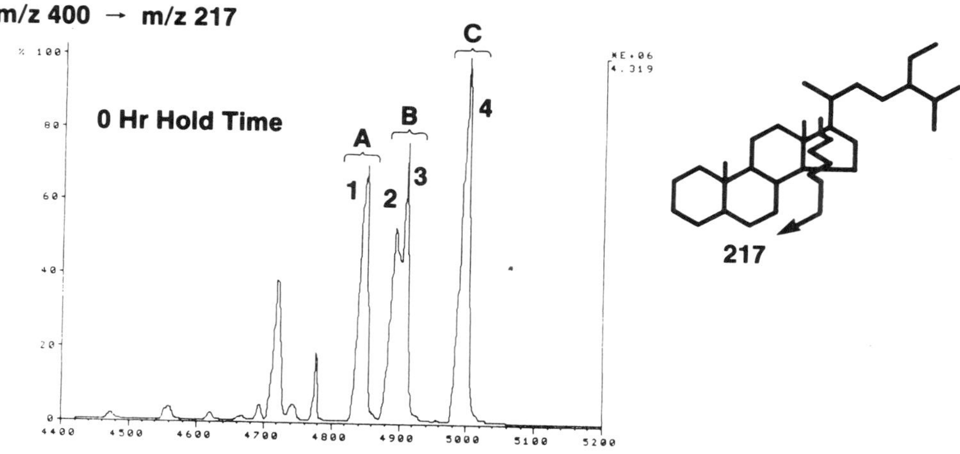

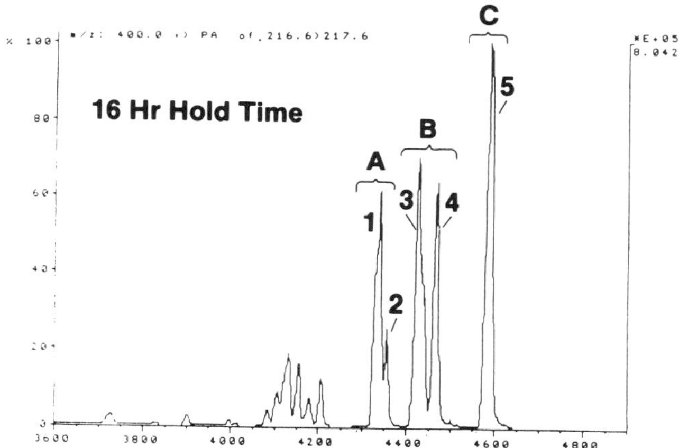

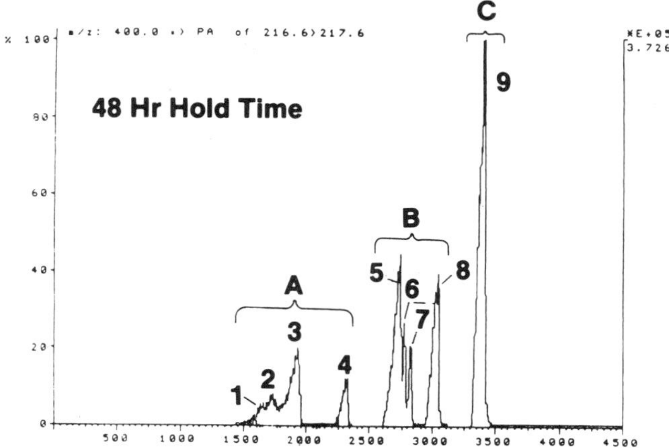

Figure 9.1 Hold time effect on GC resolution. Oven temperature, T_1, 150°C; hold times 0, 16, and 48 hours, program rate 2°C/min. to T_2, 300°C.

to *m/z* 217 daughters reaction of C_{29} steranes, that is, *m/z* 400 collision-induced decomposition to *m/z* 217 at 0-minute (top), at 16-hour hold time (middle), and 48-hour hold time (bottom). At a 0-minute hold time at 150°C there are three resolved peaks (A, B, and C), with a partially resolved doublet at B. The 16-hour hold time experiment shows A split into a doublet with a shoulder on Peak 1. B is resolved almost completely into two peaks with a shoulder on Peak 3. C remains a singlet. The 48-hour experiment shows four recognizable peaks under A, four under B, and one with a shoulder under C. Peaks 3 and 4 under A and 5 under B are broad, suggesting additional components. This suggests that there are at least a total of 14 components in this range, which under normal GC chromatographic conditions yields at best four GC peaks. Hold times of 16 and 48 hours with the column temperature at 100°C showed no effect on resolution.

A combination of factors may account for the increase in resolution using extended hold time. At 150°C continuous chromatography of the steranes is occurring. For example, all of the C_{27}-steranes and most of the C_{28}-steranes in our test mixture have already eluted from the column in 48 hours at 150°C, and when programming at 2°C per minute is started after the 48 hours hold time, the subsequent retention times of the C_{29}-steranes are considerably reduced compared to shorter hold times. (Compare scan numbers for three traces in Fig. 9.1.) Interactions between the stationary phase of the GC column and the eluent steranes may be considerably different during chromatography at 150°C compared to temperatures near 300°C where the steranes elute under normal programming conditions. However, long chromatographic procedures usually lead to band broadening, which is certainly not the case here. In fact, band sharpening was demonstrated when extended hold time experiments were applied to certain aromatic steroids (see below). This band sharpening is probably similar to that occurring during thermal diffusion experiments, which have been applied to separation of hydrocarbon mixtures (Nikolaev et al., 1971; Romero, 1970). Thus, the observed resolution improvements appear to be the result of increased chromatographic interactions plus the effects of thermal diffusion. The GC resolution increase with hold time duration was calculated using the resolution definition given in the abstract. By taking the peak width at half peak height of the singlet GC peak under C, Figure 9.1, and the adjacent GC singlet under B and their relative retention times a resolution, *Rs*, an increase of about 1.8 is calculated for the 16-hour experiment and about 5.2 times for the 48-hour experiment. Though not shown, an increase of 3.5 times was calculated for a 32-hour hold time experiment.

The nine mass spectra, Figures 9.2 through 9.10, were taken from the 48-hour hold time experiment. These mass spectra were taken with the collision gas present. Each mass spectrum was obtained by summing several scans across the GC peak of interest. The mass spectrum of Peak 1, Figure 9.1, is shown in Figure 9.2. This mass spectrum shows *m/z* 218 $>$ *m/z* 217 typical of a 14β configuration and *m/z* 149 $\gg$ *m/z* 151 which suggests 5α(H). A *m/z* 257 $\gg$ *m/z* 259 appears to confirm a 17α(H) configuration. The mass spectrum of Peak 2, Figure 9.1, is shown in Figure 9.3. This mass spectrum is similar to that of Peak 1. However, retention

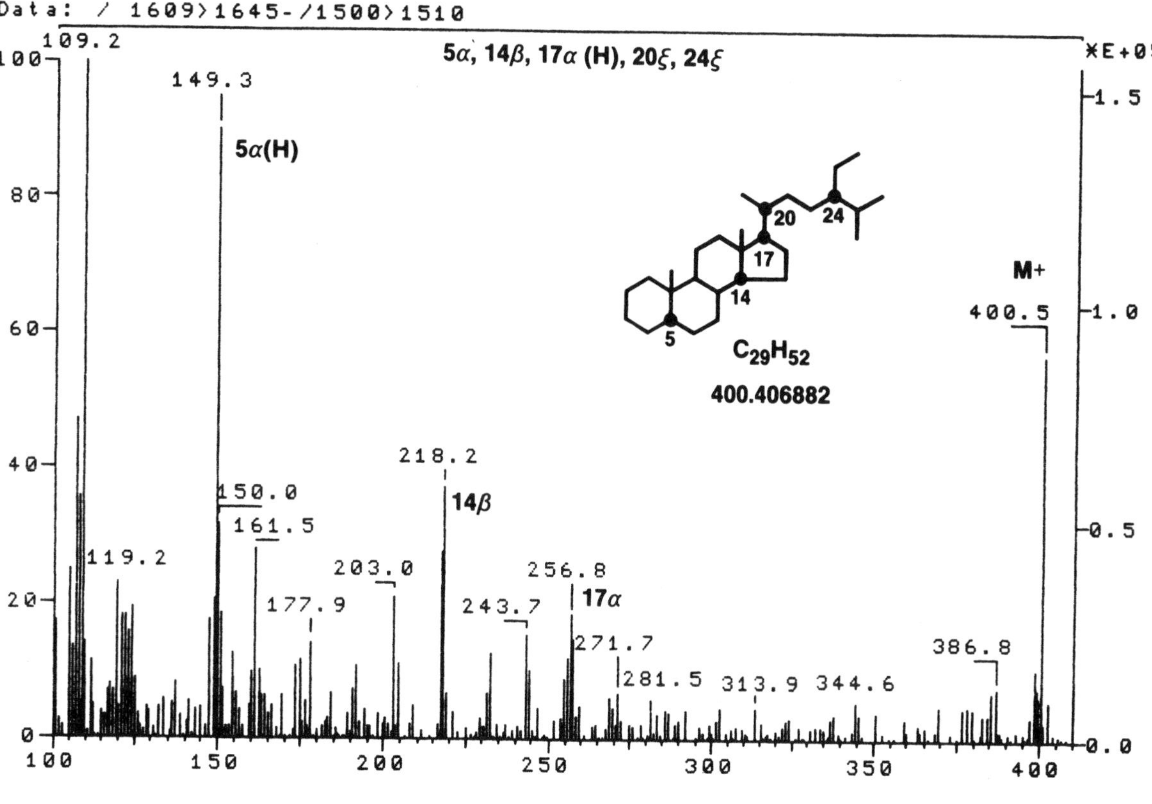

Figure 9.2 Mass spectrum one under A, see Figure 9.1.

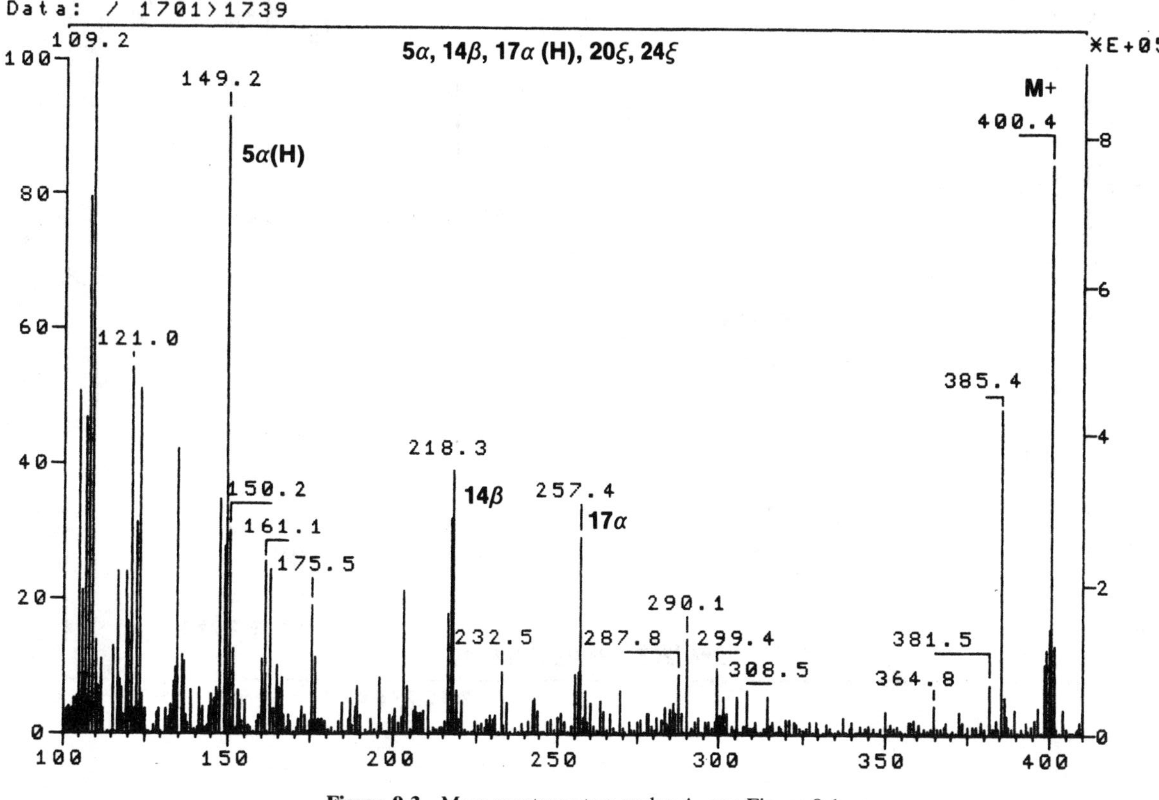

Figure 9.3 Mass spectrum two under A, see Figure 9.1.

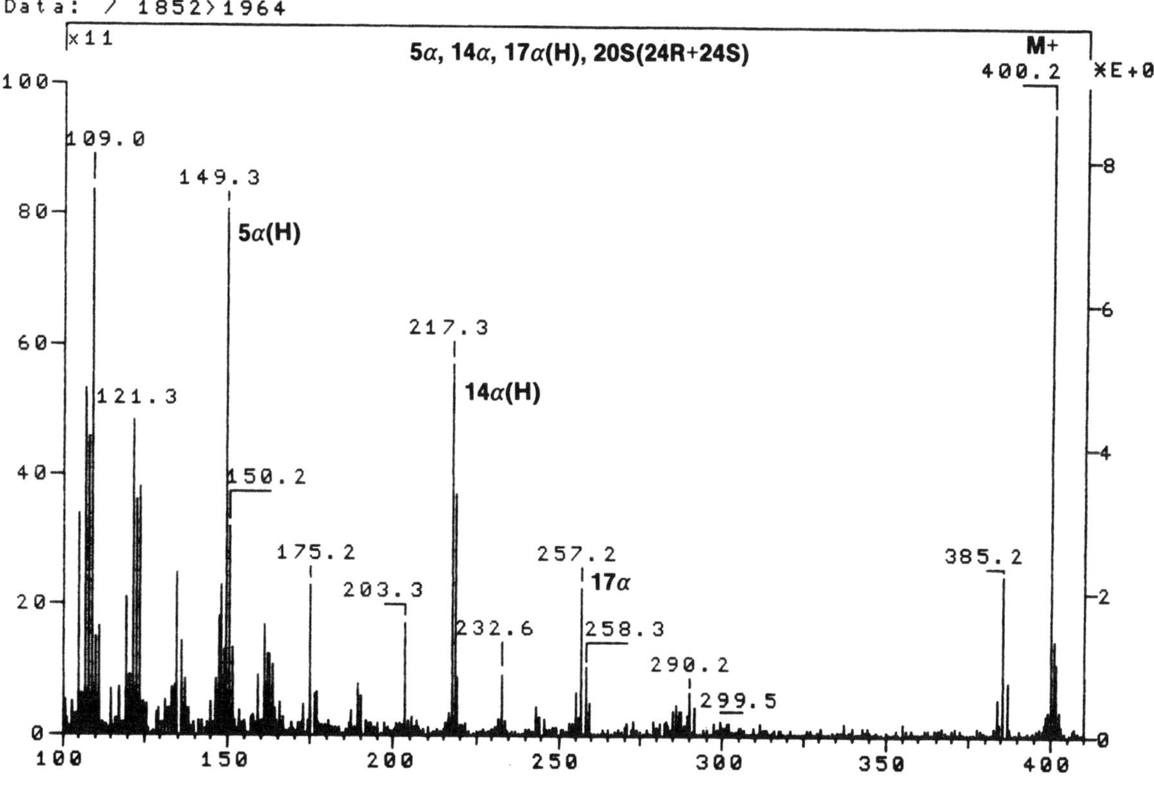

Figure 9.4 Mass spectrum three under A, see Figure 9.1.

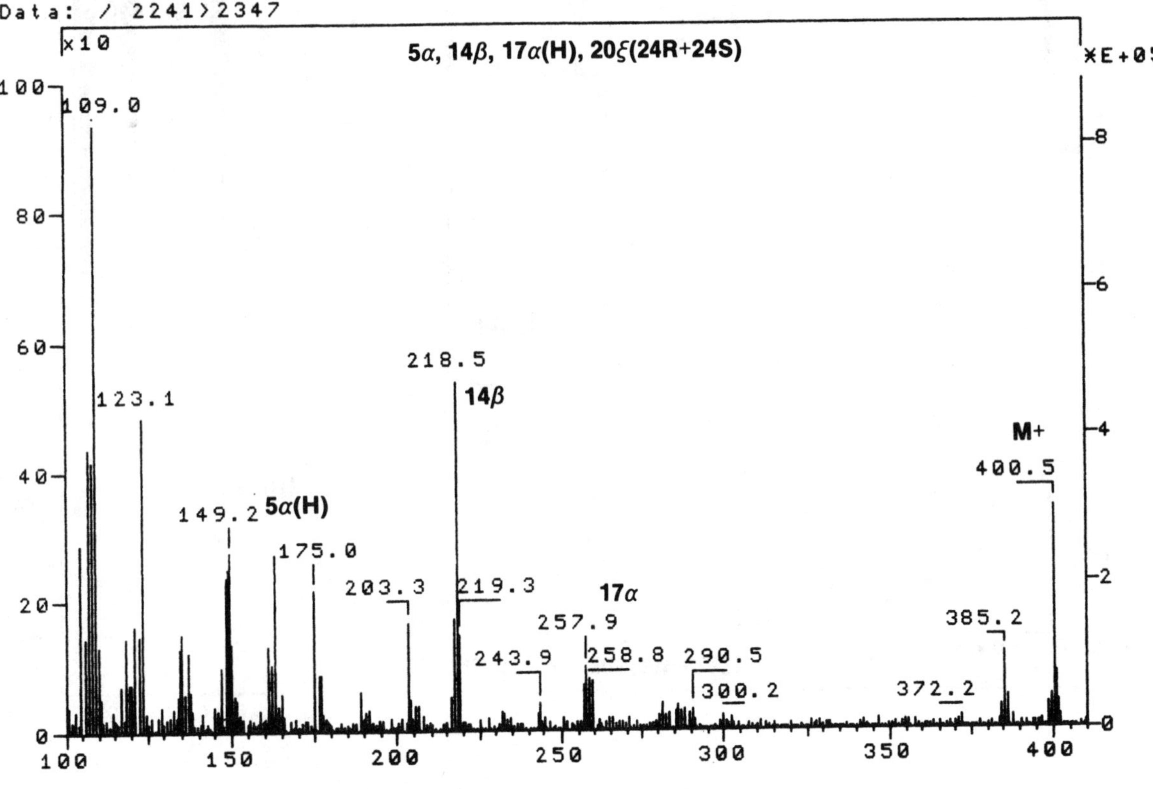

Figure 9.5 Mass spectrum four under A, see Figure 9.1.

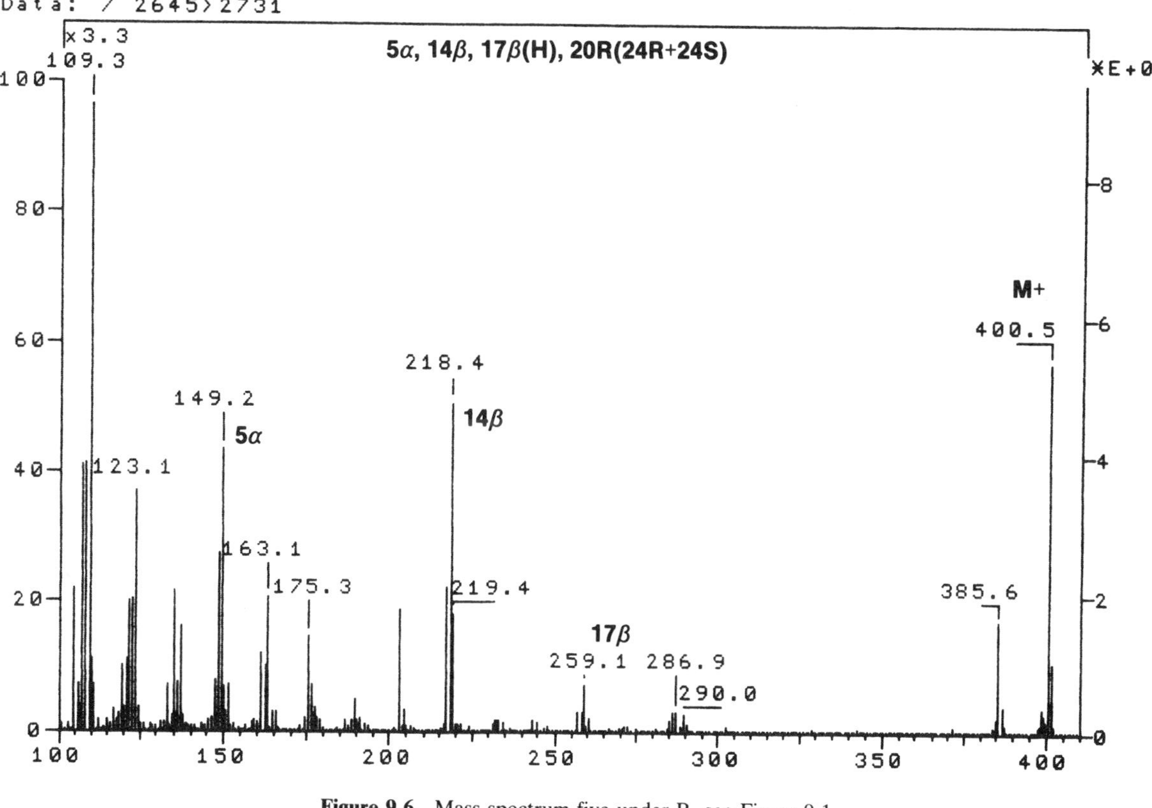

Figure 9.6 Mass spectrum five under B, see Figure 9.1.

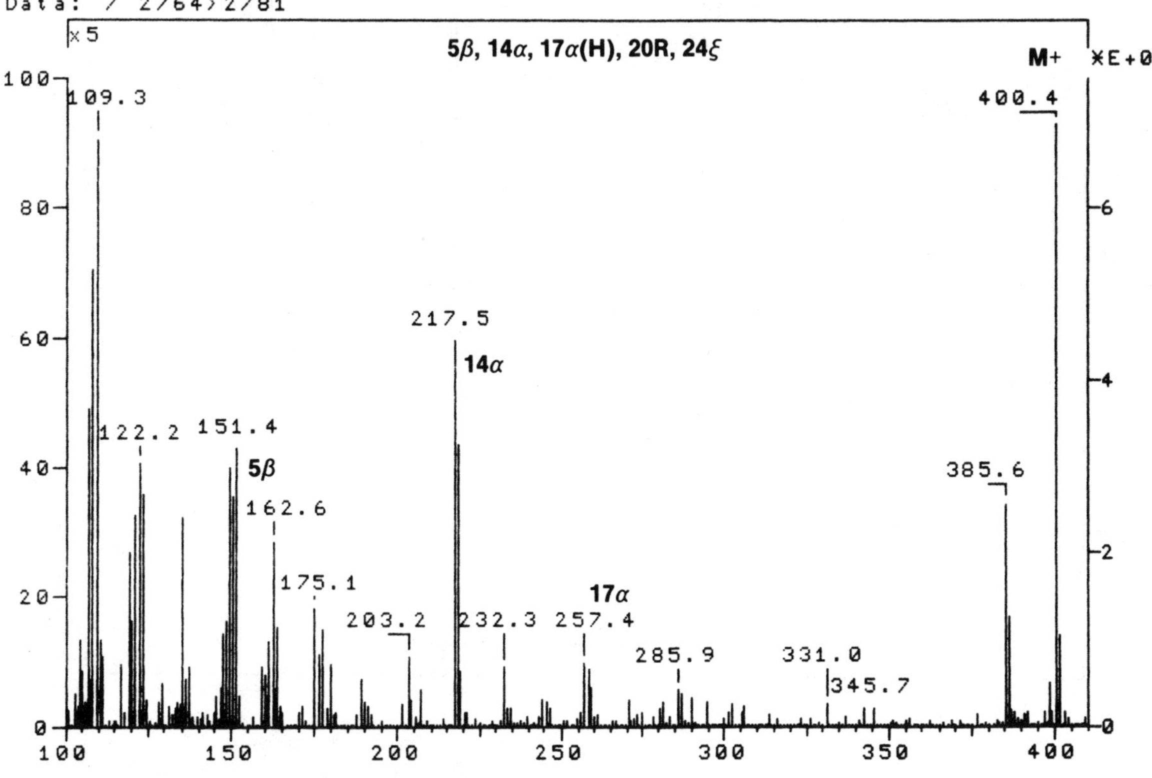

Figure 9.7 Mass spectrum six under B, see Figure 9.1.

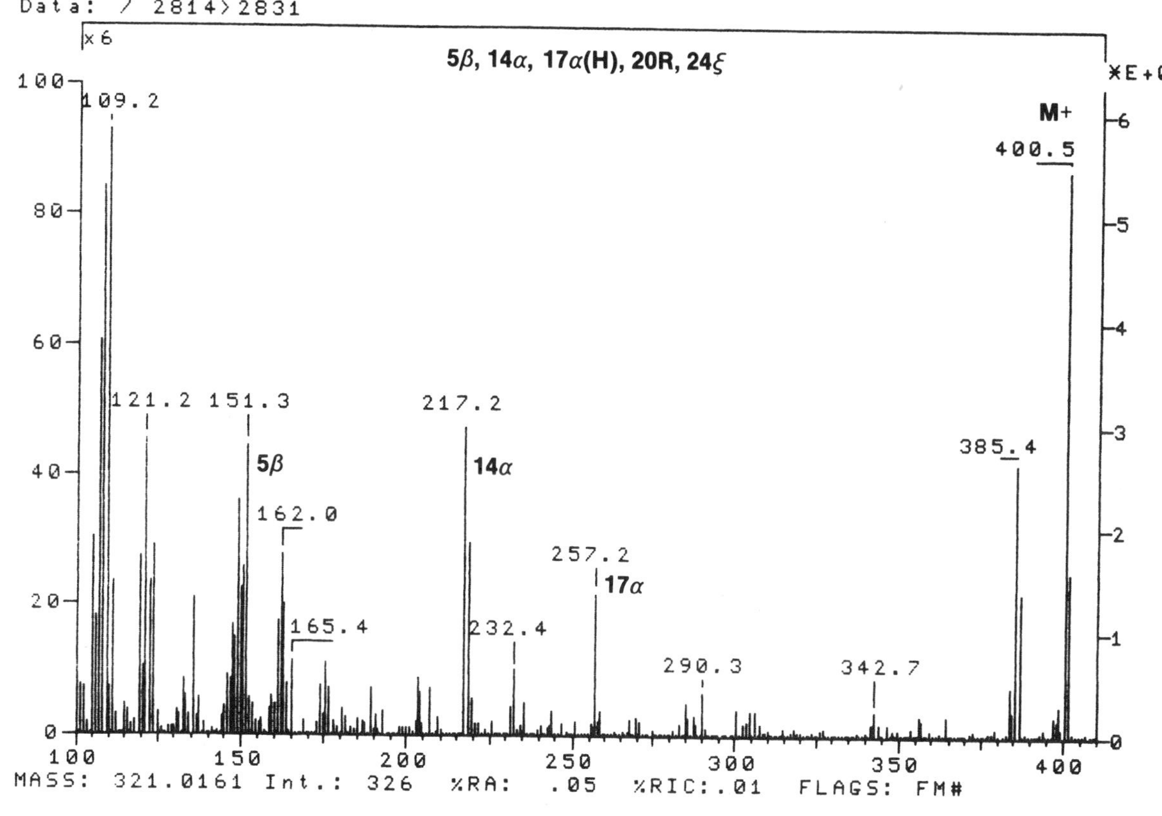

Figure 9.8 Mass spectrum seven under B, see Figure 9.1

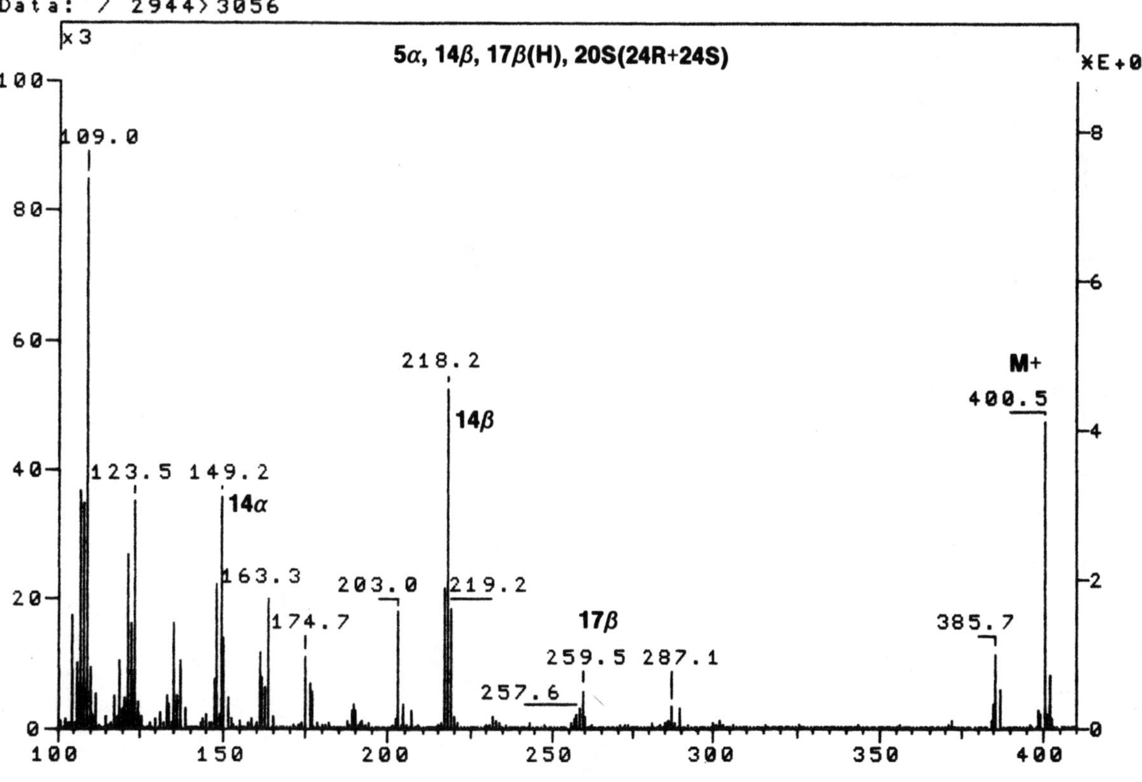

Figure 9.9 Mass spectrum eight under B, see Figure 9.1.

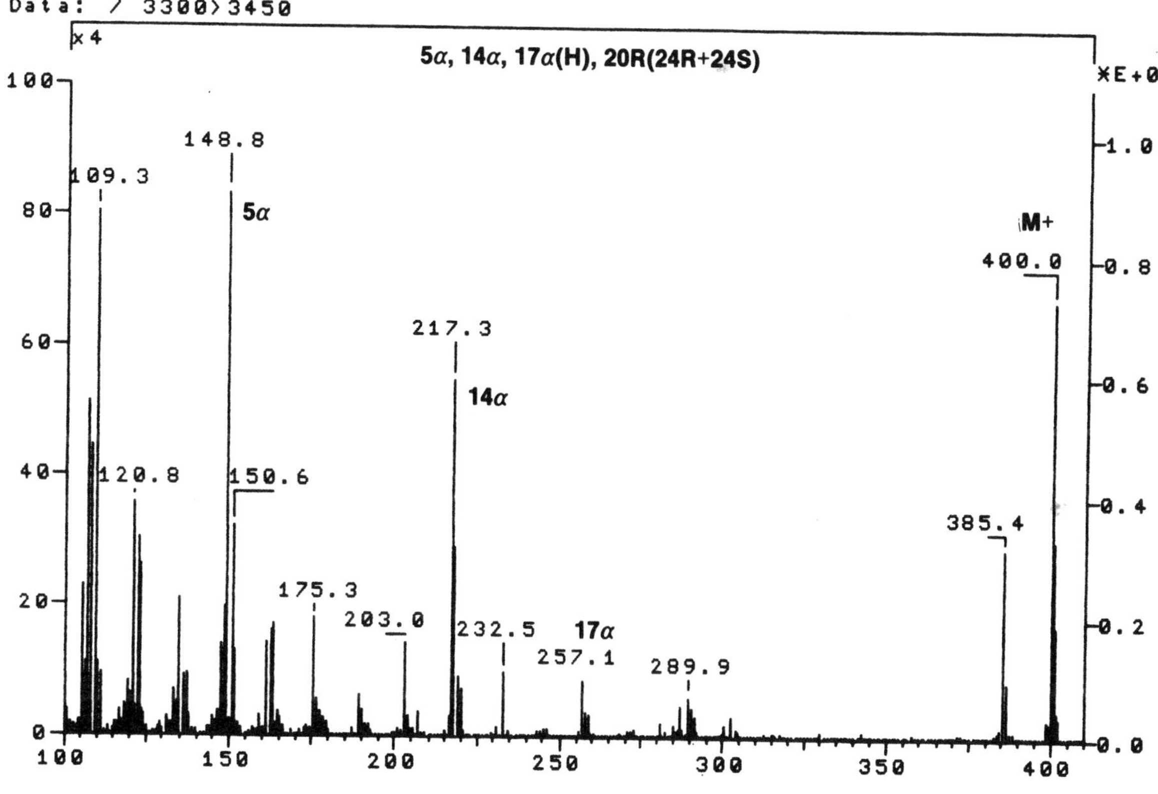

Figure 9.10 Mass spectrum nine under C, see Figure 9.1.

time and relative abundance suggest it is a C24 epimer of 1. Peaks 1 and 2 may represent 24*R* and 24*S* (order could be reversed) epimers of 5α,14β,17α(H), 20ξ,24ξ,24-ethylcholestane.

The mass spectrum of Peak 3, Figure 9.1, is shown in Figure 9.4. Here *m/z* 149 ≫ *m/z* 151 indicates 5α and *m/z* 217 ≫ *m/z* 218 suggests 14α. Again, *m/z* 257 > *m/z* 259 indicates a 17α(H) epimer. The broadness of the peak suggests a 24*R*, 24*S* epimer mix. Thus, the probable stereochemical configuration for Peak 3 is 5α,14α,17α(H),20*S*(24*R* + 24*S*).

The mass spectrum of Peak 4, Figure 9.1, is shown in Figure 9.5. Here *m/z* 149 ≫ *m/z* 151, indicating a 5α, *m/z* 218 ≫ *m/z* 217 suggests 14β, and *m/z* 257 ≫ *m/z* 259 indicates 17α(H). The broadness of the peak suggests a 24*R*, 24*S* epimer mix. The retention time is reasonable for another pair of 5α,14β,17α(H), 20ξ, 24ξ epimers.

The mass spectrum of Peak 5 from Figure 9.1 is shown in Figure 9.6. Again, *m/z* 149 ≫ *m/z* 151, indicating 5α, *m/z* 218 ≫ *m/z* 217 suggests 14β, and *m/z* 259 ≫ *m/z* 257 suggests 17β(H). The retention time is consistent with 5α,14β,17β(H)(20*R*). The broadness of the peak suggests that both the 24*R* and 24*S* epimers are present. Thus, the probable stereochemical configuration for Peak 5 is 5α,14β,17β(H),20*R*(24*R* + 24*S*).

The mass spectrum of Peak 6, Figure 9.1, is shown in Figure 9.7. Here *m/z* 151 > *m/z* 149, suggesting 5β, *m/z* 217 > *m/z* 218, suggesting 14α, and *m/z* 257 ≫ *m/z* 259 argues for 17α(H). The retention time is consistent with 5β,14α,17α(H)(20*R*). The sharpness of the peak suggests that there is only one component under this GC peak. The probable stereochemical configuration of this single component is 5β,14α,17α(H),20*R*, 24*R*, or 24*S*.

The mass spectrum of Peak 7, Figure 9.1, is shown in Figure 9.8. The mass spectrum is nearly identical to that of Peak 6. The retention time and amounts relative to Peak 6 suggest that this is the C24 epimer of Peak 6.

The mass spectrum of Peak 8, Figure 9.1, is shown in Figure 9.9. The same rationale used for Peak 5 may be used here. However, the retention time distinguishes this peak as the 20*S* epimer. The GC peak broadening suggests an epimer mix of 24*R* and 24*S*. Thus, the probable stereochemical configuration is 5α,14β,17β(H),20*S*(24*R* + 24*S*).

The mass spectrum of Peak 9, Figure 9.1, is shown in Figure 9.10. Coinjection with an authentic standard shows this to be a 5α,14α,17α(H),20*R* epimer. Both the 24*R* and 24*S* epimers elute at the same time under normal GC conditions. Peak broadening suggests an epimer mix of 24*R* and 24*S*. Thus, the probable stereochemical configuration of GC Peak 9 is 5α,14α,17α(H),20*R* (24*R* + 24*S*).

Analysis of the saturate fraction from a thermally mature petroleum gave a similar pattern of C_{29}-sterane peaks under the same extended hold time conditions. This is expected because it has been shown that sterane isomerizates prepared by heating steranes over Pd/C produce petroleum-like sterane mixtures (Seifert and Moldowan, 1979; Seifert et al., 1983). Thus, it is clear that all previously reported C_{29}-sterane 20*S*/20*R* ratios contain considerable amounts of additional coeluting

C_{29}-sterane epimers in the measurement of the 5α,14α,17α(H),20*S* epimer. The maximum or "equilibrium" value for the 20*S*/20*S* + 20*R* ratio in petroleum measured by normal chromatography GC-MS has been taken to be about 0.5 to 0.55, based on measurements of many samples (Seifert and Moldowan, 1986) in good agreement with the relative theoretical stabilities of 20*S* and 20*R* which have been calculated using molecular mechanics (van Grass et al., 1982; Petrov et al., 1976). Further, kinetic parameters for C_{29}-sterane isomerization (20*S*/20*S* + 20*R*) were calculated based on equilibrium values of 0.54 under geological conditions (Mackenzie and McKenzie, 1983). In light of the present work, these numerical parameters should be reexamined.

Hold Time Experiments on Aromatic Steroids

Additional hold time experiments were done on the mono and triaromatic steroid hydrocarbons in the aromatic fraction from a Carneros, Calif., oil. The regular MID GC-MS *m/z* 253 trace of this fraction has been reported (Moldowan and Fago, 1986). Figure 9.11a, (top traces), show the normal GC-MS/MS data of the C_{27}, C_{28}, and C_{29} parents to *m/z* 253 transitions of the monoaromatic steroids. Whereas (Fig. 9.11b), the bottom traces show the same sample and same analysis using an extended hold time of 15 hours at 150°C. Note peaks under A through I normal run (Fig. 9.11a), and hold time run (Fig. 9.11b). A in the hold time run eluted during the 15-hour hold time and was not recorded, B split from a doublet to a triplet, C from a singlet to a doublet, D singlet to a doublet, E doublet to a triplet, F singlet to a doublet, G singlet to a doublet, H stayed about the same, and I singlet to a doublet.

Analyses of triaromatic steroids from the same Carneros oil aromatic fraction are shown in Figure 9.12. The top (Fig. 9.12a) chromatographic trace shows a typical GC-MS *m/z* 231 chromatogram for C_{26}–C_{28} triaromatic steroids. The C_{26} 20*R* and C_{27} 20*S* compounds coelute under these and all previously reported GC conditions hampering the utility of triaromatic steroids for geochemical application. However, the traces from parent ions *m/z* 344, 358, 372, and 386 of *m/z* 231 resolve this analytical problem because C_{26} 20*R* is recorded on the *m/z* 344 → 231 trace and C_{27} 20*S* on the *m/z* 358 → 231 trace. This method has been applied to oil source correlation to provide C_{26}–C_{27}–C_{28} triaromatic steroid distributions (Moldowan et al., 1991). In addition, this analysis shows the C_{29}-triaromatic steroids. These are related to the 24-*n*-propylcholestanes (C_{30}-steranes) which have been used as an indicator of marine source input (Moldowan et al., 1985).

The bottom part (Fig. 9.12b) shows improved resolution on several peaks using a 66-hour extended hold time at 140°C. The peaks for C_{27} 20*S*, C_{27} 20*R*, and C_{28} 20*S* which are sharp singlets under normal GC programming conditions (Fig. 9.12a) appear as doublets in this extended hold time experiment. The doublets are assumed to be 24*R* and 24*S* epimers. The C_{29} 20*S*, 24*S*, and 24*R* isomers, separated under normal GC conditions (Fig. 9.12a), are even more widely separated under the extended hold time conditions (Fig. 9.12b).

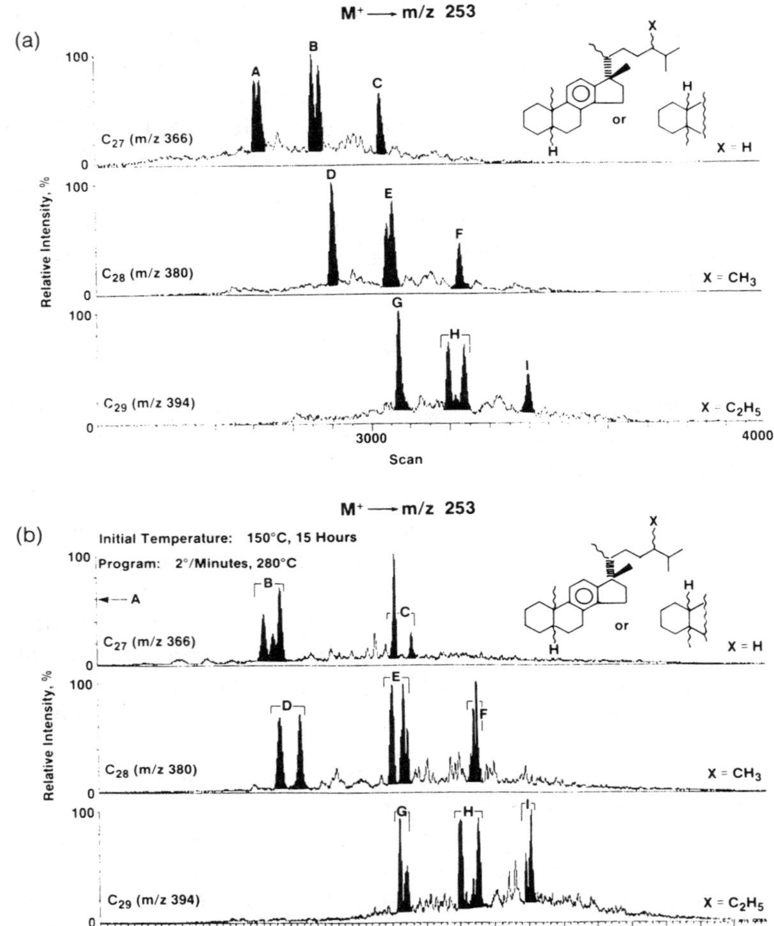

Figure 9.11 (a) Monoaromatic steroid hydrocarbons can be analyzed by GC-MSMS. (b) Extended hold time GC-MSMS resolves additional monoaromatic steroid isomers.

"Collision Gas" Full Scan Mass Spectra Experiments

The effect of collision gas on full scan mass spectra compared to normal electron impact only mass spectra of several C_{29} steranes are shown in Figures 9.13 to 9.17. Figure 9.13 compares a "collision gas" mass spectrum of a C_{29} sterane taken from the TSQ 70 with a normal mass spectrum of the same sterane taken

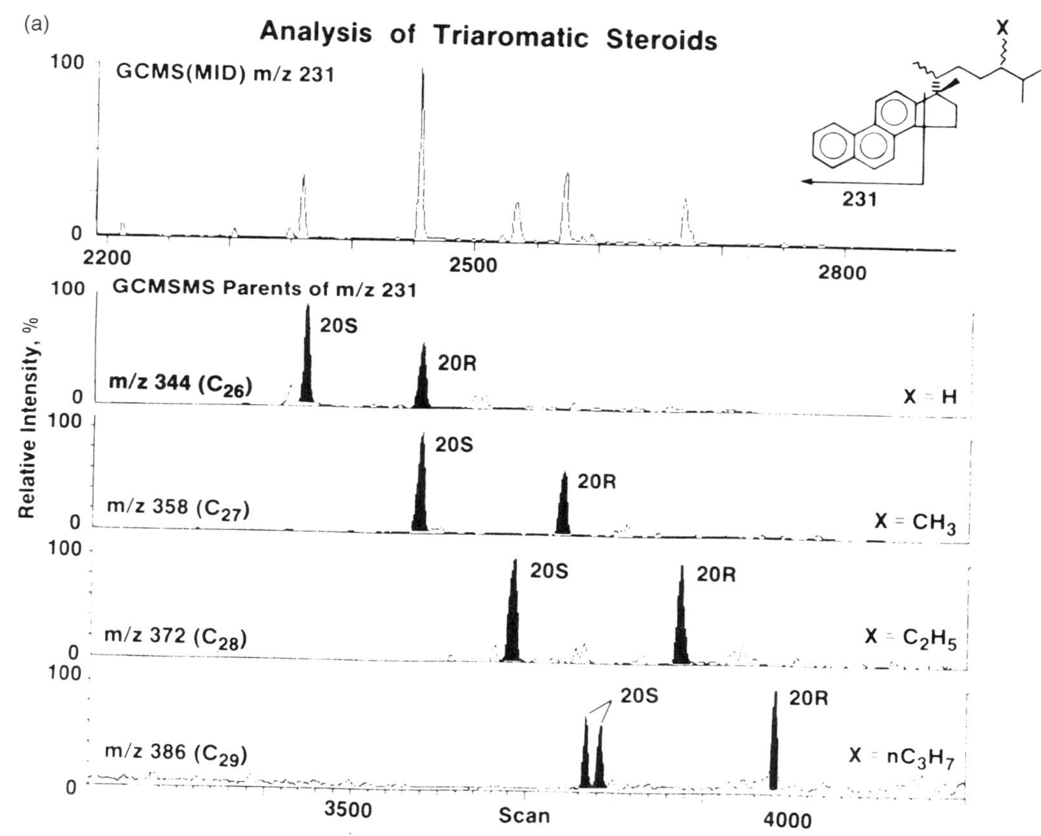

Figure 9.12 (a) Advantages in resolution and selectivity of GC-MSMS, analysis of triaromatic steroids. (b) Extended hold time effect on GC allows resolution of some triaromatic steroid epimers. (continued on next page)

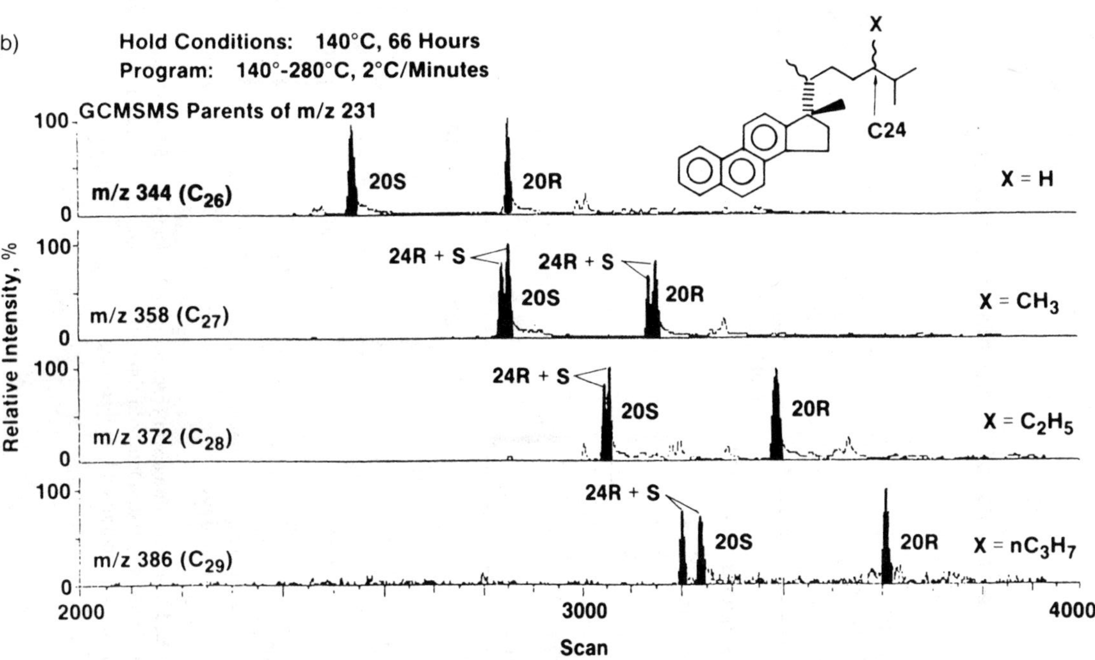

Figure 9.12 (continued)

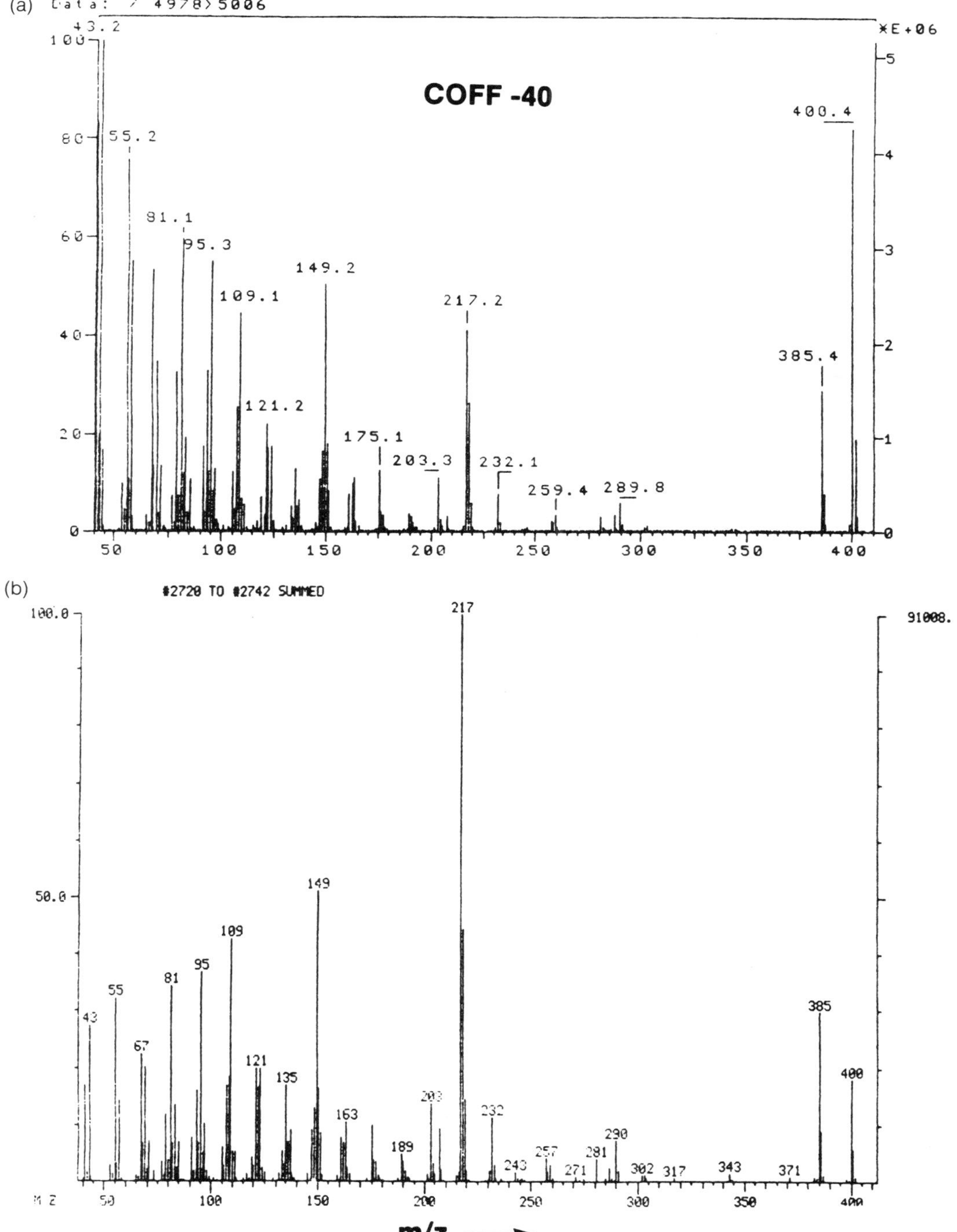

Figure 9.13 "Collision gas" versus normal mass spectrum m/z 40-410; 0 hold time graph peak one.

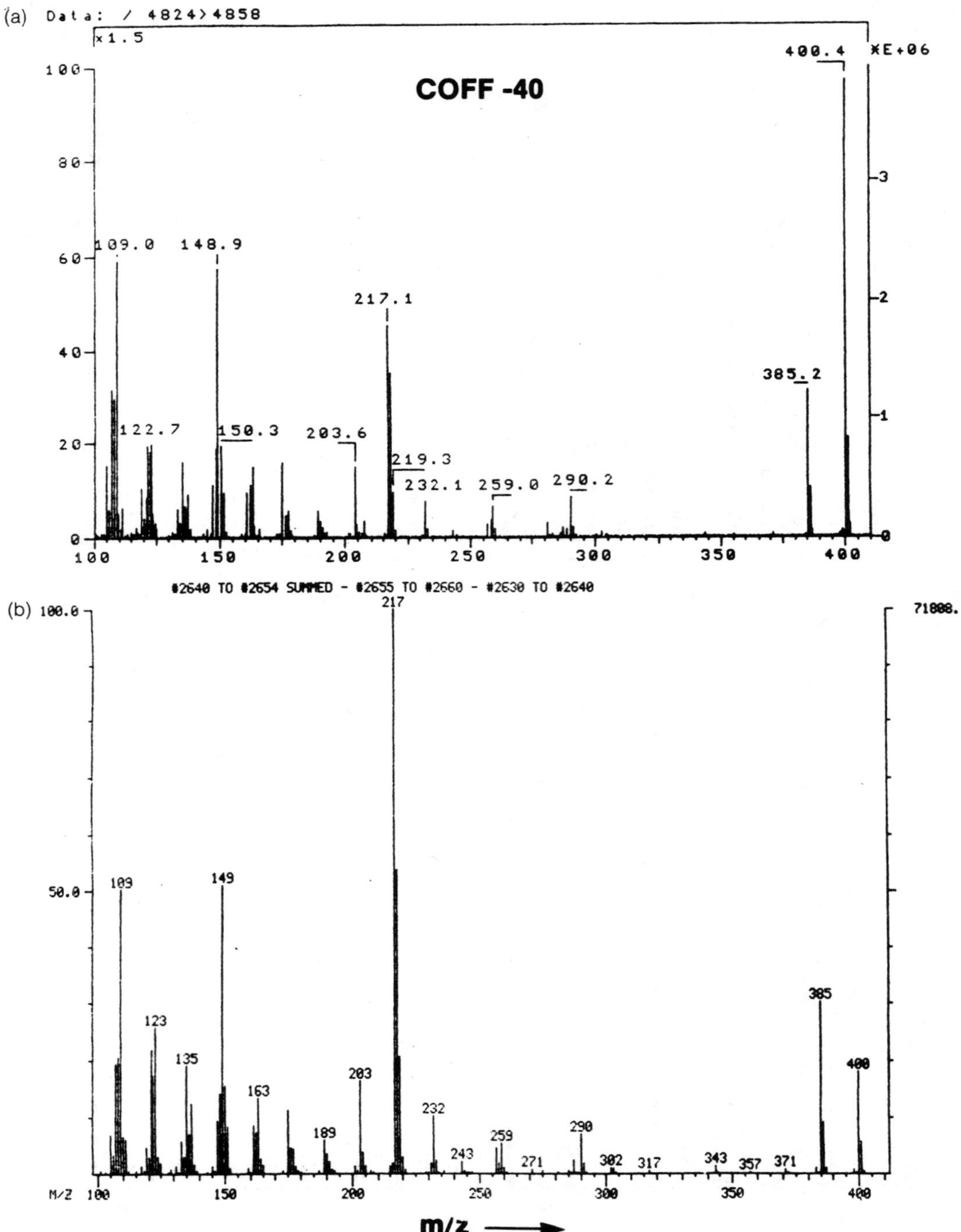

Figure 9.14 "Collision gas" versus normal mass spectrum, *m/z* 100-410; 0 hold time graph peak one.

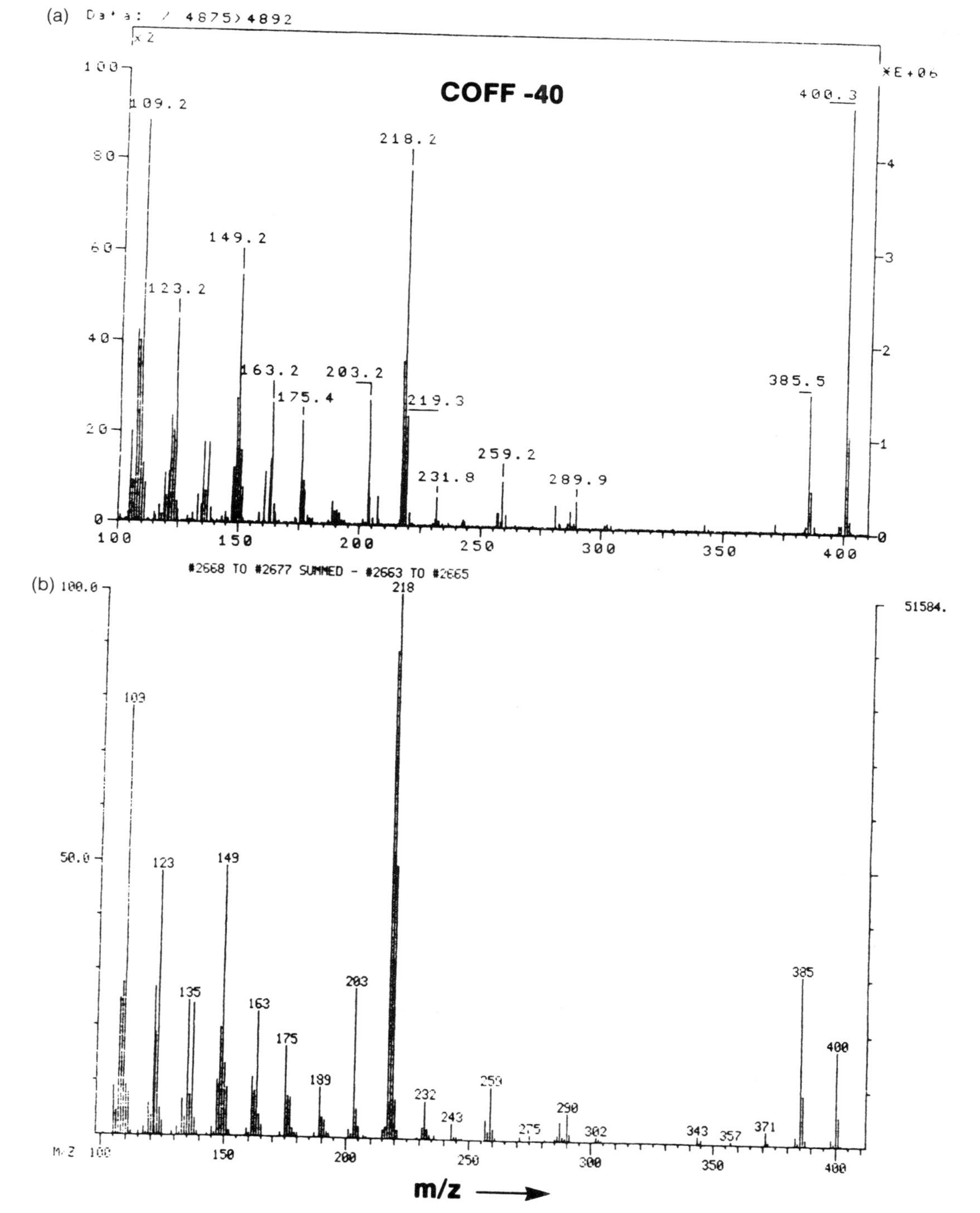

Figure 9.15 "Collision gas" versus normal mass spectrum, *m/z* 100-410; 0 hold time graph peak 2.

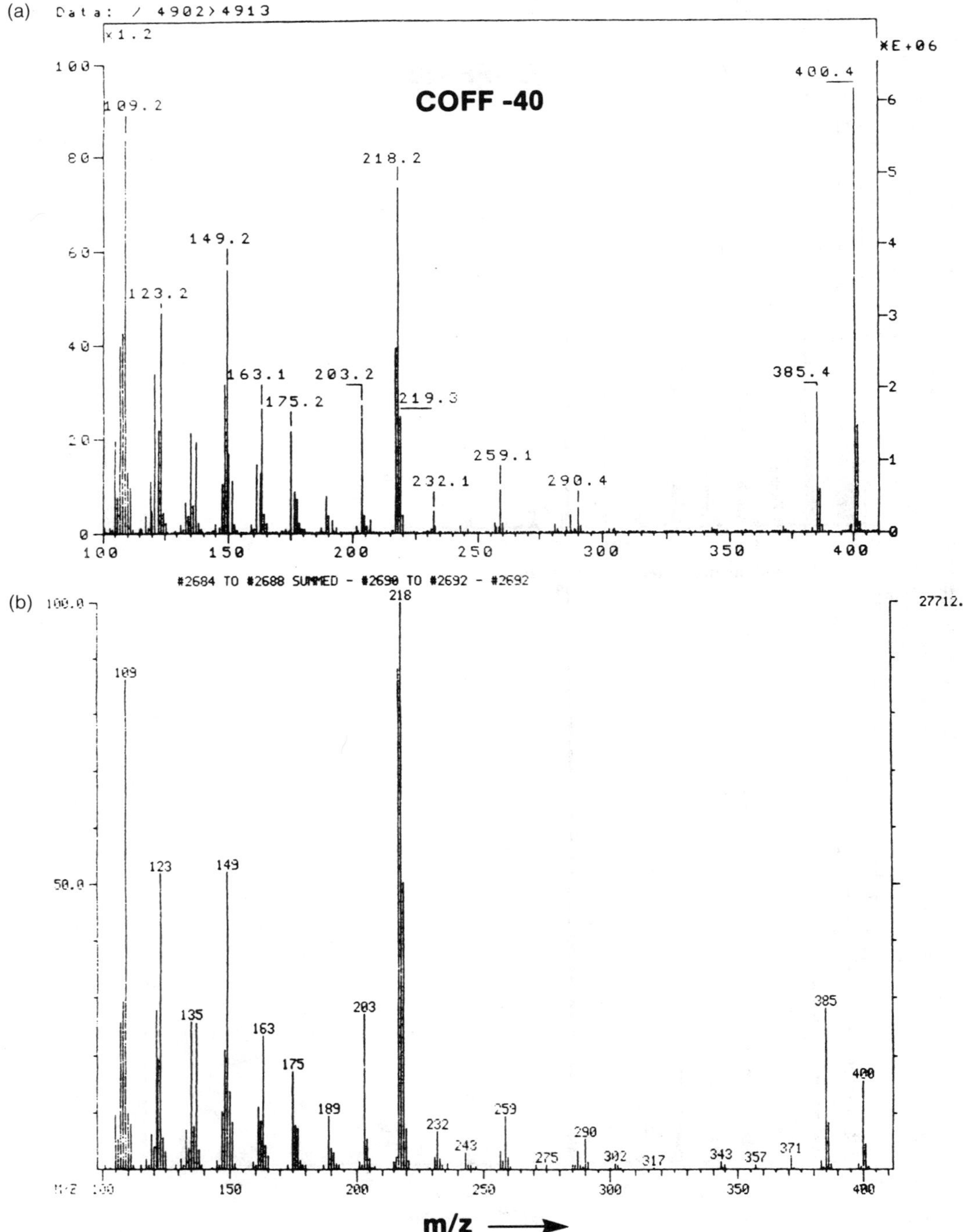

Figure 9.16 "Collision gas" versus normal mass spectrum, m/z 100-410; 0 hold time graph peak 3.

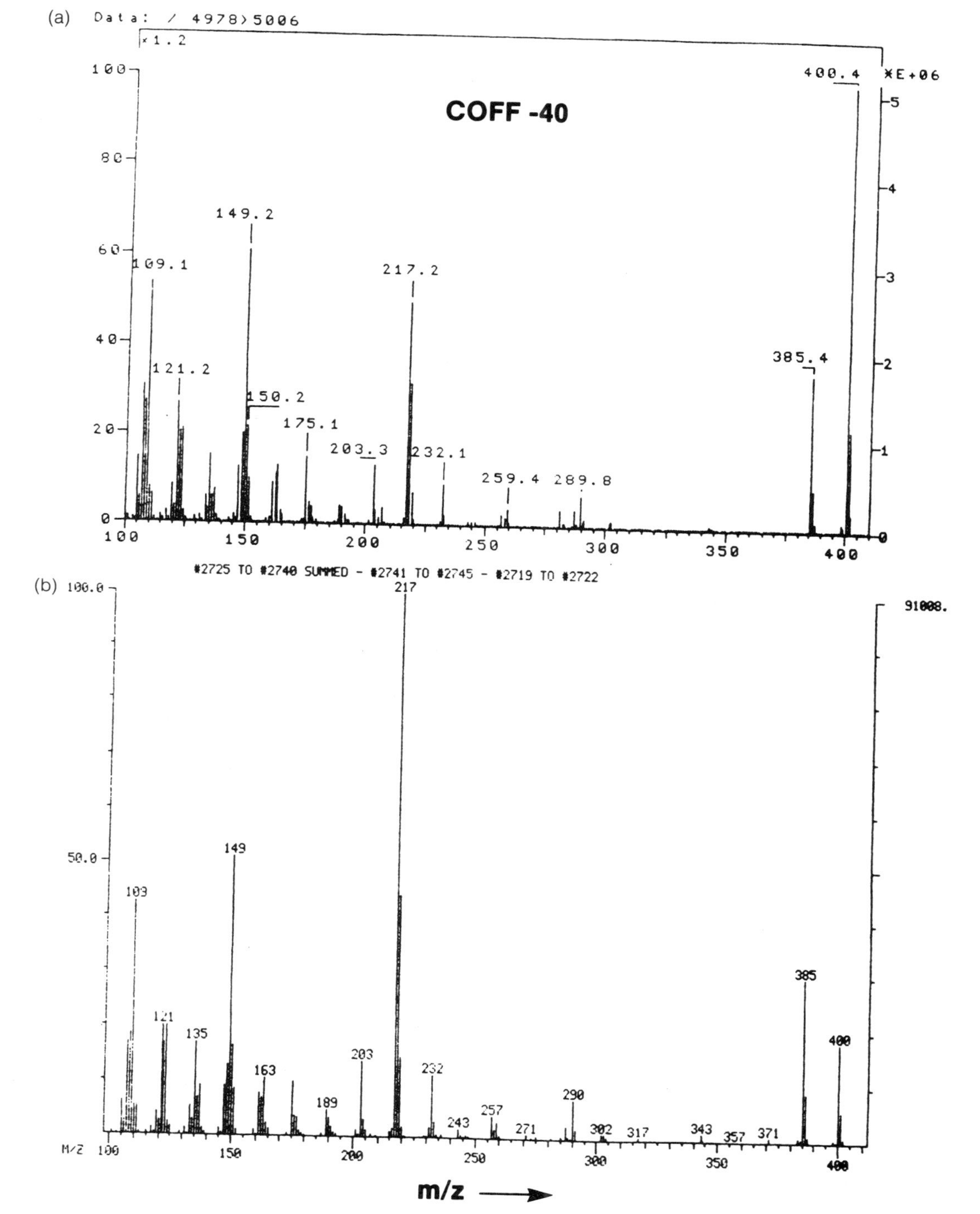

Figure 9.17 "Collision gas" versus normal mass spectrum, m/z 100-410; 0 hold time graph peak 4.

using the 7070H mass spectrometer. These data were taken from a GC run under normal conditions and the mass spectra are actually a mixture of several sterane epimers as shown above. The mass range shown is from mass 40 to the molecular ion mass. The base peak for the "collision gas" mass spectrum is *m/z* 43, whereas the base peak of the normal mass spectrum is *m/z* 218. However, by comparing the mass spectra from *m/z* 100 to the molecular ion (see Figs. 9.14 through 9.17), the mass spectra are more comparable. The diagnostic integrity of the fragments such as *m/z* 217, *m/z* 218, *m/z* 149, *m/z* 151, and so on, are retained. Generally, the "collision gas" mass spectra show a molecular ion with an intensity considerably greater than that of the parent minus a methyl group which is just the opposite from the normal mass spectra. These results demonstrate that the diagnostic value of the full scan mass spectra is not diminished with the collision gas on.

CONCLUSIONS

Increased hold time at a relatively low predetermined oven temperature can be used to improve greatly the resolving power of a capillary gas chromatographic system. However, the elapsed time is so long that this technique is mainly useful to verify the homogeneity of gas chromatographic peaks but not as a standard correlation technique.

In particular, several C_{29}-sterane diastereomers normally coeluting with 20*S*, 24-ethylcholestane were discovered using extended hold times. The presence of these coeluting isomers indicates that previously measured 20*S*/(20*S* + 20*R*) C_{29}-sterane ratios, which are widely used in geochemical correlation studies of oils and rock extracts, are too high.

This work also showed that an intensity ratio *m/z* 257 > *m/z* 259 indicates a 17α(H) sterane epimer and an intensity ratio *m/z* 257 < *m/z* 259 indicates the 17β(H) sterane epimer. The structures of several previously unrecognized minor C_{29}-sterane isomers are postulated based on their mass spectra.

Monitoring collision-induced daughter ions $M+ \rightarrow 231$ for triaromatic steroids substantially improves the analysis of their homologous series. The C_{26} 20*R* and C_{27} 20*S* triaromatic steroids are separately analyzed by this method, while single ion monitoring of *m/z* 231 fails to resolve them. Extended GC hold time also resolves additional mono- and triaromatic steroids. For example, several 24*S* and 24*R* triaromatic steroids can be partially resolved.

Finally, the presence or absence of collision gas in multiexperiment GC-MS runs does not affect the diagnostic integrity of the mass spectra obtained.

REFERENCES

MACKENZIE, A.S. and MCKENZIE, D. (1983) Isomerization and aromatization of hydrocarbons in sedimentary basins formed by extension. *Geol. Mag. 120*, 417–470.

MOLDOWAN, J.M. and FAGO, F.J. (1986) Structure and significance of a novel rearranged monoaromatic steroid hydrocarbon in petroleum. *Geochim. Cosmochim. Acta 50*, 343–351.

Moldowan, J.M., Seifert, W.K., and Gallegos, E.J. (1985) Relationship between petroleum composition and depositional environment of petroleum source rocks. *Amer. Assoc. Petrol. Geol. Bull. 69*, 1255–1268.

Moldowan, J.M., Lee, C.Y., Sundararaman, P., Salvatori, T., Alajbeg, A., Gjukić, B., Demaison, G.J., Slougui, N.E., and Watt, D.S. (1991) Source correlation and maturity assessment of select oils and rocks from the central Adriatic Basin (Italy and Yugoslavia). This text, chap. 19.

Nikolaev, B.I., Tubin, A.A., and Arnov, A.R. (1971) Use of thermal diffusion for separating liquid organic mixtures. Tr., Vses. Nauch.-Issled. Inst. Khim. Reaktivov Osobo Chist. Khim. Veschestev, No. 33, 255–260.

Petrov, A.A., Pustil'nikova, S.D., Abriutina, N.N., and Kagramonova, G.R. (1976) Petroleum steranes and triterpanes. *Neftekhimiia 16*, 411–427.

Philp, R.P., Oung, J., and Lewis, C.A. (1988) Biomarker determination in crude oils using a triple stage quadrupole mass spectrometer. *J. Chromatogr. 446*, 3–16.

Romero, J.J.B. (1970) Some technological aspects of separation by thermal diffusion in the liquid phase. Rev. Fis., Quim. Eng., Ser. A, 2(1), p. 152.

Seifert, W.K. and Moldowan, J.M. (1979) The effect of biodegradation on steranes and terpanes in crude oils. *Geochim. Cosmochim. Acta 43*, 111–126.

Seifert, W.K. and Moldowan, J.M. (1986) Use of biological markers in petroleum exploration. *Methods in Geochemistry and Geophysics* (ed. R.B. Johns), Vol. 24, pp. 261–290.

Seifert, W.K., Carlson, R.M.K., and Moldowan, J.M. (1983) Geomimetic synthesis, structure assignment, and geochemical correlation application of monoaromatized petroleum steranes. *Advances in Organic Geochemistry 1981* (eds. M. Bjorøy et al.), J. Wiley and Sons, New York, pp. 710–724.

van Graas, G., Baas, J.M.A., de Graaf, V., and de Leeuw, J.W. (1982) Theoretical organic geochemistry. I. The thermodynamic stability of several cholestane isomers calculated by molecular mechanics. *Geochim. Cosmochim. Acta 46*, 2399–2402.

Warburton, G.A. and Zumberge, J.E. (1983) Determination of petroleum sterane distributions by mass spectrometry with selective metastable ion monitoring. *Anal. Chem. 55*, 123–126.

10

Hopenes and Hopanes Methylated in Ring-A: Correlation of the Hopanoids from Extant Methylotrophic Bacteria with their Fossil Analogues

Roger E. Summons and Linda L. Jahnke

Abstract. 2α-Methyldiplopterol, 2β-methyldiplopterol (isolated from *Methylobacterium organophilum*) and a mixture of diplopterol and 3β-methyldiplopterol (isolated from *Methylococcus capsulatus*) were dehydrated using acid catalysis to yield mixtures containing the analogous hop-17(21)-ene as the major product, and with the hop-21-ene, hop-22(29)-ene, and neohop-13(18)-ene derivatives as minor products. Hydrogenation of these triterpene mixtures, also under acidic conditions, yielded the analogous 17α(H),21β(H)-hopane as the principal product in each case. Mass spectral and relative retention time correlations were then used to compare these synthetic methylhopanes with those present in bitumens.

2α-Methyl-17α(H),21β(H)-hopane and 3β-methyl-17α(H),21β(H)-hopane were found to be common to many sedimentary hydrocarbon assemblages. Plots of Ln retention time versus carbon number based on the C_{31} assignments indicate that these methylhopanes are members of C_{28}–C_{36} pseudohomologous series. 2β-Methyl-17α(H),21β(H)-hopane was identified in significant concentration in an immature ancient sediment. It was much less abundant in other sediments of intermediate maturity and absent from several mature samples, suggesting that the common sedimentary 2α-methylhopanes are probably derived from less stable 2β-methyl biogenic precursors. Mass spectra and relative gas chromatographic retention time data for the methylhopanes and the methylhopene intermediates are reported.

INTRODUCTION

Geohopanoids, long recognized as common and important constituents of fossil organic matter (Hills and Whitehead, 1966; Kimble, 1972; Ensminger et al., 1974; van Dorsselaer et al., 1974), have their principal sources in the hopanols, hopenes, and C_{35} bacteriohopane polyols of bacteria. The profound significance of the link between the fossil biomarkers and the extant bacterial source organisms became evident only after details of the structures of biological hopanoids were unequivocally determined (Gelpi et al., 1970; Bird et al., 1971; De Rosa et al., 1971; Förster et al., 1973; Langworthy and Mayberry, 1976; Rohmer and Ourisson, 1976), and after logical explanations were proposed for the stereochemical differences between the respective biological and fossil molecules (e.g., Ensminger et al., 1974, 1977; Ensminger, 1977; Seifert and Moldowan, 1978, 1980; Mackenzie et al., 1980). In subsequent work, significant structural variation in bacterial hopanoids was discovered (e.g., Ourisson et al., 1979; Rohmer et al., 1984) and their role as sterol surrogates was established (e.g., Rohmer et al., 1979; Ourisson et al., 1982, 1987). The ubiquity and abundance of hopanoids in organic-rich sediments of all ages attests to the importance of bacteria in the formation and diagenesis of the entrained organic matter.

Hopanols and bacteriohopane polyols bearing an additional methyl substituent at C-2 or C-3 in ring-A have been recognized in several classes of bacteria and cyanobacteria (Rohmer and Ourisson, 1976; Rohmer et al., 1984; Bisseret et al., 1985; Zundel and Rohmer, 1985a, 1985b), and β-orientation of the methyl substituent has been demonstrated for both types. There are numerous reports of fossil hopanoids with an additional methyl group in ring-A (e.g., Seifert and Moldowan, 1978; Brassell et al., 1980; Alexander et al., 1984; McEvoy and Giger, 1986; Summons and Powell, 1987; Summons et al., 1988; Price et al., 1988) although, to our knowledge, the specific details of their methylation patterns have not yet been established. The sedimentary methylhopanes were originally (and erroneously) thought to be 3-methylhopanes (Seifert and Moldowan, 1978; Alexander et al., 1984; McEvoy and Giger, 1986; Summons and Powell, 1987) on the basis of the earliest report of likely biogenic 3-methyl hopanoid precursors (Rohmer and Ourisson, 1976). Recent observation of an immature Ordovician sediment with two pseudohomologous series of methylhopanes each with apparent 17α(H),21β(H) and 17β(H),21α(H) stereochemistry demonstrated the need to characterize these compounds (Hoffmann et al., 1987). The isolation and structure determination of 2β-methyldiplopterol and 3β-methylbacteriohopane polyols from two types of methylotrophic bacteria (Bisseret et al., 1985; Zundel and Rohmer, 1985a) suggested a likely source of substrates from which the analogous geohopanoids could be prepared using established procedures.

Conversions of diplopterol to hopanes of the 17α(H),21β(H)-,17β(H), 21α(H)-, and 17β(H),21β(H)-series have been carried out by Tsuda et al. (1967) and Ensminger et al. (1974). The thermodynamically most stable C_{30} isomer, 17α(H), 21β(H)-hopane, was the dominant product when hop-17(21)-ene was hydrogenated under acidic conditions using Adams catalyst (Tsuda et al., 1967). Furthermore, hop-17(21)-ene is the major product formed during acid-catalyzed isomerization of diploptene or dehydration of diplopterol (Ageta et al., 1963). Hop-17(21)-ene and extended hop-17(21)-enes are also the favored hopene products during early stages of sedimentary diagenesis while unsaturated hydrocarbons are still present, prior to their complete replacement by hopanes (e.g., Ensminger, 1977; Brassell et al., 1980; ten Haven et al., 1985, 1986; McEvoy and Giger, 1986). These dehydration and hydrogenation reactions constitute a suitable route to the synthesis of 2β-methylhopane and 3β-methylhopane because their course would probably not be affected by the presence of an additional methyl group in ring-A.

In mature sediments, an extra methyl group on ring-A of steroids and hopanoids would be likely to assume the favored equatorial orientation (viz. Summons and Capon, 1988). For substitution at C-3, this is the same as the configuration found in the 3β-methyldiplopterol present in *Acetobacter pasteurianus* and *Methylococcus capsulatus*, and the 3β-methylhopan-29-ol prepared from the bacteriohopane polyols of *M. capsulatus* (Zundel and Rohmer, 1985a). For substitution at C-2, the most stable configuration is likely to be 2α-methyl, unlike the orientation of the biological analogue 2β-methyldiplopterol from *Methylobacterium organophilum* (Bisseret et al., 1985). 2α-Methyldiplopterol has, however, been synthesized and its spectral characteristics have been reported (Bisseret et al., 1985).

This chapter covers the conversion of diplopterol and its 2α-methyl-, 2β-methyl- and 3β-methyl analogues to mixtures of their respective hopenes and then to corresponding 17α(H),21β(H)-hopanes. We also discuss GC-MS comparisons between these synthetic hydrocarbons and sedimentary analogues.

MATERIALS AND METHODS

General

All solvents were distilled prior to use. Sediments were extracted and the bitumens processed and analyzed by GC-MS according to the standard procedures used in this laboratory (e.g., Summons and Powell, 1987). The instrument is a VG 7070E double focussing mass spectrometer equipped with a HP 5790 GC and VG 11-250 data system. Full scan mass spectra were collected at 2 s/decade using 70 eV and a source temperature of 250°C. Comparisons between synthetic products and components in bitumens were conducted using metastable reaction monitoring (MRM) in GC-MS using instrumental parameters previously described (Summons and Powell, 1987). Authentic standards of 17α(H),21β(H)-hopane, hop-17(21)-

ene, neohop-13(18)-ene, hop-21-ene, and hop-22(29)-ene (diploptene) were purchased from Chiron Laboratories, Trondheim, Norway. A standard of diplopterol was prepared by Wolf-Kishner reduction of 22-hydroxyhopan-3-one from Dammar resin (Bisseret et al., 1985).

Organisms and Culture Conditions

Methylococcus capsulatus (ATCC 33009) was grown in batch culture at 37°C with a nitrate-mineral salts medium and a continuous flow of a 50:50 mixture of methane : air as previously described (Jahnke and Nichols, 1986). *Methylobacterium organophilum* (ATCC 27886) was grown in batch culture at 30°C with either 0.5 percent methanol and an ammonium-mineral salts medium (Whittenbury and Dalton, 1981) or the complex medium of Hestrin and Schramm (1954). Cells from stationary phase cultures were harvested by centrifugation at 6,000 × *g* for 10 minutes. Cell pellets were washed by suspending in distilled water and centrifuging, then frozen and lyophilized.

Lipid Extraction and Isolation of Diplopterols

Total lipid extracts were prepared from the lyophilized cells by a modification of the method of Bligh and Dyer (1959) which involves a second extraction of the cell residue (Kates, 1986). The 3β- and 2β-methyldiplopterols were isolated from the total lipid extracts by thin-layer chromatography (TLC) using Merck silica gel G plates developed twice to a height of 15 cm using methylene chloride as solvent. The diplopterols were extracted from a zone (Rf 0.55) identified by the retention position of authentic diplopterol. In experiments with *Methylococcus capsulatus*, the bacteriohopane polyols and phospholipids were pre-precipitated using 0.2 percent $MgCl_2$ and cold acetone and a 3β-methylhopan-29-ol derivative was prepared by periodate oxidation and borohydride reduction according to the method of Rohmer et al. (1984). The resulting hopanols, isolated by TLC (Rf, 0.43), were eluted using methylene chloride, then acetylated, rechromatographed and analyzed by GC-MS. The products were identified as the acetates of hopan-29-ol and 3β-methylhopan-29-ol by comparison of their mass spectra with those reported by Zundel and Rohmer (1985a). The diplopterol fraction from this *M. capsulatus* culture contained a small percentage of an A-ring methyl analogue assigned as 3β-methyldiplopterol by virtue of the earlier assignment and synthesis of Zundel and Rohmer (1985a) who identified it as a minor companion to the diplopterol in both *Acetobacter pasteurianus* and *Methylococcus capsulatus*. Our assignment is supported by its identity with 3β-methyldiplopterol from *Acetobacter pasteurianus*, its cooccurrence with 3β-methylbacteriohopane polyols, and by its long GC retention time compared to diplopterol, the 2β-methyldiplopterol isolated from *M. organophilum*, and our synthetic 2α-methyldiplopterol (see below).

Synthesis of 2α-Methyldiplopterol

22-Hydroxyhopan-3-one was isolated from Dammar resin, as previously described (Dunstan et al., 1957), and then converted, using Wolff Kishner reduction, to diplopterol and to 2α-methyldiplopterol (via methylation and Wolff Kishner reduction) according to the method published by Bisseret et al. (1985). Comparison of the ^{13}C nmr spectrum of 2α-methyldiplopterol with the spectrum published by these authors confirmed its identity and that our preparation contained minor impurities of diplopterol (ca. 7%) and 2β-methyldiplopterol (ca. 7%). This latter product was also reported by Bisseret et al. (1985). In our spectrum, carbons 1, 2, and 3 were shifted by +9.1, +5, and +8.9 ppm compared to diplopterol in accord with predicted shifts for addition of a 2α-methyl substituent of +9, +6, and +9, respectively.

Conversion of Diplopterols to 17α(H),21β(H)-Hopanes

The same general procedures were used for all substrates. For example, an acidic dehydration involved refluxing (4 hr) the mixture of diplopterol and 3β-methyldiplopterol (ca. 50:1, 0.2 mg) from *M. capsulatus* in toluene (10 ml) containing conc. HCl (0.1 ml). After removal of the solvent, the hydrocarbon products were isolated by hexane elution from a silica gel column. GC-MS analysis of the mixture (comparison of retention times and mass spectra with authentic standards, Fig. 10.1a) confirmed that the hopene products (base peak *m/z* 191, Fig. 10.1b)

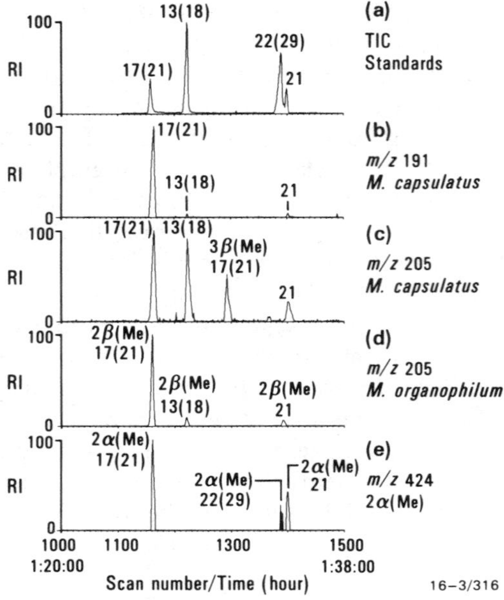

Figure 10.1 Gas chromatographic-mass spectrometric analysis of hopenes derived by HCl dehydration of diplopterols. (a) Total ion current trace for a mixture of four standard hopenes. (b) *m/z* 191 trace for the hopenes produced by dehydration of diplopterol and 3β-methyldiplopterol isolated from *M. capsulatus.* (c) *m/z* 205 ion current trace for the same mixture. (d) *m/z* 205 ion current trace of 2β-methylhopenes obtained from 2β-methyldiplopterol from *M. organophilum.* (e) *m/z* 424 ion current trace of 2α-methylhopenes obtained from 2α-methyldiplopterol. Peak notations indicate the sites of methylation and unsaturation (e.g., 3β(Me) 17(21) is 3β-methylhop-17(21)-ene.)

consisted of hop-17(21)-ene (85%), neohop-13(18)-ene (5%), and hop-21-ene (5%). Thus the major 3β-methyl analogue which only comprised about 2 percent of the mixture, eluted well after hop-17(21)-ene (base peak *m/z* 205, Fig. 10.1c) and was assigned as 3β-methylhop-17(21)-ene. Hydrogenation of this mixture using 10 percent Pd/C (5 mg) in glacial acetic acid (5 ml) and perchloric acid (50 μl) under H_2 (1 atm, 4 hr) yielded a major product identified by mass spectrum and retention time as 17α(H),21β(H)-hopane (Fig. 10.2b) and using an authentic standard (Fig. 10.2a) for comparison. The principal methylated product was, therefore, assigned as 3β-methyl-17α(H),21β(H)-hopane (Fig. 10.3b). When carried out on 2β-methyldiplopterol, the same procedure yielded a mixture of hopene products which were assigned (Fig. 10.1d), on the basis of relative retention times and the above precedent as 2β-methylhop-17(21)-ene, 2β-methylneohop-13(18)-ene, and 2β-methylhop-21-ene. On hydrogenation, this yielded a major product assigned as 2β-methyl-17α(H),21β(H)-hopane by relative retention time and mass spectrum (Fig. 10.3c) together with trace of 17α(H),21β(H)-hopane derived from some diplopterol which was present in the precursor.

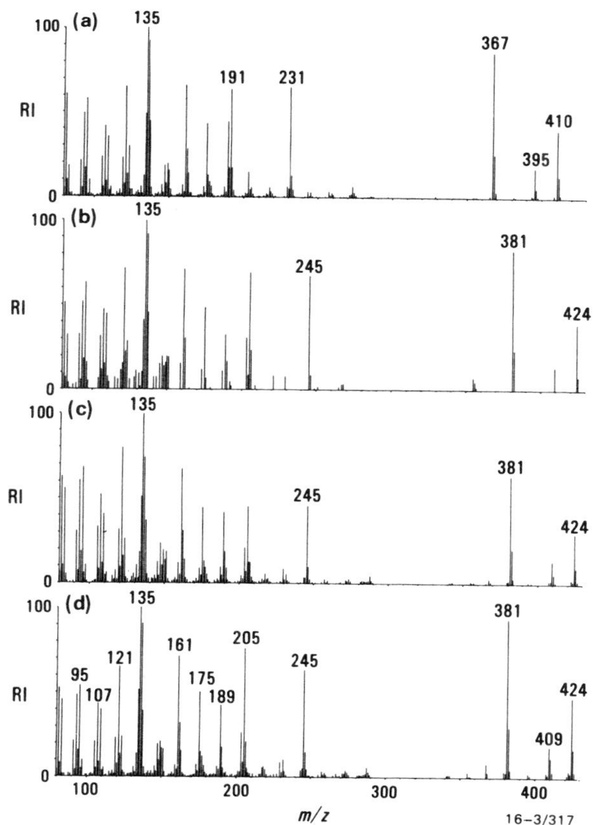

Figure 10.2 Mass spectra of standards (a) hop-17(21)-ene, (b) 3β-methylhop-17(21)-ene, (c) 2β-methylhop-17(21)-ene, and (d) 2α-methylhop-17(21)-ene.

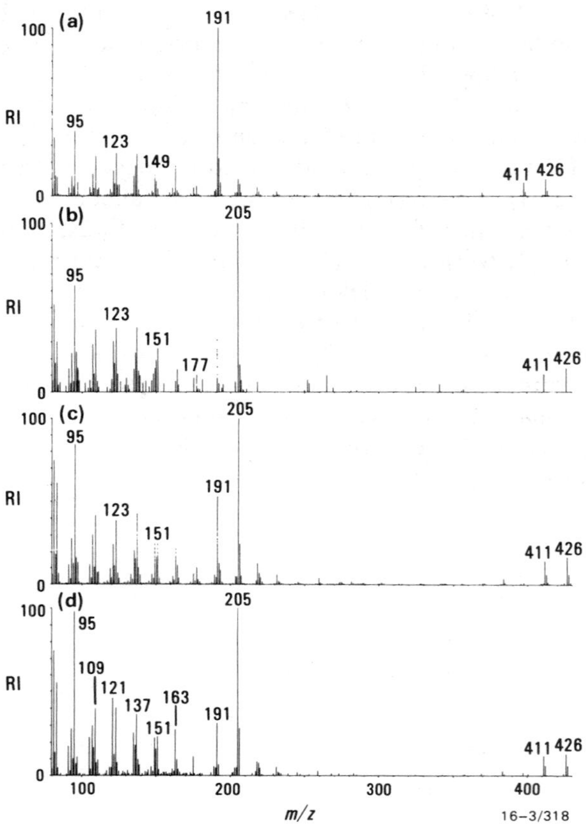

Figure 10.3 Mass spectrometric comparison of hopanes derived from acidic hydrogenation of hop-17(21)-enes. (a) 17α(H),21β(H)-hopane standard; (b) 3β-methyl-17α(H),21β(H)-hopane product *M. capsulatus*; (c) 2β-methyl-17α(H),21β(H)-hopane derived from *M. organophilum*; and (d) 2α-methyl-17α(H),21β(H)-hopane obtained from synthetic 2α-methyldiplopterol.

Synthetic 2α-methyldiplopterol yielded a similar mixture composed of 2α-methylhopenes (Fig. 10.1e) which, on hydrogenation, produced a methylhopane assigned as 2α-methyl-17α(H),21β(H)-hopane (mass spectrum Fig. 10.3d) with minor 17α(H),21β(H)-hopane (ca. 7%) and 17α(H),21β(H)-2β-methylhopane (ca. 7%) derived from impurities in the substrate. This assignment was supported by a ^{13}C nmr spectrum of the product. It showed good correlation between actual and predicted shifts of ring-A carbons due to the additional 2α-methyl substituent [i.e., +8.4, +5.8, +9 ppm for carbons 1, 2, and 3, respectively and 23.6 (predicted 23.3) for the 2α-methyl carbon)]. Signals for carbons in the DE-rings were also consistent with the 17α(H),21β(H)-stereochemistry (Balogh et al., 1973). The mass spectra of the methylhopane products were similar to those of 17α(H),21β(H)-hopane (Fig. 10.3a) except for the mass shifts due to the additional methyl group (Fig. 10.3). Neutral dehydration reactions, using benzene (10 ml) as solvent, distilled acetic anhydride (0.1 ml) and sodium acetate (20 mg) were also conducted for comparison purposes. Neutral dehydration of pure diplopterol (0.2 mg) yielded hop-22(29)-ene (diploptene) and hop-21-ene as the major products (Fig. 10.4a and d). This was repeated using the 2α-methyl- and 2α-methyldiplopterols in order to determine the relative retention times and mass spectra of these hopenes (Fig. 10.4b, c, e, and f, respectively). Although insufficient material was available to

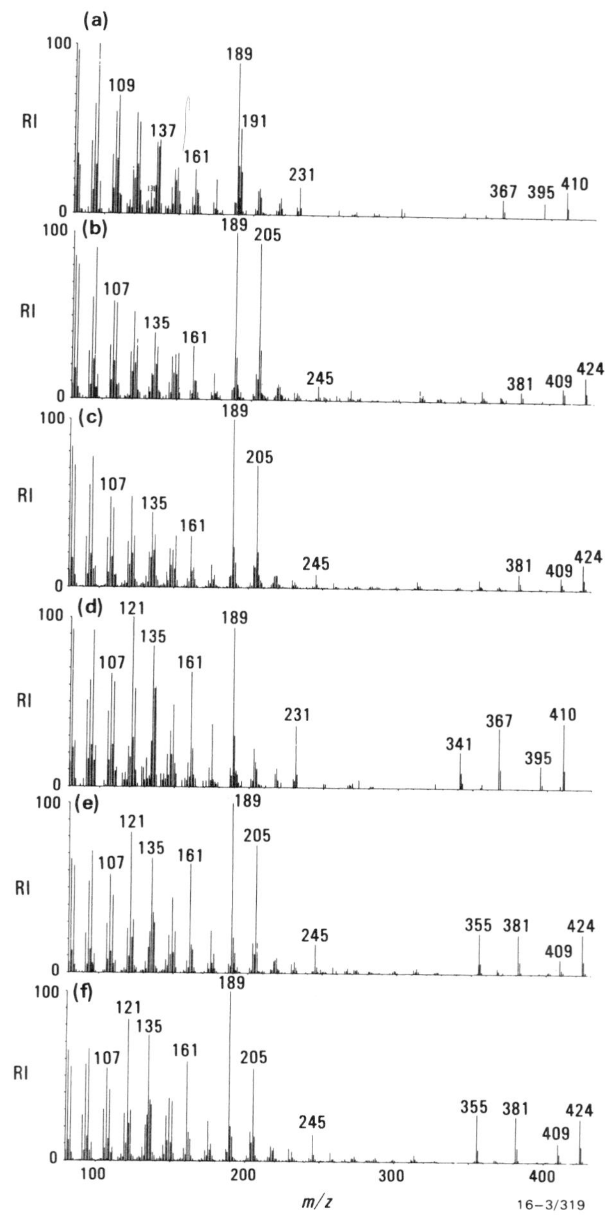

Figure 10.4 Mass spectrometric comparison of hopenes derived from neutral dehydration of diplopterols. (a) hop-22(29)-ene, (b) 2α-methylhop-22(29)-ene, (c) 2β-methylhop-22(29)-ene, (d) hop-21-ene, (e) 2α-methylhop-21-ene, and (f) 2β-methylhop-21-ene.

confirm the various olefin structures by nmr spectroscopy, the relative abundances of major fragment ions in their mass spectra, strongly support the assignments made by analogy with the diplopterol dehydration experiment. The syntheses and spectra of 3β-methylhop-21-ene and 3β-methyldiploptene have been reported by Zundel and Rohmer (1985a).

GC-MS Comparison of Synthetic and Fossil Hopanes

Two chromatographic conditions and columns of different polarity (i.e., 50 m × 0.2 mm ID apolar HP Ultra-1; 60–300°C at 3°C/min and 50 m × 0.2 mm ID moderate polarity SGE BP-10, 60–280°C at 3°C/min) were employed. Retention time plots for the extended hopanes were generated from data acquired on Ultra-1 by rapid (20°C/min) programming of the oven to 270°C for isothermal elution at that temperature. Hopanes were visualized by the MRM chromatograms for $M^{+\cdot} \rightarrow m/z$ 191.179 transitions while the methyl analogues appeared as $M^{+\cdot} \rightarrow m/z$ 205.195 transitions. In one set of comparisons using Ultra-1, a bitumen (#3050; 470 m in Santalum #1 core) extracted from the Ordovician Goldwyer Formation, Canning Basin, Western Australia (Hoffmann et al., 1987) was used as an example showing thermal maturity approaching the oil generation zone. In another experiment using BP-10, an immature bitumen (#4413; 713.6m in #1 Schoeck Errington core) extracted from the Cretaceous Greenhorn Formation, Central North American Seaway was used.

RESULTS AND DISCUSSION

Authenticity of the Precursor Diplopterols

The hopanol mixtures isolated from several species and cultures of methylotrophic bacteria had compositions consistent with data reported earlier by Zundel and Rohmer (1985a, 1985b). For instance, extraction and TLC separations of the lipids from several cultures of *M. capsulatus* yielded a diplopterol fraction which contained a minor component of 3β-methyldiplopterol which, by GC-MS analysis, eluted significantly later than diplopterol and was identical (by GC-MS analysis) to the 3β-methyldiplopterol present in *Acetobacter pasteurianus*. Periodate oxidation and borohydride reduction, acetylation and GC-MS analysis of the corresponding bacteriohopane polyols yielded mixtures with mass spectral characteristics consistent with the earlier assignments of hopan-29-ol and a later eluting 3β-methyl analogue (Zundel and Rohmer, 1985a). *M. organophilum* grown on two different media [i.e., ammonium mineral salts of Hestrin and Schramm (1954), Bisseret et al. (1985)] yielded respectively a 1:1 mixture of diplopterol and 2β-methyldiplopterol or almost pure 2β-methyldiplopterol as reported by these authors. We also repeated the procedure for isolation of 22-hydroxyhopan-3-one from Dammar resin

(Dunstan et al., 1957) and its conversion to 2α-methyldiplopterol and obtained a product with identical mass and ^{13}C nmr spectra to those reported by Bisseret et al. (1985). The 2α-methyl- and 2β-methyldiplopterols had GC retention times similar to diplopterol itself. The consistency of our results with those of the earlier workers led us to conclude that our biogenic precursors constituted appropriate standards from which to prepare hopane analogues.

Authenticity of the Hopane Products

Procedures for dehydration of diplopterols, which have 17β(H),21β(H) stereochemistry, to a variety of hopene product mixtures and their subsequent conversion to hopanes of different stereochemistry have been elaborated elsewhere (e.g., Ageta et al., 1963; Tsuda et al., 1967; Ensminger et al., 1974; van Dorsselaer et al., 1974). Since our purpose was to prepare a GC-MS standard for the ubiquitous 17α(H),21β(H)-methylhopanes found in sediments and petroleums, we utilized a strong acid to dehydrate diplopterol to a mixture containing a predominance of hop-17(21)-ene. This, in turn, is known to preferentially yield 17α(H),21β(H)-hopane ($> 55\%$) on hydrogenation with Adams catalyst in acidic solvents (Tsuda et al., 1967). Substitution of palladium (5%) on charcoal as catalyst led to products having in excess of 95 percent of the desired 17α(H),21β(H) stereochemistry (assigned because the hopane product coeluted with 17α(H),21β(H)-hopane from ancient sediments) and the minor product having 17β(H),21α(H) stereochemistry (assigned because of its coelution with moretane). Figure 10.1 shows that the hopene products from diplopterol (*ex M. capsulatus*) contain two series of compounds with the predominant hopenes having mass spectra and elution characteristics identical to authentic standards of hop-17(21)-ene, hop-21-ene, hop-22(29)-ene, and neohop-13(18)-ene. Therefore, the major coexisting methylated hopene must be the 3β-methylhop-17(21)-ene and this is supported by its retention time and by the relative abundance of major fragment ions in its mass spectrum. The 2β-methyl- and 2α-methyldiplopterols gave analogous products when dehydrated under the same conditions (Figs. 10.1 to 10.3).

Mass Spectral Analysis of Hopene Isomers

Hop-22(29)-ene is a major hydrocarbon in many bacteria including the methylotrophs reported here. We sought to determine whether methyl analogues of hop-22(29)-ene were also present (none were identified) and to obtain reference spectra that would be useful in studies of the distribution of methylhopenes in modern sediments. Hop-22(29)-ene and hop-21-ene were generally minor products when dehydration was conducted using a strong acid. We, therefore, repeated the reaction under neutral conditions because this has been reported to preferentially yield these isomers (Ageta et al., 1963). We obtained a mixture composed mainly of hop-22(29)-ene and hop-21-ene when the reaction was carried out on diplopterol, or mixtures of the 2α-methyl- and 2β-methyl analogues from the appropriate precursors. The retention times of hop-22(29)-ene, 2β-methyl-, and 2α-methylhop-22(29)-ene were almost identical and each was just resolved from the later eluting hop-21-ene (e.g., compare Fig. 10.1a and e). Mass spectral comparison of the isomers of hopenes and methyl hopenes (Figs. 10.2 and 10.4) reveal that changes to the double bond position causes marked differences in the fragmentation patterns while the position of methyl substitution causes little, if any, change apart from the 14 dalton mass shifts to predominant fragments derived from ring-A. The spectra of the three methylhop-17(21)-enes (Fig. 10.2b, c, and d) correlate with the spectrum of authentic hop-17(21)-ene (Fig. 10.2a) with regard to the relative intensity of the ions at m/z 189, 205, 245, 381, 409, and 424 compared to the corresponding fragments of m/z 189, 191, 231, 367, 395, and 410.

Chromatographic Behavior of Methylhopanes

Partial synthesis of hopane using a *M. capsulatus* diplopterol precursor, that is, one which contained 3β-methyldiplopterol as a minor component, allowed us to prepare a product identifiable as 17α(H),21β(H)-hopane (i.e., "geohopane") and carrying the 3β-methylhopane with equivalent stereochemistry. Thus, we expect geohopane and coexisting methylhopane in sedimentary bitumen, should *both* coelute with synthetic products if our approach is a valid one. In fact, all our substrates contained diplopterol as an "impurity" and consequently all methylhopane products contained some 17α(H),21β(H)-hopane. Normally, the use of such mixtures would create difficulties during chromatographic coelution experiments. However, in this case we employed metastable reaction monitoring GC-MS (Warburton and Zumberge, 1983), for comparing synthetic compounds and bitumens. This enabled us to visualize methylhopane products by virtue of their predominant $M^{+\cdot} \rightarrow m/z$ 205 transition and independently, yet simultaneously, observe the hopane products in measurement of the $M^{+\cdot} \rightarrow m/z$ 191 transition. This is analogous to the way C_{30} methyl steranes and desmethyl steranes have been differentiated even though they comprise complex, coeluting mixtures in sediments and petroleums (Moldowan et al., 1985; Summons et al., 1987; Summons and Capon, 1988). Indeed, full-scan GC-MS analysis of the hopane mixture prepared from *M. capsulatus* showed two peaks: (1) a major peak assigned as 17α(H),21β(H)-hopane on the basis of mass spectrum (Fig. 9.3a) and retention time and (2) a minor peak (ca. 2%) as 3β-methyl-17α(H),21β(H)-hopane (Fig. 9.3b). This enhanced our ability to make accurate chromatographic comparisons because the products appeared separately in the MRM chromatograms for m/z 412 → 191 and m/z 426 → 205 transitions and enabled minor alterations to chromatographic behavior to be observed. Our preparations from the 2-methyldiplopterols contained 2β-methyl-17α(H),21β(H)-hopane and 2α-methyl-17α(H),21β(H)-hopane as the major products.

Like their diplopterol and hopene precursors, both of the 2-methylhopanes eluted with similar retention times on an apolar Ultra-1 column. The 2β-methylhopane virtually coeluted with hopane, while 2α-methylhopane eluted on the trailing side and was incompletely resolved from hopane. The 3β-methylhopane eluted with a significantly longer retention time and at a point midway between 17α(H),21β(H)-homohopane (22*R*) and 17β(H),21α(H)-homohopane (22*S* + 22*R*). On a moderately polar BP-10 column, the 2β-methylhopane just preceded 2α-methylhopane and both eluted before 17α(H),21β(H)-hopane. This observation of a reversal in relative elution positions of hopanes and methylhopanes on columns of different polarity reduces the chance of error in compound identification by fortuitous coelution.

Figure 10.5 illustrates the results of coinjection experiments with the synthetic methylhopanes and the hydrocarbon fraction of a sediment extract known to contain two series of methylhopanes (#3050; see Hoffmann et al., 1987). MRM chroma-

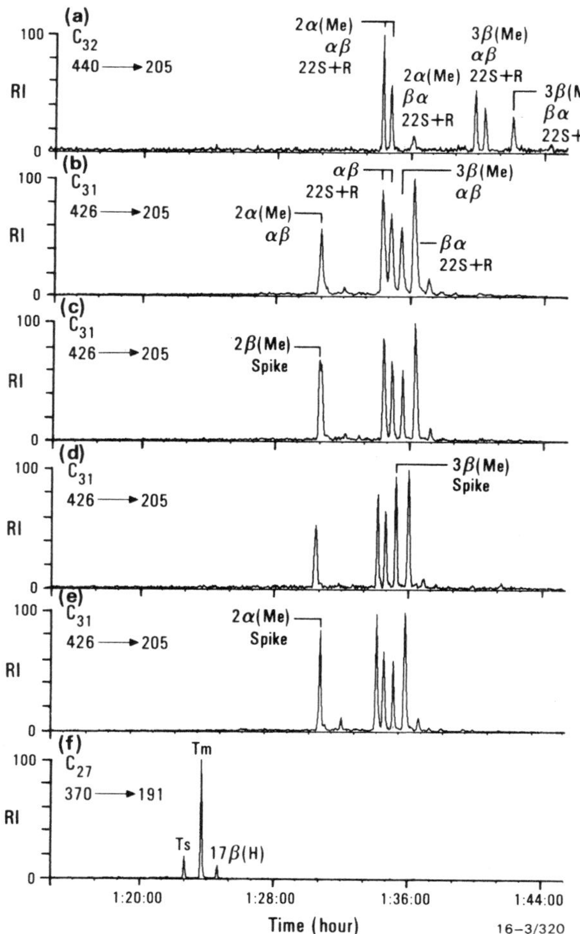

Figure 10.5 GC-MS (MRM) comparison of methylhopanes in an Ordovician sediment (#3050) extract with the chromatograms obtained by coinjection with synthetic methyl-17α(H),21β(H)-hopanes. Each chromatogram is labeled with the carbon number and reaction being monitored. αβ denotes 17α(H),21β(H) and βα denotes 17β(H),21α(H) stereochemistry respectively. Traces a, b, and f #3050 alone. Trace c with 2β-methyl-17α(H),21β(H)-hopane added. Trace d with 3β-methyl-17α(H),21β(H)-hopane added. Trace e with 2α-methyl-17α(H),21β(H)-hopane added. Chromatography is on an Ultra-1 column (60–300°C at 3°C/min).

tograms for the dominant M⁺ 426 → m/z 205 AB-ring fragment. The trace for the C_{27} hopanes is included to provide a guide to overall elution positions. The (C_{31}) homohopanes and homomoretanes appear in the M⁺ 426 → m/z 205 transition because this response arises via the alternative M⁺ → DE-ring fragmentation (Kimble, 1972; Ensminger et al., 1974; van Dorsselaer et al., 1974) which is stronger in moretanes than hopanes and corresponds to the relative abundance of AB-ring and DE-ring fragments in the main beam spectra. Trace b shows the methylhopanes as they appear in the #3050 bitumen. Trace c shows the mixture after a spike of 2β-methylhopane was added. Note that the major peak has broadened because there has been no coelution. Trace d shows the same mixture after a spike of 3β-methylhopane has been added. Coelution with the later eluting fossil methylhopane is evident because the peak is reinforced without causing any degradation in the peak shape. Trace e shows the earlier eluting fossil methylhopane reinforced by addition of synthetic 2α-methylhopane and, hence, identifies this peak. These same coinjections were repeated on a BP-10 column (data not shown) with an identical result, although in this case, the 2α- and 2β-methylhopanes were completely resolved from each other and both eluted before hopane.

Figure 10.6 shows the results of comparisons with saturated hydrocarbons from an immature sediment (#4413) using a BP-10, moderate polarity column.

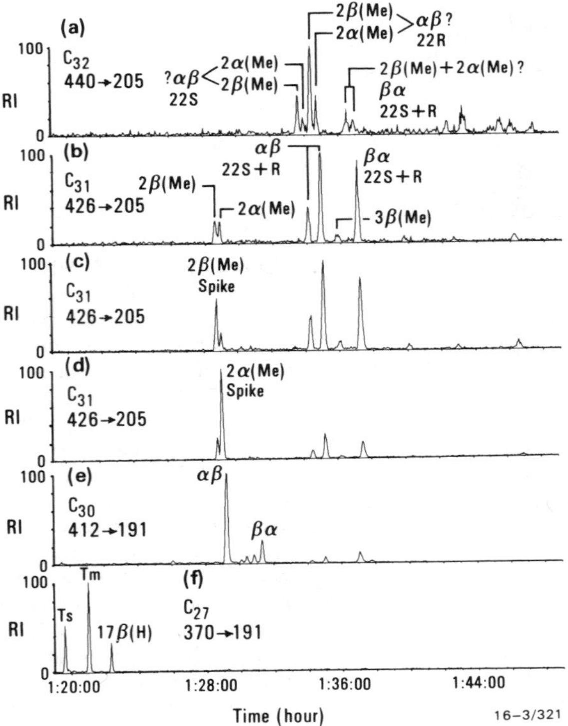

Figure 10.6 GC-MS (MRM) comparison of methylhopanes of an immature Cretaceous sediment (Greenhorn Fm. #4413) extract with the traces obtained by coinjection with methyl-17α(H),21β(H)-hopanes. Traces a, b, e, and f #4413 alone. Trace c with 2β-methyl-17α(H),21β(H)-hopane added. Trace d with 2α-methyl-17α(H),21β(H)-hopane added. Chromatography is on a BP-10 column (60–280°C at 3°C/min). "?" denotes tentative assignment.

Trace b for the bitumen alone appears to have three methylhopane peaks in addition to 17α(H),21β(H)-homohopane (22*S* + 22*R*) and 17β(H),21α(H)-homohopane (22*S* + 22*R*) in the $M^{+\cdot}$ 426 → 205 transition. We showed that the last eluting, and quantitatively minor peak was the 3β-methyl isomer by virtue of its coelution with this standard (data not shown) and this experiment sought to delineate the earlier eluting peaks for 2β and 2α-methyl isomers. This is because the 2β-methyl isomer is only likely to be abundant in very immature sediments if, as seemed probable, both 2-methyl isomers are derived from 2β-methyl-17β(H),21β(H) biogenic precursors. The low degree of maturity of this sample is evident in the homohopanes (trace a) which have not reached equilibrium with respect to the ratio of 22*S* to 22*R* isomers and in the high relative abundance of 17β(H),21α(H)-hopanes (moretanes) (Seifert and Moldowan, 1980). The data in Figure 10.6 show that the first eluting 2-methylhopane was enhanced by coinjection with the 2β-methyl-17α(H),21β(H)-hopane standard (trace c) and the second with 2α-methyl-17α(H),21β(H)-hopane standard (trace d).

In these coinjection experiments, we have only established the identities and relative elution positions of the C_{31} members of three apparent pseudohomologous C_{28}–C_{36} series. Without synthesis of other members, we cannot unambiguously assign structures of all members. However, we can tentatively identify some of them on the basis of chromatographic behavior. This is most readily achieved for the 2α- and 3β-methyl series, because the separation between these isomers is so large, and because we can make the assumption that the two series found in mature sediments will have the most stable equatorial (i.e., 2α and 3β) orientation of the ring-A methyl group. We also must assume that they behave chromatographically in the same way as the hopane and moretane series, with the additional methyl substituent in ring-A causing a uniform retention shift for each member. In immature sediments, the additional presence of 2β-methylhopanes is likely and this could complicate assignments because of the proximity of the 2α- and 2β-methyl isomers. Figure 10.7 shows Ln retention time versus carbon number plots for the C_{27}–C_{35} hopanes and the corresponding C_{28}–C_{36} methylhopanes in the moderately mature sediment #3050. The plots for each series of C_{31}–C_{35} (22*R*) hopanes and C_{32}–C_{36} (22*R*) methylhopanes are almost linear. However, it is interesting to note

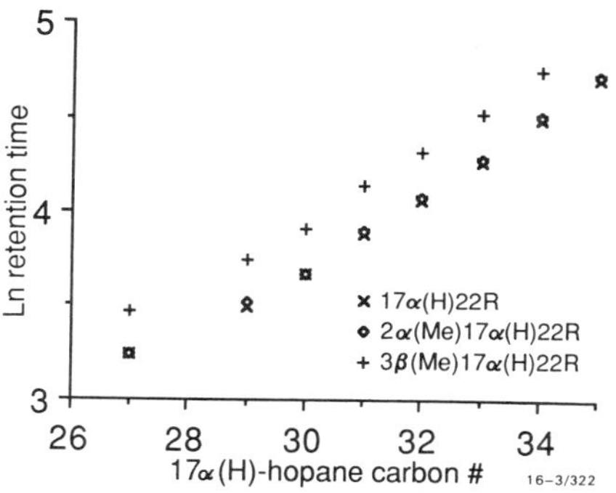

Figure 10.7 Plots of Ln retention time versus carbon number for C_{27}–C_{35} hopanes and the corresponding C_{28}–C_{36} methylhopanes from the #3050 bitumen. Chromatography is on an Ultra-1 column (60–270°C at 20°C/min and then isothermal at 270°C). Retention times were calculated from when the oven reached 270°C.

that deviation from linear behavior occurs at the C_{31} hopane (22*R*) (and the analogous C_{32} methylhopanes) in accord with the data of Larcher et al. (1987) for extended hopanes and moretanes. Moreover, these plots and others for the 17α(H),21β(H)-22*S* and the 17β(H),21α(H)-22*S* and 22*R* isomers are parallel over the range C_{28}–C_{36}. This result suggests that the early eluting series of extended methylhopanes, each of which elutes on the trailing side of the corresponding 17α(H),21β(H)-hopane on an apolar phase, comprise the 2α-methyl series while those which elute near the next higher homologue constitute the 3β-methyl series. Norhopane and trisnorhopane have methyl analogues which elute with a similar pattern suggesting the probability that they also have the same 2α-methyl and 3β-methyl substitution pattern. However, synthesis of these compounds is required for rigorous proof of identity.

Distribution of Methylhopanes in Ancient Sediments

Our results indicate that the most commonly reported methylhopanes, that is those eluting near the equivalent hopane on apolar methyl silicone columns, are 2α-methylhopanes. This series is of high to moderate abundance in Middle Proterozoic sediments of the McArthur Basin (Summons et al., 1988) and in Late Proterozoic oils of the Siberian Platform (Summons and Powell, 1989). We have observed them in the Cambrian carbonate sediments of the Officer Basin which have been discussed by McKirdy et al. (1984), in numerous Ordovician sediments, particularly those containing *G. Prisca* (Alexander et al., 1984; Hoffman et al., 1987), and other Palaeozoic sediments and oils from carbonate environments (Summons and Powell, 1987). They have also been reported from the Mesozoic (McEvoy and Giger 1986). Available data are inadequate for assignment of specific source organisms, although it is probably significant that most reports indicating high abundance of this series relate to sediments from carbonate and/or hypersaline environments. Price et al. (1988) have observed that methylhopanes are usually abundant in carbonate sediments although they are not diagnostic for this lithology. 2β-Methyl hopanoid precursors have been isolated from a cyanobacterium and from methylotrophic bacteria, but they are likely to be found in other, as yet unrecognized, prokaryote source organisms. Bisseret et al. (1985) have demonstrated that production of 2β-methyl hopanoids is dependent on specific culture conditions (also confirmed in this chapter) while Zundel and Rohmer (1985a) found that production of 3β-methyl hopanoid was enhanced by the availability of methionine. This means that additional lines of evidence will be required to identify specific origins for methylhopanes. 2α-Methylhopanes vary greatly in their abundance relative to hopanes and may comprise as much as 20 percent of the hopane load. Generally, individual 2α-methylhopanes are between 1 and 10 percent of the corresponding hopane, but there are some instances where they cannot be detected at all.

The later eluting series of 3β-methylhopanes does not appear to have been reported prior to their observation by Hoffmann et al. (1987). This is probably because they are generally present in low concentration and most easily identified by MRM analysis rather than SIR. Our initial observations are that the 3β-methyl series is ubiquitous but rarely as abundant as the 2α-methyl compounds. Improved understanding of the factors which affect the distributions of these compounds is now required. Methylhopanes will be particularly useful for oil-source rock correlation because their abundance compared to hopanes appears quite variable and, for the 2α-methyl series, at least, under some environmental control.

CONCLUSIONS

Three isomers of methylhopane were prepared from diplopterol precursors and all can be found in bitumens isolated from sedimentary rocks. The most commonly encountered isomer in mature sediments, which elutes on the trailing side of 17α(H),21β(H)-hopane on apolar phases, is 2α-methylhopane. It cooccurs in immature sediments with 2β-methylhopane, consistent with an origin from 2β-methyl hopanoid precursors. 3β-Methylhopane elutes after 17α(H),21β(H)-homohopane (22*R*) and appears to be less widely distributed than the 2-methyl isomers. Retention time correlations indicate that the 2α- and 3β-methylhopanes comprise C_{28}–C_{36} pseudohomologous series analogous to the C_{27}–C_{35} 17α(H)-21β(H)-hopanes.

Acknowledgments

We are indebted to Janet Hope and Carola Stranger for assistance with synthetic conversions, isolation, and purification of hopanoids from bacterial cultures. Zarko Roksandic assisted with GC-MS analysis and NMR spectra were run and analyzed by Robert Capon. Sediment from the Goldwyer Formation was supplied by Clinton Foster, Western Mining Corporation, Australia. John Hayes provided the sediment sample from the Greenhorn Formation. John Volkman, Christopher Boreham, Ken Peters and an anonymous reviewer contributed numerous helpful comments. Roger Summons publishes with permission of the Director, Bureau of Mineral Resources.

REFERENCES

Ageta, H., Iwata, K., and Otake, Y. (1963) A fern constituent: diplopterol, a triterpenoid isolated from *Diplopterygium glaucum* Nakai. *Chem. Pharm. Bull. 11*, 407–409.

Alexander, R., Cumbers, M., and Kagi, R.I. (1984) Geochemistry of some Canning Basin crude oils. In *The Canning Basin, W. A.* (ed. P.G. Purcell), pp. 353–358. Proc. Geol. Soc. Aust./Pet. Expl. Soc. Aust. Symposium.

Balogh, B., Wilson, D.M., Christiansen, P., and Burlingame, A.L. (1973) 17α(H)-

Hopane identified in oil shale of the Green River Formation (Eocene) by carbon-13 NMR. *Nature 242*, 603–605.

BIRD, C.W., LYNCH, J.M., PIRT, S.J., REID, W.W., BROOKS, C.J.W., and MIDDLEDITCH, B.S. (1971) Steroids and squalene in *Methylococcus capsulatus* grown on methane. *Nature 230*, 473–474.

BISSERET, P., ZUNDEL, M., and ROHMER, M. (1985) Prokaryotic triterpenoids: 2. 2β-methylhopanoids from *Methylobacterium organophilum* and *Nostoc muscorum*, a new series of prokaryotic triterpenoids. *Eur. J. Biochem. 150*, 29–34.

BLIGH, E.G. and DYER, W.J. (1959) A rapid method of total lipid extraction and purification. *Can. J. Biochem. Physiol. 37*, 911–917.

BRASSELL, S.C., COMET, P.A., EGLINTON, G., ISAACSON, P.J., MCEVOY, J., MAXWELL, J.R., THOMSON, I.D., TIBBETTS, P.J.C., and VOLKMAN, J.K. (1980) The origin and fate of lipids in the Japan Trench. In *Advances in Organic Geochemistry 1979* (eds. A.G. Douglas and J.R. Maxwell), Pergamon Press, Oxford. pp. 375–392.

DE ROSA, M., GAMBACORTA, A., MINALE, L., and BU'LOCK, J.D. (1971) Bacterial triterpanes. *J. Chem. Soc. Chem. Commun.* 619–620.

DUNSTAN, W.J., FAZAKERLEY, H., HALSALL, T.G., and JONES, E.H.R. (1957) The chemistry of the triterpenes and related compounds. Part XXXII. The chemistry of hydroxyhopanone. *Croat. Chem. Acta 29*, 173–182.

ENSMINGER, A. (1977) Evolution de composé polycycliques sédimentaires. Thèse de docteur en Sciences, Université Louis Pasteur, Strasbourg.

ENSMINGER, A., VAN DORSSELAER, A., SPYCKERELLE, CH., ALBRECHT, P. and OURISSON, G. (1974) Pentacyclic triterpenes of the hopane type as ubiquitous geochemical markers: origin and significance. In *Advances in Organic Geochemistry 1973.* (eds. B. Tissot and F. Bienner), Editions Technip, Paris. pp. 245–260.

ENSMINGER, A., ALBRECHT, P., OURISSON, G., and TISSOT, B. (1977) Evolution of polycyclic alkanes under the effect of burial (Early Toarcian shales, Paris Basin). In *Advances in Organic Geochemistry 1975* (eds. R. Campos and J. Goni), Enadimsa, Madrid. pp. 45–52.

FÖRSTER, H.J., BIEMANN, K., HAIGH, W., TATTRIE, N.H., and COLVIN, J.R. (1973) The structure of novel C_{35} triterpenes from *Acetobacter xylinum. Biochem. J. 135*, 133–143.

GELPI, E., SCHNEIDER, H., MANN, J., and ORO, J. (1970) Hydrocarbons of geochemical significance in microscopic algae. *Phytochemistry 9*, 603–612.

TEN HAVEN, H.L., DE LEEUW, J.W., and SCHENCK, P.A. (1985) Organic geochemical studies of a Messinian evaporitic basin, northern Apennines (Italy) I: Hydrocarbon biological markers for a hypersaline environment. *Geochim. Cosmochim. Acta 49*, 2181–2191.

TEN HAVEN, H.L., DE LEEUW, J.W., PEAKMAN, T.M., and MAXWELL, J.R. (1986) Anomalies in steroid and hopanoid maturity indices. *Geochim. Cosmochim. Acta 50*, 853–855.

HESTRIN, S. and SCHRAMM, M. (1954) Synthesis of cellulose by *Acetobacter xylinum. Biochem. J. 58*, 345–352.

HILLS, I.R. and WHITEHEAD, E.V. (1966) Triterpanes in optically active petroleum distillates. *Nature 209*, 977–979.

HOFFMANN, C.F., FOSTER, C.B., POWELL, T.G., and SUMMONS, R.E. (1987) Hydrocarbon biomarkers from Ordovician sediments and the fossil alga *Gloeocapsomorpha prisca Zalessky* 1917. *Geochim. Cosmochim. Acta 51*, 2681–2697.

JAHNKE, L.L. and NICHOLS, P.D. (1986) Methyl sterol and cyclopropane fatty acid composition of *Methylococcus capsulatus* grown at low oxygen tensions. *J. Bacteriol. 167*, 238–242.

KATES, M. (1986) *Techniques in Lipidology: Isolation, Analysis and Identification of Lipids.* vol. 3, Elsevier. 464 pp.

KIMBLE, B.J. (1972) The geochemistry of triterpenoid hydrocarbons. PhD Thesis, University of Bristol.

LANGWORTHY, T.A. and MAYBERRY, W.R. (1976) A 1,2,3,4-tetrahydroxy pentane-substituted pentacyclic triterpene from *Bacillus acidocaldarius. Biochim. Biophys. Acta 431*, 570–577.

LARCHER, A.V., ALEXANDER, R., and KAGI, R.I. (1987) Changes in configuration of extended moretanes with increasing sediment maturity. *Org. Geochem. 11*, 59–63.

MACKENZIE, A.S., PATIENCE, R.L., MAXWELL, J.R., VANDENBROUCKE, M., and DURAND, B. (1980) Molecular parameters of maturation in the Toarcian shales, Paris Basin, France—I. Changes in the configuration of acyclic isoprenoid alkanes, steranes and triterpanes. *Geochim. Cosmochim. Acta 44*, 1709–1721.

MCEVOY, J. and GIGER, W. (1986) Origin of hydrocarbons in Triassic Serpiano oil shales: hopanoids. In *Advances in Organic Geochemistry 1985* (eds. D. Leythaeuser and J. Rullkötter), Pergamon Press, Oxford. pp. 943–949.

MCKIRDY, D.M., KANTSLER, A.S., EMMETT, J.K., and ALDRIDGE, A.K. (1984) Hydrocarbon genesis and organic facies in Cambrian carbonates of the eastern Officer Basin, South Australia. In *Petroleum Geochemistry and Source Rock Potential of Carbonate Rocks* (ed. J.C. Palacas), Am. Assoc. Petrol. Geol. Studies in Geology 18, pp. 13–32. Tulsa, Okla.

MOLDOWAN, J.M., SEIFERT, W.K., and GALLEGOS, E.J. (1985) Relationship between petroleum composition and depositional environment of petroleum source rocks. *Am. Assoc. Pet. Geol. Bull. 69*, 1255–1268.

OURISSON, G., ROHMER, M., and ALBRECHT, P. (1979) The hopanoids. Palaeochemistry and biochemistry of a group of natural products. *Pure Appl. Chem. 51*, 709–729.

OURISSON, G., ALBRECHT, P., and ROHMER, M. (1982) Predictive microbial biochemistry: from molecular fossils to procaryotic membranes. *Trends Biochem. Sci. 7*, 233–239.

OURISSON, G., ROHMER, M., and PORALLA, K. (1987) Prokaryotic hopanoids and other polyterpenoid sterol surrogates. *Ann. Rev. Microbiol. 41*, 301–333.

PRICE, P.L., O'SULLIVAN, T.O., and ALEXANDER, R. (1988) The nature and occurrence of oil in Seram, Indonesia. In *Proc. 16th Annual Convention Indonesian Petroleum Assoc. Jakarta*, 1987. In press.

ROHMER, M. and OURISSON, G. (1976) Méthylhopanes d'*Acetobacter xylinum et d'Acetobacter rancens:* Une nouvelle famille de composés triterpéniques. *Tetrahedron. Lett.* 3641–3644.

ROHMER, M., BOUVIER, P., and OURISSON, G. (1979) Molecular evolution of biomembranes: Structural equivalents and phylogenetic precursors of sterols. *Proc. Natl. Acad. Sci. USA 76*, 847–851.

ROHMER, M., BOUVIER-NAVE, P., and OURISSON, G. (1984) Distribution of hopanoid triterpenes in prokaryotes. *J. Gen. Microbiol. 130*, 1137–1150.

SEIFERT, W.K. and MOLDOWAN, J.M. (1978) Applications of steranes, terpanes and monoaromatics to the maturation, migration and source of crude oils. *Geochim. Cosmochim. Acta 42*, 77–95.

SEIFERT, W.K. and MOLDOWAN, J.M. (1980) The effect of thermal stress on source-rock quality as measured by hopane stereochemistry. In *Advances in Organic Geochemistry 1979* (eds. A.G. Douglas and J.R. Maxwell), Pergamon Press, Oxford. pp. 229–237.

SUMMONS, R.E. and CAPON, R.J. (1988) Fossil steranes with unprecedented methylation in ring-A. *Geochim. Cosmochim. Acta 52*, 2733–2736.

SUMMONS, R.E. and POWELL, T.G. (1987) Identification of aryl isoprenoids in source rocks and crude oils: Biological markers for the green sulphur bacteria. *Geochim. Cosmochim. Acta 51*, 557–566.

SUMMONS, R.E. and POWELL, T.G. (1989) Hydrocarbon composition and the depositional environment of source rocks for the Late Proterozoic oils of the Siberian Platform. In *Early organic evolution and mineral and energy resources*. (eds. M. Schidlowski, D.M. McKirdy, and P.A. Trudinger) Springer-Verlag, Berlin. In press.

SUMMONS, R.E., VOLKMAN, J.K., and BOREHAM, C.J. (1987) Dinosterane and other steroidal hydrocarbons of dinoflagellate origin in sediments and petroleum. *Geochim. Cosmochim. Acta 51*, 3075–3082.

SUMMONS, R.E., POWELL, T.G., and BOREHAM, C.J. (1988) Petroleum geology and geochemistry of the Middle Proterozoic McArthur Basin, Northern Australia. III Composition of extractable hydrocarbons. *Geochim. Cosmochim. Acta 52*, 1747–1763.

TSUDA, Y., ISOBE, K., FUKUSHIMA, S., AGETA, H., and IWATA, K. (1967) Final clarification of the saturated hydrocarbons derived from hydroxyhopanone, diploptene, zeorin and dustanin. *Tetrahedron Lett.*, 23–28.

VAN DORSSELAER, A., ENSMINGER, A., SPYCKERELLE, C., DASTILLUNG, M., SIESKIND, O., ARPINO, P., ALBRECHT, P., OURISSON, G., BROOKS, P.W., GASKELL, S.J. KIMBLE, B.J., PHILP, R.P. MAXWELL, J.R., and EGLINTON, G. (1974) Degraded and extended hopane derivatives (C_{27}–C_{35}) as ubiquitous geochemical markers. *Tetrahedron Lett.* 1349–1352.

WARBURTON, G.A. and ZUMBERGE, J.E. (1983) Determination of petroleum sterane distributions by mass spectrometry with selective metastable ion monitoring. *Anal. Chem. 55*, 123–126.

WHITTENBURY, R. and DALTON, H. (1981) The methylotrophic bacteria. In: *The Prokaryotes* (eds. M.P. Starr, H. Stolp, H.G. Truper, A. Balows, and H.G. Schegel) vol. 1, Springer, Berlin. pp. 894–902.

ZUNDEL, M. and ROHMER, M. (1985a) Prokaryotic triterpenoids.1. 3β-methylhopanoids from *Acetobacter* sp. and *Methylococcus capsulatus*. *Eur. J. Biochem. 150*, 23–27.

ZUNDEL, M. and ROHMER, M. (1985b) Prokaryotic triterpenoids.3. The biosynthesis of 2β-methylhopanoids and 3β-methylhopanoids of *Methylobacterium organophilum* and *Acetobacter pasteurianus* spp. *pasteurianus*. *Eur. J. Biochem. 150*, 35–39.

11

An Oil-Source Correlation Study Using Age-Specific Plant-Derived Aromatic Biomarkers

Robert Alexander, Alfons V. Larcher, Robert I. Kagi, and Peter L. Price

Abstract. Whether or not the sediments in the Eromanga Basin of Australia have generated petroleum is a problem of considerable commercial importance which remains contentious as it has not yet been resolved unequivocally. Sediments of the underlying Cooper Basin were deposited throughout the Permian and much of the Triassic and deposition in the overlying Eromanga Basin commenced in the Early Jurassic and extended into the Cretaceous. As Araucariaceae (trees of the kauri pine group) assumed prominence for the first time in the Early to Middle Jurassic and were all but absent in older sediments, a promising approach would seem to be using the presence or absence of specific araucarian chemical marker signatures as a means of distinguishing oils formed from source rocks in the Eromanga Basin from those derived from the underlying Cooper Basin sediments.

The aromatic hydrocarbon compositions of the sediment extracts from the Cooper and Eromanga Basins have been examined to identify their distinctive fossil hydrocarbon markers. Sediments from the Eromanga Basin, which contain abundant microfossil remains of the araucarian plants, contain diterpane hydrocarbons and aromatic hydrocarbons which bear a strong relationship to natural products in modern members of the Araucariaceae. Sediments from the Permo-Triassic Cooper

Basin, which predate the Jurassic araucarian flora, have different distributions of aromatic hydrocarbons.

Many oils found in the Cooper/Eromanga region do not have the biological marker signatures of the Jurassic sediments and appear to be derived from the underlying Permian sediments; however, several oils contained in Jurassic to Cretaceous reservoirs show the araucarian signature of the associated Jurassic to Early Cretaceous source rock sediments. It is likely, therefore, that these oils were sourced and reservoired within the Eromanga Basin and have not migrated from the Cooper Basin sequences below.

INTRODUCTION

Terpenoid and steroidal natural products derived from plants undergo transformations during diagenesis and maturation of the sediments to produce saturated and aromatic hydrocarbons which are structurally related to their natural product precursors. These biomarkers can be identified by detailed chemical analysis of the aromatic and saturated hydrocarbon fractions of crude oils and sediment extracts. In cases where a modern representative of a plant type is available, the biomarkers which might result from incorporation of such material into sediments can be inferred from the natural product composition of the modern plant. Further, when the evolutionary history of the plant family is known, the presence or absence of such biomarkers may serve as a useful indicator for the age of a sediment sample.

Sediments of the Cooper Basin were deposited throughout the Permian and much of the Triassic and deposition in the Eromanga Basin commenced in the Early Jurassic and extended into the Cretaceous (Kantsler et al., 1983). Evolution of the Araucariaceae (trees of the Kauri pine group) occurred in the Early to Middle Jurassic (Stewart, 1983; Miller, 1977, 1982), so the presence or absence of biomarkers derived from resins of the araucarian flora provides a convenient method for distinguishing oils derived from Eromanga Basin sediments from those derived from the underlying Cooper Basin formations.

Biomarkers from Araucariaceae

The bicyclic and tricyclic natural products shown in Figure 11.1 have been reported to be present in the resins and fossil resins of the Araucariaceae (Thomas, 1969). In particular, resins from various species of Agathis have been investigated in detail, and in most cases the components that are the most abundant are the bicyclic and tricyclic skeletal types (Thomas, 1969). In contrast, resins derived from other conifer families contain a predominance of compounds with different skeletal types (Aplin et al., 1963; Cambie et al., 1971; Carman and Sutherland, 1979). For example, resins derived from Podocarpaceae contain an abundance of diterpenoids with tetracyclic skeletons (Aplin et al., 1963).

The relative abundances of the aromatic and saturated biomarkers derived

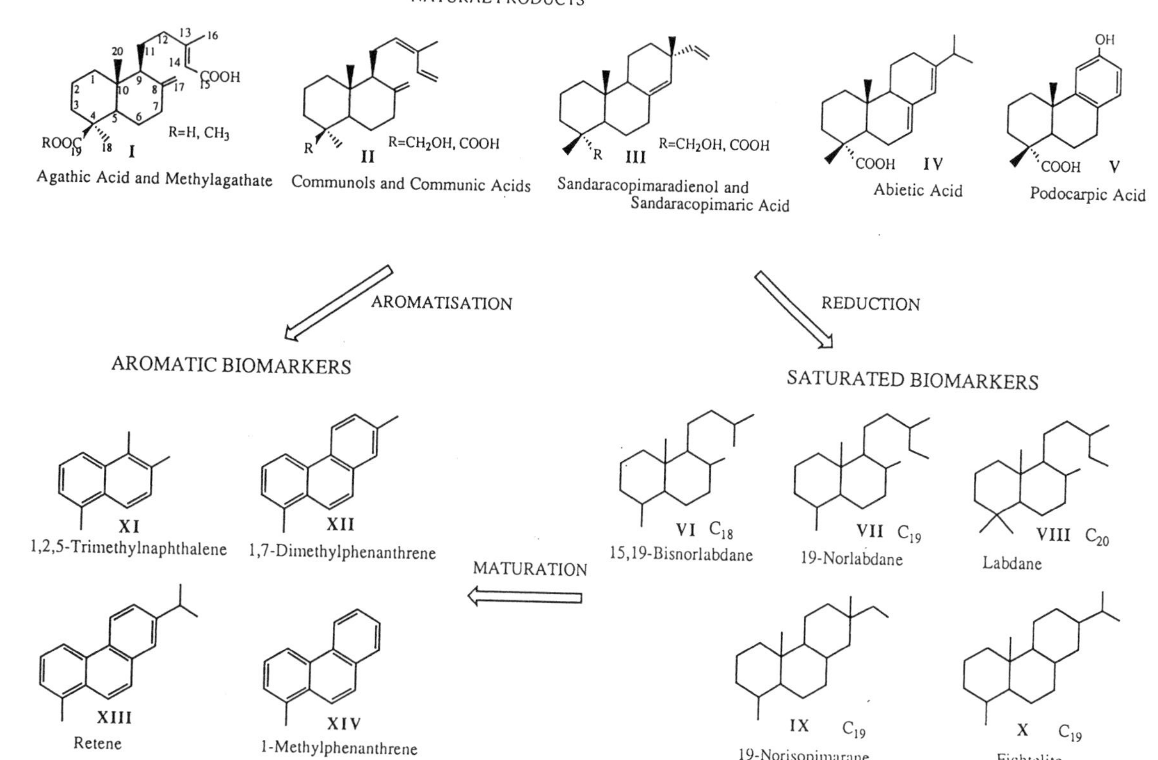

Figure 11.1 Predicted relationships between aromatic biomarkers, saturated biomarkers, and their natural product precursors.

from resin source materials depends upon the conditions experienced during deposition, diagenesis and maturation of the sediments (Wakeham et al., 1980; Barnes and Barnes, 1981; Alexander et al., 1987). In depositional environments which favor coal formation, aromatization of the plant-derived terpenoids is strongly favored relative to reduction and preservation of saturated biomarkers (Strachan et al., 1988; Alexander, Hazai et al., 1987). In the case of sediments from the Eromanga Basin system, the abundance of saturated biomarkers is extremely low in many crude oils making crude oil-source correlations difficult (Alexander et al., 1988). In contrast, aromatic compounds with resin-derived natural product precursors are abundant and provide a more convenient method of oil-source correlation.

Most of the aromatization reactions shown in Figure 11.1 have been shown to occur in laboratory experiments. For example, Ruzicka and Hosking (1930) conducted laboratory experiments and found that aromatization of bicyclic components of resins resulted in formation of agathalene (1,2,5-trimethylnaphthalene) and aromatization of the tricyclic compounds yielded pimanthrene (1,7-dimethylphenanthrene) and retene.

Reports of the cooccurrence of these aromatic compounds along with their partially aromatized natural products in coals (Simoneit et al., 1986; Radke et al., 1986; Alexander, Hazai et al., 1987; Puttman and Villar, 1987) is strong evidence that the aromatization processes also occur in sediments under geological conditions. The formation of 1-methylphenanthrene from abietic acid-type and sandaracopimaric acid-type natural product precursors has not yet been demonstrated. However, the widespread occurrence of biomarkers which are derived from natural product precursors by loss of an isopropyl, or more generally by loss of a 1-methylalkyl substituent (the relationship between C_{27} and C_{30} hopanes is one example), suggests that 1-methylphenanthrene is a likely aromatic biomarker for abietic acid-type natural products. Another likely source of 1-methylphenanthrene is the monoaromatic podocarpic acid (Fig. 11.1) which is also a common constituent of conifer resins (Campbell and Todd, 1942). Other unknown sources could also introduce this compound.

The collective presence of the group of aromatic compounds shown in Figure 11.1 indicates that a range of natural products, typical of those in modern Agathis resins, were present during diagenesis and maturation of the sediments. It must be emphasized that the individual compounds do not have a unique natural product source. For example, other sources of 1,2,5-trimethylnaphthalene (XI) are the pentacyclic triterpenoids (Strachan et al., 1988) and retene (XIII) can be derived from phyllocladane (Alexander, Hazai et al., 1987). However, the presence of the full suite of four aromatic compounds in a sample provides strong evidence for the range of precursor natural products which are characteristic of resins from Araucariaceae.

Saturated biomarkers could be formed by defunctionalization and reduction of natural product precursors. In the case of the natural products (I–IV) present in Agathis resins, loss of the oxygenated C-19 functional group at C-4 produces a group of C_{19} bicyclic and tricyclic biomarkers. Additional loss of the C-15 carboxyl group from agathic acid (I) gives the C_{18} biomarker 15,19-bisnorlabdane (VI). Although fichtelite (X) can be formed from abietic acid (IV) by an analogous series of defunctionalization and reduction steps to those which produce 19-norisopimarane (IX) and the norlabdanes, this process would be expected to operate only when conditions were very favorable for reduction. The presence of two conjugated double bonds within the ring system of the abietic acid molecule suggests that it will readily undergo aromatization and that only aromatized biomarkers such as retene (XIII) and possibly 1-methylphenanthrene (XIV), will result in most cases. Compounds VI, VII, and IX have recently been reported (Alexander et al., 1988) to occur in some sediments from this region, however, their low-relative abundance limits their usefulness for correlation purposes.

In summary, we propose that suites of aromatic biomarkers and suites of saturated biomarkers are derived from organic matter containing Agathis resins. The proportions of each suite and the composition of the suite itself depend on depositional environment and maturation: in coaly and more mature sediments, the aromatic biomarkers would be expected to be more abundant than the saturated biomarkers.

Occurrence of Fossil Araucariaceae in Cooper/Eromanga Basin Sediments

The effectiveness of the oil to source correlation strategy adopted in this study depends on araucariacean plants being prominent in the source rocks of the Eromanga Basin but absent in the Cooper Basin source rocks. The distribution of Araucariaceae in the Cooper/Eromanga Basin sequence is set out in a generalized way on Figure 11.2 and is based primarily on the distribution of araucariacean pollen together with megafloral data including that from Gould (1975), Miller (1977, 1982), and Stockey (1982).

The Permian floras recovered in source rocks of the Merrimelia Formation and the Gidgealpa Group belong to the distinctive Glossopterid floras whose megafloral elements have been described by Gould (1975), Gould and Shibaoka (1980), and Retallack (1980). The floras are dominated by pteridosperms (seed ferns) including the distinctive Glossopterid plants, with only a minor proportion of conifers: none of these can be considered to have close affinities with extant families or genera of conifers and in particular, the araucariacean conifers. The microfloras recovered from the Permian include no pollen which can be clearly identified with modern conifers, although a rare component of the basaccate pollen bears some similarity to podocarpacean pollen. The pteridosperms maintained their dominance throughout the Permian and Triassic although changes in the prominence of various pteridosperm groups did occur.

Near the close of the Permian, more or less corresponding with the change in the Cooper Basin sedimentation from the coal measures of the Toolachee Formation to the "red bed" sediments of the Nappamerri Formation, the pteridosperms

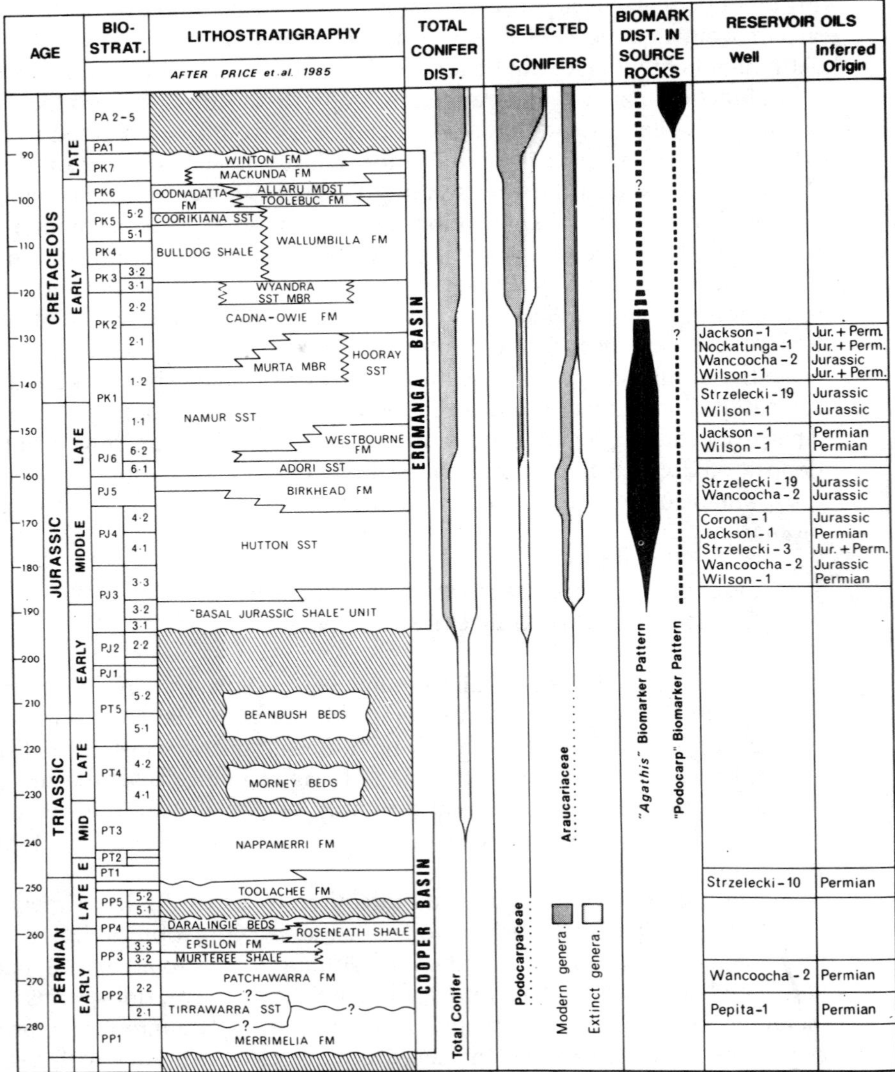

Figure 11.2 Stratigraphy and biomarker distribution of the Cooper and Eromanga Basin sediments.

underwent an abrupt change with the introduction of the *Dicroidium* flora and the associated spore/pollen defined *Falcisporites* microflora. The *Dicroidium* flora as a whole included only a low proportion of conifers (see Gould, 1975; Gould and Shibaoka, 1980; Retallack, 1980; and Townrow, 1969) including *Voltziopsis* and *Rissikia* although the conifers may be locally abundant, for example, basal Narrabean Group, Sydney Basin (Retallack, 1980). Of the Triassic conifers, only *Rissikia* which is considered to be podocarpacean by Townrow (1969), is related to a modern family; no araucariacean forms have been positively identified. In terms of the Triassic microfloras, which are dominated (up to 90% of the total assemblage) by the pteridosperm pollen *Falcisporites*, a very low proportion of pinacean-like and podocarpacean-like pollen occur. In addition, isolated specimens of inaperturate pollen, which may represent araucariacean-affinities, occur in widely separated samples of Middle to Late Triassic age.

With the commencement of sedimentation in the Great Artesian Basin at about the beginning of the Jurassic, the flora underwent another major change. Conifers attained dominance over the pteridosperms for the first time and pollen records include a low but significant proportion of araucariacean forms. The earliest Jurassic microflora assemblages (PJ1 lower PJ3 1), such as those from the "Basal Jurassic Unit," are strongly dominated by *Classopollis* (al. *Carollina*) pollen believed to be derived from the Cheirolepidiaceae, an extinct conifer group (Price et al., 1985). Cretaceous members of this family have been allied to extant Cupressaceae (e.g., Miller, 1977), which are widely represented in Australian floras.

Concomitant with a reduction in *Classopollis* numbers, araucarian-like pollen assume prominence and remain common throughout the overlying Jurassic microfloras (upper PJ3.1 PK1.1) that are associated with the upper "Basal Jurassic Unit" to Namur Sandstone sequence. Saccate pollen (probably largely podocarpean) are prominent also, and a distinctive trisaccate pollen type (*Microcachryidites*), attributed to the podocarp *Microcachrys*, appears in the Adori Sandstone (PJ6) and attains prominence in the section above. These bi- and trisaccate elements apparently displace the araucariacean forms, which are no longer common in sediments overlying the Namur Sandstone. The other modern podocarpacean genera, Dacrydium and Phyllocladus, occur initially in the Cretaceous and, together with Podocarpus, give rise to the distinctive diterpane biomarkers from the Cretaceous Gippsland Basin described by Alexander et al., (1987).

Megafloral remains of Araucariaceae abound in the Middle and Late Jurassic and are more closely allied to the genus *Araucaria* than to *Agathis*, whose fossil record extends back only into the Tertiary (Stockey, 1982). No megafloral remains of Araucariaceae have been recorded in Australian Triassic sediments (Townrow, 1969). Thus, while there is some tentative evidence of sparse araucariacean forms in the Triassic, they are not a significant part of the Australian flora before late Early Jurassic.

EXPERIMENTAL

Samples

The crude oils and sediment samples used in this study are listed in Tables 11.1 and 11.2. A well location map is shown in Figure 11.3. The sediment samples were recovered as cores and cuttings and were selected on the basis of a genetic

Table 11.1 GEOCHEMICAL DATA FOR SEDIMENT SAMPLES

Well	Depth (m)	Age	Formation	Maturity			Source			
				DNR-1	TNR-1	R_c	1,2,5-TMN / 1,3,6-TMN	1-MP / 9-MP	1,7-DMP / X	Retene / 9-MP
Arrabury 1	1958	Jurassic	Birkhead	1.1	0.27	0.57	18	3.7	6.6	1.3
	2010	Jurassic	Birkhead	1.4	0.45	0.61	12	3.3	2.3	1.6
	2205	Jurassic	B.Jurassic	2.3	0.68	0.65	3.1	1.4	1.2	0.6
	2622	Permian	Toolachee	4.3	0.77	0.80	0.10	0.58	0.27	0.01
	2757	Permian	Merrimelia	5.2	0.84	0.75	0.85	0.68	0.39	0.05
Challum 1	1836	Jurassic	Birkhead	0.25	0.39	0.72	18	2.7	3.1	0.73
	1855	Jurassic	Hutton	1.8	0.35	0.60	3.9	2.6	4.3	0.26
	2327	Permian	Toolachee	6.3	0.82	0.85	0.31	0.65	0.32	0.01
	2370	Permian	Patchawarra	7.5	0.69	0.92	0.19	0.57	0.24	0.01
Chookoo 1	1669	Jurassic	Birkhead	5.3	0.81	0.84	1.2	1.1	1.0	0.38
	1800	Jurassic	B.Jurassic	6.3	0.69	0.88	0.75	0.69	0.50	0.14
	1830	Permian	Toolachee	4.7	0.75	0.75	1.3	0.76	0.51	0.02
	1870	Pre-Permian	Pre-Permian	5.1	0.71	0.80	1.3	0.77	0.49	0.06
Naccowlah East 1	1702	Jurassic	Birkhead	2.7	0.57	0.62	9	3.2	5.2	2.8
	2044	Permian	Patchawarra	3.6	0.78	0.53	2.4	1.0	0.73	0.16
	2099	Permian	Patchawarra	3.2	0.69	0.64	3.5	0.88	0.68	0.08
	2257	Permian	Patchawarra	5.1	1.2	0.78	1.4	0.72	0.37	0.03
Pepita 1	1791	Jurassic	Birkhead	3.5	0.58	0.77	3.5	2.0	2.3	0.54
Strzelecki 5	1352	Jurassic	Murta	3.9	0.65	0.77	1.9	1.3	0.77	0.32
	1417	Jurassic	Namur	4.1	0.68	0.74	2.2	1.4	0.87	0.38
	1672	Jurassic	Birkhead	3.8	0.74	0.80	1.8	1.5	1.3	0.57
	1673	Jurassic	Birkhead	3.2	0.61	0.75	3.0	1.3	1.2	0.68
	1684	Jurassic	Birkhead	2.8	0.52	0.71	4.4	1.4	1.1	0.68
	1847	Permian	Toolachee	7.3	0.90	0.80	0.75	0.70	0.35	0.01
	1851	Permian	Toolachee	5.4	0.91	0.86	0.87	0.66	0.39	0.01
	1857	Permian	Toolachee	6.1	0.81	0.75	0.65	0.65	0.24	0.01
Strzelecki 1	1960	Permian	Murteree	6.0	1.0	0.85	0.48	0.67	0.31	0.02
	1992	Permian	Patchawarra	6.3	0.90	0.78	0.33	0.75	0.36	0.02
	2005	Permian	Patchawarra	6.1	0.77	0.86	0.29	0.68	0.39	0.01
Thargomindah 2	1253	Jurassic	Westbourne	1.9	0.38	0.54	6.7	7.8	2.9	7.9
Wancoocha 4	1566	Jurassic	Birkhead	2.0	0.29	0.53	6.2	1.8	1.8	4.5
Wancoocha 1	1766	Permian	Epsilon	3.2	0.47	0.51	1.8	0.67	0.35	0.27
	1776	Permian	Murteree	2.9	0.48	0.62	4.5	1.1	0.89	1.7
	1878	Permian	Patchawarra	3.7	0.60	0.66	2.6	0.72	0.47	0.10

Table 11.2 GEOCHEMICAL DATA FOR CRUDE OILS

	Reservoir			Maturity			Source			
Well	Depth (m)	Age	Formation	DNR-1	TRN-1	R_c	1,2,5-TMN / 1,3,6-TMN	1-MP / 9-MP	1,7-DMP / X	Retene / 9-MP
Corona 1		Jurassic	Hutton	12.8	0.39	0.60	7.6	2.4	1.6	5.3
Jackson 1	DST2	Jurassic	Murta	5.9	0.56	0.64	1.2	1.1	0.88	3.0
	DST5	Jurassic	Westbourne	6.9	0.52	0.87	0.68	0.88	0.32	0.47
	DST7	Jurassic	Hutton	5.2	0.67	0.99	0.98	1.0	0.42	1.3
Nockatunga 1		Jurassic	Murta	5.8	0.54	0.61	1.21	1.8	1.6	5.4
Pepita 1		Permian	Tirrawarra	5.9	0.66	0.71	0.31	0.76	0.38	0.13
Strzelecki 19	DST3	Jurassic	Namur	4.3	0.63	0.80	3.2	1.3	0.80	0.71
	DST1	Jurassic	Birkhead	4.0	0.60	0.87	3.0	1.2	1.3	0.75
Strzelecki 3	DST3	Jurassic	Hutton	4.3	0.66	0.82	1.9	1.6	0.63	0.98
Strzelecki 10	DST3	Permian	Toolachee	3.9	0.63	0.74	2.8	0.95	0.53	0.09
Wancoocha 2	DST1	Jurassic	Murta	1.9	0.28	0.56	3.2	1.4	1.0	2.6
	DST4	Jurassic	Birkhead	2.4	0.33	0.57	6.0	2.7	1.9	3.9
	DST5	Jurassic	Hutton	2.8	0.39	0.60	3.4	1.6	1.4	1.7
	DST3	Permian	Patchawarra	5.4	0.54	0.80	1.3	0.75	0.61	0.19
Wilson 1	DST2	Jurassic	Murta	3.9	0.67	0.71	1.1	1.1	0.69	3.2
	DST6	Jurassic	Namur	2.5	0.35	0.70	3.1	1.5	1.2	4.2
	DST4	Jurassic	Westbourne	6.5	0.66	0.75	0.83	1.8	0.33	1.3
	DST5	Jurassic	Hutton	6.3	0.79	0.97	0.97	0.4	0.36	0.98

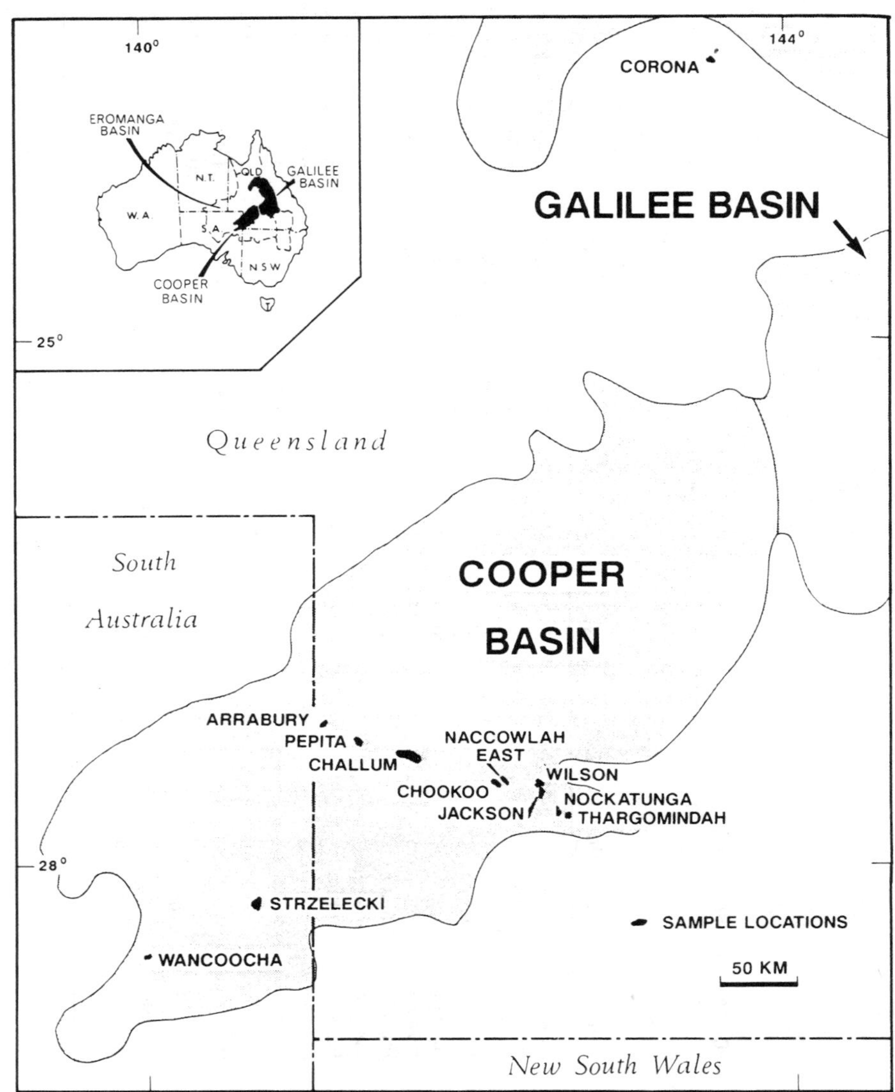

Figure 11.3 Map of study area showing location of wells.

potential, as measured by the Rock-Eval pyrolysis technique, of greater than 2 kg per ton.

Isolation of Aromatic Fractions

Aromatic fractions were isolated from sediments and crude oils by techniques previously described in detail (Alexander et al., 1985). This procedure involves recovery of an aromatic fraction from the silicic acid column after elution of the saturate fraction. This aromatic fraction is then subjected to thin-layer chromatography on alumina and the fraction containing dinuclear and trinuclear aromatic compounds is isolated.

Gas Chromatography (GC) and Gas Chromatography-Mass Spectrometry (GC-MS)

GC and GC-MS analyses were carried out using a HP 5890 gas chromatograph and a HP 5895B capillary GC—quadrupole MS—computer data system. Both systems were equipped with OCI3 on-column injectors (SGE, Australia) and 50 m × 0.2 mm i.d. WCOT fused silica columns with immobilized stationary phases. Aromatic fractions were analyzed using a 5 percent phenylmethylsilicone column (Hewlett-Packard) and the oven temperature was programmed at 1.5°C/min to 300°. Aromatic compound ratios measured from mass fragmentograms were corrected for differences in molar responses. Samples for analysis were dissolved in hexane and injected with an oven temperature of 50°C. Hydrogen was used as carrier gas at a linear velocity of 28 cm/s. In GC-MS, samples were analyzed in the data acquisition mode, by scanning from 50 to 450 Dalton in 1.3 s cycles, or by selected ion monitoring, using dwell times of 10 ms for each ion monitored. Typical MS operating conditions were: EM voltage 2200 V; electron energy 70 eV; ion source temperature 250°C.

RESULTS AND DISCUSSION

Table 11.1 shows geochemical data for the sediment samples together with parameters based on aromatic hydrocarbons which provide information about the maturity of the samples and their biological origins. The corresponding data for crude oils are shown in Table 11.2. Table 11.3 shows the definition of each of the abbreviations used in the tables of data and the measurement techniques used to obtain them.

The maturity and biomarker parameters based on aromatic compounds were obtained by GC-FID and GC-MS analysis of the samples. The two sets of partial chromatograms shown in Figure 11.4 are typical of those of samples of Jurassic and Permian age. The dimethylnaphthalene components in this pair of samples

Table 11.3 KEY TO TABLES 11.1 AND 11.2 INCLUDING MEASUREMENT TECHNIQUES USED TO OBTAIN DATA

Abbreviation	Definition	Technique[a]
DNR-1	$\frac{[2,6\text{-DMN}] + [2,7\text{-DMN}]}{[1,5\text{-DMN}]}$ DMN = dimethylnaphthalene	GC-FID
TNR-1	$\frac{[2,3,6\text{-TMN}]}{[1,4,6\text{-TMN}] + [1,3,5\text{-TMN}]}$ TMN = trimethylnaphthalene	m/z 170
R_c	$0.6\,\frac{1.5 \times ([2\text{-MP}] + [3\text{-MP}])}{[\text{P}] + [1\text{-MP}] + [9\text{-MP}]} + 0.4$ P = phenanthrene MP = methylphenanthrene	m/z 178, 192
$\frac{1,2,5\text{-TMN}}{1,3,6\text{-TMN}}$	$\frac{[1,2,5\text{-Trimethylnaphthalene}]}{[1,3,6\text{-Trimethylnaphthalene}]}$	m/z 170
$\frac{1\text{-MP}}{9\text{-MP}}$	$\frac{[1\text{-Methylphenanthrene}]}{[9\text{-Methylphenanthrene}]}$	m/z 192
$\frac{1,7\text{-DMP}}{X}$	$\frac{[1,7\text{-Dimethylphenanthrene}]}{[\text{Coeluting mixture of DMP isomers}]}$	m/z 206
$\frac{\text{Retene}}{9\text{-MP}}$	$\frac{[\text{Retene}]}{[9\text{-Methylphenanthrene}]}$	m/z 219, 192

[a]Ions used in mass fragmentogram.

show only slight differences, indicating that the samples have similar maturities. There are, however, marked differences in the chromatograms of the trimethylnaphthalene and alkylphenanthrene components of the two samples: these differences reflect differences in the original source material.

Maturity of Sediments and Crude Oils

Three maturity parameters based on the relative abundances of aromatic hydrocarbons were used in this study (see Table 11.3). One is based on the relative abundances of isomeric dimethylnaphthalenes (DNR-1, Radke, 1987), a second is based on trimethylnaphthalenes (TNR-1, Alexander et al., 1985; Radke, 1987), and the third is based on phenanthrene and the methylphenanthrenes (R_c, Radke, 1987). The calculated vitrinite reflectance (R_c) values are likely to be least reliable in the low-maturity zone ($R_c < 0.7$) because of anomalously high concentrations of 1-methylphenanthrene in samples of Jurassic age. In chromatograms of more mature samples, such as that of the Jurassic sample shown in Figure 11.4, the predominance of 1-methylphenanthrene (1-MP) relative to 9-methylphenanthrene (9-MP) is less pronounced. The net result of this effect is for low-maturity samples of Jurassic age to have an artificially low R_c value. Notwithstanding this effect, the R_c values shown in Table 11.1 indicate that the sediments have a range of maturities within the oil window and that this range coincides with the values shown in Table

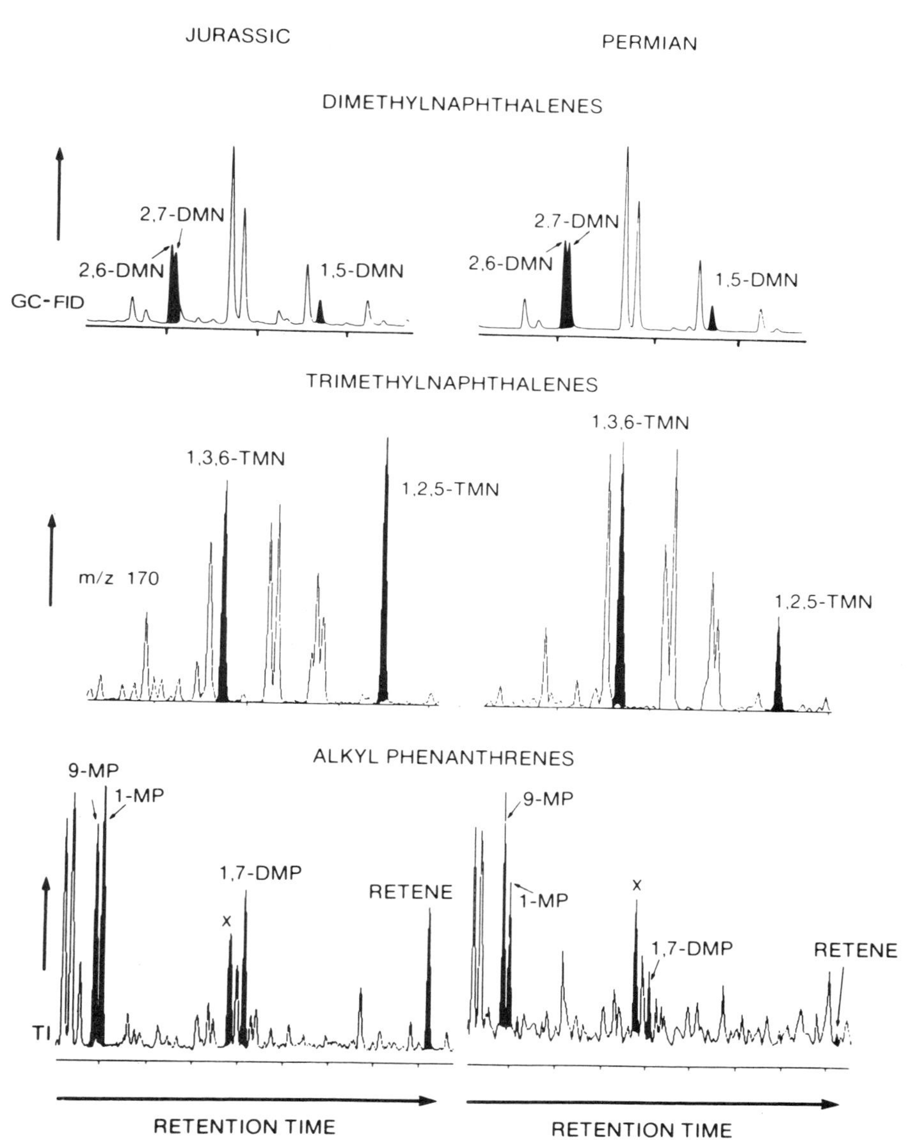

Figure 11.4 Partial GC-FID traces and mass chromatograms of aromatic fractions from typical Jurassic and Permian samples.

11.2 for the crude oils. The maturity ranges of the sediments and crude oils are further illustrated in Figure 11.5 where the DNR-1 and TNR-1 maturity indicators are shown on a plot. This plot shows that DNR-1 and TNR-1 have an approximately linear relationship for sediment samples. The samples of Jurassic age range from

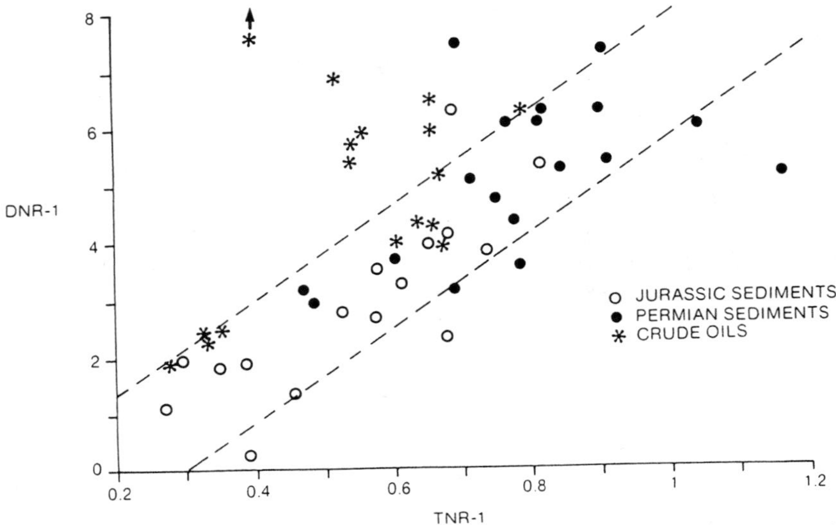

Figure 11.5 A plot of DNR-1 against TNR-1 for Jurassic sediments, Permian sediments and crude oils. For definitions refer to Table 11.3.

low maturity (TNR-1 < 0.5; DNR-1 < 5) to moderate maturity (TNR-1 ≈ 0.8; DNR-1 ≈ 6) and the samples of Permian age are mainly of moderate maturity, although a few samples have higher values. The values for the crude oils span the low- and moderate-maturity zones, but a significant number of samples do not conform to the DNR-1, TNR-1 relationship shown by the sediment samples and have relatively high DNR-1 values. The reasons for this effect are unclear but it could be associated with differences in heating rates (Alexander et al., 1986) or with differences in source material leading to alteration in input of one of the compounds used in the maturity indicator.

Aromatic Biomarkers

The abundances of the key aromatic biomarker compounds have been expressed as ratios and are shown in Table 11.1. The ratio 1,2,5-TMN/1,3,6-TMN is dependent on maturity since 1,2,5-TMN is the hydrocarbon which is formed initially in the aromatization of the natural product precursors, and this isomer subsequently reacts to give 1,3,6-TMN and other isomeric trimethylnaphthalenes (Strachan et al., 1988). Typically, what is observed is that as maturity increases with depth in a sediment sequence, a maturity zone is eventually reached where it is no longer possible to recognize enhanced 1,2,5-TMN concentrations. The values shown in Table 11.1 show that the thermal regime experienced by this basin system has apparently been such that it is possible to recognize samples with high 1,2,5-TMN input into the zone of moderate maturity. For example, the samples of Jurassic age from Strzelecki 5 have a higher value than most of the Permian samples. A cut-off value of 1.8 has been adopted to indicate that samples with lower values either did not receive an enhanced input of 1,2,5-TMN from the source material or that the samples are now too mature for these values to be used to make a meaningful assessment.

The relative abundance of 1-methylphenanthrene has been expressed as a ratio to that of 9-methylphenanthrene: 9-methylphenanthrene was selected because its stability is similar to that of 1-methylphenanthrene (Radke et al., 1986). As with 1,2,5-TMN, enhanced levels of 1-methylphenanthrene persist in these sediments at moderate maturity: in this case, a value of the ratio greater than 1 has been taken to indicate input from araucarian flora.

Figure 11.6 is a plot of the methylphenanthrene and the trimethylnaphthalene biomarker parameters designed to differentiate between samples on the basis of their contents of 1,2,5-TMN and 1-methylphenanthrene. All samples of Jurassic age except one have values for these two parameters which locate them in the upper right quadrant, indicating high input of both of the aromatic biomarkers derived from araucarian flora. Most samples of Permian age fall within the lower left quadrant, although some lie outside this region because of their higher contents of 1,2,5-TMN, which we attribute to sources other than araucarian flora.

Two other aromatic biomarker ratios have been calculated. The ratio of the concentration of 1,7-DMP to that of a peak labeled x in Figure 11.4 which is due to an unresolved mixture of 1,3-DMP, 3,9-DMP, 2,10-DMP, and 3,10-DMP (Radke et al., 1986), and the ratio of the concentration of retene to that of 9-methylphenanthrene. Values of the retene-based parameter and the 1,7-DMP-based parameter for the sediment and crude oil samples are shown in Figure 11.7. When benchmark values for these two ratios of 0.3 and 0.8, respectively are taken, it is apparent

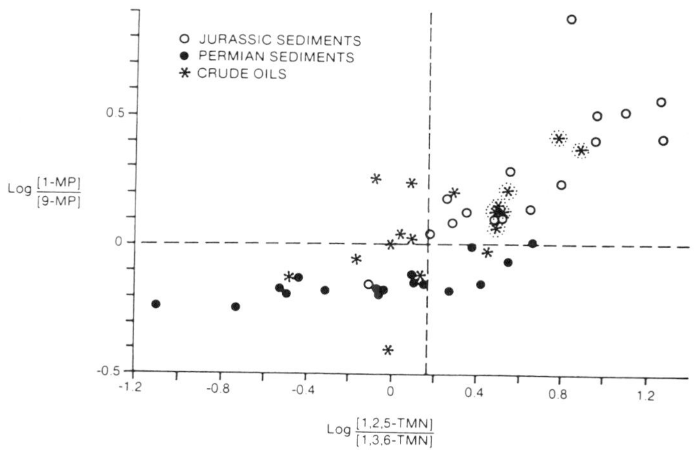

Figure 11.6 A log-log plot of source parameters based on alkylphenanthrenes for Jurassic sediments, Permian sediments, and crude oils.

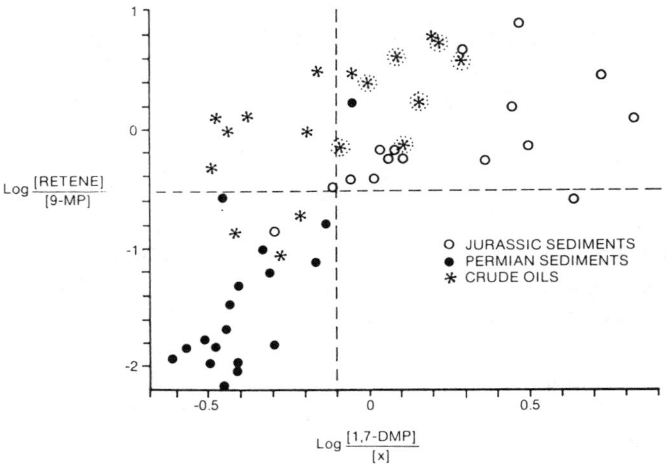

Figure 11.7 A log-log plot of source parameters based upon alkylnaphthalenes for Jurassic sediments, Permian sediments, and crude oils.

from the plot that sediments of Jurassic age are located predominantly in the upper right quadrant and those of Permian age are predominantly in the lower left quadrant. In both Figures 11.6 and 11.7, one sample of Jurassic age is located in the lower left quadrant. This sample—Chookoo 1, 1800 m—from the basal Jurassic contains abundant *Classopollis* but little evidence of araucariacean fossils. Although it is of Jurassic age, its position on these plots is consistent with that of a sample without an input from araucariacean flora. The position of the crude oil on these plots can be used to identify those samples that contain significant concentrations of the suite of aromatic biomarkers, which are proposed as indicators of *Agathis* resin input. Such samples will be located in the upper right quadrant in both plots; they are Corona 1, Strzelecki 19 DST3, Strzelecki 19 DST1, Wancoocha 2 DST1, Wancoocha 2 DST4, Wancoocha 2 DST5, and Wilson 1 DST6. The symbols representing these samples have been circled by dots in Figures 11.6 and 11.7.

Crude Oil: Source Rock Relationships

Table 11.4 shows values of the aromatic-hydrocarbon-based maturity indicators measured on crude oils and sediments recovered from the Wancoocha field. The striking feature of these data is the progressive increase in maturity with increasing age (depth) of the formation irrespective of the type of sample. The maturity indicators show that the crude oils in reservoirs have similar maturities to those of indigenous hydrocarbons contained in shales in similar stratigraphic locations. Such a situation could arise by either (a) entrapment of low-maturity crude oils from a deeper source followed by in-reservoir maturation, or (b) recent

Table 11.4 VALUES OF MATURITY PARAMETERS BASED ON AROMATIC HYDROCARBONS FOR OILS AND SEDIMENTS RECOVERED FROM THE WANCOOCHA FIELD

Age	Formation	Sample type	Maturity indicator		
			DNR-1	TNR-1	R_c
Jurassic	Murta	Crude oil	1.9	0.28	0.56
Jurassic	Birkhead	Source rock	2.0	0.29	0.53
Jurassic	Birkhead	Crude oil	2.4	0.33	0.57
Jurassic	Hutton	Crude oil	2.8	0.39	0.60
Permian	Epsilon	Source rock	3.2	0.47	0.51
Permian	Patchawarra	Source rock	3.7	0.60	0.66
Permian	Patchawarra	Crude oil	5.4	0.57	0.80

entrapment of crude oil from source beds with similar maturities to those of the reservoir.

The presence of the suite of aromatic biomarkers indicative of a source of Jurassic age in each of the crude oils from reservoirs of Jurassic age (Wancoocha 2 DST1, DST4, DST5) and their absence from the crude oil (Wancoocha 2 DST3) in the reservoir of Permian age, indicates that the second of these two scenarios is the more likely.

In contrast to the Wancoocha #2 crude oils, those from Wilson #1 do not show a consistent relationship between maturity and their location in the stratigraphic sequence. Judging by the three maturity indicators shown in Table 11.2, it is apparent that the crude oil located in the Namur is the least mature. The crude oil located in the shallower Murta Member is somewhat more mature, and this in turn is less mature than the deeper crudes from the Westbourne and Hutton Formations. The aromatic biomarkers indicative of a source of Jurassic age are strongly represented in the crude oil from the Namur and virtually absent from the more mature crudes from the Westbourne and Hutton Formations. Although the values of the aromatic biomarker parameters of the crude oil from the Murta do not place it directly with the samples of Jurassic age, nevertheless its borderline values for these parameters, together with its low-maturity values suggest strongly that it has some contribution from a source of Jurassic age. We, therefore, suggest that it is probably a mixture comprised of crude oils derived from low-maturity material of Jurassic age and more mature material of Permian age.

The crude oils from Strzelecki 10 and Strzelecki 19 show strong relationships between the age of reservoir and the age of source. The reservoirs of Jurassic age contain crude oils with a strong representation of aromatic biomarkers derived from Jurassic sediments, whereas the crude oil from the reservoir of Permian age lacks these compounds. We conclude from this that a local source from within the Jurassic has been the predominant source of the crude oils reservoired within the formations of Jurassic age. The crude oil from the Hutton Sandstone at Strzelecki 3 fails to qualify as a crude oil from a source of Jurassic age because the value for one of the four aromatic source parameters is marginally too low. Such an effect can result from mixing of crude oils from sources of Jurassic age and Permian age. The value of the 1,7DMP/X parameter in the sample of Jurassic age is influenced more than the other aromatic parameters on mixing because of the high-relative abundance of compounds X in the crude oils from sources of Permian age. The higher relative abundance of compounds X in samples of Permian age compared with samples of Jurassic age is illustrated in the partial chromatograms shown in Figure 11.4. We suggest that the crude oil from Strzelecki 3 has resulted from accumulation of crude oils from sources of both Jurassic and Permian ages.

The crude oils from Jackson #1 appear to be derived predominantly from organic matter which lacked the biomarker precursors characteristic of sediments of Jurassic age. The crude oil reservoired within the Murta Member may have some contribution from a source of Jurassic age. The lower maturity of this crude oil (based on R_c values in Table 11.2), together with its higher abundance of aromatic biomarkers relative to the crude oils from the Westbourne and Hutton Formations, may indicate a minor contribution from a local source within the formations of Jurassic age.

The inferred origins of the oils as outlined and the lithostratigraphic relationships of the reservoirs are summarized in Figure 11.3.

BIOSTRATIGRAPHICAL CONSIDERATIONS

In the present study, the sediment suites indicate the absence of *Agathis* markers in the Permian but their presence in the Jurassic, corresponding to the observed distribution of araucariacean fossils; however, the use of chemotaxonomic methods for characterization of fossil material is subject to a number of limitations. For example, it is uncertain whether or not the compositions of resins found in modern *Agathis* are a reliable guide to that of earlier members. Given the observed distribution of the *Agathis* biomarker pattern in the sediment extracts from the Cooper and Eromanga Basins relative to the distribution of araucariacean micro and mega fossils (see Fig. 11.3), this appears not to be a problem. It should be noted, however, that the genus *Agathis* does not extend back in time beyond the Tertiary and hence the *Agathis* biomarker pattern relates to other araucariacean forms as well as *Agathis*.

Dispersal and accumulation mechanisms may introduce further uncertainties. The diterpanes and aromatics have their origins in the resins which are distributed in the wood in specialized cells and resin canals and in the vascular tissue of the leaves. Resin is also exuded from the wood, sometimes accumulating in deposits of considerable thickness on the forest floor. Thus, the resin can enter the sediments via the wood and leaves, being transported to the depositional site and on occasions, be present as autochthonous deposits as in the Kauri forests of New Zealand. In some depositional systems, however, only the wind and surface water transported pollen and leaf cuticle (often only the thicker, less perforated, upper surface cuticle) reach the depositional site. In such circumstances it is possible that, while araucariacean microfossils may be represented, the araucariacean chemical fossils derived from the wood and leaf conductive tissue may not be present.

In the present study only one Jurassic shale sample was found to be entirely devoid of the araucariacean fingerprint of diterpanes and aromatic hydrocarbons. This was the *Araucariacea*-poor, *Classopollis*-rich, basal Jurassic sample from Chookoo-1, 1800 m, which may represent a case where Araucariaceae were not established in a particular ecosystem, being at a very early stage of their development immediately prior to their widespread distribution in the mid Jurassic. Such lower "Basal Jurassic Unit" shales are known from localized isolated parts of the Eromanga Basin and may be good source rocks which could yield Jurassic-sourced oils lacking the araucariacean markers.

In view of the two, albeit somewhat exceptional circumstances just described, one must allow for the possibility that some Jurassic source rocks may lack the diterpane/aromatic markers. Thus, emphasis must be placed on the use of the presence of these chemical markers to indicate a Jurassic source in the Cooper Basin region, as opposed to the converse. That is, the use of the absence of the diterpane/aromatic fingerprint as evidence of a Permian source for a Jurassic reservoired oil, must be approached with some caution, especially if the *Classopollis*-dominated, *Araucariacea*-sparse, basal Jurassic sediments are implicated as the source. Nevertheless, the near-ubiquitous distribution of the araucariacean diterpane/aromatic markers and araucariacean microfossils for the postbasal Jurassic shales (Birkhead, Westbourne, Murta) would suggest that if oil were sourced from such sediments there would be a strong probability that the *Agathis* markers should be represented in any such oil.

CONCLUSIONS

Suites of aromatic compounds that are indicative of plant resins from *Araucariaceae* have been identified in widely distributed sediments of Jurassic age from the Eromanga Basin. Similar biomarker assemblages have been shown not to be present in sediments of Permian age from the Cooper Basin.

Some crude oils contained in reservoirs of Jurassic to Cretaceous age have been shown to contain the biomarker assemblage characteristic of the sediments of Jurassic age, and therefore, appear to have been derived from sediments within the Eromanga Basin. Other crude oils in reservoirs of this age, together with all crude oils from reservoirs of Permian age do not contain the biomarker signature of the sediments of Jurassic age, and are therefore presumed to have been derived from the Permian sediments of the Cooper Basin.

Some crude oils recovered in Jurassic reservoirs contain some of the biomarkers which indicate a Jurassic source, but the match with the biomarkers of the Jurassic sediments is not strong. Close examination of the detailed composition of the aromatic components of these oils suggests that they are comprised of mixtures of material derived from sources within both the Cooper Basin and the Eromanga Basin.

Acknowledgments

The authors gratefully acknowledge CSR Petroleum, Delhi Petroleum, the Geological Survey of Queensland, and Santos Petroleum for access to samples and permission to publish. Financial support for the project of which this work forms a part was provided under the National Energy Research Development and Demonstration Program.

REFERENCES

Alexander, R., Kagi, R.I., Rowland, S.J., Sheppard, P.N., and Chirila, T.V. (1985) The effects of thermal maturity on distributions of dimethylnaphthalenes and trimethylnaphthalenes in some Ancient sediments and petroleums. *Geochimica et Cosmochimica Acta 49*, 385–395.

Alexander, R., Strachan, M.G., Kagi, R.I., and van Bronswijk, W. (1986) Heating rate effects on aromatic maturity indicators. *Organic Geochemistry 10*, 997–1003.

Alexander, R., Noble, R.A., and Kagi, R.I. (1987) Fossil resin biomarkers and their application in oil to source-rock correlation, Gippsland Basin, Australia. *Australian Petroleum Exploration Association Journal 27*(1), 63–72.

Alexander, G., Hazai, I., Grimalt, J., and Albaiges, J. (1987) Occurrence and transformation of phyllocladanes in brown coals from Nograd Basin, Hungary. *Geochimica et Cosmochimica Acta 51*, 2065–2073.

Alexander, R., Larcher, A.V., Kagi, R.I., and Price, P.L. (1988) The use of plant-derived biomarkers for correlation of oils with source rocks in the Cooper/Eromanga Basin system, Australia. *Australian Petroleum Exploration Association Journal 28*, 310–324.

Aplin, R.T., Cambie, R.C., and Rutledge, P.S. (1963) The taxonomic distribution of some diterpene hydrocarbons. *Phytochemistry 2*, 205–14.

Barnes, M.A. and Barnes, W.C. (1981) Oxic and anoxic diagenesis of diterpenes in lacustrine sediments. In *Advances in Organic Geochemistry 1981*, (eds. Bjorøy, M. et al.) John Wiley and Sons, Chichester, 289–298.

Cambie, R.C., Madden, R.J., and Parnell, J.C. (1971) Chemistry of the *Podocarpaceae*. XXVIII. Constituents of some *Podocarpus* and other species. *Australian Journal of Chemistry 24*, 217–21.

Campbell, W.P. and Todd, D. (1942). The structure and configuration of resin acids. Podocarpic acid and ferruginol. *Journal of the American Chemical Society 64*, 928–935.

Carman, R.M. and Sutherland, M.D. (1979) Cupressene and other diterpenes of *Cupressus* species. *Australian Journal of Chemistry 32*, 1131–42.

GOULD, R.E. (1975) The succession of Australian pre-Tertiary megafossil floras. *The Botanical Review 41*, 453–483.

GOULD, R.E. and SHIBAOKA, M. (1980) Some aspects of the formation and petrofabric features of coal members in Australia, with special reference to the Tasman orogenic zone. *Coal Geology 2*, 1–29.

KANTSLER, A.J., PRUDENCE, T.J.C., COOK, A.C., and ZWIGULIS, M. (1983) Hydrocarbon habitat of the Cooper/Eromanga Basin, Australia. *Australian Petroleum Exploration Association Journal 23*, 75–92.

MILLER, C.N. JR. (1977) Mesozoic Conifers. *The Botanical Review 43*(2), 217–280.

MILLER, C.N. JR. (1982) Current status of Paleozoic and Mesozoic conifers. *Review of Palaeobotany and Palynology 37*, 99–114.

PRICE, P.L., FILATOFF, J., WILLIAMS, A.J., PICKERING, S.A., and WOOD, G.R. (1985) Late Palaeozoic and Mesozoic palynostratigraphical units. CSR Oil & Gas Division Palynological Facility Report No. 274/25 1.20 (Queensland Department of Mines Open File Report No. 14012).

PUTTMAN, W.I. and VILLAR, H. (1987) Occurrence and geochemical significance of 1,2,5,6-tetramethylnaphthalene. *Geochimica et Cosmochimica Acta 51*, 3023–3029.

RADKE, M., WELTE, D.H., and WILLSCH, H. (1986) Maturity parameters based on aromatic hydrocarbons: influence of the organic matter type. *Organic Geochemistry 10*, 51–63.

RADKE, M. (1987) Organic geochemistry of aromatic hydrocarbons. In *Advances in Petroleum Geochemistry 2*, 141–207.

RETALLACK, G.J. (1980) Late Carboniferous to Middle Triassic megafloras from the Sydney Basin. In *A Guide to the Sydney Basin.* (eds. C. Herbert & R. Helby) New South Wales Geological Survey, Bulletin 26, 384–430.

RUZICKA, L. and HOSKING, J.R. (1930) Higher terpenoid compounds XLII: dehydrogenation and isomerisation of agathic acid. *Helvetica Chimica Acta 13*, 1402–23.

SIMONEIT, B.R.T., GRIMALT, J.O., WANG, T.G., COX, R.E., HATCHER, P.G., and NISSENBAUM, A. (1986) Cyclic terpenoids of contemporary resinous plant detritus and of fossil woods, ambers and coals. *Organic Geochemistry 10*, 877–889.

STEWART, W.N. (1983) Paleobotany and the evolution of plants. Cambridge University Press, Cambridge, 348.

STOCKEY, R.A. (1982) The araucariaceae: An evolutionary perspective. *Review of Paleobotany and Polynology 37*, 133–154.

STRACHAN, M.G., ALEXANDER, R., and KAGI, R.I. (1988) Trimethylnaphthalenes in crude oils and sediments: Effects of source and maturity. *Geochimica et Cosmochimica Acta 52*, 1255–1264.

THOMAS, B.R. (1969) Kauri resins—modern and fossil. In *Organic Geochemistry—Methods and Results* (eds. G. Eglinton and M.T.J. Murphy) Springer-Verlag, Berlin, pp. 599–618.

TOWNROW, J.A. (1969) Some Lower Mesozoic Podocarpaceae and Araucariaceae. In Gondwana Stratigraphy, IUGS Symposium, Buenos Aires. UNESCO, Paris, 159–184.

WAKEHAM, S.G., SCHAFFNER, C., and GIGER, W. (1980) Polycyclic aromatic hydrocarbons in recent lake sediments-11. Compounds derived from biogenic precursors during early diagenesis. *Geochimica et Cosmochimica Acta 44*, 415–429.

12

Quantitative Analysis of Triterpane and Sterane Biomarkers: Methodology and Applications in Molecular Maturity Studies

Adolfo G. Requejo

Abstract. With the increasing commercial availability of high purity, authentic steroid and triterpenoid hydrocarbon standards, the ability to accurately quantify the concentrations of biological markers in geologic samples is being facilitated. This chapter discusses aspects of the analytical methodology employed in the quantitative analysis and its application in the study of sterane-derived parameters for thermal maturity assessment in oils and source rocks. Quantitative biomarker data suggest that, while sterane aromatization appears to involve direct conversion of ring-C monoaromatic to triaromatic compounds, little evidence exists for epimerization of saturated steranes at C-20. It is possible that the latter reaction does not occur to a significant extent in natural systems, which is contrary to the commonly held view that steranes isomerize at the C-20 position with increasing thermal maturity. The absolute concentrations of steranes and hopanes in two families of oils generated from marine source facies (Type II and II-S) were found to be highly correlated with the degree of sterane aromatization. A generalized reaction scheme linking absolute concentrations to aromatization is proposed to account for these observations.

INTRODUCTION

Traditionally, results of triterpenoid and steroid biomarker analyses of geologic samples have been presented using one of two formats: (1) mass-chromatogram "fingerprints," illustrating the distribution of individual components within a specific biomarker compound class (e.g., *m/z* 217 for the steranes), or (2) ratios of chromatographic peak areas, which are most often used to depict the relative abundance of precursors and products in a geochemical reaction (e.g., the percent 20*S* sterane isomerization parameter) or the proportions of different biomarker compounds present in a sample (e.g., hopane/sterane ratios). These approaches possess several inherent limitations. For example, it is often difficult to relate distributions of biomarkers isolated in different liquid chromatographic fractions (e.g., aromatic steranes versus saturated steranes). Comparisons of the quantities of these compounds in different samples can also be complicated.

With the increasing commercial availability of high purity, authentic steroid and triterpenoid hydrocarbon standards, the ability to accurately quantify the concentrations of biological marker compounds in geologic samples is being facilitated. Determination of absolute concentrations adds a new dimension to biomarker analysis which could not be adequately addressed using the conventional fingerprinting approaches. This chapter discusses aspects of the analytical methodology employed in the quantitative analysis and its application in the study of sterane-derived maturity parameters in oils and source rocks.

ANALYTICAL METHODS

Quantitation of absolute biomarker concentrations does not require deviation from the conventional liquid chromatographic isolation of these compounds followed by gas chromatography/mass spectrometry analysis. The principal requirement is the addition of an "internal" or "surrogate" standard at some stage of the analytical scheme. A "surrogate" standard is defined as a model compound which is added at the initial stages of an analytical scheme (for example, either prior to or immediately after precipitation of asphaltenes). The term internal standard is used when the model compound is added immediately prior to instrumental analysis. The difference in terminology is illustrated in Figure 12.1. All of the quantitative data presented here have been determined using the internal standard method.

Previous studies have employed a similar quantitative approach to biomarker characterization, using various standard compounds (Seifert and Moldowan, 1979; Rullkötter et al., 1984; Abbott et al., 1984; Dahl et al., 1985; Mackenzie et al., 1985). The quantitative data reported in this study differs from most of the earlier work in the incorporation of *response factors*, which compensate for differences in detector response between the various analytes and the model compound. These

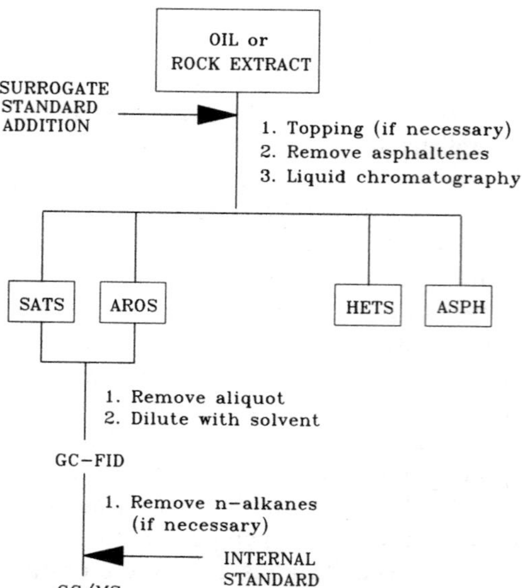

Figure 12.1 Sample analytical scheme showing the stage at which different types of quantification standards are added to samples.

are calculated from parallel analysis of authentic external standard mixtures by comparing the *m/z* 217 response per unit amount of the internal standard, in this case 5α(H)-androstane, to that of individual steranes (*m/z* 217) or triterpanes (*m/z* 191). Examples of the external standard mixtures are shown in Figure 12.2. For quantitation of aromatic sterane compounds, the synthetic aromatic hydrocarbon o-terphenyl was used as an internal standard. At the time this work was initiated, individual aromatic steranes were not readily available for use in external standard mixtures, hence reported concentrations of these compounds are not response-corrected.

Subsequent work has shown that 5α(H)-androstane may not be the most suitable standard for triterpane and sterane quantitation. In addition to 5α(H)-androstane, two other compounds were evaluated as potential standards: 5β(H)-cholane and 3,3-d_2,5α(H),14α(H),17α(H)-cholestane (20*R*). Table 12.1 lists the calculated response factors for two epimers of 5α(H),14α(H),17α(H)-ethyl cholestane (20*S* and 20*R*) and for 17α(H),21β(H)-hopane relative to all three model compounds. This comparison shows that both cholane and cholestane exhibit instrumental responses more similar to the biomarkers of interest than does androstane (this is especially notable in the case of hopane), and hence may be more representative standards. More recent publications reporting quantitative data for biomarker compounds have employed deuterated steranes as internal standards (Eglinton and Douglas, 1988; Mello et al., 1988). More than 10 years ago, Seifert and Moldowan (1979) reported biomarker concentrations in crude oils using an almost identical analytical approach, including the use of "sensitivity factors" to

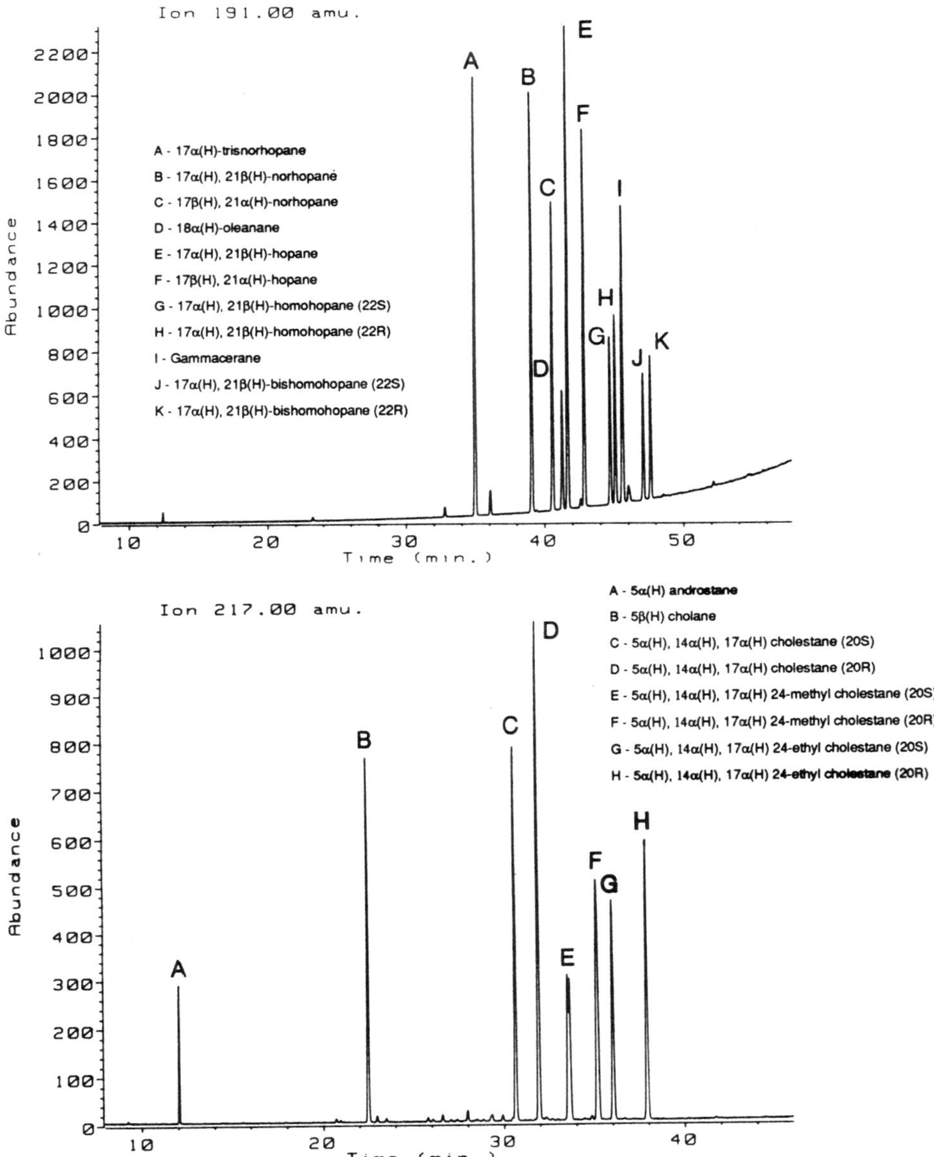

Figure 12.2 Mass chromatograms of m/z 191 and 217 depicting distributions of individual hopanes and steranes in external standard mixtures.

Table 12.1 TYPICAL RESPONSE FACTORS FOR COMMON TRITERPANE AND STERANE BIOMARKERS CALCULATED USING THREE STANDARD COMPOUNDS

	Analyte		
Standard	Ethyl-cholestane (20*R*) (*m/z* 217)	Ethyl-cholestane (20*S*) (*m/z* 217)	Hopane (*m/z* 191)
Androstane	2.1–2.8	1.7–2.3	7.1–8.0
5β-Cholane	0.6–0.7	0.5–0.6	1.4–1.6
3,3-d2-Cholestane	0.6	0.5	1.3–1.5

Data are based on 30 SIM GC-MS analyses of the standards shown in Figure 12.2 over a period of six months.

compensate for differences in instrument response relative to added 5β(H)-cholane standard.

The precision of the quantitative data has been evaluated through replicate analyses of oils. These data are summarized in Table 12.2 for two oils derived from "marine" and "terrestrial" source facies. Relative standard deviations for calculated concentrations of 17α(H),21β(H)-hopane are ± 28% and ±5%, respectively, with the greater values corresponding to the lower concentrations found in the "marine" oil (Table 12.2). This trend may indicate lesser precision in samples containing lower biomarker concentrations. It may also result in part from the large response factors used when 5α(H)-androstane is used to calculate hopane concentrations (Table 12.1). The data in Table 12.2 also indicate that absolute concentrations of individual compounds can be determined with less precision using the internal standard technique than conventional biomarker parameters determined from peak area ratios. Use of the "surrogate" approach with standards other than 5α(H)-androstane has resulted in higher analytical precision (Requejo, unpublished data).

A QUANTITATIVE APPROACH TO ASSESSING MOLECULAR PARAMETERS

Quantitative biomarker data have received limited application in petroleum geochemistry, chiefly because the technique had been used sparingly until recently. Rullkötter et al. (1984) used biomarker concentrations together with other molecular characteristics to classify families of oils in the Williston Basin. The same authors found that absolute biomarker concentrations were also useful in recognizing mixing of oils of different maturities (cf. Mackenzie et al., 1985). Concentrations of biomarkers have been shown to increase as a result of biodegradation relative to unaltered samples (Seifert and Moldowan, 1979; Requejo et al., 1989). Geochemical applications have recently been extended to include classification of

Table 12.2 ASSESSMENT OF ANALYTICAL PRECISION FOR CALCULATED CONCENTRATIONS OF HOPANE AND FOR SEVERAL COMMONLY USED TRITERPANE AND STERANE PARAMETERS IN "MARINE" AND "TERRESTRIAL" OILS

			Triterpane Parameters				Concentration[e] of	Sterane Parameters					
										Carbon Number[h]			
Oil	Replicate Number		Ts/*Tm*[a]	C_{29}/C_{30}[b]	% C_{30}-αβ[c]	%22*S*[d]	Hopane (ppm)	%20*S*[f]	%ββ[g]	C_{27}	C_{28}	C_{29}	Hopane/Sterane[i]
Marine	1		2.05	0.38	0.94	58.1%	123	44.9%	64.1%	30.8%	23.7%	45.5%	1.6
Marine	2		2.05	0.38	NC	57.2%	70	46.4%	61.9%	30.2%	24.9%	44.9%	1.5
Marine	3		1.97	0.36	0.94	57.1%	94	45.0%	62.7%	32.2%	23.4%	44.4%	1.7
Marine	4		1.83	0.37	0.94	58.2%	68	46.3%	62.3%	29.6%	26.1%	44.3%	1.6
Marine	5		2.02	0.38	0.93	56.7%	123	44.2%	61.9%	31.1%	23.5%	45.4%	1.7
		Mean	1.98	0.37	0.94	57.5%	96	45.4%	62.6%	30.8%	24.3%	44.9%	1.6
		SD	0.09	0.01	0.01	1.0%	27	0.9%	0.9%	1.0%	1.2%	0.6%	0.1
Terrestrial	1		0.50	1.34	0.95	55.5%	248	34.3%	63.8%	30.7%	17.7%	51.6%	2.9
Terrestrial	2		0.49	1.33	0.95	55.5%	274	33.8%	63.8%	30.6%	17.4%	52.0%	2.9
Terrestrial	3		0.51	1.35	0.95	55.4%	253	34.3%	63.7%	30.9%	17.4%	51.7%	2.8
Terrestrial	4		0.47	1.33	0.95	55.4%	244	34.2%	63.3%	30.4%	17.7%	51.9%	2.9
Terrestrial	5		0.48	1.36	0.95	55.2%	240	34.4%	63.0%	30.5%	17.2%	52.3%	2.8
		Mean	0.49	1.34	0.95	55.4%	252	34.2%	63.5%	30.6%	17.6%	51.9%	2.9
		SD	0.02	0.01	0.00	0.1%	13	0.2%	0.4%	0.2%	0.2%	0.3%	0.1

NC = Not Calculated

[a]Ts/*Tm* = 18α(H)-trisnorneohopane/17α(H)-trisnorhopane

[b]C_{29}/C_{30} = C_{29} 17α(H),21β(H)-norhopane/C_{30} 17α(H),21β(H)-hopane

[c]%C_{30}αβ = C_{30} 17α(H),21β(H)-hopane/C_{30} 17α(H),21β(H)-hopane + C_{30} 17β(H),21α(H)-hopane

[d]%22*S* = C_{31} 17α(H),21β(H)-homohopane (22*S*)/C_{31} 17α(H),21β(H)-homopane (22*S* + 22*R*)

[e]Calculated relative to the weight of whole (untopped) oil

[f]%20*S* = C_{29} 5α(H),14α(H),17α(H)-sterane (20*S*)/C_{29} 5α(H),14α(H),17α(H)-sterane (20*S* + 20*R*)

[g]%ββ = 2[C_{29} 5α(H),14β(H),17β(H)-sterane (20*S* + 20*R*)]/C_{29} 5α(H),14α(H),17α(H)-steranes (20*S* + 20*R*) + 2[C_{29} 5α(H),14β(H),17β(H)-sterane (20*S* + 20*R*)]

[h]Percent of C_{27}, C_{28}, and C_{29} 5α(H),14α(H),17α(H)-steranes (20*R*)

[i]Hopane/Sterane = C_{30} 17α(H),21β(H)-hopane/C_{29} 5α(H),14α(H),17α(H)-steranes (20*S* + 20*R*) + C_{29} 5α(H),14β(H),17β(H)-steranes (20*S* + 20*R*)

depositional environments (Mello et al., 1988) and assessment of the relative contributions of biomarkers originating from bitumen and kerogen during simulated maturation of source rocks (Eglinton and Douglas, 1988). Quantitation has also been employed in the analysis of other classes of biomarker compounds, such as porphyrins (Baker et al., 1987).

In this study, the absolute concentrations of sterane and triterpane biomarkers are used to quantitatively assess variations in the individual compounds which make up biomarker-derived maturity parameters. The concept underlying this approach is shown diagrammatically in Figure 12.3. Distributions of biological markers are frequently correlated with bulk characteristics of samples, such as API gravity and sulfur contents in the case of oils (Curiale et al., 1985) or thermal maturity in the case of sediments (Mackenzie et al., 1980). These correlations frequently involve ratios of peak areas corresponding to individual biomarker compounds (percent 20*S* is a good example), resulting in a crossplot such as the one shown in Figure 12.3. However, absolute concentrations of the individual compounds which make up the formulated ratio can vary widely and still satisfy the observed relationship (Fig. 12.3). How the absolute concentrations of individual components in the ratio vary could greatly influence our interpretation of the geochemical processes which result in the observed correlation. Yet, the traditional fingerprinting approach is inadequate in discerning these variations, as it considers only the relative quantities.

This approach has been used to examine the common sterane isomerization parameter, percent 20*S*. It has conventionally been held that, with increasing maturity, the 20*R* isomer of ethyl cholestane is *converted* to its geologically stable 20*S* epimer and, indeed, this conversion has been demonstrated in the laboratory (Seifert et al., 1983; Abbott et al., 1984). However, concentrations of ethyl cholestane

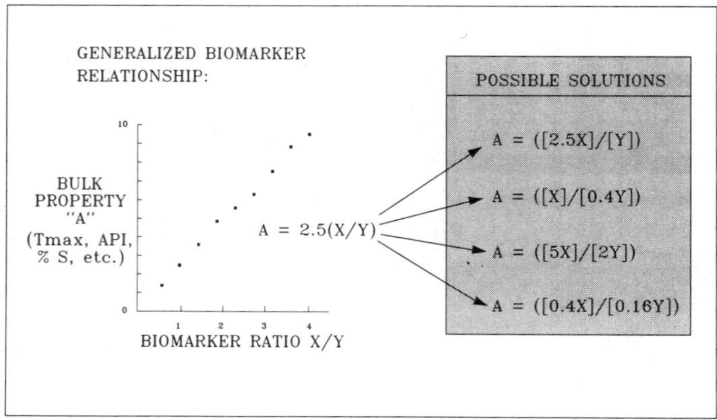

Figure 12.3 Schematic diagram illustrating the concept behind the use of quantitative data in the assessment of biomarker parameters. Concentrations of individual components comprising a parametric ratio can vary widely and still satisfy the observed relationship.

isomers (20*S* and 20*R*) in both source rocks and oils representing several organic matter types at varying maturity levels do not support a *net* conversion of 20*R* to 20*S* with increasing maturity. This is illustrated in Figures 12.4 and 12.5, which show plots of the absolute concentrations of the two ethyl cholestane epimers versus the percent $20S_Q$ parameter (the Q subscript signifies that the ratio is based on absolute concentrations rather than peak areas) for source rock bitumens and biodegraded sandstone oil stains from the Jameson Land Basin, East Greenland

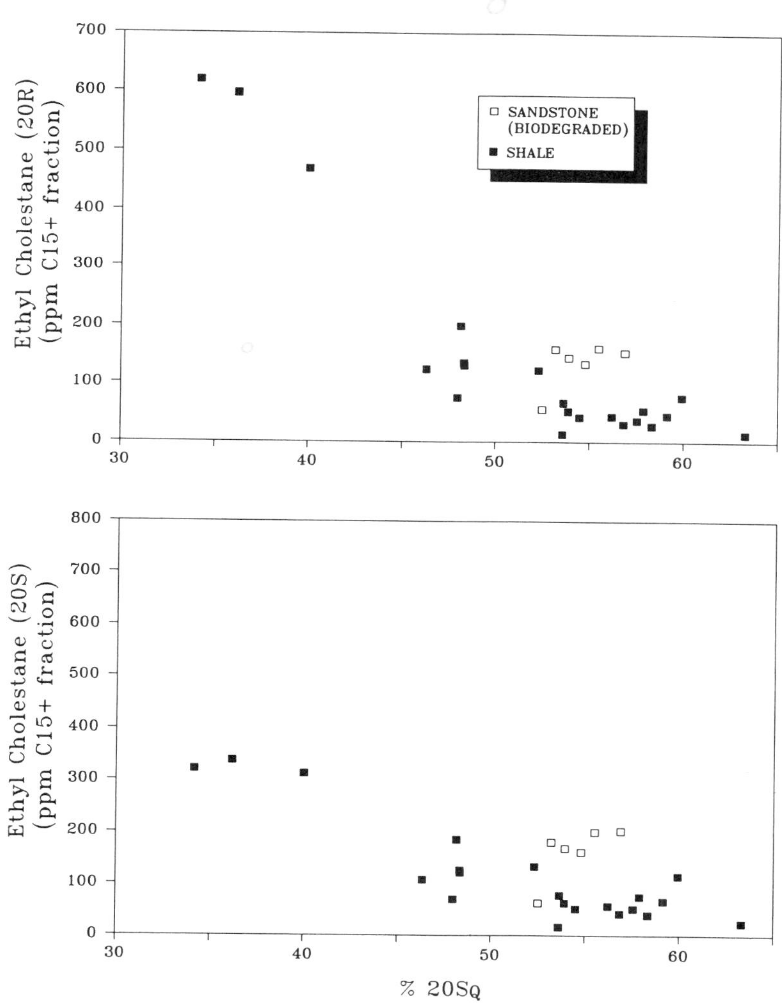

Figure 12.4 Plots of ethyl cholestane concentrations (20*S* and 20*R*) versus extent of isomerization at C-20 (% $20S_Q$) in bitumens (filled symbols) and biodegraded oil extracted from sandstones (open symbols), Jameson Land, East Greenland (Requejo et al., 1989).

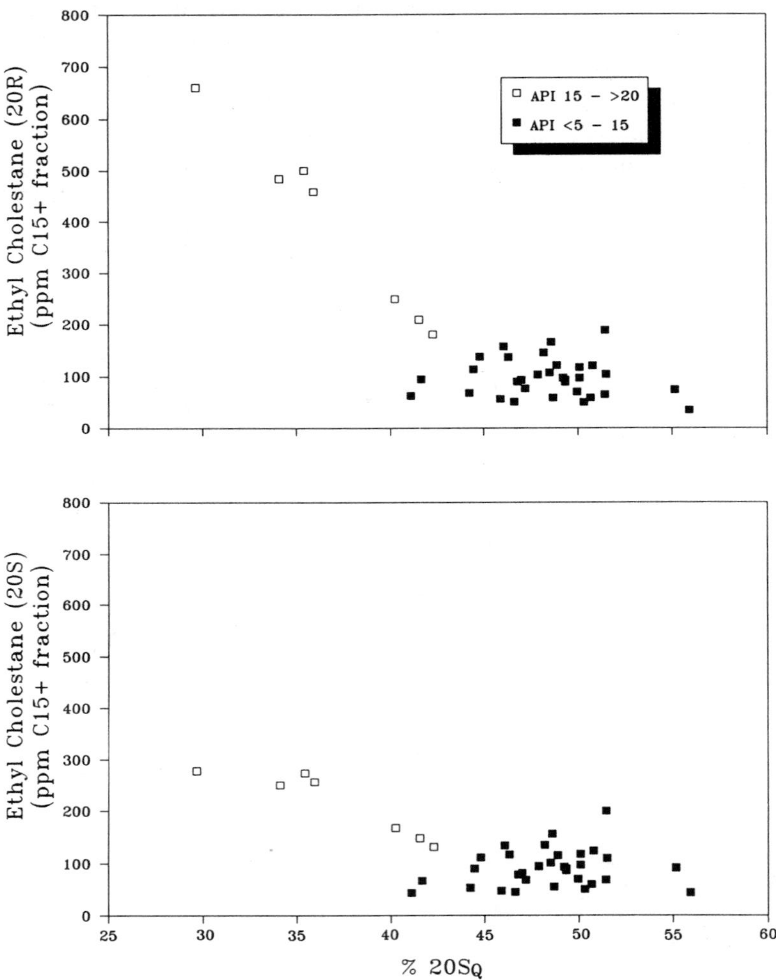

Figure 12.5 Plots of ethyl cholestane concentrations (20*S* and 20*R*) versus extent of isomerization at C-20 (% $20S_Q$) in Monterey oils of varying API gravity (filled symbols < 15° API; open symbols > 15° API).

(Requejo et al., 1989), and also for a series of 38 oils generated from the Monterey Formation, respectively. Concentrations are reported per unit C_{15+} fraction, in order to facilitate comparison of data from bitumens and oils. These results show that both the 20*S* and 20*R* isomers decrease in concentration with increasing maturity, but at slightly different rates, such that the 20*S* epimer becomes prevalent at higher percent $20S_Q$ values. If a net conversion of 20*R* to 20*S* were occurring, the concentration of the latter compound (product) would be expected to increase as the former (reactant) decreased. Although on the surface this would appear intuitively obvious, that is, higher molecular-weight components in oils and bitumens generally decrease in abundance with increasing maturity, references in the literature commonly refer to a "conversion" of 20*R* to 20*S*. A more accurate representation would state that the 20*S* epimer increases in abundance relative to 20*R*, while both decrease in concentration with increasing level of maturity. Curiously, Monterey oils possessing higher API gravities have the lowest percent $20S_O$ values (Fig. 12.5), suggesting that additional factors beside maturation are affecting the isomerization values in oils generated from Type II-S kerogens.

These data challenge the traditionally held view regarding sterane isomerization and thermal maturity. They imply that either (1) epimerization of steranes at the C-20 position does not occur to a significant extent with increasing thermal maturation, or (2) that other reactions (aromatization or other degradative reactions) effectively outcompete epimerization at increasing levels of maturation (Fig. 12.6), thereby preventing the accumulation of products. Previous studies have pointed out that steranes possessing geologically mature configurations at C-20 occur in immature sediments and can arise through mechanisms other than direct isomerization of the 20*R* configuration (ten Haven et al., 1986; Peakman and Maxwell, 1988; Peakman et al., 1989). For example, Peakman and Maxwell (1988) have shown that reduction of Δ7 sterenes can result in the formation αββ (20*R*) and (20*S*) steranes commonly used in maturity assessments, and might account for their abundance in an immature marl from a hypersaline deposit. It is possible that such pathways are more significant than previously acknowledged in the formation of 20*S* sterane isomers in geologic samples.

An alternative interpretation of the trends illustrated in Figures 12.4 and 12.5 is that sterane concentrations decrease due to dilution by hydrocarbons generated at higher maturity. This interpretation assumes that the hydrocarbons generated at elevated thermal maturities become increasingly depleted in biomarkers. This view is not supported by the results of laboratory thermal maturation experiments reported by Eglinton and Douglas (1988). These authors found that maximum yields of triterpanes and steranes from hydrous pyrolysis of kerogens are obtained over roughly the same temperature intervals as the maximum yields of bulk py-

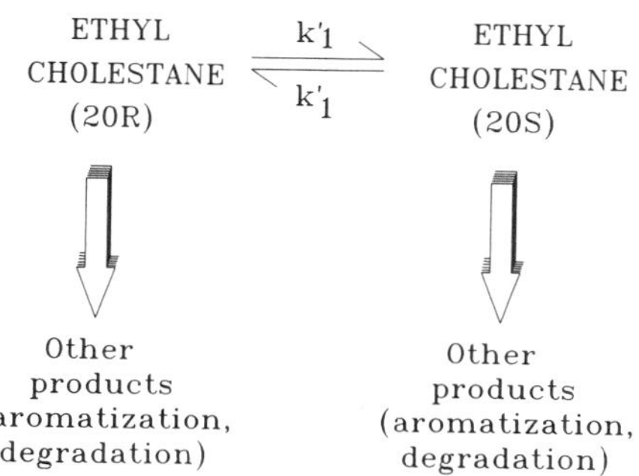

Figure 12.6 Schematic diagram depicting the fate of ethyl cholestane. Quantitative data suggests that reactions other than isomerization (bold arrows) are the major catagenic pathway for steranes.

rolysates and C_{15}–C_{30} *n*-alkanes, although maximum yields of steranes were obtained at a slightly lower temperature than the triterpanes. Hence, from a volumetric standpoint, dilution at higher maturities would not be favored. The most likely explanation for the trends shown in Figures 12.4 and 12.5 is that either conversion of the 20*R* to the 20*S* epimer is not as significant a reaction as previously believed or that it is part of a reaction sequence, which removes the products and prevents their accumulation. Clearly, the quantitative data offer valuable insight into the geologic fate of biomarker compounds which could not be addressed using the conventional fingerprinting techniques.

A similar approach can be extended to the study of sterane aromatization. Figures 12.7 and 12.8 show plots of the calculated concentrations of the sum of the ring-C monoaromatic steranes 5α(H)-$C_{28}H_{44}$ (20*R*) + 5β(H)-$C_{29}H_{46}$ (20*R*) (*m*/*z* 253) and the triaromatic sterane $C_{28}H_{38}$ (20*R*) (*m*/*z* 231) versus the sterane aromatization parameter percent TRI_E (the subscript E signifies that the ratio is empirical, based on peak areas rather than absolute concentrations). These components represent the numerator and denominator, respectively, in the percent TRI_E parameter (Mackenzie, 1984) and correspond to postulated reactant and product in the aromatization of ring-C monoaromatics to triaromatics (with exception of the coeluting C_{28} monoaromatic compound). The data sets shown correspond to the Monterey oils discussed earlier (Fig. 12.7) and a series of 23 oils from the South Pass 61 field, offshore Louisiana (Fig. 12.8). The latter oils represent a single genetic family, derived from siliciclastic, marine source facies at a similar level of maturity (Requejo and Halpern, 1990). Variations in the bulk properties of these oils have been attributed to postgenerative alteration, primarily biodegradation and evaporative fractionation associated with tertiary migration.

In contrast to the saturated sterane data, the concentrations of these aromatic compounds exhibit a product-precursor relationship with respect to the degree of aromatization; that is, the monoaromatic steranes decrease in concentration while the triaromatic compound concomitantly increases with increasing aromatization (Figs. 12.7 and 12.8). Several of the offshore Louisiana oils have aromatized sterane concentrations which fall above the main trendline, which is highlighted by the shaded area in Figure 12.8. These samples all show evidence of biodegradation (Requejo and Halpern, 1990), which has the net effect of increasing absolute biomarker concentrations through selective removal of more labile components (Seifert and Moldowan, 1979; Requejo et al., 1989). A cursory attempt at a mass balance to determine whether there is direct conversion of the C_{29} mono- to the C_{28} triaromatic compound reveals generally good agreement. For example, in the Monterey oils, values of approximately 20 percent TRI_E correspond to C_{28} triaromatic sterane concentrations of approximately 50 ppm, while C_{28} + C_{29} monoaromatic sterane concentrations have decreased by approximately 100 ppm relative to the least aromatized samples (Fig. 12.7). The difference may be due to contributions from the coeluting C_{28} monoaromatic compound or to the subsequent reaction of the triaromatic compound to form as yet uncharacterized products. Immature Monterey Formation bitumens and heavy oils are known to contain high-

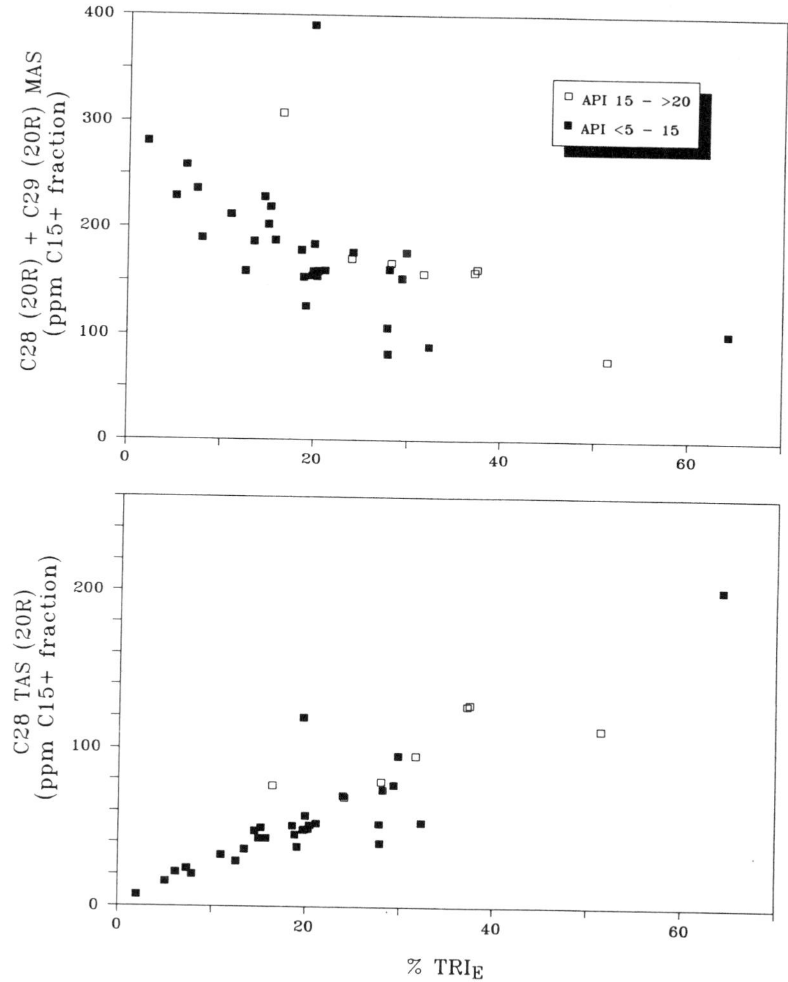

Figure 12.7 Plots of ring-C monoaromatic sterane [MAS; 5α(H)-$C_{28}H_{44}$ (20*R*) + 5β(H)-$C_{29}H_{46}$ (20*R*)] and triaromatic sterane [TAS; $C_{28}H_{38}$ (20*R*)] concentrations versus the sterane aromatization parameter percent TRI_E in Monterey oils.

monoaromatic sterane contents (Curiale et al., 1985), although the relative proportion of mono- and triaromatic steranes in bitumens can vary according to lithofacies (Curiale and Odermatt, 1989). A similar mass balance of the aromatic sterane concentrations in the unaltered offshore Louisiana oils also shows favorable agreement.

Another advantage of the quantitative approach is the ability to compare parameters determined from analysis of saturated and aromatic hydrocarbon subfractions obtained after liquid chromatographic separation. In the offshore Lou-

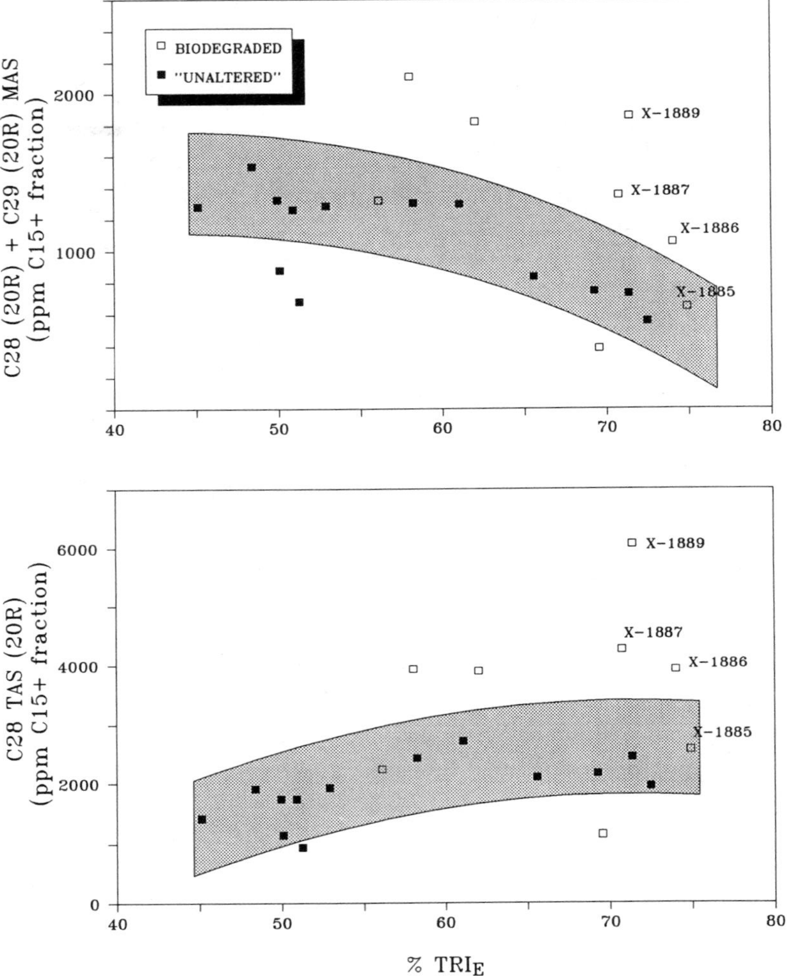

Figure 12.8 Plots of ring-C monoaromatic sterane [MAS; 5α(H)-$C_{28}H_{44}$ (20*R*) + 5β(H)-$C_{29}H_{46}$ (20*R*)] and triaromatic sterane [TAS; $C_{28}H_{38}$ (20*R*)] concentrations versus the sterane aromatization parameter percent TRI_E in oils from the South Pass 61 field, offshore Louisiana (Requejo and Halpern, 1990). The shaded region highlights the trends in concentration with increasing aromatization. Biodegraded samples are identified by the open symbols.

isiana oils, it was initially noted that, despite the uniformity in biomarker composition of the oils, concentrations of hopane and ethyl cholestane (20*R*) varied widely (236–1848 ppm and 93–422 ppm, respectively) (Requejo and Halpern, 1990). The degree of sterane aromatization also varied over a range greater than expected for a family of oils at similar maturity levels (45–75%). A crossplot of these parameters for both the Louisiana and Monterey oils (Figs. 12.9 and 12.10) reveals that the absolute concentrations of these biomarkers are highly correlated with the degree of sterane aromatization. In the case of the Monterey oils, the correlation is better among the heavy oils (those with API gravities in the range 5 to 15°). Biomarker concentrations in Monterey oils with API gravities > 15° are more variable. Some of the more extensively biodegraded Louisiana oils (e.g., X-1889, X-1887; Fig. 12.10) deviate from the main trend, most likely as a result of

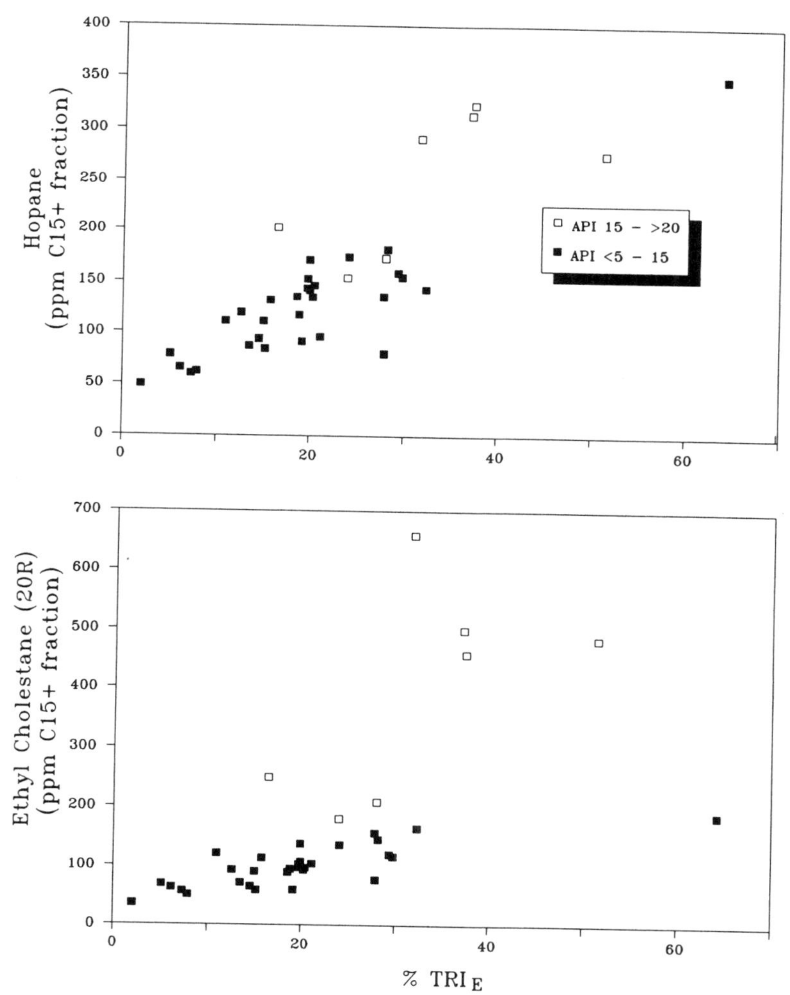

Figure 12.9 Plots of 17α(H),21β(H)-hopane and 5α(H),14α(H),17α(H)-ethyl cholestane (20*R*) versus the sterane aromatization parameter percent TRI_E in Monterey oils.

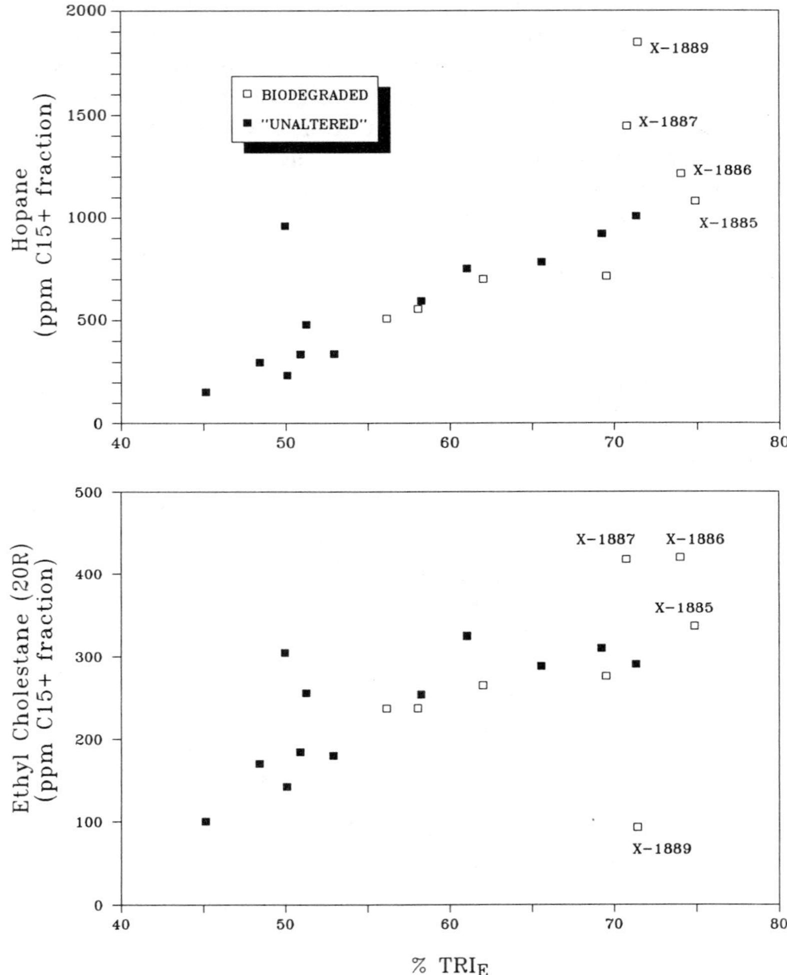

Figure 12.10 Plots of 17α(H),21β(H)-hopane and 5α(H),14α(H),17α(H)-ethyl cholestane (20*R*) versus the sterane aromatization parameter percent TRI_E in oils from the South Pass 61 field, offshore Louisiana.

the "concentration" effect described earlier. One severely biodegraded oil (X-1889), has been altered to the extent that ethyl cholestane (20*R*) has been degraded and its concentration significantly reduced (Fig. 12.10). These findings are particularly noteworthy in the case of the Louisiana oils, as conventional maturity indicators (alkane/isoprenoid ratios, gasoline-range parameters, etc.) indicate that the oils are of uniform maturity, yet the aromatized sterane parameter varies considerably.

Why the absolute concentrations of these compounds covary with sterane aromatization is unclear. Sterane aromatization has been unquestionably linked to thermal maturation (Mackenzie, 1984), although additional processes appear to influence the relative proportion of mono- and triaromatic steranes in oils (Hoffman et al., 1984). One explanation which attempts to reconcile the observations presented here is depicted schematically in Figure 12.11. According to this scheme, higher concentrations of the saturated steranes (ethyl cholestane in the example shown) are accompanied by a higher degree of sterane aromatization because aromatization would be the principal pathway governing the fate of steroidal hydrocarbons. Hence, the higher the concentration of saturated steranes, the further the aromatization reaction would proceed (shaded region in Fig. 12.11). Isomerization becomes a relatively minor catagenic process. The aromatization is presented as a progression starting with the conversion of the fully saturated compound to the monoaromatized homologue. In actual fact, the evidence for the occurrence of this initial step in natural systems is limited and its incorporation in the scheme is based primarily on the results of laboratory thermal maturation experiments

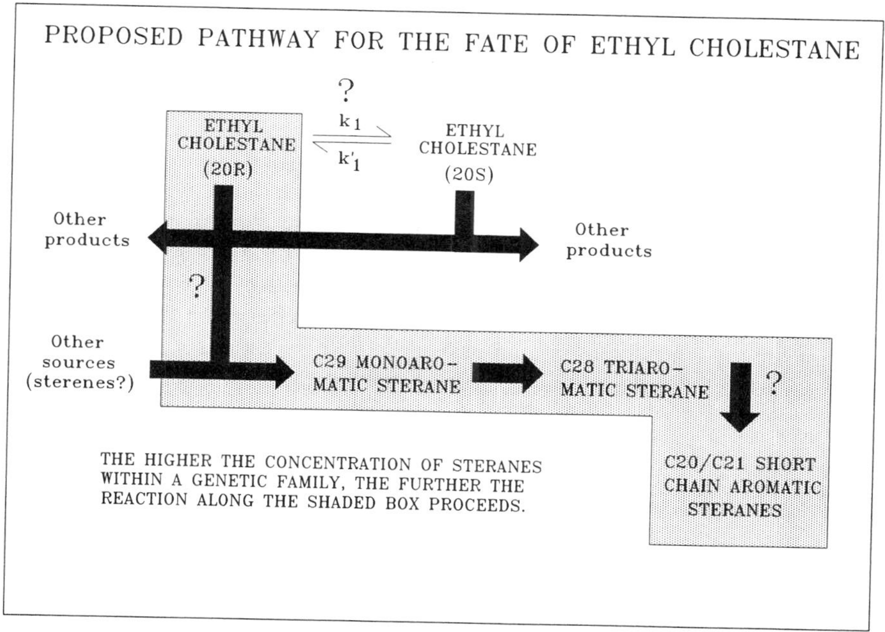

Figure 12.11 Schematic diagram of a proposed catagenic pathway for ethyl cholestane. Progressive aromatization is the principal reaction mechanism, while isomerization is a lesser process. Question marks refer to reactions for which evidence is lacking or marginal at present. At advanced stages in this sequence aromatization would go to completion and concentrations of both the saturated and aromatic steranes would be expected to decrease with any further thermal maturation. See text for further caveats and details.

(Seifert et al., 1983). There may exist sources for monoaromatic steranes other than the saturate compounds (Fig. 12.11). Similarly, the evidence for conversion of the triaromatic compound to the short-chain homologue is questionable, based on recent reports (Beach et al., 1989). However, the aromatization of ring-C monoaromatic steranes is suggested both by the quantitative data presented here for actual samples and by the results of laboratory experiments (Seifert et al., 1983; Abbott and Maxwell, 1988). At more advanced stages, aromatization would go to completion and concentrations of both the saturated and aromatic steranes would be expected to decrease with any further thermal maturation. It should be emphasized that the supporting data is limited to *cogenetic families of oils*, in which variations associated with level of maturity and organic matter type are limited. The proposed scheme is speculative, but generally consistent with the measured data and amenable to testing through further empirical observations and quantitative laboratory maturation experiments.

One curious aspect of the data in Figures 12.9 and 12.10 is the large difference in the concentrations of hopane and ethyl cholestane in the offshore Louisiana and Monterey oils. Previous work has shown that Monterey oils tend to be enriched in biomarker compounds (Curiale et al., 1985). However, the data in Figures 12.7 and 12.8 show that, per unit C_{15+} fraction, the offshore Louisiana oils contain up to threefold greater absolute concentrations of hopane and ethyl cholestane than the Monterey oils. This anomaly can be explained by the differences in the overall composition of C_{15+} components in the respective oils. The offshore Louisiana oils generally consist of < 3 percent asphaltenes and up to 73 percent saturated hydrocarbons (Requejo and Halpern, 1990), while the heavier Monterey oils generally contain 25 to > 50 percent asphaltenes and < 25 percent saturated hydrocarbons (Requejo, unpublished data). Thus, the C_{15+} fraction of the Louisiana oils is composed in greater part of the saturated hydrocarbon "compartment" which contains the biomarker compounds. The Monterey oils contain saturated hydrocarbon compartments enriched in biomarkers, but this compartment is an overall smaller portion of the overall C_{15+} fraction, hence the lower concentrations.

SUMMARY

Quantitative biomarker data have been used to examine the fate of individual compounds which are used in sterane-derived molecular maturity parameters. The results suggest that, while aromatization appears to involve direct conversion of ring-C monoaromatic to triaromatic compounds, little evidence exists for epimerization of saturated steranes at C-20. It is possible that the latter reaction does not occur to a significant extent in natural systems, which is contrary to the commonly held view that steranes isomerize at the C-20 position with increasing thermal maturity. A correlation is observed between the absolute concentrations of saturated steranes and the degree of sterane aromatization. A generalized reaction scheme is proposed to account for these observations, but the specific molecular pathway(s) by which these reactions proceed are uncertain at the present time.

These results illustrate the importance of quantitative data in the study of biomarker geochemistry. As exemplified by this study, interpretations based on quantitative results can lead to insights which were not possible when using the conventional fingerprinting approach. A quantitative approach to the study of molecular parameters should gain in popularity as additional applications are documented.

Acknowledgments

I thank the management of Atlantic Richfield for permission to publish these results. Jim Nicola and Rick Tharpe are acknowledged for their technical support. Jackie Reed contributed to the development of the analytical methods and the data interpretation.

REFERENCES

Abbott, G.D. and Maxwell, J.R. (1988) Kinetics of the aromatization of rearranged ring-C monoaromatic steroid hydrocarbons. *Org. Geochem. 13*, 881–885.

Abbott, G.D., Lewis, C.A., and Maxwell, J.R. (1984) Laboratory simulation studies of steroid aromatization and alkane isomerization. *Org. Geochem. 6*, 31–38.

Baker, E.W., Louda, J.W., and Orr, W.L. (1987) Application of metalloporphyrin biomarkers as petroleum maturity indicators: The importance of quantification. *Org. Geochem. 11*, 303–309.

Beach, F., Peakman, T.M., Abbott, G.D., Sleeman, R., and Maxwell, J.R. (1989) Laboratory thermal alteration of triaromatic steroid hydrocarbons. *Org. Geochem., 14*, 109–111.

Curiale, J.A. and Odermatt, J.R. (1989) Short-term biomarker variability in the Monterey Formation, Santa Maria Basin. *Org. Geochem. 14*, 1–13.

Curiale, J.A., Cameron, D., and Davis, D.V. (1985) Biological marker distribution and significance in oils and rocks of the Monterey Formation, California. *Geochim. Cosmochim. Acta. 49*, 271–288.

Dahl, B., Speers, G.C., Steen, A., Talnaes, N., and Johansen, J.E. (1985) Quantification of steranes and triterpanes by gas chromatographic-mass spectrometric analysis. In *Petroleum Geochemistry in Exploration of the Norwegian Shelf* (eds., B.M. Thomas et al.), Norwegian Petroleum Society, Graham & Trotman, London, 303–307.

Eglinton, T.I. and Douglas, A.G. (1988) Quantitative study of biomarker hydrocarbons released from kerogens during hydrous pyrolysis. *Energy & Fuels 2*, 81–88.

Hoffman, C.F., A.S. Mackenzie, C.A. Lewis, J.R. Maxwell, J.L. Oudin, B. Durand, and M. Vandenbroucke (1984) A biological marker study of coals, shales and oils from the Mahakam Delta, Kalimantan, Indonesia. *Chem. Geol. 42*, 1–23.

MACKENZIE, A.S. (1984) Applications of biological markers in petroleum geochemistry. In *Advances in Petroleum Geochemistry*, V. 1 (eds., J. Brooks and D. Welte), London, Academic Press, 115–213.

MACKENZIE, A.S., RULLKÖTTER, J., WELTE, D.H., and MANKIEWICZ, P. (1985) Reconstruction of oil formation and accumulation in North Slope, Alaska, using quantitative gas chromatography-mass spectrometry. In Alaska North Slope Oil/Rock Correlation Study (eds., L.B. Magoon and G. Claypool), *AAPG Studies in Geology #20*, AAPG, Tulsa, Oklahoma, 319–377.

MACKENZIE, A.S., PATIENCE, R.L., MAXWELL, J.R., VANDENBROUCKE, M., and DURAND, B. (1980) Molecular parameters of maturation in the Toarcian shales, Paris Basin, France-I. changes in the configurations of acyclic isoprenoid alkanes, steranes and terpanes. *Geochim. Cosmochim. Acta. 44*, 1709–1721.

MELLO, M.R., GAGLIANONE, P.C., BRASSELL, S.C., and MAXWELL, J.R. (1988) Geochemical and biological marker assessment of depositional environment using Brazilian offshore oils. *Mar. Pet. Geol. 5*, 205–223.

PEAKMAN, T.M., TEN HAVEN, H.L., RECHKA, J.R., DE LEEUW, J.W., and MAXWELL, J.R. (1989) Occurrence of (20*R*)- and (20*S*)-$\Delta^{8(14)}$ and Δ^{14} 5α(H)-sterenes and the origin of 5α(H),14β(H),17β(H)-steranes in an immature sediment. *Geochim. Cosmochim. Acta. 53*, 2001–2009.

PEAKMAN, T.M. and MAXWELL, J.R. (1988) Early diagenetic pathways of steroid alkenes. *Org. Geochem. 13*, 583–592.

REQUEJO, A.G. and HALPERN, H.I. (1990) A geochemical study of oils from the South Pass 61 field, offshore Louisiana. In *Geochemistry of Gulf Coast Oils and Gases: Their Characteristics, Origin, Distribution and Exploration and Production Significance* (eds. D. Schumacher and B.F. Perkins), SEPM, pp. 219–235.

REQUEJO, A.G., HOLLYWOOD, J., and HALPERN, H.I. (1989) Recognition and source correlation of migrated hydrocarbons in the Upper Jurassic Hareelv Formation, Jameson Land, East Greenland. *AAPG Bulletin. 73*, 1065–1088.

RULLKÖTTER, J., MACKENZIE, A.S., WELTE, D.H., LEYTHAEUSER, D., and RADKE, M. (1984) Quantitative gas chromatography-mass spectrometry analysis of geological samples. *Org. Geochem. 6*, 817–827.

SEIFERT, W.K. and MOLDOWAN, J.M. (1979) The effect of biodegradation on steranes and terpanes in crude oils. *Geochim. Cosmochim. Acta. 43*, 111–126.

SEIFERT, W.K., CARLSON, R.M.K., and MOLDOWAN, J.M. (1983) Geomimetic synthesis, structure assignment, and geochemical correlation application of monoaromatized petroleum steroids. In *Advances in Organic Geochemistry 1981* (eds. M. Bjorøy and John Chichester), Wiley, pp. 710–724.

TEN HAVEN, H.L., DE LEEUW, J.W., PEAKMAN, T.M., and MAXWELL, J.R. (1986) Anomalies in steroid and hopanoid maturity indices. *Geochim. Cosmochim. Acta. 50*, 853–855.

13

Oil-to-Source-Rock Correlation Using Carbon-Isotopic Data and Biological Marker Compounds, Cook Inlet-Alaska Peninsula, Alaska

Leslie B. Magoon and Donald E. Anders

Abstract. Rock and oil samples from the Cook Inlet-Alaska Peninsula area were analyzed to determine the source of the commercial hydrocarbons produced in the Cook Inlet basin from lower Tertiary nonmarine sandstone reservoirs. Rock Eval (hydrogen index, HI) analysis and organic-carbon (OC) content were used to identify the most favorable rock samples for solvent extraction and carbon-isotopic, gas-chromatographic (GC), and gas-chromatographic/mass-spectrometric (GC-MS) analyses. On the basis of organic-matter richness, 5 Tertiary nonmarine coal and shale samples and 12 Mesozoic (Upper Triassic and Middle Jurassic) marine shale samples were selected. A total of 28 oil and condensate samples from producing wells, drill-stem tests, field separators, and seeps were used for oil-to-oil and oil-to-source-rock correlation. On the basis of biomarker and carbon-isotopic data, 4 of the shallower oils and condensates are from nonmarine source rocks, and 24 of the deeper oils are sourced from marine shales. Geochemical and regional geologic considerations indicate the following conclusions. The Upper Tertiary nonmarine oils and condensates associated with commercial microbial gas accumulations are geochemically similar to the immature organic matter in the Tertiary nonmarine rocks. In the upper Cook Inlet, marine oils in lower Tertiary nonmarine reservoirs originated from Middle Jurassic rocks that matured during

Pliocene to Holocene time. In the lower Cook Inlet-Alaska Peninsula area, oils migrated from both Upper Triassic and Middle Jurassic source rocks during Late Cretaceous to early Tertiary time. Although four possible source rocks are identified, this oil-to-source-rock study shows that biomarkers are necessary to determine whether the Upper Triassic marine shale contributed hydrocarbons to the migrated oil in the lower Cook Inlet and the Alaska Peninsula.

INTRODUCTION

Previous work on the origin of the Cook Inlet oil was by Kelly (1963), Osment et al. (1967), Young et al. (1977), and Magoon and Claypool (1981). Various theories for the origin of Cook Inlet-Alaska Peninsula hydrocarbons have been proposed, the most common of which are that the oil originated (1) from the Tertiary nonmarine rocks, (2) from the Tertiary marine rocks to the south and east, or in the center, of the basin, or (3) from the Jurassic or Cretaceous rocks.

Kelly (1963) indicated on the basis of geologic evidence that the Jurassic oil on the Iniskin Peninsula is indigenous; however, he also considered the oil in the Swanson River oil field to be indigenous to the basal Tertiary rocks or Upper Cretaceous strata. Osment et al. (1967) preferred the Jurassic rocks as the source of the oil because (1) oil seeps from the Jurassic rocks on the lower west side of the Cook Inlet, (2) this seep oil is analytically similar to oil produced from Tertiary reservoirs, and (3) concentrations of several trace elements in both oils correspond closely. Young et al. (1977) used changes in C_5–C_7 and C_{15+} hydrocarbon composition to calculate the age of the oils. Cook Inlet oils from Tertiary reservoirs were calculated to have ages suggesting that the hydrocarbons are indigenous.

Magoon and Claypool (1981) presented general crude-oil characteristics showing that API gravity decreases from 55 to 28° as sulfur content increases from less than 0.05 to greater than 0.20 weight percent. On the basis of U.S. Bureau of Mines Hempel distillation analysis (Blasko et al., 1972, p. 23), all the oils from Tertiary reservoirs are similar (Magoon and Claypool, 1981). Except for the Kenai condensate and biodegraded oils, C_7 hydrocarbon compositions and gas chromatograms of C_{15+} saturated hydrocarbons also indicate that the oils in Tertiary reservoirs are all similar. Carbon-isotopic ratios ($\delta^{13}C$ values) of the saturated-hydrocarbon fraction in these oils range from -32 to -30 permil except for the Kenai condensate, which has -26.7 permil. Rock analyses of the Middle Jurassic and Cretaceous marine, and Tertiary nonmarine rocks included gas chromatograms of C_{15+} hydrocarbons and carbon-isotopic data on saturated hydrocarbons. The gas chromatograms of saturated hydrocarbons from rocks of the Middle Jurassic Tuxedni Group and of oils from Tertiary reservoirs were judged similar enough to indicate a reasonable oil/source-rock pair (Magoon and Claypool, 1981). Finally, on the basis of δD and $\delta^{13}C$ values of whole-oil and condensate samples from five producing fields, including the Kenai gas field, Yeh and Epstein (1981) suggested that all the oils are of marine origin.

Using data from biological markers, *n*-paraffin content, and carbon and hydrogen isotopes on oils from around the world, Moldowan et al. (1985) and Peters et al. (1986) developed parameters to assess the oil from marine shales, marine carbonates, and nonmarine rocks. Included in their studies were three hydrocarbon samples from the Cook Inlet that, on the basis of their parameters, originated from either marine shales or nonmarine rocks.

The purpose of this report is to determine: (1) whether the lower Tertiary oils derive from a common source and to what degree are they similar, (2) whether the Kenai condensate is similar to the nonmarine oils of Moldowan et al. (1985) and Peters et al. (1986), (3) whether the nonmarine organic matter in the Tertiary rocks can be distinguished from the marine organic matter in the Upper Triassic and Middle Jurassic rocks, and (4) whether the oil or condensate compares favorably with the organic matter in the Tertiary nonmarine or Mesozoic marine rocks.

SAMPLE SELECTION

Rock Samples

A total of 153 rock samples were analyzed by using Rock Eval and organic-carbon (OC) content to select the most favorable material for solvent extraction and carbon-isotopic, gas-chromatographic (GC), and gas-chromatographic/mass-spectrometric (GC-MS) analyses: 140 samples from 13 Cook Inlet wells, and 13 samples from the Puale Bay section. Previous results on these samples were reported by Magoon and Claypool (1981) and Claypool (1986). The samples used in this study were selected on the basis of availability, organic richness, and thermal maturity (Fig. 13.1). Rock samples from the Puale Bay area were included because the oil seeping from the Mesozoic rocks on the Alaska Peninsula is thought to be similar to Cook Inlet oils (Magoon and Claypool, 1981).

In general, the Mesozoic rocks were deposited in a marine environment, and the Tertiary rocks in a nonmarine environment (Figs. 13.2 and 13.3; Magoon and Egbert, 1986). The oldest organic-rich unit is found in unnamed Upper Triassic rocks of the Puale Bay area (samples 15–17, Table 13.1). Triassic rocks with source-rock potential are unknown in the Cook Inlet area. The Lower Jurassic Talkeetna Formation is volcaniclastic and uncharacteristic of a petroleum source rock (Magoon and Egbert, 1986). On the basis of previous studies (Magoon and Claypool, 1981), the Middle Jurassic rocks that underlie the entire Cook Inlet–Alaska Peninsula area are good petroleum sources. Nine Middle Jurassic rock samples (6–14) are included in this study: two outcrop samples from the Kialagvik Formation in the Puale Bay section, four conventional core samples from the Tuxedni Group (samples 9–12) recovered from below the Swanson River oil field in the Soldatna Creek 33–33 well, and chips from three conventional cores from the Tuxedni Group (samples 6–8) recovered from the Beal 1 well drilled on the Iniskin Peninsula.

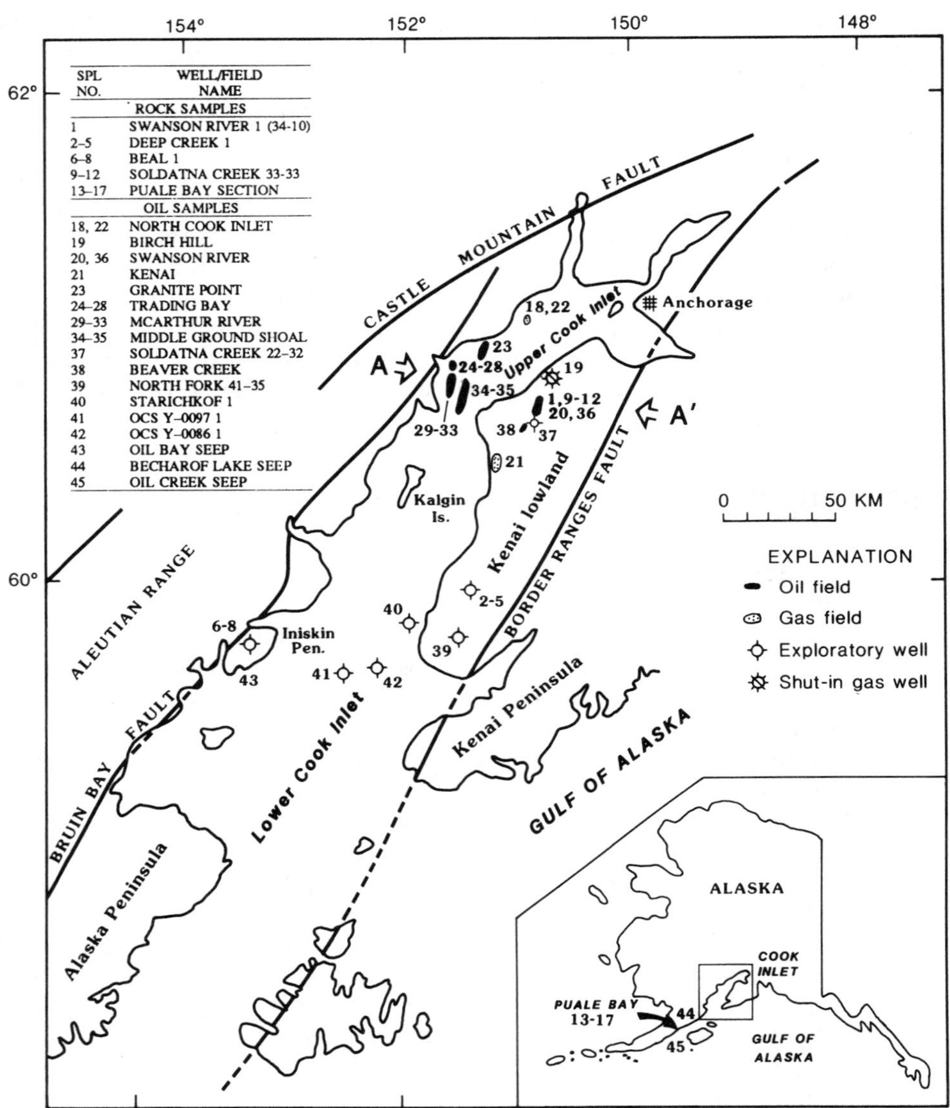

Figure 13.1 Cook Inlet–Alaska Peninsula area, showing locations of rock and oil samples and of cross section A–A′ (Fig. 13.2). Numbers refer to samples in Table 13.1.

Everywhere, the Cretaceous rocks are poor petroleum sources, whenever analyzed (Magoon and Claypool, 1981; Magoon, 1986; Claypool, 1986).

The Tertiary rocks are more than 6,000 m (20,000 ft) thick in the upper Cook Inlet but are almost absent on the Alaska Peninsula north of Puale Bay. In the

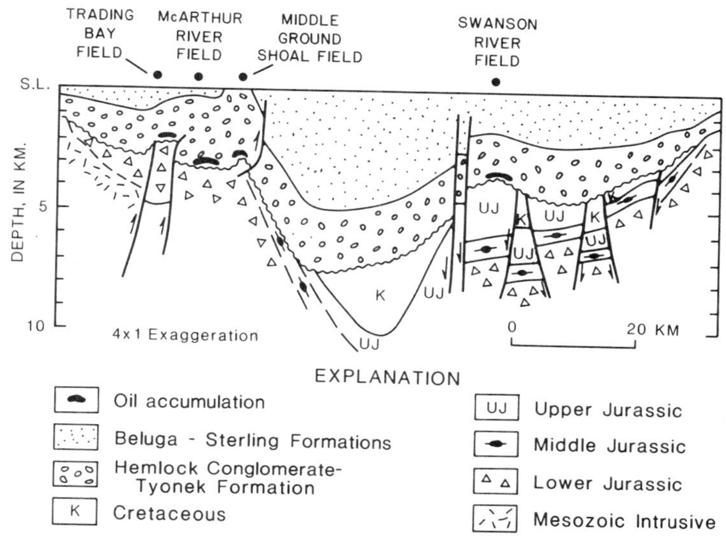

Figure 13.2 Cross section A–A′, showing stratigraphic units and oil fields in the Cook Inlet–Alaska Peninsula area (modified from Boss et al., 1976).

upper Cook Inlet, the oldest Tertiary unit is the West Foreland Formation, which is represented by two conventional core samples (4–5) from the Deep Creek 1 well drilled on the Kenai lowland. On the basis of its lithologic description, the Hemlock Conglomerate, a major reservoir unit, contains very little organic matter. The Tyonek Formation, which contains abundant coal, is represented by three conventional cores: from the Deep Creek 1 well (samples 2–3) and the Swanson River 1 well (sample 1). Even though the Beluga and Sterling Formations contain coal, they were not sampled because their organic matter is of low-thermal maturity and probably similar to that of the Tyonek Formation.

Oil Samples

Oil production is restricted to the Tertiary rocks in the Cook Inlet (Fig. 13.3); more than a billion barrels has been produced, of which 80 percent is from the Hemlock Conglomerate, 18 percent from the Tyonek Formation, and the remaining 2 percent from the West Foreland Formation (Alaska Oil and Gas Conservation Commission, 1988).

A total of 28 oil and condensate samples from reservoirs or seeps that range in age from Jurassic to Pliocene are included in this study (samples 18–45, Table 13.1). Three oil-seep samples, all from the Jurassic Naknek Formation (Oil Bay, sample 43, on the Iniskin Peninsula; Becharof Lake, sample 44, and Oil Creek, sample 45, on the Alaska Peninsula) were studied. Two oils (samples 37 and 42, recovered by drill-stem tests) from the Cretaceous Matanuska Formation are also

System	Series, sub-series	Cook Inlet (Lower — Upper)	Maximum thickness (meters)	Petroleum Production	Samples: Rock	Samples: Oil
TERTIARY	Pliocene U, L	Sterling Formation (Kenai Group)	1050	☼		21 (C)
	Miocene U	Beluga Fm. (Kenai Group)	1525	☼		
	Miocene M, L; Oligocene U	Tyonek Formation (Kenai Group)	2135	☼ ● 18%	1-3 (T_1)	18,19 (C) 20 (N) 23 (C) 24,29,34,38 (N) 25,27,40 (B)
	Oligocene L	Hemlock Conglomerate (Kenai Group)	450	● 80%		22 (C) 30,31,35,36,39 (N) 26,28 (B)
	Eocene U, M					
	Eocene L; Paleocene U	West Foreland Fm.	1000	● 2%	4-5 (T_2)	32,33 (N)
	Paleocene L					
CRETACEOUS	Senonian	Kaguyak Fm.; Upper part of Matanuska Fm.	2600			37 (N) 42 (B)
	Upper; Lower					
	Neocomian	Unnamed rocks	571			
JURASSIC	Upper	Naknek Fm.	2185			
		Chinitna Fm.	715			
	Middle	Tuxedni Group	2960		6-14 (J)	43,44,45 (B)
	Lower	Talkeetna Fm.	2575			
TRIASSIC	Upper	Unnamed rocks	395		15-17 (T̶R)	
	Middle					
	Lower					

Figure 13.3 Stratigraphic column of the Cook Inlet area, showing intervals for each sample (numbers). Percentages indicate relative amounts of oil recovered from the lower Tertiary rocks. Tr, Upper Triassic; J, Middle Jurassic; T_2, Paleocene West Foreland Formation; T_1, Miocene and Oligocene Tyonek Formation; C, condensate; N, normal oil; B, biodegraded oil. Geologic time scale modified from van Eysinga (1975); rock units and thicknesses from Fisher and Magoon (1978).

included. Oil samples from oil fields include: two samples from the West Foreland Formation (samples 32–33) in the McArthur River field, seven samples from the Hemlock Conglomerate (samples 22, 26, 28, 30, 31, 35, 36) in five fields, two samples from the Hemlock Conglomerate and Tyonek Formation (samples 23–24) in the Granite Point and Trading Bay fields, and five samples from the Tyonek Formation (samples 25, 27, 29, 34, 38) in four fields. Drill-stem test samples are from the North Cook Inlet State 1 well (sample 18), the North Fork 41-35 well (sample 39), the Starichkof State 1 well (sample 40), the OCS Y-0097 well (sample 41), and the OCS Y-0086 well (sample 42). A condensate, presumably from the Sterling Formation, collected from a separator (sample 21) in the Kenai gas field, is included, along with the analytical results of Moldowan et al. (1985) and Peters et al. (1986) for three samples (19–20, 37). Two samples (19–20) are of light oil/condensate recovered from Miocene sandstone reservoirs in the Birch Hill gas field and above the Swanson River oil field, respectively. Sample 37 is from an Upper Cretaceous sandstone reservoir just south of the Swanson River oil field.

EXPERIMENTAL PROCEDURE

The rock and oil analyses used in this study are as follows (Table 13.1). Rock analyses include Rock Eval, organic carbon content, and Soxhlet extraction. Carbon-isotopic analyses of the C_{15+} saturated- and aromatic-hydrocarbon fractions, and GC and GC-MS analyses of the C_{15+} saturated-hydrocarbon fraction, for both rock and oil are used for oil-to-oil and oil-to-source-rock correlations. GC-MS information includes *m/z* 191 and 217 fragment-ion traces. Also included from Magoon and Claypool (1981) are API gravity and sulfur content.

Rock samples were powdered and Soxhlet extracted in chloroform for 16 hours. Elemental sulfur was removed by using elemental copper. Asphaltenes were removed from the oils and rock extracts by precipitation at 22°C in *iso*-octane. The deasphaltened oils and bitumens were fractionated into saturates, aromatics, and resins (NSO) via silica-gel/alumina column chromatography, using *iso*-octane to elute the saturated hydrocarbons, benzene to elute the aromatic hydrocarbons, and benzene/Methanol (50/50) to remove the resins.

Gas chromatograms shown were obtained by using a Hewlett-Packard 5880 gas chromatograph[1] equipped with a DB-5 fused-silica capillary column (50 m × 0.327 mm) programmed from 50 to 320°C at 4°C/min (2 minutes initial, 15 minutes final). The carrier gas was hydrogen at a linear velocity of 27 cm/s.

Gas chromatography-mass spectrometry was used to monitor selected fragment ions at *m/z* 191 and 217, characteristic of tricyclic and pentacyclic terpane and sterane biomarkers, respectively. A Hewlett-Packard 5880 gas chromatograph was equipped with a DB-5 fused-silica capillary column (50 m × 0.327 mm) pro-

[1]Any use of trade, product, or firm names in this chapter is for descriptive purposes only and does not imply endorsement by the U.S. Government.

Table 13.1 PETROLEUM GEOCHEMISTRY OF ROCK AND OIL SAMPLES FROM THE COOK INLET–ALASKA PENINSULA, ALASKA

Spl. no.	Field name	Well name	API no.	Location S-T-R/Block	Depth interval (ft)	Age	Rock unit
			Nonmarine rocks				
1	Swanson River	Swanson River 1 (34-10)	5013310136	10–8N–9W	8504–8512	Olig.–Mio.	Tyonek Fm.
2	-	Deep Creek 1	5013310004	15–2S–13W	7001–7002	Olig.–Mio.	Tyonek Fm.
3	-	Deep Creek 1	5013310004	15–2S–13W	10,288–10,289	Olig.–Mio.	Tyonek Fm.
4	-	Deep Creek 1	5013310004	15–2S–13W	13,688–13,690	Paleocene	West Foreland Fm.
5	-	Deep Creek 1	5013310004	15–2S–13W	13,696–13,697	Paleocene	West Foreland Fm.
			Marine rocks				
6	-	Beal 1	5012110008	17–5S–23W	6410–6415	M. Jurassic	Tuxedni Gp.
7	-	Beal 1	5012110008	17–5S–23W	6420–6424	M. Jurassic	Tuxedni Gp.
8	-	Beal 1	5012110008	17–5S–23W	8683–8707	M. Jurassic	Tuxedni Gp.
9	Swanson River	Soldatna Creek 33-33	5013320293	33–8N–9W	15,040	M. Jurassic	Tuxedni Gp.
10	Swanson River	Soldatna Creek 33-33	5013320293	33–8N–9W	15,050	M. Jurassic	Tuxedni Gp.
11	Swanson River	Soldatna Creek 33-33	5013320293	33–8N–9W	15,060	M. Jurassic	Tuxedni Gp.
12	Swanson River	Soldatna Creek 33-33	5013320293	33–8N–9W	15,074	M. Jurassic	Tuxedni Gp.
13	-	Puale Bay Section	5012995007	29–28S–37W	Surface	M. Jurassic	Kialagvik Fm.
14	-	Puale Bay Section	5012995007	29–28S–37W	Surface	M. Jurassic	Kialagvik Fm.
15	-	Puale Bay Section	5012995007	5–29S–37W	Surface	U. Triassic	Unnamed rocks
16	-	Puale Bay Section	5012995007	5–29S–37W	Surface	U. Triassic	Unnamed rocks
17	-	Puale Bay Section	5012995007	5–29S–37W	Surface	U. Triassic	Unnamed rocks
			Nonmarine oils				
18	North Cook Inlet	North Cook Inlet State 1	5028310018	29–12N–9W	11,007–11,026	Olig.–Mio.	Tyonek Fm.
19	Birch Hill	Birch Hill 22–25	5013310029	25–9N–9W	8190–8220	Olig.–Mio.	Tyonek Fm.
20	Swanson River	Swanson River 212–27	5013310153	27–8N–9W	7600	Olig.–Mio.	Tyonek Fm.
21	Kenai	Kenai KSU–2	5013395001	8–4N–11W	3858–4210	Pliocene	Sterling Fm.

			Marine oils				
22	North Cook Inlet	Cook Inlet State 1–A	5028310013	6–11N–9W	12,404–12,477	Oligocene	Hemlock Cgl.
23	Granite Point	Granite Point State 20	5013320012	31–11N–11W	8690–9620	Olig.–Mio.	Hem–Ty
24	Trading Bay	Trading Bay State 2	5013320122	34–10N–13W	9016–9808	Olig.–Mio.	Hem–Ty
25	Trading Bay	Trading Bay State A–7	5013320036	4–9N–13W	4605–4625	Olig.–Mio.	Tyonek Fm.
26	Trading Bay	Trading Bay 1–A	5013310052	4–9N–13W	5362–5444	Oligocene	Hemlock Cgl.
27	Trading Bay	Trading Bay State A–8	5013320043	4–9N–13W	5758–6535	Olig.–Mio.	Tyonek Fm.
28	Trading Bay	Trading Bay State A–6	5013320020	4–9N–13W	6069–6370	Oligocene	Hemlock Cgl.
29	McArthur River	Trading Bay K–30	5013320211	21–9N–13W	10,876–11,268	Olig.–Mio.	Tyonek Fm.
30	McArthur River	Trading Bay K–17	5013320248	21–9N–13W	11,423–11,771	Oligocene	Hemlock Cgl.
31	McArthur River	Trading Bay G–21	5013320267	29–9N–13W	—	Oligocene	Hemlock Cgl.
32	McArthur River	Trading Bay G–33	5013320212	29–9N–13W	10,086–10,625	Eocene	West Foreland Fm.
33	McArthur River	Trading Bay K–30	5013320211	17–9N–13W	12,008–12,480	Eocene	West Foreland Fm.
34	Middle Ground Shoal	Middle Ground Shoal 6	5013310073	31–9N–12W	5782–6607	Olig.–Mio.	Tyonek Fm.
35	Middle Ground Shoal	Middle Ground Shoal 13	5013320039	31–9N–12W	7870–8425	Oligocene	Hemlock Cgl.
36	Swanson River	Swanson River 12–22	5013310147	22–8N–9W	10,800–10,895	Oligocene	Hemlock Cgl.
37	-	Soldatna Creek 22–32	5013310022	32–7N–9W	14,175–14,200	Cretaceous	Matanuska Fm.
38	Beaver Creek	Beaver Creek 4	5013320239	33–7N–10W	15,060–15,100	Olig.–Mio.	Tyonek Fm.
39	-	North Fork 41–35	5023110004	35–4S–14W	10,808–10,859	Oligocene	Hemlock Cgl.
40	-	Starichkof 1	5023110002	33–3S–15W	6920–6930	Olig.–Mio	Tyonek Fm.
41	-	OCS Y–0097 1	5522000002	LCI Bl 401	—	—	—
42	-	OCS Y–0086 1	5522000003	LCI Bl 318	9400	Cretaceous	Matanuska Fm.
43	-	Oil Bay seep	-	12–6S–24W	Surface	M. Jurassic	Naknek Fm.
44	-	Becharof Lake seep	5012995001	Unknown	Surface	M. Jurassic	Naknek Fm.
45	-	Oil Creek seep	5012995001	10–29S–40W	Surface	M. Jurassic	Naknek Fm.

Data for samples 19, 20, and 37 from Moldowan et al., (1985) and Peters et al., (1986).

Samples 29 and 30 contain a large amount of gammacerane.

[-, no information; API, American Petroleum Institute; Arom, aromatic hydrocarbons; Asph, asphaltenes; Cgl., conglomerate; ^{13}C, carbon isotope; CPI, carbon-preference index; Fm., Formation; GCMS, gas-chromatographic/mass spectrometric analysis; Gp., Group Hem–Ty, Hemlock Conglomerate and Tyonek Formation, undivided; HI, hydrogen index; OC, organic carbon; OEP, odd-even-predominance index; Olig.–Mio., Oligocene and Miocene; PI, production index; Res, resins; Sat, saturated hydrocarbons; S-T-R/Block, section-township-range/Outer Continental Shelf block; S, sulfur, T_{max}, Rock Eval maximum temperature; Bit, total bitumen; x, sample contains abundant 5β20*R* sterane, hopenes, ββ hopanes, and numerous other immature biomarkers; xx, tricyclic content too low for accurate measurement; Tri, C_{19} tricyclic/C_{23}; Hop, Hopane/C_{19} tricyclic; Mig, migration distance; 32*S*/*R*, 22*S*-17αH,21βH, 30,31-bishomohopane/22*R*-17αH,21βH, 30,31-bishomohopane; 33*S*/*R*, 22*S*-17αH,21βH, 30,31,32-trishomohopane/22*R*-17αH,21βH, 30,31,32-trishomohopane; *Tm*/*Ts*, 17αH-22,29,30-trisnorhopane/18αH-22,29,30-trisnorneohopane; 1, C_{29} ααα 20*S*/ααα 20*S* + 20*R*; 2, C_{29} αββ 20*S* + 20*R*/(αββ 20*S* + 20*R* + ααα 20*S* + 20*R*); 3, C_{29} ααα 20*S*/ααα 20*R*; 4, C_{29} αββ 20*R*/ααα 20*R*; T_1, Miocene and Oligocene; T_2, Paleocene; J, Middle Jurassic; Tr, Upper Triassic; C, condensate; N, normal oil; B, biodegraded oil]

Table 13.1 PETROLEUM GEOCHEMISTRY OF ROCK AND OIL SAMPLES FROM COOK INLET–ALASKA PENINSULA, ALASKA—Continued

		Rock eval				Solvent extraction											
Spl. no.	OC (wt %)	HI	OI	PI	T_{max} °C	Sat (ppm)	Arom (ppm)	Res (ppm)	Asph (ppm)	Bit (ppm)	S/A	Sat (%)	Arom (%)	Res (%)	Asph (%)	API (°)	S (%)
Nonmarine rocks																	
1	2.6	303	46	0.04	427	264	627	879	108	2265	0.42	14	33	47	6	-	-
2	71.7	100	43	0.03	426	642	1110	1587	1776	8380	0.58	13	22	31	35	-	-
3	41.6	192	29	0.03	427	1944	1508	1520	2766	8200	1.29	25	19	20	36	-	-
4	19.7	356	16	0.02	431	1992	5308	2361	1506	11,397	0.36	18	48	21	13	-	-
5	10.6	400	17	0.02	434	4591	5233	3387	2817	17,695	0.88	29	33	21	18	-	-
Marine rocks																	
6	0.9	137	12	0.09	437	234	209	118	79	685	1.12	37	33	18	12	-	-
7	1.1	154	20	0.01	442	327	274	144	275	969	1.19	32	27	14	27	-	-
8	0.7	112	21	0.17	449	339	143	98	72	697	2.36	52	22	15	11	-	-
9	1.9	349	16	0.03	433	226	240	196	74	873	0.94	31	33	27	10	-	-
10	2.1	357	21	0.03	429	570	295	243	106	1116	1.93	47	24	20	9	-	-
11	0.8	153	32	0.08	431	168	121	94	32	416	1.39	40	29	23	8	-	-
12	1.9	325	21	0.03	432	298	234	233	108	882	1.27	34	27	27	12	-	-
13	1.5	411	47	0.04	435	666	391	245	86	1484	1.70	48	28	18	6	-	-
14	1.4	304	43	0.05	436	388	437	301	99	1349	0.89	32	36	25	8	-	-
15	1.7	347	106	0.07	443	611	336	229	99	1343	1.82	48	26	18	8	-	-
16	2.8	463	58	0.09	440	1219	782	637	188	3380	1.56	43	28	23	7	-	-
17	1.3	226	84	0.07	442	429	300	192	48	1114	1.43	44	31	20	5	-	-

Nonmarine oils																	
18	-	-	-	-	-	-	-	-	-	-	10.8	61	6	32	1	46.9	0.03
19	-	-	-	-	-	-	-	-	-	-	-	-	-	-	-	37.4	-
20	-	-	-	-	-	-	-	-	-	-	-	-	-	-	-	25.4	-
21	-	-	-	-	-	-	-	-	-	-	9.6	73	13	14	0	28.6	0.05
Marine oils																	
22	-	-	-	-	-	-	-	-	-	-	6.4	73	11	12	4	37.3	0.04
23	-	-	-	-	-	-	-	-	-	-	9.6	79	8	12	1	40.8	0.06
24	-	-	-	-	-	-	-	-	-	-	-	-	-	-	-	28.6	0.12
25	-	-	-	-	-	-	-	-	-	-	3.7	63	17	15	4	24.5	0.09
26	-	-	-	-	-	-	-	-	-	-	7.2	72	10	14	4	29.4	0.07
27	-	-	-	-	-	-	-	-	-	-	6.9	73	11	11	5	28.6	0.10
28	-	-	-	-	-	-	-	-	-	-	3.8	58	15	20	7	30.1	0.07
29	-	-	-	-	-	-	-	-	-	-	6.1	69	11	16	4	33.4	0.07
30	-	-	-	-	-	-	-	-	-	-	7.5	74	10	14	3	33.3	0.07
31	-	-	-	-	-	-	-	-	-	-	-	-	-	-	-	34.0	0.09
32	-	-	-	-	-	-	-	-	-	-	6.8	69	10	14	7	31.7	0.08
33	-	-	-	-	-	-	-	-	-	-	5.0	66	13	14	6	32.1	0.09
34	-	-	-	-	-	-	-	-	-	-	7.7	75	10	14	2	36.2	0.04
35	-	-	-	-	-	-	-	-	-	-	5.4	72	13	14	1	35.4	0.07
36	-	-	-	-	-	-	-	-	-	-	5.2	67	13	16	4	32.2	0.09
37	-	-	-	-	-	-	-	-	-	-	-	-	-	-	-	-	0.23
38	-	-	-	-	-	-	-	-	-	-	9.4	78	8	11	3	36.6	0.04
39	-	-	-	-	-	-	-	-	-	-	5.3	62	12	18	9	25.5	0.12
40	-	-	-	-	-	-	-	-	-	-	4.2	56	13	17	14	25.6	0.21
41	-	-	-	-	-	-	-	-	-	-	3.3	50	15	16	19	26.0	0.21
42	-	-	-	-	-	-	-	-	-	-	7.9	77	10	10	3	27.8	0.22
43	-	-	-	-	-	-	-	-	-	-	-	-	-	-	-	-	-
44	-	-	-	-	-	-	-	-	-	-	3.4	60	18	17	5	16.4	0.20
45	-	-	-	-	-	-	-	-	-	-	3.5	60	17	18	5	16.6	0.21

Table 13.1 PETROLEUM GEOCHEMISTRY OF ROCK AND OIL SAMPLES FROM COOK INLET–ALASKA PENINSULA, ALASKA–Continued

	$\delta^{13}C$		Gas chromatography				Gas-Chromatographic/Mass-spectrometric ratios											
Spl.	Sat (o/oo)	Arom (o/oo)	OEP	CPI	Pr/Ph	C_{17}/Pr	Tri	Hop	32S/R	33S/R	Tm/Ts	1	2	3	4	Sym	Mig (km)
							Nonmarine rocks										
1	−28.6	−29.7	2.9	1.8	2.5	1.1	0.4	9.0	x	x	x	x	x	x	x	T_1	-
2	−27.2	−25.6	2.4	1.6	3.6	0.8	0.5	12.5	x	x	x	0.1	x	0.1	x	T_1	-
3	−28.4	−27.0	1.9	1.5	6.5	0.2	xx	xx	0.8	x	x	0.1	x	0.1	x	T_1	-
4	−27.3	−26.3	1.7	1.4	10.0	0.2	11.7	3.4	0.7	0.7	0.3	0.2	x	0.1	x	T_2	-
5	−28.0	−26.5	1.2	1.0	8.6	0.2	12.0	9.2	0.7	0.6	0.4	0.2	x	0.1	x	T_2	-
							Marine rocks										
6	−29.6	−28.2	1.0	1.0	2.8	1.4	0.7	13.1	1.2	1.4	0.7	0.3	0.6	0.2	1.1	J	-
7	−29.3	−28.1	1.0	1.0	2.8	1.3	0.5	14.4	1.5	1.2	2.0	0.4	0.6	0.5	1.7	J	-
8	−30.6	−29.5	1.0	1.0	2.8	2.0	0.7	5.8	1.0	1.8	0.6	0.3	0.5	0.7	1.0	J	-
9	−29.7	−29.0	1.6	1.2	4.3	0.3	0.2	60.5	0.9	1.0	4.9	0.2	x	0.4	x	J	-
10	−29.5	−29.0	1.6	1.3	4.0	0.4	0.2	24.2	1.1	1.1	3.5	0.2	0.3	0.3	0.4	J	-
11	−28.9	−28.6	1.4	-	3.6	0.6	0.2	7.4	0.7	0.7	2.8	0.2	0.3	0.2	0.4	J	-
12	−29.7	−29.6	1.5	1.2	3.9	0.4	0.2	11.9	1.4	1.1	2.6	0.2	0.3	0.3	0.4	J	-
13	−30.4	−30.0	1.4	-	3.5	0.5	0.2	30.3	1.0	1.1	2.1	0.2	0.3	0.2	0.3	J	-
14	−31.2	−30.2	1.2	1.1	2.7	0.6	0.3	60.0	1.1	1.1	2.3	0.3	0.3	0.3	0.4	J	-
15	−29.6	−29.1	1.1	1.0	2.7	0.7	0.1	40.0	1.2	1.5	1.3	0.3	0.3	0.5	0.4	Tr	-
16	−30.0	−29.8	1.1	1.0	2.9	0.7	0.2	30.0	1.1	1.2	1.9	0.3	0.3	0.5	0.3	Tr	-
17	−29.2	−28.8	1.1	1.0	2.9	0.8	0.4	7.5	1.3	0.9	1.8	0.4	0.4	0.6	0.5	Tr	-

								Nonmarine oils									
18	−29.1	−26.3	-	1.8	4.2	2.0	7.1	1.8	0.8	1.0	2.4	0.3	0.5	0.4	0.7	C	45
19	−28.8	−26.5	-	-	-	-	-	-	-	-	-	-	-	-	-	C	25
20	−28.8	−25.8	-	-	-	-	-	-	-	-	-	0.5	0.4	0.4	0.5	N	19
21	−26.2	−25.7	-	2.5	4.9	0.1	-	-	-	-	-	-	-	-	-	C	-
								Marine oils									
22	−29.9	−27.4	-	1.0	3.3	2.3	1.3	4.3	1.3	1.4	0.9	0.5	0.6	1.0	1.5	C	41
23	−29.7	−28.0	-	1.0	3.4	2.0	2.4	3.9	1.2	1.6	1.1	0.4	0.6	0.8	1.2	C	32
24	−30.2	−28.5	-	-	-	-	-	-	-	-	-	-	-	-	-	N	30
25	−29.6	−28.4	-	-	2.1	0.3	0.5	32.7	1.4	1.5	0.6	0.5	0.6	1.1	1.7	B	30
26	−30.1	−28.7	-	-	2.2	0.5	0.5	32.9	1.4	1.5	0.8	0.5	0.6	1.2	1.7	B	30
27	−30.0	−28.1	-	-	2.6	0.9	0.5	30.7	1.4	1.6	0.7	0.5	0.6	1.0	1.5	B	30
28	−30.0	−28.7	-	-	2.7	0.3	0.7	24.8	1.5	1.5	0.8	0.5	0.6	1.2	1.7	B	30
29	−30.1	−28.7	-	1.0	3.0	1.5	0.6	29.6	1.4	1.4	0.8	0.5	0.6	1.0	1.5	N	30
30	−30.1	−28.7	-	1.0	3.1	1.6	0.5	24.7	1.5	1.3	0.8	0.5	0.6	1.0	1.5	N	30
31	−30.1	−28.8	-	1.0	-	-	-	-	-	-	-	-	-	-	-	N	30
32	−30.5	−28.9	-	1.0	2.9	1.4	0.3	49.7	1.6	1.5	1.0	0.5	0.6	1.0	1.5	N	30
33	−30.4	−29.1	-	1.0	3.1	1.5	0.5	16.6	1.3	1.4	1.0	0.5	0.6	1.1	1.5	N	30
34	−30.0	−27.8	-	1.0	3.0	1.7	0.9	19.5	1.4	1.4	0.9	0.5	0.6	1.0	1.5	N	21
35	−30.0	−28.2	-	1.0	3.0	1.6	0.8	16.2	1.5	1.4	0.9	0.5	0.6	1.0	1.5	N	21
36	−30.0	−29.4	-	1.0	2.8	1.5	0.7	24.6	1.5	1.6	0.9	0.5	0.6	1.1	1.6	N	19
37	−30.6	−29.9	-	1.0	3.3	-	0.3	46.6	1.5	1.6	1.0	0.5	0.6	0.9	1.1	N	19
38	−30.1	−28.7	-	1.0	2.8	1.8	0.6	14.5	1.5	1.6	0.6	0.5	0.6	1.1	1.8	N	10
39	−30.3	−28.0	-	1.6	2.8	1.7	0.5	19.6	1.3	1.4	1.6	0.4	0.6	0.8	1.2	N	-
40	−30.1	−29.3	-	-	2.3	1.0	0.3	37.3	1.4	1.4	1.2	0.5	0.6	1.0	1.3	B	-
41	−30.8	−29.9	-	-	2.6	1.6	0.2	9.8	1.4	1.3	0.9	0.5	0.6	0.9	1.3	B	-
42	−30.2	−29.0	-	-	-	-	0.4	18.3	1.2	1.2	0.5	0.5	0.6	1.0	1.3	B	-
43	−29.4	−28.4	-	-	-	-	-	-	-	-	-	-	-	-	-	B	-
44	−29.8	−29.0	-	-	1.5	-	0.2	13.1	1.5	1.4	0.9	0.5	0.5	0.9	1.2	B	-
45	−30.2	−29.5	-	-	1.7	-	0.3	5	1.4	1.4	0.7	0.5	0.5	1.1	1.5	B	-

grammed from 50 to 170°C at 30°C/min and from 170 to 320°C at 3°C/min (2 minutes initial, 9 minutes final). The carrier gas was helium at a linear velocity of 27 cm/s. A Kratos/AEI MS-30 ion detector was run in multiple-ion mode, using a Fluke programmable power supply to switch the accelerating voltage. Source temperature was 250°C, electron-impact voltage 70 V, dynamic resolution 3,000 amu, and source pressure 1×10^{-6} torr.

Rock Eval results were obtained on an automated Rock-Eval II instrument (Delsi/Nermag), using the IFP Fina process.

Carbon-isotopic ratios ($\delta^{13}C$ values) of saturated- and aromatic-hydrocarbon fractions from whole oils and rock extracts were determined by using a Finnigan/Mat 251 mass spectrometer. The hydrocarbons were combusted to CO_2 in an atmosphere of oxygen, and the resulting CO_2 scrubbed and cryogenically dehydrated under vacuum. All delta values are reported relative to the Peedee belemnite (PDB) marine carbonate standard.

GEOLOGIC SETTING

Almost 60 years of exploration preceded the discovery of the first commercial oil field in the Cook Inlet area, the Swanson River field. Exploration started on the Iniskin Peninsula (Fig. 13.1) in 1902, where seven wells were drilled. Between 1921 and 1957, only nine exploratory wells were drilled before the Swanson River oil field was discovered. By the end of 1987, almost 1.1 billion bbl of oil and 5.3 trillion cubic feet (tcf) of gas had been produced, leaving 90 million bbl of oil and 3.3 tcf of gas yet to be extracted. The history of exploration and the framework and petroleum geology of this area have been discussed by many workers (Kelly, 1963, 1968; Detterman and Hartsock, 1966; Crick, 1971; Blasko et al., 1972; Kirschner and Lyon, 1973; MacKevett and Plafker, 1974; Blasko, 1974, 1976; Boss et al., 1976; Hite, 1976; Magoon et al., 1976; Fisher and Magoon, 1978; Claypool et al., 1980; Magoon et al., 1980; Magoon and Claypool, 1981; Reed et al., 1983; Magoon, 1986; Magoon and Egbert, 1986).

The Cook Inlet-Alaska Peninsula area is bounded on the northwest by the Alaska-Aleutian Range and the Talkeetna Mountains, and on the southeast by the Kenai Mountains. The pre-Upper Cretaceous sequence and correlative plutonic rocks of the Alaska-Aleutian Range batholith constitute the Peninsular terrane. The Border Ranges fault separates this terrane from the accreted Chugach terrane on the southeast.

In the Puale Bay on the Alaska Peninsula, unnamed Upper Triassic rocks, as much as 800 m (2,625 ft) thick, consist of fossiliferous limestone and volcanic agglomerate intruded by basaltic dikes and sills (Imlay and Detterman, 1977). In their fossil content, especially *Monotis sp.*, and fetid odor when broken, these rocks are similar to the Shublik Formation on the North Slope. The same Triassic rocks are presumed to underly much of the Cook Inlet, but nowhere are petroleum source-rock characteristics exhibited in adjacent outcrops.

The Lower Jurassic Talkeetna Formation, a volcaniclastic sequence, and the Alaska-Aleutian batholith are considered by most workers to be economic basement for the Cook Inlet petroleum province.

The Middle Jurassic rocks include the Tuxedni Group in the Cook Inlet area and the Kialagvik Formation on the Alaska Peninsula. The Tuxedni Group is thick and widespread, and contains potential petroleum source rocks. This group includes as many as six formations and ranges in thickness from 536 to 2,748 m (1,760–9,015 ft) in the Iniskin-Tuxedni area (Detterman and Hartsock, 1966) where the section is abundantly fossiliferous and consists of 35 percent sandstone, 65 percent siltstone, and minor amounts of tuff. All the sediment was deposited within a shallow-marine-shelf environment. The Kialagvik Formation crops out along the north shore of Puale Bay on the Alaska Peninsula (Imlay and Detterman, 1977). The unit is at least 730 m (2,400 ft) thick, is sparsely fossiliferous, and consists of 10 percent sandstone conglomerate and 90 percent siltstone representing submarine slope-channel fill and slope deposits, respectively.

The Upper Jurassic Naknek Formation contains a high proportion of feldspathic sandstone and conglomerate, but because of laumontite cementation it is a poor reservoir (Franks and Hite, 1980; Bolm and McCulloh, 1986). The Upper Cretaceous Matanuska Formation is predominantly siltstone containing small amounts of organic matter. Under the lower Cook Inlet, potential sandstone reservoirs in this unit were penetrated.

The Cenozoic rocks in the Cook Inlet area overlap the Alaska-Aleutian batholith on the northwest and the Border Ranges fault on the southeast. Calderwood and Fackler (1972) defined and named the critical Cenozoic stratigraphic units—West Foreland Formation, Hemlock Conglomerate, Tyonek Formation, Beluga Formation, and Sterling Formation—that are regionally correlated (Alaska Geological Society, 1969a–d, 1970a,b) and mapped (Hartman et al., 1972). These rock units were all deposited in a nonmarine forearc-basin setting. The provenance for the conglomerate, sandstone, siltstone, shale, and volcaniclastic debris was local highs flanking the basin and interior Alaska. Each of these rock units is a reservoir for oil or gas somewhere in the basin. Throughout the section, numerous large coal deposits formed and were preserved (Barnes and Payne, 1956; Barnes and Cobb, 1959).

The Tertiary units included in this study are the West Foreland Formation, Hemlock Conglomerate, and Tyonek Formation. The West Foreland Formation is as much as 600 m (2,000 ft) thick (Hartman et al., 1972). Its depositional environment is nonmarine fluvial, and sediments was derived from local source terranes (Kirschner and Lyon, 1973). Conglomerate on the west flank of the Cook Inlet indicates proximity to the source terrane. Coalescing alluvial fans from the northwest flank spread southeastward. Ash from volcanic eruptions covered the entire area, in some places burying trees in growth position; locally, coal swamps developed. The depositional environments of the Hemlock Conglomerate are fluvial deltaic to estuarine (Hartman et al., 1972; Kirschner and Lyon, 1973; Boss et al., 1976; Hite, 1976); the unit consists of conglomerate, conglomeratic sandstone, some siltstone, and a few thin coal seams. The depositional environment of the Tyonek Formation is fluvial, deltaic, and estuarine (Hartman et al., 1972; Kirschner and Lyon, 1973), or alluvial (Boss et al., 1976; Hite, 1976); the unit consists of massive sandstones and thick coal beds.

ANALYTICAL RESULTS

To simplify referencing the rock or oil samples listed in Table 13.1 to the figures, the following categories are used. "Nonmarine and marine rocks" refers to the depositional environment of the rock material; letters refer to the age of the rock unit. "Nonmarine and marine oils" refers to the type of organic matter (provenance) that sourced the hydrocarbons on the basis of biomarker (Moldowan et al., 1985; Peters et al., 1987) and carbon-isotopic data; letters refer to condensate (API gravity $> 37°$), normal oils (presence of *n*-alkanes), and biodegraded oils (diminished *n*-alkanes).

Richness, Kerogen Type, and Thermal Maturity

Rock Eval and OC data on 17 samples indicate that the Upper Triassic, Middle Jurassic, and Tertiary strata have the highest potential as source rocks. Analytical results for 12 core samples from 4 wells and 5 outcrop samples from the Puale Bay area indicate that 14 contain more than 1 weight percent OC and 11 samples have hydrogen-index (HI) values greater than 200 (Table 1). On the basis of T_{max} values, nine samples are immature ($< 435°C$), and the remaining eight samples are mature (435–460°C).

Middle Jurassic samples from the Beal 1 well at 1,950 to 2,650 m (6,400–8,700 ft) are slightly more mature than Middle Jurassic samples from the Soldatna Creek 33–33 well at 4,600 m (15,000 ft), as indicated by T_{max} values from Rock Eval analyses. If the thermal gradients that caused this level of maturity are similar on both sides of the Cook Inlet, then 2,100 ± 300 m (7,000 ± 1,000 ft) of uplift and erosion is indicated at the Beal 1 well. Because the Jurassic section is exposed near the well on the Iniskin Peninsula, this amount of uplift is reasonable.

The HI and oxygen-index (OI) values plotted on a modified van Krevelen diagram (Fig. 13.4) indicate that the samples from the Tertiary Tyonek and West Foreland Formations, Middle Jurassic rocks (Soldatna Creek 33–33 well and Puale Bay section), and Upper Triassic strata (Puale Bay section) contain type II/III kerogens, with HI values that range from 300 to 400. The OC content in two coal samples from the Tertiary rocks exceeds 40 weight percent.

Extract (C_{15+} hydrocarbon) and OC data indicate that 10 samples are fair oil-source rocks, 5 are good oil-source rocks, and 2 are coal (Fig. 13.5). On the basis of these data, the two West Foreland Formation samples (T_2) have the best oil-source-rock qualities. All samples lack migrated hydrocarbons that would affect

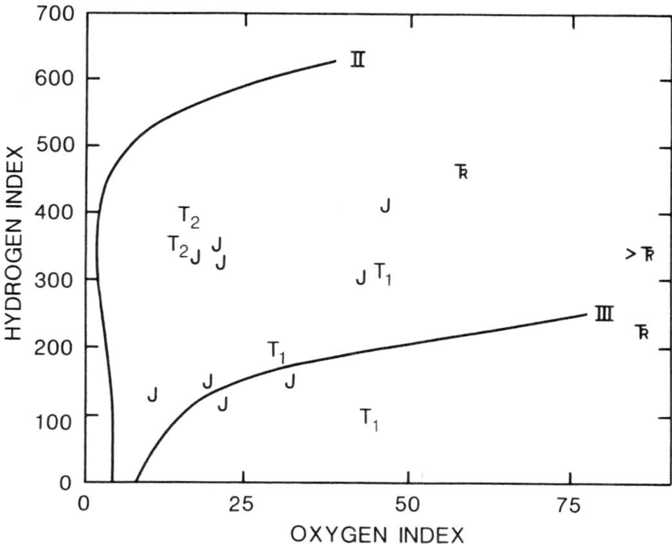

Figure 13.4 Rock Eval and organic-carbon data plotted on a van Krevelen diagram to show kerogen type. See Figure 13.3 for explanation of symbols.

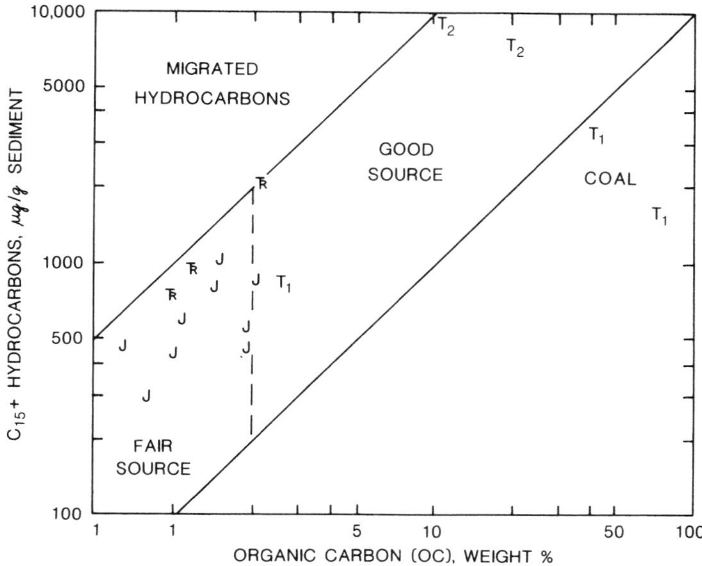

Figure 13.5 Hydrocarbon versus organic-carbon contents, showing source-rock richness. See Figure 13.3 for explanation of symbols.

or contaminate the samples for oil-to-source rock correlation (Fig. 13.5). The saturated-to-aromatic (S/A) hydrocarbon ratio is less than 3.0 for all the rocks analyzed (Table 13.1). The Upper Triassic and Middle Jurassic samples are fair to good source rocks, and the Tertiary samples are good source rocks.

On the basis of these results, three Tertiary nonmarine samples, five Middle Jurassic marine shale samples, and two Upper Triassic samples are possible source rocks for the oil recovered from the Mesozoic rocks and the oil being produced from Tertiary reservoirs.

Hydrocarbon and Nonhydrocarbon Fractions

Hydrocarbon- and nonhydrocarbon-fraction concentrations in rocks and oils differ mainly in the proportion of saturated hydrocarbons (Table 13.1, Fig. 13.6). In the rock samples, saturated hydrocarbons, aromatic hydrocarbons, and nonhydrocarbons are in nearly equal proportions; but in the oil samples, the proportion of saturated hydrocarbons is highest.

Nonpolar saturated hydrocarbons are concentrated preferentially in the expulsion (primary) and migration (secondary) process over more polar nonhydrocarbons, a phenomenon recognized by Leythaeuser et al. (1988). The average composition of the nonmarine-rock extract is unlike that of the marine-rock extract: the nonmarine rocks contain more nonhydrocarbons (35–66%) than do the marine rocks (24–39%).

On the basis of hydrocarbon- and nonhydrocarbon-fraction concentrations, the marine rocks compare more favorably to the marine oils than do the nonmarine

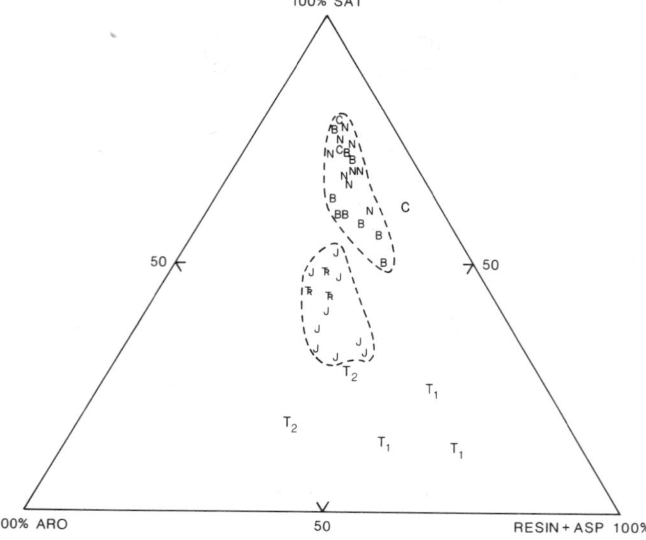

Figure 13.6 Triangular diagram comparing contents of saturated hydrocarbons, aromatic hydrocarbons, and nonhydrocarbons (resins and asphaltenes) in rock and oil samples from the Cook Inlet–Alaska Peninsula area. See Figure 13.3 for explanation of symbols.

rocks. The consistent difference between the percentages of saturated hydrocarbons in the marine-rock extract and the marine oils can be explained by a preferential retention of polar compounds in the source-rock during expulsion and migration.

Carbon-Isotopic Data

A reasonable correlation exists between the carbon-isotopic compositions of the saturated- and aromatic-hydrocarbon fractions in the rocks and oils. The $\delta^{13}C$ value of the nonmarine oil and condensates ranges from −29.1 to −28.8 permil for saturated hydrocarbons and from −26.3 to −25.8 for aromatic hydrocarbons. The higher $\delta^{13}C$ value (−26.2 permil) for saturated hydrocarbons in the Kenai condensate (sample 21) is due to (1) a condensate from an unknown source rock or (2) physical fractionation in the separator.

Four of the five nonmarine-rock samples have isotopic ratios close to those of the three condensates. Aromatic hydrocarbons in the fifth nonmarine-rock sample (1) have a higher $\delta^{13}C$ value. The $\delta^{13}C$ value of the four nonmarine-rock samples ranges from −28.4 to −27.2 for saturated hydrocarbons and from −27.0 to −25.6 permil for aromatic hydrocarbons. Sample 1 is 2–4 permil lighter in aromatic hydrocarbons than the other four nonmarine-rock samples. The isotopic difference between the four nonmarine rocks (samples 2–5) and the nonmarine oil and condensates (samples 18–21) is about 1 permil; the oil and condensates are isotopically lighter than the rock extracts.

The apparent similarity between the three light oils/condensates and the four rocks suggest they may be related genetically. First, because these particular samples are immature, they themselves are not the source; but if additional, mature samples were obtained, they may more closely match the oils. However, wherever the Tertiary rocks are penetrated, they are found to be immature. Second, because these oils are associated with natural-gas accumulations, the natural gas might have stripped out heavier hydrocarbons from the immature rocks, thereby giving the impression that these liquids originated from coal or closely associated shales (Philp and Gilbert, 1982).

The carbon-isotopic compositions of the marine oils and marine rocks are nearly identical. The $\delta^{13}C$ value of the saturated hydrocarbons in 12 rocks ranges from −31.2 to −28.9 permil, a 2.3-permil difference, and for 27 oils from −30.8 to −29.4, a 1.4-permil difference; the $\delta^{13}C$ value of the aromatic hydrocarbons in these rocks ranges from −30.2 to −28.1, a 2.1-permil difference, and for oils from −29.9 to −27.4 permil, a 2.5-permil difference. The range in isotopic ratio for saturated hydrocarbons in the rocks partly overlaps that of the oils, and the range for aromatic hydrocarbons completely overlaps that of the oils (Fig. 13.7). Isotopic data indicate that the average $\delta^{13}C$ value of the saturated hydrocarbons in the oils is about 0.5 permil lighter than in the rocks.

The apparent similarity of the marine oils and marine rock extracts suggests a genetic relation. The slight shift in isotopic ratios may be related to maturity or expulsion. Because $\delta^{13}C$ value is fixed at deposition, this shift would indicate that

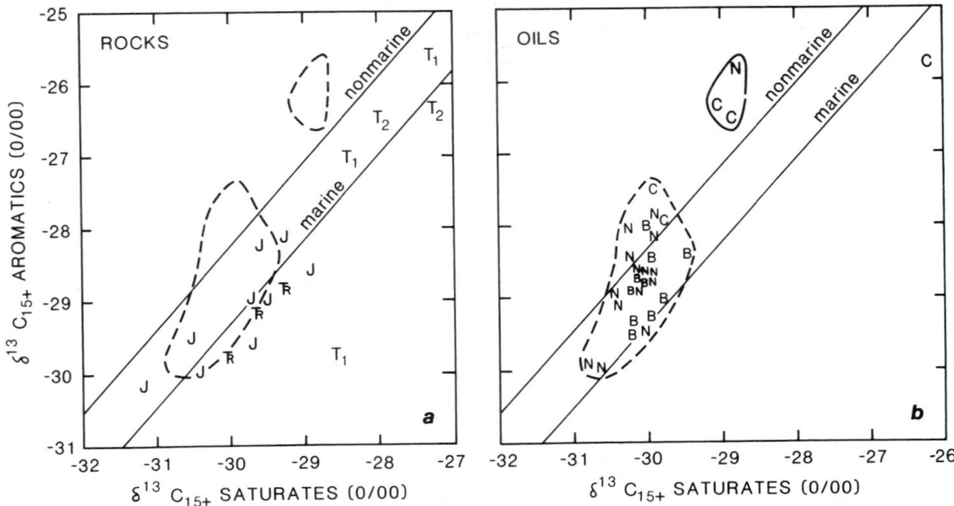

Figure 13.7 Carbon-isotopic ratios of saturated versus aromatic hydrocarbons in rock(a) and oil(b) samples from the Cook Inlet–Alaska Peninsula area. Diagonal lines (from Sofer, 1984) refer to the isotopic composition of waxy nonmarine oils and nonwaxy marine oils. See Figure 13.3 for explanation of symbols.

saturated hydrocarbons containing ^{12}C rather than ^{13}C are more likely to be expelled from the kerogen during thermal maturation.

Gas-Chromatographic/Mass-Spectrometric Analysis

Rock data. Gas chromatograms and two fragmentograms (*m/z* 191 and 217) for a representative sample from each rock unit are shown in Figure 13.8. The biomarker peaks are identified on Table 13.2.

The GC signatures of saturated hydrocarbons in rocks of the Tyonek Formation (samples 1–3) are represented by data from the Swanson River 1 well (sample 1), and of saturated hydrocarbons in rocks of the West Foreland Formation (samples 4–5) by data from the Deep Creek 1 well (sample 4). All five Tertiary rock samples have a strong odd-carbon *n*-alkane preference, high pristane-to-phytane (Pr/Ph) ratios, and low *n*-C^{17}/Pr ratios, all indicative of immature terrestrial organic matter. Of all the rock units sampled, the West Foreland Formation contains the most extractable hydrocarbon per unit of organic carbon and yield among the best samples with respect to residual-hydrocarbon potential (HI = 356–400 mg HC/g OC, Table 13.1).

The *m/z* 191 and 217 fragment-ion traces—tricyclic, tetracyclic, and pentacyclic terpanes and steranes, respectively—of organic matter from rocks of the Tyonek Formation show an abundance of compounds indicative of low-thermal

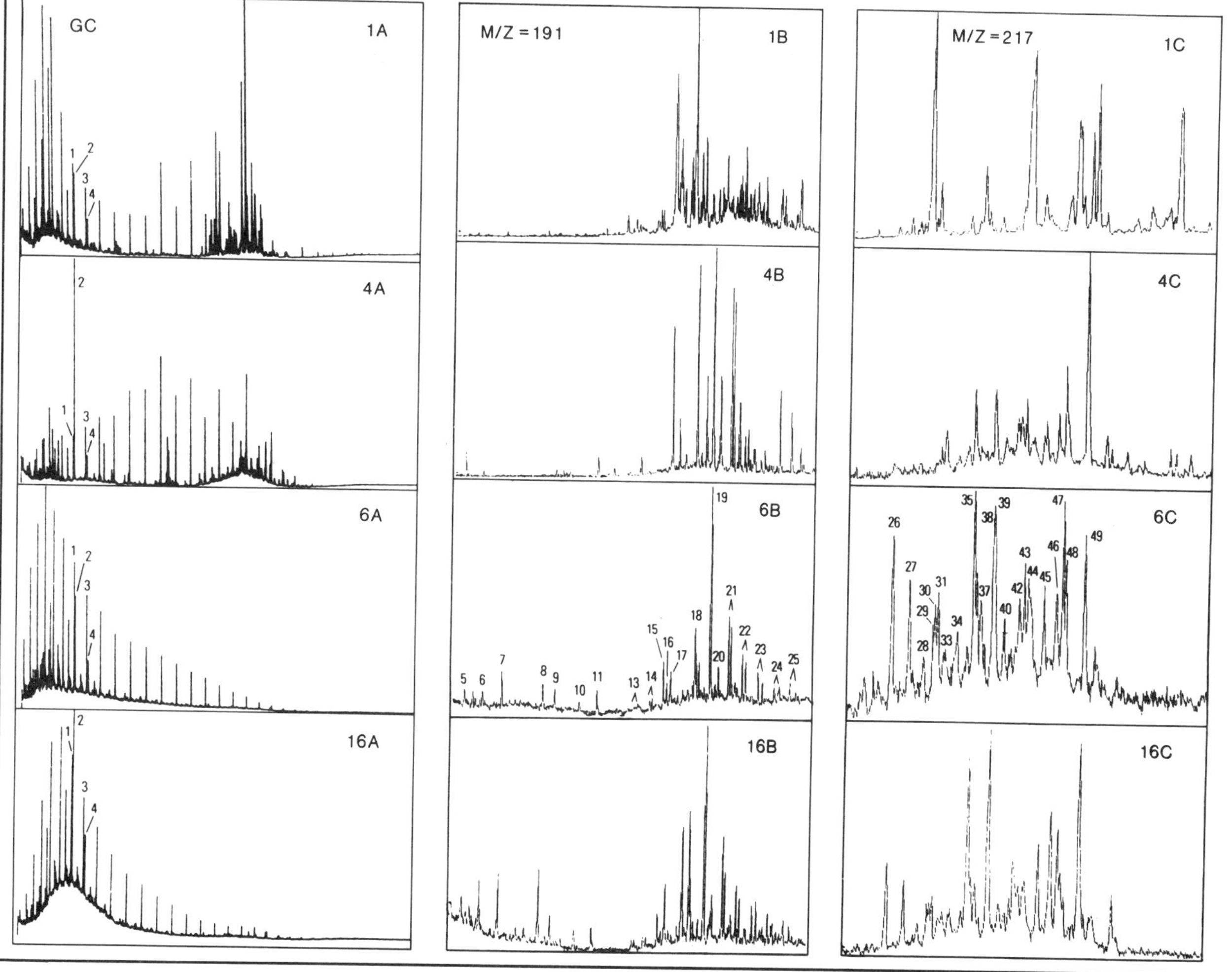

Figure 13.8 Gas chromatograms and two GC-MS (*m/z* 191 and 217) fragmentograms of saturated hydrocarbons in extracts from Tertiary nonmarine and Mesozoic marine units. Note the tricyclic hydrocarbons present in the Triassic rock extract.

Table 13.2 IDENTIFICATION FOR NUMBERED PEAKS ON FIGURES 13.8–13.10

1. heptadecane (n-C_{17})
2. pristane
3. octadecane (n-C_{18})
4. phytane
5. C_{19} tricyclic terpane
6. C_{20} tricyclic terpane
7. C_{21} tricyclic terpane
8. C_{23} tricyclic terpane
9. C_{24} tricyclic terpane
10. C_{25} tricyclic terpane
11. C_{24} tetracyclic terpane
12. C_{26} tricyclic terpanes
13. C_{28} tricyclic terpanes
14. C_{29} tricyclic terpanes
15. 18α(H)-trisnorneohopane (Ts)
16. C_{30} pentacyclic (unknown)
17. 17α(H)-trisnorhopane (Tm)
18. norhopane (C_{29})
19. hopane (C_{30})
20. moretane (C_{30})
21. C_{31} hopanes (S&R)
22. C_{32} hopanes (S&R)
23. C_{33} hopanes (S&R)
24. C_{34} hopanes (S&R)
25. C_{35} hopanes (S&R)
26. 13β,17α-diacholestane (20S)
27. 13β,17α-diacholestane (20R)
28. 13α,17β-diacholestane (20S)
29. 13α,17β-diacholestane (20R)
30. unidentified
31. 24-methyl-13β,17α-diacholestane (20S)
32. unidentified
33. unidentified
34. 24-methyl-13β,17α-diacholestane (20R)
35. 24-methyl-13α,17β-diacholestane (20S) + 14α,17β-cholestane (20S)
36. 24-ethyl-13β,17α-diacholestane (20S) + 14β,17β-cholestane (20R)
37. 24-methyl-13α,17β-diacholestane (20R) + 14β,17β-cholestane (20S)
38. 14α,17α-cholestane (20R)
39. 24-ethyl-13β,17α-diacholestane (20R)
40. 24-ethyl-13α,17β-diacholestane (20S)
41. 24-methyl-14α,17α-cholestane (20S)
42. 24-methyl-14β,17β-cholestane (20R) + 24-ethyl-13α,17β-diacholestane (20R)
43. 24-methyl-14β,17β-cholestane (20S)
44. unknown
45. 24-methyl-14α,17α-cholestane (20R)
46. 24-ethyl-14α,17α-cholestane (20S)
47. 24-ethyl-14β,17β-cholestane (20R)
48. 24-ethyl-14β,17β-cholestane (20S)
49. 24-ethyl-14α,17α-cholestane (20R).

maturity (e.g., 5β, 20*R*-steranes, hopenes, ββ-hopanes, etc., Fig. 13.8, sample 1). Terpane and sterane biomarkers in rocks of the West Foreland Formation, represented in Figure 13.8 by sample 4 (see *m/z* 191 and 217 traces), are slightly more mature than those in rocks of the Tyonek Formation, as evidenced by the disappearance of hopenes and ββ-hopanes and a decreasing content of 5β-steranes.

The Middle Jurassic marine rocks of the Tuxedni Group are buried deeper in the north than in the south but are less thermally mature in the north than stratigraphically equivalent units in the south. The more mature samples (samples 6–9, Table 13.1) have T_{max} values that range from 437 to 439°C. Gas chromatograms of the saturated hydrocarbons indicate an abundance of light hydrocarbons, no odd- or even-carbon preference in the *n*-alkanes, low cyclic-biomarker content (terpanes and steranes), Pr/Ph ratios of 2.8, and *n*-C_{17}/Pr ratios in the range 1.3–2.0 (sample 6, Fig. 13.8). Organic matter associated with the less mature (T_{max} = 429–433°C) rocks of the Tuxedni Group (samples 9–12, Table 13.1) has higher Pr/Ph ratios (3.6–4.3), lower *n*-C_{17}/Pr ratios (0.3–0.6), and a strong odd-carbon preference in the higher carbon number *n*-alkanes, suggesting some terrestrial input into the Middle Jurassic rocks.

The most mature Upper Triassic marine rocks (T_{max} = 440–443°C) came from outcrops in the Puale Bay area (samples 15–16). On the basis of GC peaks, hydrocarbons in the mature Upper Triassic rocks are nearly identical to those in the mature Middle Jurassic samples (6–8), showing an abundance of light hydrocarbons, little or no odd- or even-carbon preference in the *n*-alkanes, low-cyclic biomarker content (terpanes and steranes), and high Pr/Ph ratios (avg 2.8). The only observed GC differences between the mature Upper Triassic and Middle Jurassic samples are in their *n*-C_{17}/Pr ratios. *n*-C_{17}/Pr ratios in the Upper Triassic rocks (0.7–0.8) are lower than in the Middle Jurassic rocks because of the lower thermal maturity of the Upper Triassic rocks.

The *m/z* 191 and 217 fragment-ion traces of hydrocarbons in the mature Middle Jurassic Tuxedni Group (samples 6–8) and the slightly less mature Upper Triassic strata (samples 15–17) are represented in Figure 13.8 by samples 6 and 16, respectively. The *m/z* 191 traces (tricyclic, tetracyclic, and pentacyclic terpanes) are quite similar in both sample sets. The principle differences in the *m/z* 191 fragment-ion traces are as follows: (1) the Middle Jurassic samples all show an unidentified pentacyclic compound (M^+412 = 30%, *m/z* 191 = 100%, *m/z* 109 = 50%, *m/z* 81 = 90%) eluting between the C_{27} hopanes (*Ts* and *Tm*) whose *m/z* 191 peak is at least 75 percent as intense as the *m/z* 191 fragment-ion peak from *Tm*—this unidentified pentacyclic compound is negligible or absent in the Upper Triassic rock samples; and (2) the Upper Triassic rocks all contain C_{28} tricyclic terpanes whose *m/z* 191 peaks are at least 75 percent as intense as the *m/z* 191 peaks of the C_{29} tricyclic terpanes—the C_{28} tricyclic terpanes are negligible or absent in the Middle Jurassic rocks. These two differences appear to be source related and, therefore, useful for correlation purposes. All of the Upper Triassic and Middle Jurassic rock samples are dominated by C_{27} and C_{29} regular steranes, a result consistent with the Rock Eval HI and OI data, suggesting both a mix of terrestrial and marine organic matter. Both sample sets also show an abundance of C_{27} diasteranes.

Condensate data. Five samples (18–19, 21–23) can be classified as condensates. On the basis of API gravity (28.6°), the Kenai sample would not be classified as a condensate, except that it was collected at the field separator. In addition, its low viscosity, clear straw color, and absence of asphaltenes make it appear to be a condensate. Isotopically, the Kenai hydrocarbons are the heaviest of all the oils or condensates examined ($\delta^{13}C$ value of saturates, −26.2, and of aromatics, −25.7 permil). The gas chromatogram of the saturated-hydrocarbon fraction in the Kenai sample indicates approximately 47 percent hydrocarbons ($< C_{15}$), abundant isoprenoids and isoalkanes, a carbon-preference index (CPI) of 2.5 (n-C_{28}–C_{30}), a Pr/Ph ratio of 4.9, a low n-C_{17}/Pr ratio (0.1), and a low concentration of n-alkanes. This near-absence of n-alkanes suggests that the Kenai sample may be biodegraded. The m/z 191 fragment-ion trace shows the presence of diterpanes but very low concentrations of triterpanes; no steranes were observed. Several two- and three-ring compounds were observed by mass-scanning the GC peaks, but they were not identified.

The true northern Cook Inlet condensates (samples 18, 22, and 23) each show a smooth exponential decline in n-alkane intensities with increasing carbon number (see GC sample 18, Fig. 13.9). North Cook Inlet condensate (sample 18) appears to have picked up some immature hydrocarbons during migration or in the reservoir. Unlike samples 22 and 23, sample 18 has an odd-carbon preference (CPI = 1.8, n-C_{28}–C_{30}), the less thermally stable C_{27} hopane (Tm) is more intense than the more thermally stable C_{27} hopane (Ts)(Tm/Ts ratio, 2.4), the extended hopanes (C_{32} and C_{33}) have S/A ratios < 1, and the C_{29} sterane-maturation parameters are low (Fig. 13.9, Table 13.1). The other two condensates from North Cook Inlet (sample 22) and Granite Point (sample 23) oil fields do not have these immature characteristics (Table 13.1). The most notable differences in biomarkers between these two condensates are the relative intensities of the C_{19} and C_{20} tricyclic terpanes and in the intensity of an unknown diterpane peak eluting between the C_{19} and C_{20} tricyclic terpanes (similar to peak in sample 18, Table 13.1, Fig. 13.9). The m/z 191 fragment-ion peaks representing C_{19} and C_{20} tricyclic terpanes and the unknown compound are twice as intense in sample 18 as in samples 22 and 23. The intense C_{19} and C_{20} tricyclic-terpane peaks in sample 18 may indicate a dominant terrestrial contribution in this sample but a lesser terrestrial contribution in samples 22 and 23.

Normal-oil data. The normal oils are represented by 13 samples (20, 24, 29–39). Gas chromatograms of the normal oils (samples 29–39) are represented in Figure 13.9 by sample 30. Except for a slight break in the curve at n-C_{18}, the relative distributions of n-alkanes decline linearly with increasing carbon number. The relative concentrations of cyclic biomarker are low, Pr/Ph ratios range from

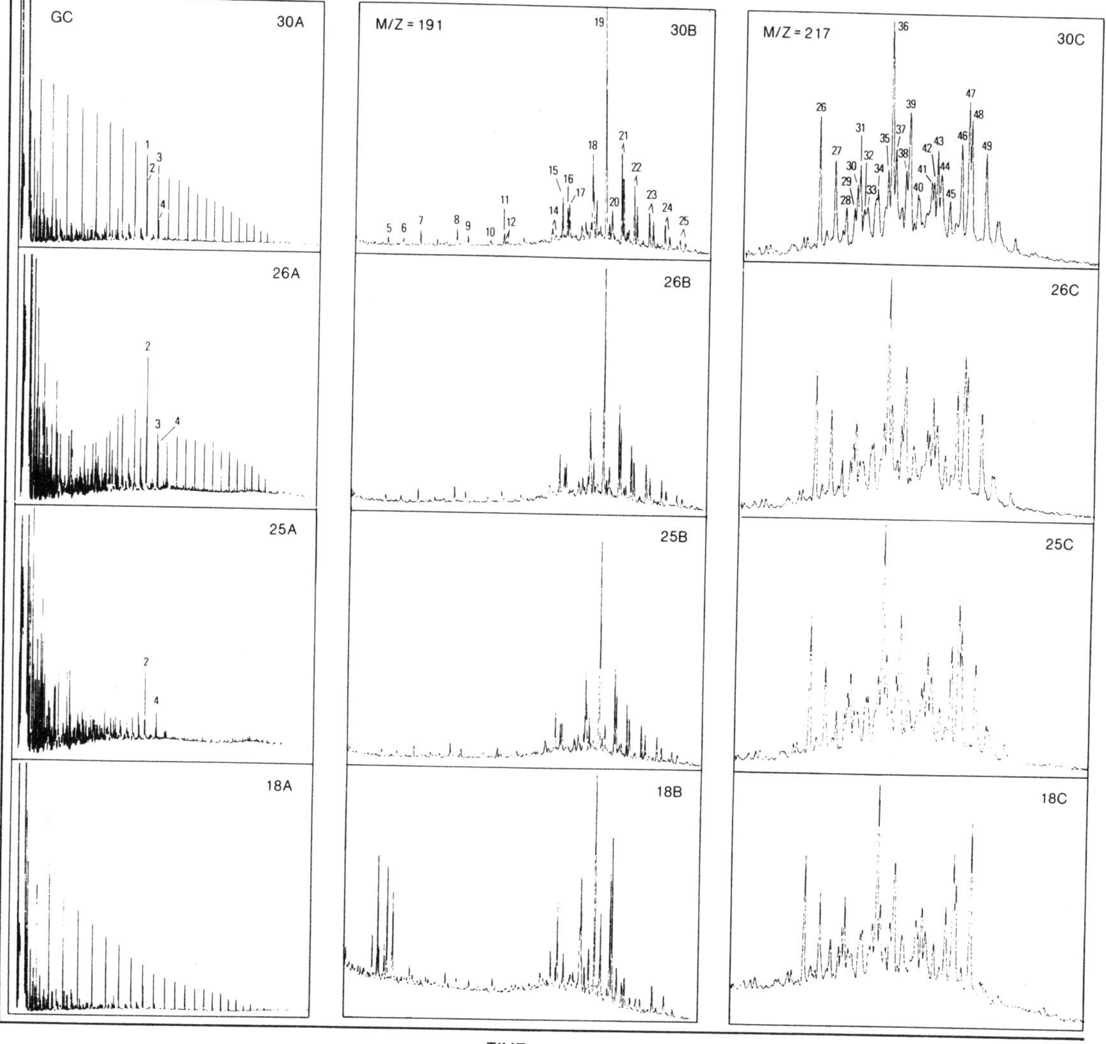

Figure 13.9 Gas chromatograms and two GC-MS (m/z 191 and 217) fragmentograms of saturated hydrocarbons for normal and several stages of biodegraded oils.

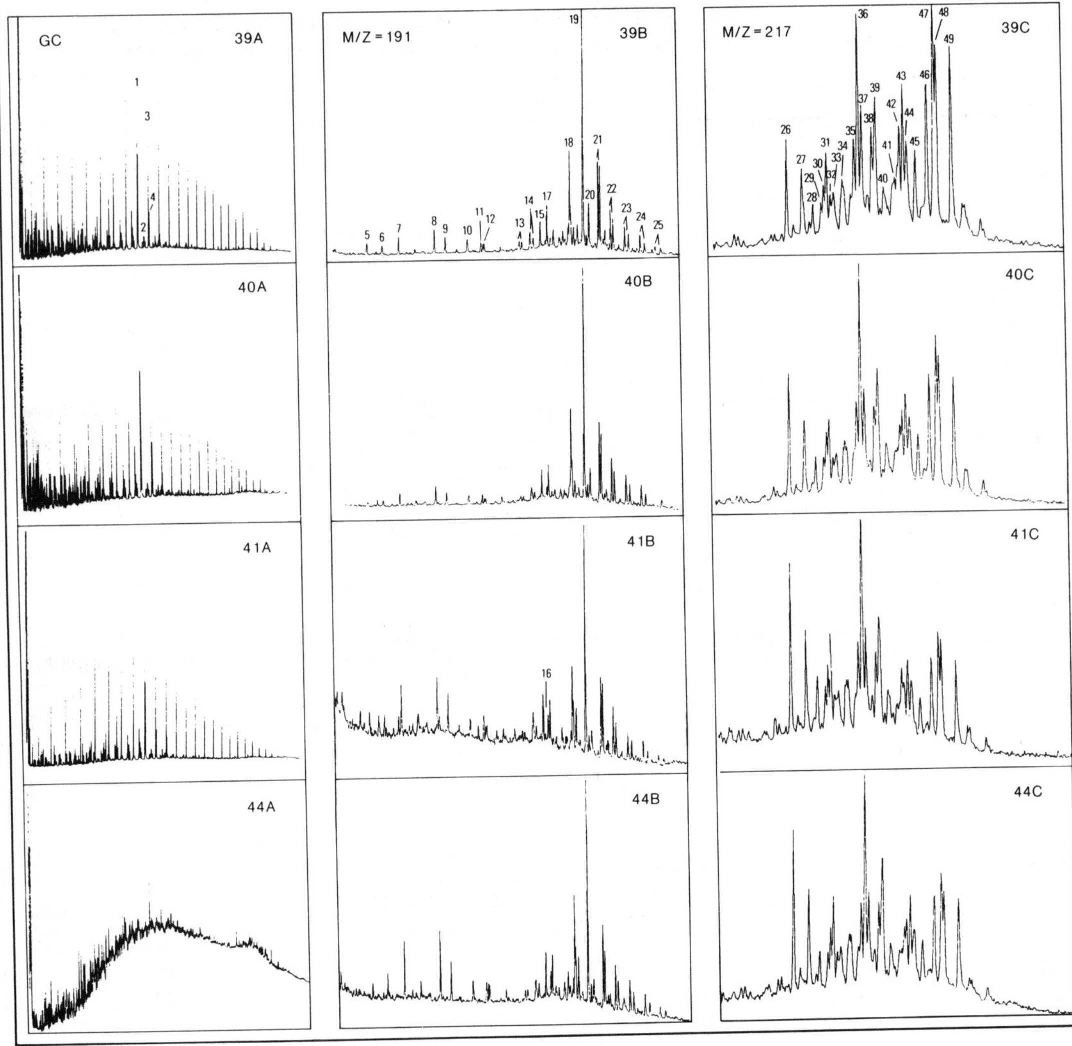

Figure 13.10 Gas chromatograms and two GC-MS (m/z 191 and 217) fragmentograms of saturated hydrocarbons for immature and biodegraded oils. Note the tricyclic hydrocarbons present in these oils.

2.6 to 3.3, and n-C_{17}/Pr ratios range from 1.5 to 1.8. Cyclic biomarkers characterized by the m/z 191 and 217 fragment-ion traces (tricyclic, tetracyclic, and pentacyclic terpanes and steranes, respectively) show an exponential decline in extended hopanes with increasing carbon number (C_{31}–C_{35}); the C_{27} hopanes *Ts* and *Tm* have an unidentified pentacyclic triterpane (M^+412 = 30%, m/z 191 = 100%, m/z 109 = 50%, m/z 81 = 90%) eluting between them; the tricyclic-terpane distributions (C_{19}–C_{26}) show C_{23} to be the dominant tricyclic terpane, with C_{21} tricyclic terpane running a close second; the steranes are generally dominated by C_{27} and C_{29} regular steranes; and C_{27} diasteranes are also abundant.

Some of the southernmost oils (sample 39, Fig. 13.10) show subtle but significant differences from the more northern oils (samples 29–38) that could indicate a second, supplementary source. Sample 39 has a CPI of 1.6 (n-C_{28}–C_{30}), whereas oil samples 29–38 have CPI's closer to 1.0. The unidentified pentacyclic compound (M^+412) observed eluting between *Ts* and *Tm* in samples 29–38 is negligible in sample 39, and C_{28} and C_{29} tricyclic terpanes are more prevalent in sample 39 then in samples 29–38. This second, supplementary source could be from Tertiary organic matter.

Biodegraded-oil data. A total of 10 oils are biodegraded (samples 25–28, 40–45). Surface seeps containing no n-alkanes and altered isoprenoids (samples 43–45, Table 13.1; sample 44, Fig. 13.10) are considered to be the most intensely degraded. Degradation of subsurface oils ranges from mild (samples 26, 27, 40, 41) to extensive (samples 25, 28).

The terpane (m/z 191) and sterane (m/z 217) biomarker traces in samples 25–28 are, for all practical purposes, identical to the cyclic-biomarker traces of the major oil type (samples 29–38) and probably are genetically related (compare samples 25 and 26 with sample 30, Fig. 13.9). Degraded oils from the southern Kenai Peninsula (represented by samples 40 and 41, Fig. 13.10) and the Iniskin Peninsula (represented by sample 44, Fig. 13.10) have m/z 191 and 217 mass chromatograms resembling both the northern Cook Inlet oils (samples 29–38) and the slightly different oil (sample 39) from the Kenai Peninsula. Each of these degraded oils (samples 40–45) contains a considerable amount of the unidentified pentacyclic compound (M^+412) common to the northern Cook Inlet oils (samples 29–38), and of the more pronounced C_{28} and C_{29} tricyclic terpanes common to the Kenai Peninsula oil (sample 39). These more southern, degraded oils (samples 40–45) may have a dual source.

OIL-TO-OIL CORRELATIONS

The Kenai condensate (sample 21) is so unusual in its biomarker traces that it appears to be genetically unrelated to any of the other oils or condensates from the Cook Inlet.

North Cook Inlet condensate (sample 18) is unusually rich in C_{19} and C_{20}

tricyclic terpanes relative to the other terpanes and contains an unknown component of equal intensity to the C_{19} and C_{20} tricyclic terpanes eluting between them. The *m/z* 191 and 217 traces provided by Moldowan for sample 20 show these same peaks in approximately the same relative abundance as found in sample 18. The C_{29} sterane maturation parameters show sample 20 to be slightly less mature than sample 18. The carbon-isotopic ratios for the hydrocarbons in sample 18 are similar to those for the hydrocarbons in condensates (samples 19–20) reported by Moldowan et al. (1985) and Peters et al. (1986). On the basis of isotopic and biomarker similarities, samples 18 to 20 could be genetically related.

Condensate samples 22 and 23, oil samples 29–38, and degraded-oil samples 25–28 (Table 13.1) have nearly identical *m/z* 191 and 217 mass chromatogram traces and so are probably genetically related.

Normal-oil sample 39 contains relatively little of the unidentified pentacyclic compound (M^+412, eluting between *Ts* and *Tm*) that is prevalent in samples 22, 23, and 25–38, and it shows well-defined peaks for C_{28} and C_{29} tricyclic terpanes (J.M. Moldowan, oral commun.; Seifert et al., 1979) that are absent or less well defined in samples 22, 23, and 25–38. Also evident in sample 39 but not in samples 22, 23, and 25–38 is an odd-carbon preference. These differences suggest another, less mature supplementary source for the North Fork oil (sample 39) than for oil samples 22, 23, and 25–38.

Finally, the degraded oils (samples 40–45, Table 13.1) from wells and surface seeps at least 100–160 km (60–100 miles) farther south than the Trading Bay, McArthur River, Middle Ground Shoal, and Swanson River oil types but in the same vicinity as the North Fork 41-35 (sample 39) oil type have cyclic-biomarker characteristics common to both oil types. This similarity suggests a mixture of oils from two sources.

OIL-TO-SOURCE-ROCK CORRELATIONS

Because of the immaturity of the nonmarine rocks examined in this text, it is difficult to arrive at any conclusions as to how their alkanes, isoalkanes, or cyclic biomarkers would appear if the rocks were thermally mature, or whether these rocks are the source of any of the condensates or oils from the Cook Inlet. Isotopically, the hydrocarbons in the nonmarine rocks are isotopically heavier than those in the marine rocks. These higher isotopic ratios for the hydrocarbons in the nonmarine rocks are much closer to those for the hydrocarbons in the North Cook Inlet (sample 18), Birch Hill (sample 19), and Swanson River (sample 20) oil fields than are any of the isotopic ratios for the hydrocarbons in the marine rocks. Although the isotopic ratios for condensate and oil samples 18–20 and the nonmarine-rock samples are similar, no rocks (marine or nonmarine) have as high relative concentrations of C_{19} and C_{20} tricyclic terpanes and the unknown component that elutes between them as does oil sample 18 from the North Cook Inlet oil field. Therefore, the source of the petroleum in samples 18–20 cannot be assigned with certainty, although the odd-carbon preference and abundance of C_{19} and C_{20} tricyclic terpanes in sample 18 suggest a possible terrestrial source.

On the basis of similarities in stable isotopes and the relative abundances of *n*-alkanes, pristane and phytane, tricyclic terpanes (C_{19}–C_{26}), hopanes, and steranes, little difference can be seen between the hydrocarbons in the mature Upper Triassic and Middle Jurassic rocks. The only observed organic-chemical differences between the Upper Triassic and Middle Jurassic rocks are quite subtle. The Middle Jurassic rocks are generally distinguishable from the Upper Triassic rocks by the presence of an unidentified pentacyclic compound (M^+412) eluting between *Ts* and *Tm* in the Middle Jurassic rocks but negligible or absent in the Upper Triassic rocks. C_{28} and C_{29} tricyclic terpanes are generally more prevalent in the Upper Triassic than in the Middle Jurassic rocks (J.M. Moldowan, oral commun.; Seifert et al., 1979). In fact, these two distinctions were used to correlate the rock extracts with the oils and condensates because of the similarities of all the other correlation parameters. Oil and condensate samples 22, 23, and 25–38 all contain the unidentified pentacyclic component characteristic of the Middle Jurassic rocks but only negligible C_{28} tricyclic terpanes. Oil sample 39 contains almost none of the unidentified pentacyclic component common to the Middle Jurassic rocks but well-defined C_{28} and C_{29} tricyclic terpanes common to the Upper Triassic rocks. Oil samples 40–45 have a moderate concentration of C_{28} and C_{29} tricyclic terpanes plus the unidentified pentacyclic component. This phenomenon suggests two different cosources, the Upper Triassic and Middle Jurassic source rocks, for samples 40–45.

CONCLUSIONS

The source of the Kenai condensate could not be determined because of its unusual hydrocarbon signature. The North Cook Inlet and Birch Hill condensates and the Swanson River oil (samples 18–20) have carbon-isotopic ratios closer to those of the hydrocarbons in the Cook Inlet nonmarine rocks than in the marine rocks—also, unusually high-relative abundances of C_{19} and C_{20} tricyclic terpanes in the North Cook Inlet condensate (sample 18) could indicate a possible terrestrial source or contribution. Condensate samples 22 and 23, normal-oil samples 24, 29–38, and degraded-oil samples 25–28 have bulk and molecular parameters that are nearly identical to those in the Middle Jurassic rocks. Farther south, near the tip of the Kenai Peninsula, occurs an oil, North Fork 41–35 (sample 39), whose biomarkers appear to be more like those in the Upper Triassic than in the Middle Jurassic rocks. This conclusion is based on the near-absence in oil sample 39 of the unidentified C_{30} pentacyclic component prevalent in the Middle Jurassic rocks but absent or negligible in the Upper Triassic rocks, and on the increase in this oil of the relative intensities of C_{28} and C_{29} tricyclic terpanes, more prevalent in the Upper Triassic than in the Middle Jurassic rocks. The rest of the oils (sample 40–45), from the lower Kenai Peninsula and the Iniskin Peninsula, display biomarker characteristics common to both the Upper Triassic and Middle Jurassic rock, suggesting a possible dual source for these oils.

The major conclusion that can be drawn from the regional geology, and the carbon-isotopic and biomarker data for the Cook Inlet–Alaska Peninsula oils and rocks is that the biomarker data were essential to identify four petroleum systems.

1. The lower Tertiary oils correlate with source rocks in the Middle Jurassic Tuxedni Group. All of the commercial oil being produced from the West Foreland Formation, the Hemlock Conglomerate, and the lower part of the Tyonek Formation originated from marine siltstone of the Tuxedni Group of Middle Jurassic age. Geologic and thermal-maturity data from the Soldatna Creek 33–33 well in the Cook Inlet area indicate that the depth to maturity exceeds 4,500 m (15,000 ft). In the subsurface between the Middle Ground Shoal field and the Swanson River field, the Middle Jurassic source rocks are at least this deep. The onset of oil generation occurred during late Pliocene and Pleistocene time (Magoon and Claypool, 1981). Petroleum emanating from this depocenter appears to be restricted to the immediate area and has migrated less than 45 km (Table 13.1).
2. The Tertiary light oils/condensates compare favorably with the Tertiary nonmarine organic matter. All the commercial natural-gas deposits are microbial rather than thermal (Claypool et al., 1980), but the liquids associated with this gas have a different origin. Even though the biologic markers from light oil/condensate compare favorably with organic matter in the nonmarine Tertiary rocks, these same rocks are immature. Because the thermal and microbial natural gases probably originated during Pliocene and Pleistocene time, the fluids also originated simultaneously. These fluids could have originated from immature kerogen or been naturally extracted from organic matter by migrating natural gas, but the precise geologic and geochemical mechanism is unclear.
3. The oils on the Alaska Peninsula compare favorably with those from the unnamed Upper Triassic marine shales and the Middle Jurassic Kialagvik Formation. The timing and method of generation of this oil are uncertain, but rather than overburden, a magmatic heat source is suspected because of the proximity and timing of intrusion for plutonic bodies (Hudson, 1986). The timing of petroleum generation is suspected to be Late Cretaceous. This petroleum system is interpreted to be separate and distinct from a similar system in the lower Cook Inlet.
4. The oils in the lower Cook Inlet that were recovered from the Upper Cretaceous Matanuska Formation and the lower Tertiary Hemlock Conglomerate compare favorably with those in the unnamed Upper Triassic shales on the Alaska Peninsula and the Middle Jurassic Tuxedni Group in the Cook Inlet. The Middle Jurassic rocks on the Iniskin Peninsula are marginally mature and suspected to be the depocenter responsible for the oil shows in the lower

Cook Inlet south of Kalgin Island. These source rocks probably matured during Late Cretaceous and early Tertiary time. On the basis of the biomarker composition of the oil, which suggests that two source rocks are involved—Upper Triassic and Middle Jurassic—organic-rich Upper Triassic rocks are suspected to underlie the Lower Jurassic Talkeetna Formation beneath the lower Cook inlet. Hydrocarbons generated in these Upper Triassic rocks migrated up faults in the lower Cook Inlet into the Middle Jurassic and younger rocks to mix with the Tuxedni Group oil. The expelled hydrocarbons migrated into fractures and available reservoirs. During uplift of the lower Cook Inlet, the trapped hydrocarbons seeped upward into the Tertiary reservoirs, such as at the North Fork well. If this scenario is true, it explains the absence of commercial accumulations south of Kalgin Island, the Cretaceous oil shows, and the biodegradation of these oils.

Acknowledgments

We thank Don Blasko, U.S. Bureau of Mines, for his help in collecting many oil samples. Geologists from the Chevron, Marathon, and Mobil Oil Cos. generously supplied crude oil and core material from various wells critical to this study. Dave King, Sister Carlos Lubeck, Eric Michael, Chuck Threlkeld, and April Vuletich of the U.S. Geological Survey did the analytical work. M.A. Pytte of Chevron U.S.A. provided important information about samples included in this study. J.A. Curiale and W.B. Hughes reviewed the manuscript and provided many helpful suggestions. Special thanks to Michael Moldowan for his invitation to contribute to this memorial volume.

REFERENCES

Alaska Geological Society (1969a) Northwest to southeast stratigraphic correlation section, Drift River to Anchor River, Cook Inlet basin, Alaska. Anchorage, vertical scale 1 inch = 500 feet.

——— (1969b) South to north stratigraphic correlation section, Anchor Point to Campbell Point, Cook Inlet basin, Alaska. Anchorage, vertical scale 1 inch = 500 feet.

——— (1969c) South to north stratigraphic correlation section, Kalgin Island to Beluga River, Cook Inlet basin, Alaska. Anchorage, vertical scale 1 inch = 500 feet.

——— (1969d) West to east stratigraphic correlation section, West Foreland to Swan Lake, Cook Inlet basin, Alaska. Anchorage, vertical scale 1 inch = 500 feet.

——— (1970a) South to north stratigraphic correlation section, Campbell Point to Rosetta, Cook Inlet basin, Alaska. Anchorage, vertical scale 1 inch = 500 feet.

——— (1970b) West to east stratigraphic correlation section, Beluga River to Wasilla, Cook Inlet basin, Alaska. Anchorage, vertical scale 1 inch = 500 feet.

Alaska Oil and Gas Conservation Commission, 1988 (1987) Statistical Report. Alaska Oil and Gas Conservation Commission [available from 3001 Porcupine Drive, Anchorage, AK 99501-3192].

Barnes, F.F. and Cobb, E.H. (1959) Geology and coal resources of the Homer district, Kenai coal field, Alaska. U.S. Geological Survey Bulletin 1058-F, F217-F260.

Barnes, F.F. and Payne, T.G. (1956) The Wishbone district, Matanuska coal field, Alaska. U.S. Geological Survey Bulletin 1016, 88 pp.

Blasko, D.P. (1974) Natural gas fields—Cook Inlet basin, Alaska. Bureau of Mines Open-File Report 35-74, 29 pp.

Blasko, D.P. (1976) Oil and gas seeps in Alaska—Alaska Peninsula, Western Gulf of Alaska. U.S. Bureau of Mines Report of Investigations 8122, 78 pp.

Blasko, D.P., Wenger, W.J., and Morris, J.C. (1972) Oilfields and crude oil characteristics, Cook Inlet basin, Alaska. U.S. Bureau of Mines Report of Investigations 7688, 44 pp.

Bolm, J.G. and McCulloh, T.H. (1986) Sandstone diagenesis, In Geologic studies of the lower Cook Inlet COST No. 1 well, Alaska Outer Continental Shelf. (ed. L.B. Magoon.) *U.S. Geological Survey Bulletin 1596*, 51–53.

Boss, R.F., Lennon, R.B., and Wilson, B.W. (1976) Middle Ground Shoal oil field, Alaska, In North American oil and gas fields. (ed. Jules Braunstein) American Association of Petroleum Geologists Memoir 24, 1–22.

Calderwood, K.W. and Fackler, W.C. (1972) Proposed stratigraphic nomenclature for Kenai Group, Cook Inlet basin, Alaska. *Am. Assoc. Pet. Geol. Bull. 56*, 739–754.

Claypool, G.E. (1986) Petroleum geochemistry, In Geologic studies of the lower Cook Inlet COST No. 1 well, Alaska Outer Continental Shelf. (ed. L.B. Magoon) *U.S. Geological Survey Bulletin 1596*, 33–39.

Claypool, G.E., Threlkeld, C.N., and Magoon, L.B. (1980) Biogenic and thermogenic origins of natural gas in Cook Inlet basin, Alaska. *Am. Assoc. Pet. Geol. Bull. 64*, 1131–1139.

Crick, R.W. (1971) Potential petroleum reserves, Cook Inlet, Alaska, In Future petroleum provinces of the United States—their geology and potential. (ed. I.H. Cram) American Association of Petroleum Geologists Memoir 15, Vol. 1, 109–119.

Detterman, R.L. and Hartsock, J.K. (1966) Geology of the Iniskin-Tuxedni Region, Alaska. *U.S. Geological Survey Professional Paper 512*, 78 pp.

Fisher, M.A. and Magoon, L.B. (1978) Geologic framework of lower Cook Inlet Alaska. *Am. Assoc. Pet. Geol. Bull. 62*, 373–402.

Franks, S.G. and Hite, D.M. (1980) Controls of zeolite cementation in Upper Jurassic sandstones, lower Cook Inlet, Alaska [abs.]. *Am. Assoc. Pet. Geol. Bull. 64*, 708–709.

Hartman, D.C., Pessel, G.H., and McGee, D.L. (1972) Preliminary report on stratigraphy of Kenai Group, upper Cook Inlet, Alaska. Alaska Division of Geological Surveys, Special Report No. 5, 4 p., 7 maps, scale 1:500,000, 1 pl.

Hite, D.M. (1976) Some sedimentary aspects of the Kenai Group, Cook Inlet, Alaska, In Recent and ancient sedimentary environments in Alaska. (ed. T.P. Miller) Alaska Geological Society, Anchorage, Alaska, I1–I23.

Hudson, T. (1986) Plutonism and provenance—implications for sandstone composition, In Geologic studies of the lower Cook Inlet COST No. 1 well, Alaska Outer Continental Shelf. (ed. L.B. Magoon) *U.S. Geological Survey Bulletin 1596*, 55–60.

Imlay, R.W. and Detterman, R.L. (1977) Some Lower and Middle Jurassic beds in Puale

Bay–Alinchak Bay area, Alaska Peninsula [geologic notes]. *Amer. Assoc. Pet. Geol. Bull. 61*, 607–611.

KELLY, T.E. (1963) Geology and hydrocarbons in Cook Inlet basin, Alaska, In Backbone of the Americas. (eds. O.E. Childs and B.W. Beebe) American Association of Petroleum Geologists Memoir 2, 278–296.

KELLY, T.E. (1968) Gas accumulations in nonmarine strata, Cook Inlet basin, Alaska, In Natural gases of North America. (eds. W.B. Beebe and B.F. Curtis) American Association of Petroleum Geologists Memoir 9, Vol. 1, 49–64.

KIRSCHNER, C.E. and LYON, C.A. (1973) Stratigraphic and tectonic development of Cook Inlet petroleum province, In Arctic geology. (ed. M.G. Pitcher) American Association of Petroleum Geologists Memoir 19, 396–407.

LEYTHAEUSER, D., SCHAFFER, R.G., and RADKE, M. (1988) Geochemical effects of primary migration of petroleum in Kimmeridge source rocks from Brae field area, North Sea, I: Gross composition of C_{15+} solid organic matter and molecular composition of C_{15+} saturated hydrocarbons. *Geochim. Cosmochim. Acta 52*, 701–713.

MACKEVETT, E.M., JR. and PLAFKER, G. (1974) The Border Ranges fault in south-central Alaska. *U.S. Geological Survey Journal of Research*, Vol. 2, 323–329.

MAGOON, L.B., ed. (1986) Geologic studies of the lower Cook Inlet COST No. 1 well, Alaska Outer Continental Shelf. *U.S. Geological Survey Bulletin 1596*, 99 pp.

MAGOON, L.B., ADKISON, W.L., and EGBERT, R.M. (1976) Map showing geology, wildcat wells, Tertiary plant-fossil localities, K-Ar age dates, and petroleum operations, Cook Inlet area, Alaska. *U.S. Geological Survey Miscellaneous Investigations Map I-1019*, 3 sheets, scale 1:250,000.

MAGOON, L.B. and CLAYPOOL, G.E. (1981) Petroleum geology of Cook Inlet basin, Alaska—an exploration model. *Am. Assoc. Pet. Geol. Bull. 65*, 1043–1061.

MAGOON, L.B. and EGBERT, R.M. (1986) Framework geology and sandstone composition, In Geologic studies of the lower Cook Inlet COST No. 1 well, Alaska Outer Continental Shelf. (ed. L.B. Magoon) *U.S. Geological Survey Bulletin 1596*, 65–90.

MAGOON, L.B., GRIESBACH, F.R., and EGBERT, R.M. (1980) Nonmarine Upper Cretaceous rocks, Cook Inlet, Alaska. *Am. Assoc. Pet. Geol. Bull. 64*, 1259–1266.

MOLDOWAN, J.M., SEIFERT, W.K., and GALLEGOS, E.J. (1985) Relationship between petroleum composition and depositional environment of petroleum source rocks. *Am. Assoc. Pet. Geol. Bull. 69*, 1255–1268.

OSMENT, F.C., MORROW, R.M., and CRAIG, R.W. (1967) Petroleum geology and development of the Cook Inlet basin of Alaska (with French abs.), In Origin of oil, geology, and geophysics. 7th World Petroleum Congress Proceedings, Mexico, London, Elsevier Publishing Co., Vol. 2, 141–150.

PETERS, K.E., MOLDOWAN, J.M., SCHOELL, M., and HEMPKINS, W.B. (1986) Petroleum isotopic and biomarker composition related to source rock organic matter and depositional environment. In *Advances in organic geochemistry, petroleum geochemistry, part I.* (Leythaeuser, D., and Rullkötter, J.) *Org. Geochem. 10*, no. 1–3, 17–27.

PHILP, R.P. and GILBERT, T.D. (1982) Unusual distribution of biological markers in an Australian crude oil. Nature 299, 245–247.

REED, B.L., MIESCH, A.T., and LANPHERE, M.A. (1983) Plutonic rocks of Jurassic age in the Alaska-Aleutian Range batholith: Chemical variations and polarity. *Geological Society of America Bulletin* 94, 1232–1240.

SEIFERT, W.K., MOLDOWAN, J.M., and JONES, J.W. (1979) Application of biological marker chemistry to petroleum exploration. World Petroleum Congress, 10th, Bucharest, Proceedings: London, Heyden and Son, Ltd. 425–440.

SOFER, Z. (1984) Stable carbon isotope compositions of crude oils: Application to source depositional environments and petroleum alteration. *Am. Assoc. Pet. Geol. Bull. 68*, 31–49.

VAN EYSINGA, F.W.B. (1975) *Geologic time table.* Amsterdam, Elsevier.

YEH, H-W. and EPSTEIN, S. (1981) Hydrogen and carbon isotopes of petroleum and related organic matter. *Geochim. Cosmochim. Acta 45*, 753–762.

YOUNG, A., MONAGHAN, P.H., and SCHWEISBERGER, R.T. (1977) Calculation of ages of hydrocarbons in oils—Physical chemistry applied to petroleum geochemistry I. *Am. Assoc. Pet. Geol. Bull. 61*, 573–600.

14

Molecular Maturity Parameters Within a Single Oil Family: A Case Study from the Sverdrup Basin, Arctic Canada

Joseph A. Curiale

Abstract. Oils and thermal condensates from Jurassic and Triassic reservoirs of the Sverdrup Basin, Arctic Canada, have been analyzed using elemental, isotopic, and molecular techniques. The samples range from 21 to 49° API gravity, and are derived from marine organic matter. The oils from the three shallowest reservoirs are heaviest and most biodegraded. Molecular data indicate that at least 11 of the 12 samples are from a single source unit (the Triassic Schei Point Group), although molecular and isotopic data for the oils also provide evidence for organic facies variations within the Schei Point Group.

The Sverdrup oils encompass a wide range of maturities, with $C_{21}/(C_{21} + C_{28})$-20*R*-triaromatic steroid hydrocarbon ratios ranging from 0.20 to 0.78. This ratio covaries with API gravity and with several other molecular parameters, including rearranged/regular steranes, 18α(H)-trisnorneohopane/17α(H)-trisnorhopane, C_{23}-tricyclic terpane/hopane, $n\text{-}C_{20}/n\text{-}C_{29}$, and $n\text{-}C_{29}/(1/2)(n\text{-}C_{28} + n\text{-}C_{30})$. All of these molecular parameters are shown to be valid maturity indicators for mature and postmature crude oils and condensates in this basin. It is proposed that maturation variability for oils of the Sverdrup Basin is a function of the thermal maturity of the Schei Point shales at the time of expulsion. The maturity levels of the oils do not correspond geographically to known present-day Schei Point Group shale maturity. This is attributed to uplift and possible remigration of hydrocarbons and to the occurrence of known nonregional sites of elevated maturity caused by local igneous intrusions and emplacement of salt structures.

INTRODUCTION

The assessment of thermal maturity in petroleum source rocks is a necessary step in the economic evaluation of a sedimentary basin. Numerous methods for the determination of maturity in both the soluble and insoluble fractions of organic matter in source rocks have been developed. These range from bulk determinations involving visual examination of kerogen to specific stereochemical isomerizations at a single chiral center (Naeser and McCulloh, 1989). Despite the evolution of this wide range of maturity determinants, most routine assessments still reflect the historical perspective derived from coal analysis, and utilize gross optical (microscopic) measurements of kerogen, including reflectance of the vitrinite maceral or color of organic material.

The advent of methods for molecular determination of thermal maturity provided an opportunity to cross-correlate detailed molecular indicators with more conventional kerogen parameters, increasing confidence in both (Mackenzie et al., 1981, 1983; Mackenzie and Maxwell, 1981; Lewan et al., 1986; Curiale et al., 1989). In addition, the availability of molecular maturity parameters made possible maturity assessments of petroleum and solid bitumens. These assessments are often useful in oil-source rock correlations and calculation of migration histories (Curiale, in press).

Most of the molecular maturity parameters already defined are only applicable to source rocks (and oils generated from source rocks) that are in immature or mature stages of their thermal cycle. Postmature source rocks (and their analogous oils and condensates) are more difficult to thermally assess on this basis (Seifert and Moldowan, 1986). For example, of the 11 molecular ratios examined by Mackenzie and Maxwell (1981), only ratios involving side-chain length of triaromatic steroid hydrocarbons remain "unequilibrated" above a vitrinite reflectance level of about 1.0–1.1% R_o. Recently, other molecular maturity parameters, including other aromatic hydrocarbons, have been proposed for use with mature rocks and oils (e.g., Alexander et al., 1985; Abbott and Maxwell, 1988).

The problems associated with finding useable molecular parameters for postmature sections are often of limited interest to the petroleum explorationist. As a consequence, previous studies of molecular maturity parameters in crude oils have focused on petroleum generated at vitrinite reflectance equivalents of about 1.0% R_o or less. Yet, most such molecular measurements have already equilibrated at this maturity (Mackenzie and Maxwell, 1981). In the case of thermally mature oils and thermal condensates, however, direct molecular maturity assessment can be useful in deconvoluting the structural history of a sedimentary basin, and in reconstructing past petroleum migration pathways in the subsurface. Such assessments require understanding the behavior of a wide range of molecular maturity parameters in oils and condensates generated at elevated levels of thermal stress. Parameters which have not yet equilibrated at higher maturity levels include those which involve selective carbon-carbon bond breaking (e.g., ratios which monitor side-chain length of selected steranes; Beach et al., 1989; Sakata et al., 1988) and those which measure the preferential loss of a distinctive class of biomarkers (e.g., tricyclic/pentacyclic terpane ratio; Kruge et al., 1989). Such indices can be used to order suites of cogenetic mature and postmature oils according to thermal maturity (Leenheer and Zumberge, 1987). In addition, the impact of excessive thermal stress on biomarker parameters which are often considered to be maturity-invariant can then be investigated in detail (Sakata et al., 1988).

In this study, a set of 12 oils and condensates from the Sverdrup Basin, Arctic Canada (Fig. 14.1; Table 14.1), has been used to evaluate and cross-correlate a group of molecular maturity parameters. Results will be compared and contrasted with bulk measurements published previously (Curiale, 1989). As will be shown, this oil set is genetically constant and covers a wide range of thermal maturities and is, therefore, ideal for an investigation of molecular maturity parameters in petroleum.

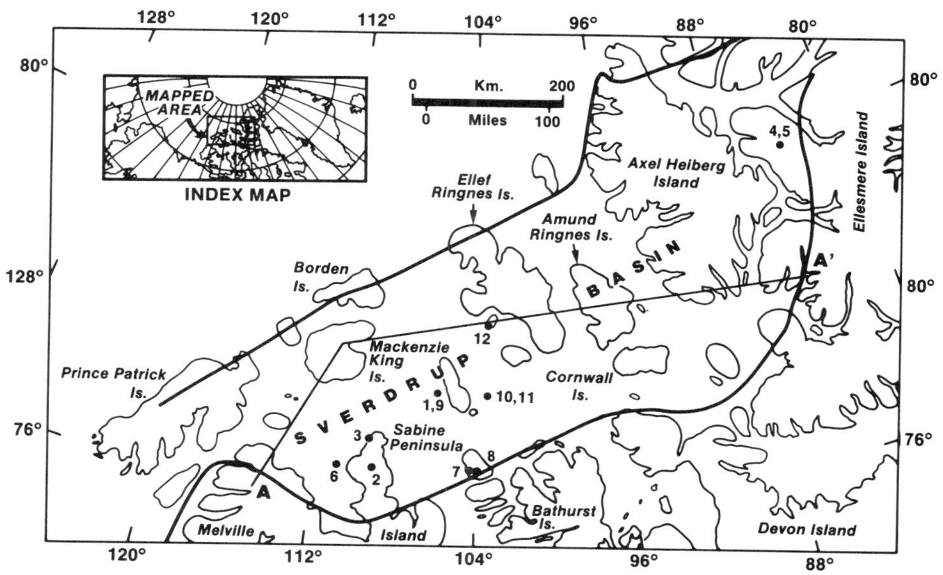

Figure 14.1 Map of Sverdrup Basin, Arctic Canada, showing oil and condensate sample locations (Table 14.1). Bold lines represent the edge of Mesozoic outcrops in the basin. Adapted from Powell (1978). Cross section of A-A′ shown in Figure 14.2.

Table 14.1 SVERDRUP BASIN OIL SAMPLE LOCATION AND DEPTH INFORMATION

Sample	Well name	DST	Depth (m)	Reservoir	API
1	Cisco B-66	1	2097-2105	King Christian	—
2	Drake Point L-67	3	1402-1444	Bjorne Fiord	20.0
3	Roche Point O-43	3	2737-2754	L. Schei Point	49.3
4	Romulus C-42	4	1032-1044	Deer Bay/Savik	27.8
5	Romulus C-42	13	2842-2865	L. Schei Point	48.1
6	West Hecla P-62	2	1065-1077	Bjorne Fiord	21.0
7	Bent Horn F-72A	3	3060-3260	Eid/Blue Fiord	41.6
8	Bent Horn N-72	9	3203-3212	Blue Fiord	43.5
9	Cisco B-66	4	2104-2108	King Christian	—
10	MacLean I-72	6	1769-1773	Heiberg/Blaa Mt.	—
11	MacLean I-72	13	1792-1796	Heiberg/Blaa Mt.	—
12	Thor P-38	3	3870-3890	Heiberg	32.1

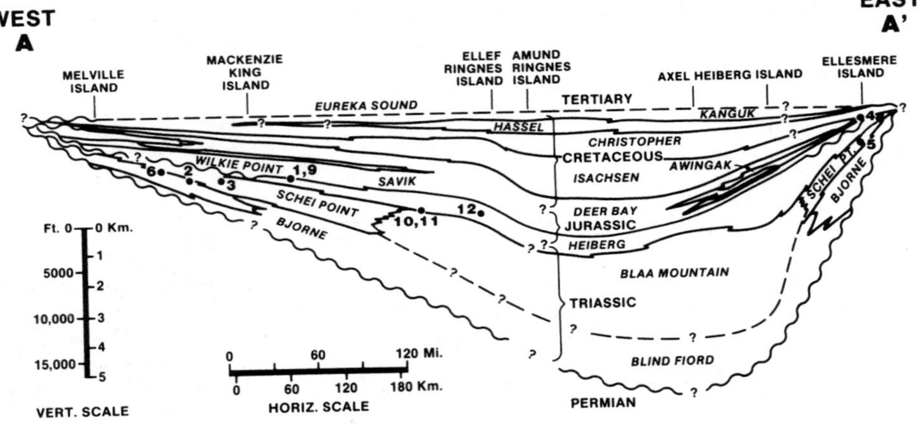

Figure 14.2 West-east cross section through Sverdrup Basin, showing Mesozoic section, asymmetric character of basin, and projected oil and condensate sample locations. Line of section shown in Figure 14.1. From Snowdon and Roy (1975).

Geologic History of the Sverdrup Basin

The late Paleozoic to Tertiary rocks of the Sverdrup Basin and the early Paleozoic rocks of the Franklinian Basin constitute two distinct sedimentary troughs of the Canadian Arctic Islands. The Sverdrup Basin is the northernmost of the two basins, and covers over 500,000 km^2, including offshore areas (Fig. 14.1). The basin contains up to 15 km of carbonates, evaporites and marine and nonmarine clastics (Fischer et al., 1980), and potential reserves of almost three billion barrels of oil (Proctor et al., 1984).

After late Devonian-early Mississippian time, the Sverdrup Basin began to develop immediately north of the Franklinian Basin, and collected sediments at various depocenters through the Miocene (Stuart-Smith and Wennekers, 1977). The basin subsided asymmetrically, and the sediment column is thickest in the east (Snowdon and Roy, 1975). Figure 14.2 shows an idealized cross section through the basin, indicating (approximately) projected locations of the samples in the present study. A generalized stratigraphic section is presented in Figure 14.3. The Sverdrup oil reservoirs that were sampled for this study are indicated on this section (note that the Bent Horn oils, samples 7 and 8 in Table 14.1, are strictly considered part of the Franklinian succession, and are thus not shown in Figs. 14.2 and 14.3).

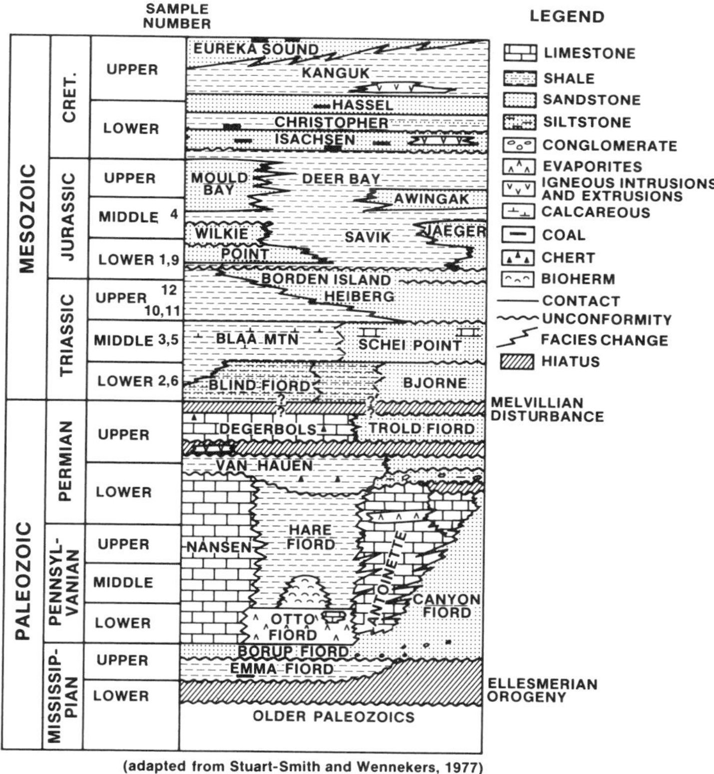

Figure 14.3 Upper Paleozoic through Cretaceous stratigraphic section for Sverdrup Basin. Stratigraphic location of samples is indicated by sample number. From Stuart-Smith and Wennekers (1977).

A more detailed review of the structural and stratigraphic history of the basin has been presented by Stuart-Smith and Wennekers (1977).

Igneous intrusions and salt diapirs are present at various locations throughout the Sverdrup Basin (Balkwill and Fox, 1982; Stuart-Smith and Wennekers, 1977; Brooks et al., in press). In particular, both intrusions and salt structures are recognized on (and adjacent to) the Sabine Peninsula (site of oils 2, 3, and 6; Figs. 14.1 and 14.3), in the late Mississippian through early Tertiary section (Stuart-Smith and Wennekers, 1977). Salt diapirs are also present in the central portion of the basin, between Melville and Mackenzie King Islands (Balkwill and Fox, 1982), whereas igneous intrusions are common in the northwest (Axel Heiberg Island). Such thermal events (unrelated to the smooth thermal gradients characteristic of simple subsidence) must be considered in any interpretation of source rock and petroleum maturity in the basin. Brooks et al. (in press), for example, feel that intrusions in Jurassic and Triassic rocks of the Blaa Mountain Group have contributed to the anomalous thermal maturity often observed in the central and eastern portions of the basin. Although such intrusions must be pervasive to have significantly effected hydrocarbon generation, these authors also note that certain reservoirs that are very close to salt diapirs contain natural gas and condensate, but no oil.

Several studies have addressed the question of thermal maturity of sedimentary strata in the Sverdrup Basin. Fischer et al. (1980) studied spore coloration, vitrinite reflectivity and electron spin resonance response in Mesozoic kerogens, and concluded that maximum paleotemperatures were reached in these sediments during late Cretaceous and Tertiary times. Earlier work by Snowdon and Roy (1975), utilizing cuttings gas data, suggests that the "oil window" occurs between 2000 and 3400 m below the base of the Tertiary. This is generally consistent with conclusions of Henao-Londono (1977) for the western portion of the basin. For the eastern Sverdrup, however, Henao-Londono (1977) suggested that thermal destruction of oil may have occurred in rocks of the early Mesozoic.

Powell (1978) provided the most comprehensive study of Canadian Arctic petroleum source rock potential to date. He concluded that mature and overmature source rocks in the Sverdrup Basin are found at depths below 3000 m and 4500 m, respectively, and that the associated gases of Drake and Hecla Fields (Table 14.1 and Fig. 14.1) were produced at low levels of thermal maturation. This latter conclusion is in contrast to suggestions of Brooks et al. (in press) that the gas in these fields is probably derived from mature to overmature Schei Point sediments. Powell (1978) also presents thermal maturity maps for the major units of the basin; his map for the Schei Point Group was subsequently updated by Brooks et al. (in press), and is discussed later.

Petroleum Source Rock Occurrences in the Sverdrup Basin

The source of the oil in this basin has been a subject of dispute since Trettin and Hills (1967) published their first data on the Triassic tar sands of Melville Island (Fig. 14.1). These workers concluded that oil sourced from Pennsylvanian to Jurassic beds originally had gravities of 16 to 31° API, and had subsequently altered (due to exposure) to form tars at the surface. Powell (1978; p. 80) specifically cites the Schei Point Formation (Figs. 14.2 and 14.3) as the "source for the Melville Island tar sands and the oil shows in the vicinity of the Sabine Peninsula on Melville Island." However, he considers the Schei Point to be an unlikely source for large quantities of oil in the Sverdrup Basin. Powell (1978) considered the Heiberg Formation to be the probable source for the Thor Island oils (Table 14.1), and the Weatherall/Bird Fiord Formation (a Devonian unit of the Franklinian succession) is the suggested source for the Bent Horn oils (Table 14.1). Powell also concludes that maximum hydrocarbon generation in the Franklinian Basin occurred in the Middle to Late Devonian.

Detailed molecular source rock analyses and oil-source rock correlations involving six oils and numerous samples of the Schei Point Group of the Sverdrup Basin were recently completed by Brooks et al. (in press). Their data indicate that the Schei Point Group contains good to excellent petroleum source rocks. Whereas the bulk geochemical data (Rock-Eval pyrolysis yields) presented by these authors suggest that organic facies within the Schei Point is variable, this variation is not evident in their biomarker data. They concluded that the six oils examined constitute a single genetic family, and that the shales of the Schei Point Group (undifferentiated) are the source for this family.

The classification of a number of Sverdrup Basin oils into a single family, as indicated by data of Brooks et al. (in press), is also supported by data presented below. The ability to invoke a single source for the Sverdrup oils (including, interestingly, the Bent Horn oils that were considered previously to be sourced from the Franklinian succession), indicates that chemical variability within the oil set is caused by alteration processes that are largely source-invariant, such as maturation, migration or in-reservoir alteration. This will be useful in assessing the effect of maturation on the distribution of specific molecular suites.

METHODS

Whole oils were analyzed by gas chromatography (1 μl injection; split) under conditions outlined in Curiale et al. (1985). Gas chromatography-high resolution mass spectrometry was used for biomarker determination, in order to provide increased specificity and to permit whole oil injection (Mackenzie, Disko, and Rullkötter, 1983; Lin et al., 1989). Prior to GC-MS analysis, all oils were diluted approximately 1:10 with dichloromethane. Samples were injected (injector temperature: 315°C) in solution (1 μl) into an HP5890A gas chromatograph via an HP 7673A autosampler. Samples were chromatographed on a 30m DB-5 fused silica column. The oven was temperature programmed from 40°C (held 2 min, then ramped at 20°C/min) to 150°C, then to 310°C (ramped at 3°C/min). The chromatograph was interfaced to a VG 70-250SE high-resolution mass spectrometer, which was operated at 10,000 R.P. Source temperature was 250°C. The mass spectrometer was set in selected ion recording mode to detect the following eight-ion suite: *m/z* 177.1638, 191.1798, 217.1955, 218.2034, 231.1174, 231.2113, 253.1956, and 253.2895. Further injections were made, where necessary, to monitor sterane parents (C_{26}–C_{30}) and methylphenanthrene base peaks using alternative ion suites, all at exact masses. Data were processed using the VG 11-250 data system.

All molecular distributions and ratios discussed in this chapter are measured from peak heights (using automatic processing algorithms constructed within the data system), with the exception of methylphenanthrene indices (Radke, 1987), which are calculated as peak areas. The results of all automated peak detection algorithms were individually inspected, and baselines were manually selected where appropriate (less than 5% of all cases). In cases where coelution problems were severe, data were not used. Two samples (5 and 8) contained extremely low quantities of tetracyclic and pentacyclic hydrocarbons, and were not amenable to whole oil GC-MS analysis. Subsequent liquid chromatographic processing designed to concentrate these compounds prior to analysis yielded fractions that were still too depleted in biomarkers for reasonable determination.

RESULTS

Ten Sverdrup Basin oils reservoired in Jurassic and Triassic rocks (Figs. 14.2 and 14.3), and two oils reservoired in Devonian rocks of the Franklinian succession, were studied. Oil gravities range from 21 to 49° API, and reservoir depths vary from 1032 m to 3870 m (Table 14.1). Bulk data for the 12 oils, including stable carbon isotope ratios and sulfur and vanadium concentrations, were presented by Curiale (1989). Carbon isotope ratios of −30.7 0/00 to −29.1 0/00 (PDB; untopped whole oil), sulfur and vanadium concentrations of less than 0.7 percent and 6 ppm (respectively), and the occurrence of C_{30}-desmethylsteranes (Brooks et al. in press; Curiale, 1989), collectively indicate a mature marine source for the organic matter (Curiale, 1989). This conclusion is consistent with previous results of Brooks et al. (in press) for a set of six Sverdrup Basin oils.

Biodegradation, as interpreted from selective loss of *n*-alkanes, is apparent in samples 2, 4, and 6 (Table 14.1; Fig. 14.4). These three oils have the lowest API gravities and presently reside in the shallowest reservoirs of the entire sample set (Table 14.1). They are also among the highest in sulfur and vanadium concentrations (Curiale, 1989). The relationship among biodegradation, API gravity and depth is discernible in Figure 14.4, which shows whole oil gas chromatograms for

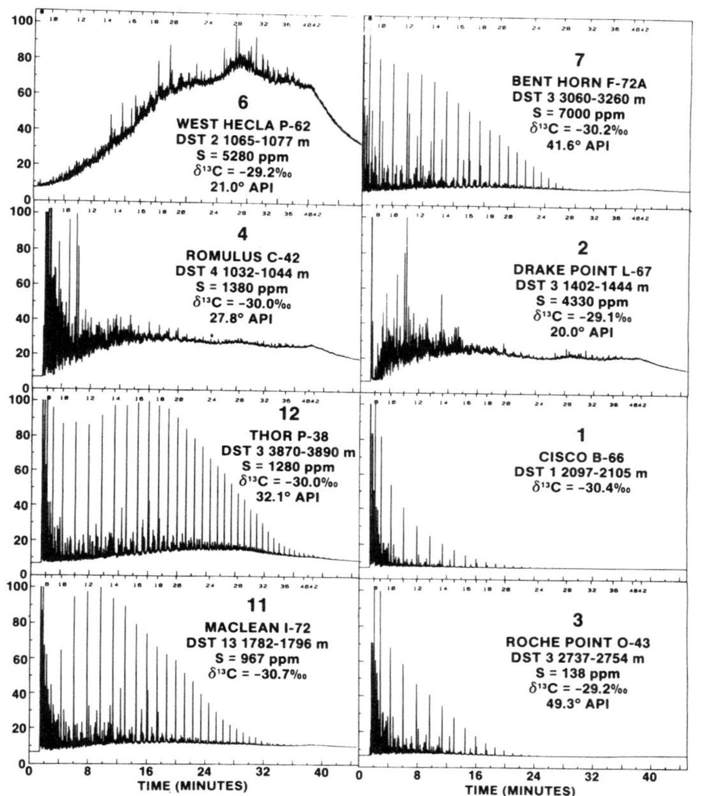

Figure 14.4 Whole oil gas chromatograms for 8 of the 12 oil samples. Numbers above each chromatogram indicate (approximate) elution location of *n*-alkane of that carbon number. Analytical methods are described in Curiale et al. (1985). Sulfur and isotopic data from Curiale (1989). Samples 2, 4, and 6 are inferred to be biodegraded, based solely on the depletion of *n*-alkanes. Samples are ordered (top left to bottom left; top right to bottom right) according to increasing thermal maturity, as determined from $C_{21}/(C_{21} + C_{28})$-20*R*-triaromatic steroid hydrocarbons (see Fig. 14.10 and discussion in text).

8 of the 12 Sverdrup oils. Aside from oils 2, 4, and 6, the oil set is generally free of severe biodegradation.

Molecular data obtained from whole oil analysis by gas chromatography-(high resolution) mass spectrometry are listed in Table 14.2 for 10 of the 12 samples (samples 5 and 8 are excluded because of their extremely low C_{15+} hydrocarbon content). The distribution of $5\alpha,14\alpha,17\alpha,20R$-$C_{27-29}$ steranes by carbon number is shown in Figure 14.5. This plot suggests that the sterane-producing biota which contributed to the source rocks of all (but one) of these oils is very similar. The

Table 14.2 MOLECULAR PARAMETERS, SVERDRUP BASIN OILS[a]

	N-Alkanes		Hopanes			Steranes			Aromatics	
1[b]	2[c]	3[d]	4[e]	5[f]	6[g]	7[h]	8[i]	9[j]	10[k]	11[l]
1	4.17	0.90	0.59	0.59	7.65	38:22:40	0.51	0.11	0.76	0.73
2	4.55	0.92	0.58	0.59	8.23	38:23:40	0.53	0.11	0.74	0.71
3	3.45	1.02	0.59	0.62	4.99	38:21:41	0.49	0.08	0.78	0.62
4	0.74	nd	0.60	0.60	4.91	35:21:45	0.48	0.06	0.28	0.66
6	0.51	0.60	0.59	0.60	5.99	35:20:44	0.48	0.07	0.26	nd
7	nd	nd	0.59	0.59	8.37	39:20:41	0.53	0.13	0.61	nd
9	2.86	0.74	0.57	0.60	8.38	38:21:41	0.53	0.11	0.55	0.76
10	nd	nd	0.57	0.61	11.39	27:15:58	0.42	nd	0.20	0.81
11	3.57	0.92	0.56	0.59	11.29	35:22:43	0.54	0.09	0.53	0.81
12	1.12	nd	0.59	0.60	6.44	36:21:43	0.47	0.05	0.33	1.34

[a]Samples 5 and 8 excluded due to extremely low C15+ hydrocarbon content. Parameters designated nd were not determined due to either low concentration or insufficient accuracy. All measurements are based on heights calculated from mass chromatograms, as indicated in column number designations; relative precision ranges from 5 to 20%. Results differ to some extent from those reported by Curiale (1989), due to differing sample preparation and peak integration methods. Further biomarker ratios for this sample set are reported in Curiale (1989).

[b]Sample number (see Table 14.1).

[c]n-C_{20}/n-C_{29} (from m/z 253.2895)

[d]n-C_{29}/(1/2) (n-C_{28} + n-C_{30}) (from m/z 253.2895)

[e]22*S*/(22*S* + 22*R*)-17α(H),21β(H)-homohopane (from m/z 191.1798)

[f]22*S*/(22*S* + 22*R*)-17α(H),21β(H)-bishomohopane (from m/z 191.1798)

[g]Hopane/5α,14α,17α,20*R*-ethylcholestane (from m/z 191.1798 and m/z 217.1955)

[h]5α,14α,17α,20*R*-C_{27}:C_{28}:C_{29} regular sterane (from m/z 217.1955)

[i]20*S*/(20*S* + 20*R*)-5α,14α,17α-ethylcholestane (from m/z 217.1955)

[j](5α,20*R*-C_{29}-monoaromatic steroid hydrocarbon)/(5α,20*R*-C_{29}-monoaromatic steroid hydrocarbon + 20 *R*-C_{28}-triaromatic steroid hydrocarbon) (from m/z 253.1956 and m/z 231.1174)

[k]C_{21}/(C_{21} + C_{28})-20*R*-triaromatic steroid hydrocarbons (from m/z 231.1174)

[l]Methylphenanthrene Index (MPI) I of Radke (1987) (from m/z 192.0936)

SVERDRUP OILS

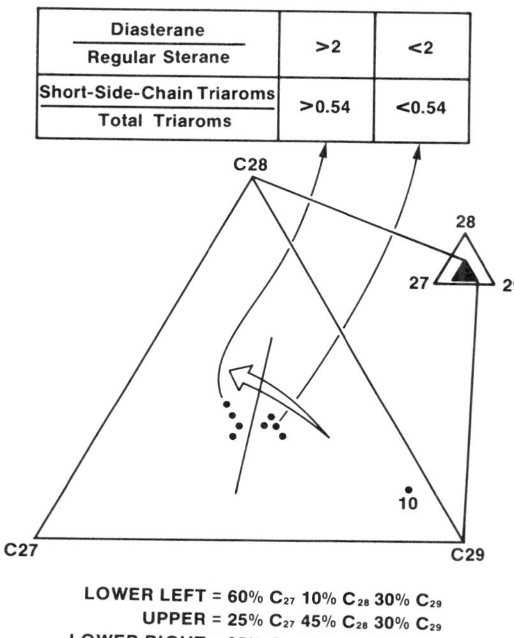

Figure 14.5 Ternary diagram depicting the distribution of 5α,14α,17α,20*R*-C_{27-29} steranes. Sample 10 is considered an outlier and is possibly derived from a different combination of sources than samples 1–9, 11, 12. Arrow indicates direction of increasing maturation for the sample set. Samples to the left of the vertical line have relatively high diasterane/regular sterane and $C_{21}/(C_{21} + C_{28})$-20*R*-triaromatic steroid hydrocarbon ratios (see text and Table 14.2 for specific components of these ratios). Vertices of large triangle are: lower left = 60% C_{27}, 10% C_{28}, 30% C_{29}; upper = 25% C_{27}, 45% C_{28}, 30% C_{29}; lower right = 25% C_{27}, 10% C_{28}, 65% C_{29}.

sole exception is sample 10, DST 6 from MacLean I-72 (1769–1773 m), which contains a significantly higher content of 5α,14α,17α,20*R*-ethylcholestane, relative to the C_{27} and C_{28} homologues. A typical *m*/*z* 217.1955 sterane distribution for the majority of the oils is shown in Figure 14.6 (top), whereas the C_{29}-biased distribution for oil 10 is shown at the bottom of this figure. It is noted that, while internal distributions of regular steranes and internal distributions of rearranged (dia-) steranes are very similar for all samples of the major oil group, the ratio of regular to rearranged steranes differs significantly. This difference is attributed to maturation effects, as discussed.

The similarity of 5α,14α,17α,20*R* sterane carbon number distribution for all oils except sample 10 also extends to the C_{26} and C_{30} members of the sterane family (not shown). Typical parent mass chromatograms (at exact masses *m*/*z* 372.3744, 386.3900, 400.4056, and 414.4212) of a nonbiodegraded, relatively low-maturity oil are shown for sample 12 in Figure 14.7, depicting the C_{27-30} homologues. Again, when differences in biodegradation and maturity (see below) are accounted for, these distributions are similar for oils 1 through 9 and 11 and 12, suggesting a single source facies with respect to steroid contributions. The pairwise similarity of the $C_{27}:C_{29}$ parent distributions and the $C_{28}:C_{30}$ parent distributions in oil 12 of Figure 14.7 is of interest, and may indicate the presence of C_{28} and C_{30} methylsteranes in

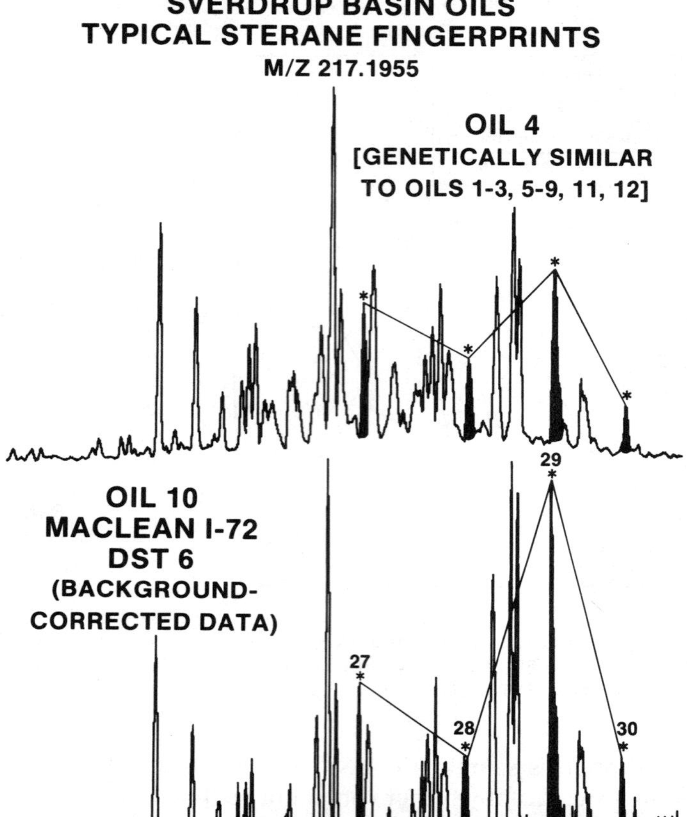

Figure 14.6 *m/z* 217.1955 sterane distribution for oil 4 (top), shown as a typical example of the sterane distributions of oils 1–9, 11, 12; and the sterane distribution for oil 10 (bottom). 5α,14α,17α,20*R*-C_{27-29} steranes are highlighted with an asterisk and their carbon number. Note that lower mass chromatogram is background-corrected to remove rising baseline features.

these samples. The potential occurrence of methylsteranes is supported by the presence of specific peaks in the *m/z* 231.2113 mass chromatogram, and is inferred by the occurrence of dinoflagellates in the Schei Point shales (Brooks et al., in press). The distribution of methylsteranes in these oils and condensates is currently being investigated through metastable ion monitoring experiments.

The typical *m/z* 191.1798 mass chromatogram for the major oil family is shown at the top of Figure 14.8. Whereas the tricyclic/pentacyclic terpane ratio within the sample set changes with maturity (as discussed below), the internal distribution of

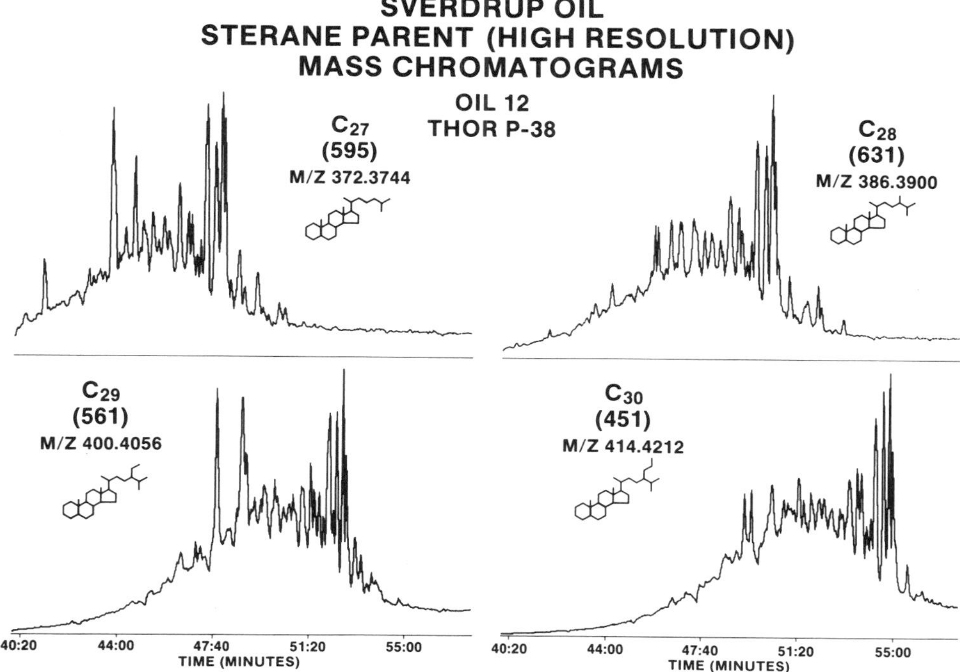

Figure 14.7 Parent mass chromatograms (m/z 372.3744, 386.3900, 400.4056, 414.4212) for C_{27-30} steranes of Oil 12 (Thor P-38). The relative ion intensity of the maximum peak in each chromatogram is shown in parentheses.

tricyclic and the internal distribution of the pentacyclic compounds are constant throughout the oil family. Also shown in Figure 14.8 (bottom) is the (background-corrected) m/z 191.1978 chromatogram of the outlier oil.

The similarity among internal distributions of steranes and hopanes indicates that the biota contributing each of these molecular types to the oil's source rocks is distinctive. However, variable contributions of steroid-containing and hopanoid containing organisms (e.g., phytoplankton and bacteria, respectively) cannot be monitored by separate examination of m/z 217.1955 and m/z 191.1798 chromatograms. In an effort to examine detailed differences in the source facies of these oils, the ratio of hopane to 5α,14α,17α,20*R*-ethylcholestane has been calculated. Figure 14.9 shows this ratio plotted against the C_{30}/C_{29}-5α,14α,17α,20*R*-desmethylsterane ratio (left axis) and the $\delta^{13}C$ ratio of the aromatic hydrocarbons (right axis; data from Curiale, 1989). The isotope ratio of the *aromatic* hydrocarbons was used in order to minimize alteration effects common with whole oil or aliphatic hydrocarbon isotope ratios. With the exception of outlier oil 10, the hopane/sterane ratio increases with increasing C_{30}/C_{29} ratio, and with decreasing $\delta^{13}C$ value. These relationships suggest that the source facies contributions for the major oil family

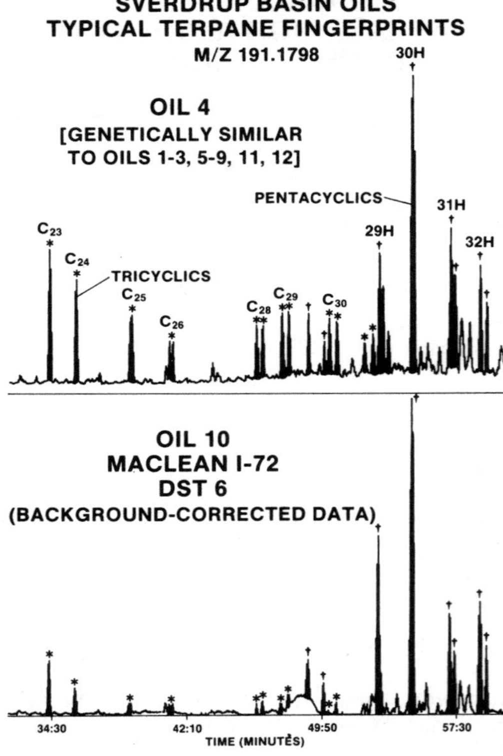

Figure 14.8 *m/z* 191.1798 terpane distribution for oil 4 (top), shown as a typical example of the terpane distributions of oils 1–9, 11, 12; and the terpane distribution for oil 10 (bottom). Tricyclic terpanes and pentacyclic triterpanes (hopanes) are highlighted with asterisks and daggers, respectively. Note that lower mass chromatogram is background-corrected to remove rising baseline features.

are not constant from oil to oil. Instead, increasing hopane/sterane ratios (increasing relative bacterial input to the source?) are accompanied by a relative increase in the C_{30} desmethylsteranes and a decrease in $\delta^{13}C$ of the aromatic hydrocarbons.

The conclusions drawn from Figure 14.9 reflect comments of Brooks et al. (in press), who noted that Rock-Eval data from the Schei Point Group source rocks for these oils indicate the presence of different "kerogen types," whereas biomarker data (based on sterane carbon number distributions) suggested a single organic facies. Such a conflict is reconciled here by noting that differing ratios of two different biomarker *classes* (hopanes and steranes) probably reflect differing organic facies within the (Schei Point Group) source rocks. In any event, the Rock-Eval values of Brooks et al. (in press) were undoubtedly overprinted to some extent by maturity effects, which casts some doubt on kerogen type differences deduced solely from Rock-Eval data. In summary, it would appear that these oils do indeed represent a single oil family (excepting oil 10) in the sense that a single stratigraphic group (Schei Point) is responsible for them, as previously deduced by Brooks et al. (in press). However, organic facies changes within the Schei Point source shales are measurable, and can be recognized in both the molecular and the isotopic data.

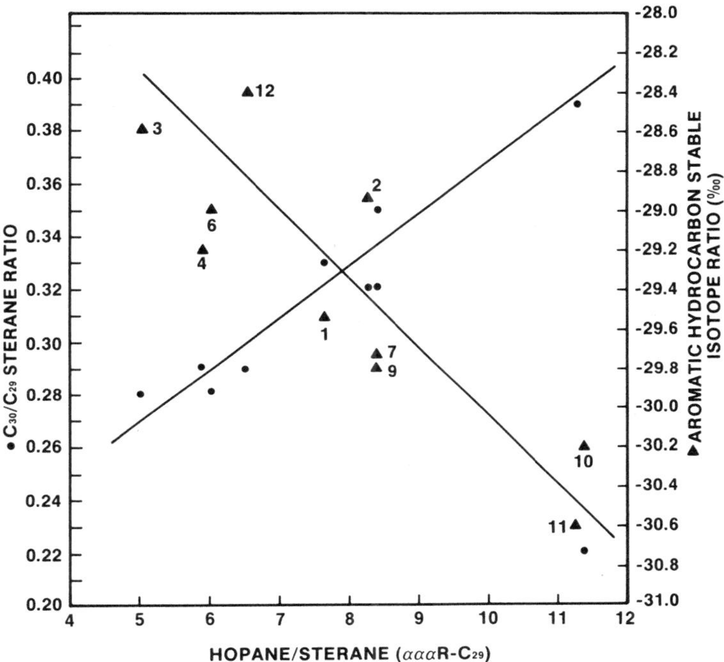

Figure 14.9 Plot of the hopane/5α,14α,17α,20*R*-ethylcholestane ratio (x-axis) versus the C_{30}/C_{29}-5α,14α,17α,20*R*-desmethylsterane ratio (left, circles) and the aromatic hydrocarbon stable carbon isotope ratio (right, triangles; data from Curiale, 1989). Sloping lines are best fits to the data.

Molecular Maturity Determinations

Conventional molecular maturity parameters for the oil set are listed in Table 14.2. Based on 22*S*/(22*S* + 22*R*)-17α(H),21β(H)-homohopane, 22*S*/(22*S* + 22*R*)-17α(H),21β(H)-bishomohopane, and 20*S*/(20*S* + 20*R*)-5α,14α,17α-ethylcholestane ratios (columns 4, 5, and 8 in Table 14.2) 11 of the 12 samples are thermally mature (i.e., these oils are all at thermal equilibrium with respect to these ratios). Oil 10 is measurably less mature, as indicated by a marginally lower 20*S*/(20*S* + 20*R*)-5α,14α,17α-ethylcholestane ratio.

Examination of conventional biomarker distributions and the molecular data listed in Table 14.2 indicates that certain consistent trends are present in the Sverdrup oil set. Most conspicuous among these trends is the relative change in triaromatic steroid hydrocarbon distribution by carbon number. As shown in Table 14.2

(column 10), the ratio of $C_{21}/(C_{21} + C_{28})$-20*R*-triaromatic steroid hydrocarbons ranges from 0.20 (oil 10) to 0.78 (oil 3). This range covers virtually the entire thermal window for oil generation, as measured previously in petroleum source rocks (e.g., Mackenzie et al., 1981, 1983). Figure 14.10 shows the *m/z* 231.1174 chromatograms, ordered by $C_{21}/(C_{21} + C_{28})$ ratio, for 8 of the 12 oils (excluding oil 10, based on the possibility that it may not be fully cogenetic with the major

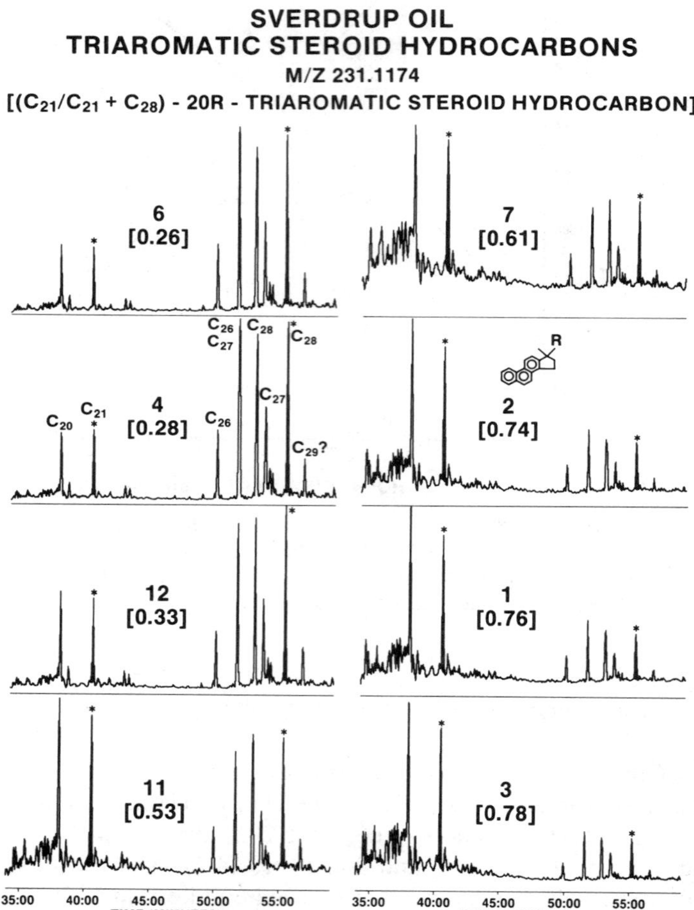

Figure 14.10 *m/z* 231.1174 mass chromatograms for 8 of the 12 oil samples, showing the distribution of triaromatic steroid hydrocarbons. Samples are ordered (top left to bottom left; top right to bottom right) according to increasing thermal maturity, as determined by the $C_{21}/(C_{21} + C_{28})$-20*R*-triaromatic steroid hydrocarbon ratio (peaks highlighted with asterisk; value given in brackets; see text). Data are derived from 1 μl (dichloromethane-diluted) whole oil injections.

oil family). This order, oils 6, 4, 12, 11, 7, 2, 1, and 3, respresents the maturity trend of these oils from least to greatest.

Comparison of the $C_{21}/(C_{21} + C_{28})$ ratio in these oils with other maturity-affected molecular ratios indicates that molecular maturity trends in these oils can be monitored using a large suite of parameters. Figure 14.11 shows the triaromatic chain length ratio plotted against (A) the diasterane/regular sterane ratio (13β,17α,20*R*/14α,17α,20*R*-ethylcholestane), (B) the 18α(H)-22,29,30-trisnorneo-hopane/17α(H)-22,29,30-trisnorhopane ratio, (C) the C_{23}-tricyclic terpane/hopane ratio, and (D) the *n*-C_{20}/*n*-C_{29} ratio. In all cases, there is a monotonic increase in each ratio with increasing $C_{21}/(C_{21} + C_{28})$ ratio. Figures 14.12 and 14.13 show the sterane (*m*/*z* 217.1955) and hopane (*m*/*z* 191.1798) chromatograms for eight of the oils, in the same order as in Figure 14.10. These figures graphically depict (respectively) the increase in the diasterane/regular sterane and C_{23}-tricyclic terpane/hopane ratios with increasing maturity.

It was concluded previously that most of the samples in this oil set are commonly sourced, based on sterane carbon number distributions. Yet minor but systematic changes in this distribution are visible in Figure 14.5, where the more mature oils (as defined by diasterane/regular sterane ratios and $C_{21}/(C_{21} + C_{28})$ ratios) are shown to be slightly depleted in the 5α,14α,17α,20*R*-ethylcholestane

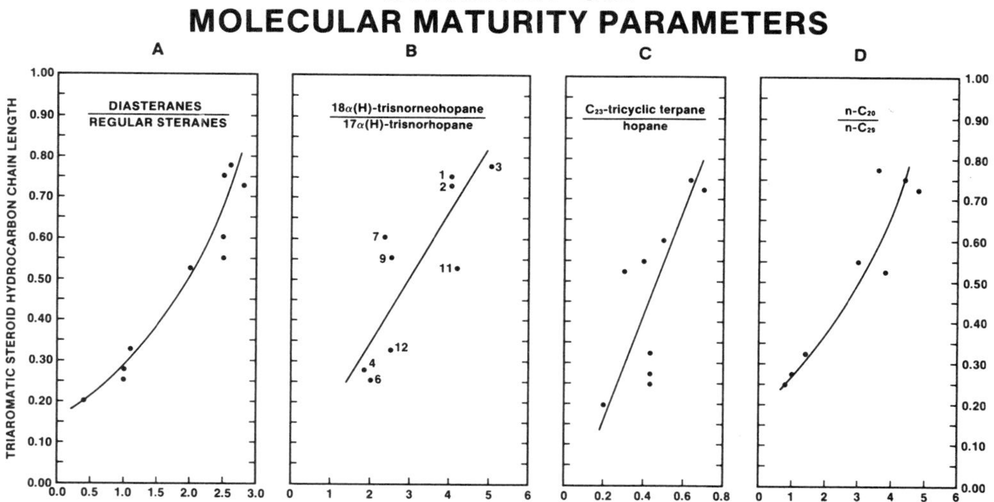

Figure 14.11 Plots of the $C_{21}/(C_{21} + C_{28})$-20*R*-triaromatic steroid hydrocarbon ratio (y-axis) versus four other molecular maturity parameters for the Sverdrup oils. Plate A: 13β,17α,20*R*/14α,17α,20*R*-ethylcholestane; Plate B: 18α(H)-22,29,30-trisnorneohopane/17α(H)-22,29,30-trisnorhopane ratio; Plate C: C_{23}-tricyclic terpane/hopane ratio; and Plate D: *n*-C_{20}/*n*C_{29} alkane ratio. Curves on each plate are interpreted best fit lines.

SVERDRUP OIL
STERANES
M/Z 217.1955

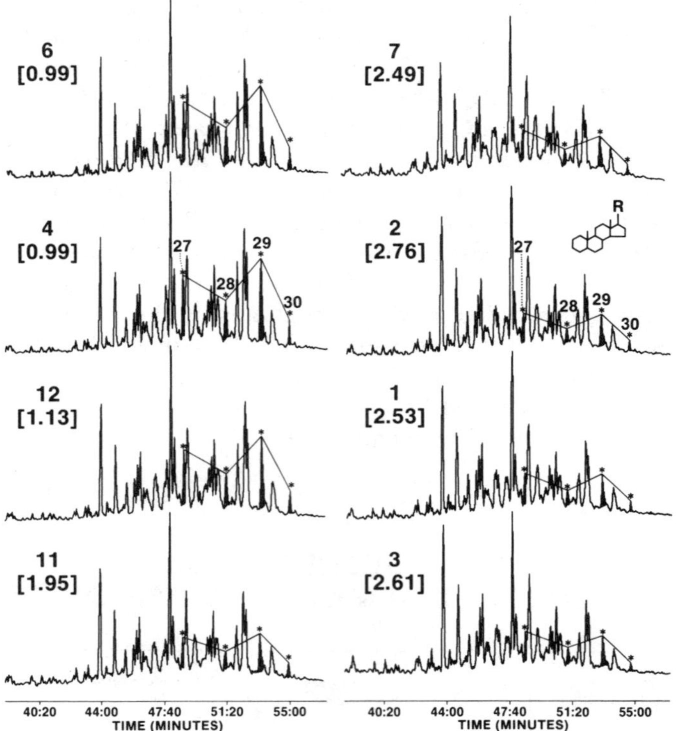

Figure 14.12 m/z 217.1955 mass chromatograms for 8 of the 12 oil samples, showing the distribution of rearranged and regular steranes. $5\alpha,14\alpha,17\alpha,20R$-$C_{27-30}$ steranes are highlighted with an asterisk. Samples are ordered (top left to bottom left; top right to bottom right) according to increasing thermal maturity, as determined by the $C_{21}/(C_{21} + C_{28})$-20R-triaromatic steroid hydrocarbon ratio (as in Fig. 14.10). Value given in brackets is the $13\beta,17\alpha,20R/14\alpha,17\alpha,20R$-ethylcholestane ratio. Data are derived from 1 μl (dichloromethane-diluted) whole oil injections.

component. The possibility of relative depletion of this sterane with increasing maturity has been noted and discussed elsewhere (Mackenzie, 1984, p. 191; Curiale, 1986; Sakata et al., 1988).

The presence of a slight even-carbon predominance among the n-alkanes in the Sverdrup oils was observed previously by Brooks et al. (in press). Our data suggest that the extent of this predominance is quite variable, and that this variability is maturity-related. Figure 14.14 shows the relationship between the n-C_{29}/$(1/2)(n$-C_{28} + n-$C_{30})$ ratio (y-axis) and the $C_{21}/(C_{21} + C_{28})$-20R-triaromatic steroid

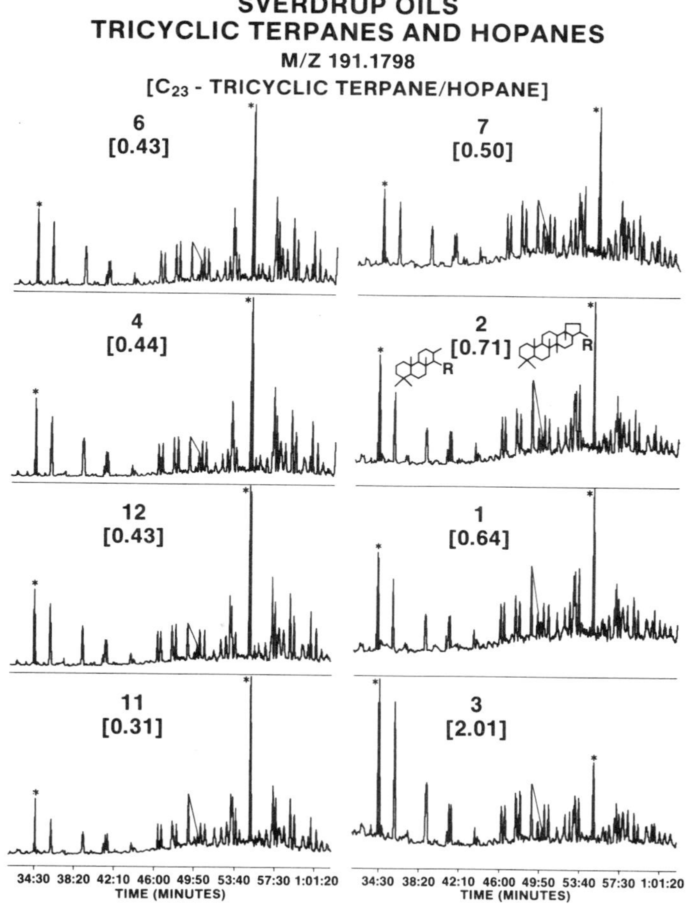

Figure 14.13 *m/z* 191.1798 mass chromatograms for 8 of the 12 oil samples, showing the distribution of tricyclic and pentacyclic terpanes. Samples are ordered (top left to bottom left; top right to bottom right) according to increasing thermal maturity, as determined by the $C_{21}/(C_{21} + C_{28})$-20*R*-triaromatic steroid hydrocarbon ratio (as in Fig. 14.10). Value given in brackets is the C_{23}-tricyclic terpane/hopane ratio (each of these compounds is highlighted with an asterisk). Data are derived from 1 μl (dichloromethane-diluted) whole oil injections.

hydrocarbon (left) and 18α(H)-22,29,30-trisnorneohopane/(17α(H)-22,29,30-trisnorhopane (right) ratios. The carbon number predominance disappears in the most mature oils.

In addition to the relationship between thermal maturity and molecular parameters, the oil set also exhibits a minor correlation between maturity and API gravity. Generally, API gravity of the oils, with the exception of the biodegraded

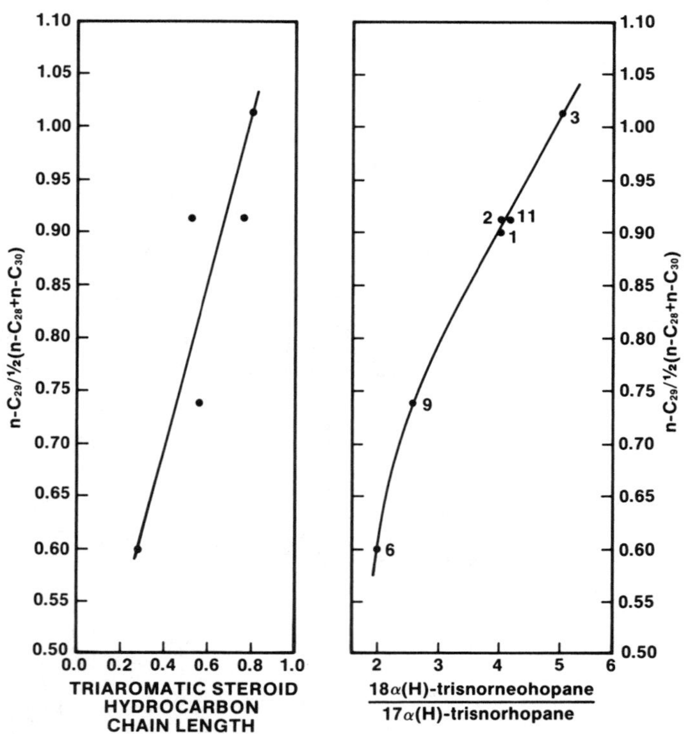

Figure 14.14 Plots of the n-C_{29}/(1/2)(n-C_{28} + n-C_{30}) ratio (y-axis) versus the C_{21}/(C_{21} + C_{28})-20*R*-triaromatic steroid hydrocarbon ratio (left) and the 18α(H)-22,29,30-trisnorneohopane/17α(H)-22,29,30-trisnorhopane ratio (right).

oil 2, increases with increasing C_{21}/(C_{21} + C_{28}) triaromatic steroid hydrocarbon ratio. This relationship, while not conclusive, is consistent with the observation that most of the correlative molecular maturity parameters discussed are dependent on carbon-carbon cracking mechanisms. An increase in API gravity would be expected with increasing extent of cracking, leading to thermal condensates such as those present in the Sverdrup Basin.

The consistent trend among molecular parameters within this oil set suggests that the effects of thermal maturity are observable over a wide range of molecular types, even in oils whose C-20 sterane and C-22 hopane epimers are already thermally equilibrated. The ability to classify all but one of the oils/condensates as sourced from the Schei Point shales (Brooks et al., in press) provides an opportunity to investigate the relationship between present-day source rock maturity and present-day reservoired oil/condensate maturity.

DISCUSSION

Present-day thermal maturity of the Schei Point Group has been determined by Powell (1978), and Powell's results were subsequently updated by Brooks et al. (in press). The distribution of immature, mature, and overmature Schei Point shales is shown in Figure 14.15. Maturity levels generally follow basin outlines, with the Schei Point being least mature on the southern and northern flanks of the basin.

Oil sample locations from this study are superimposed onto Figure 14.15. Comparison of geographic location of the oils with the maturity of the Schei Point source shales indicates only a very minor correspondence between present-day source maturity and present-day oil maturity. The low-maturity oils 6 and 4 are located along the basinal flanks, and the high-maturity oil 3 is located adjacent to

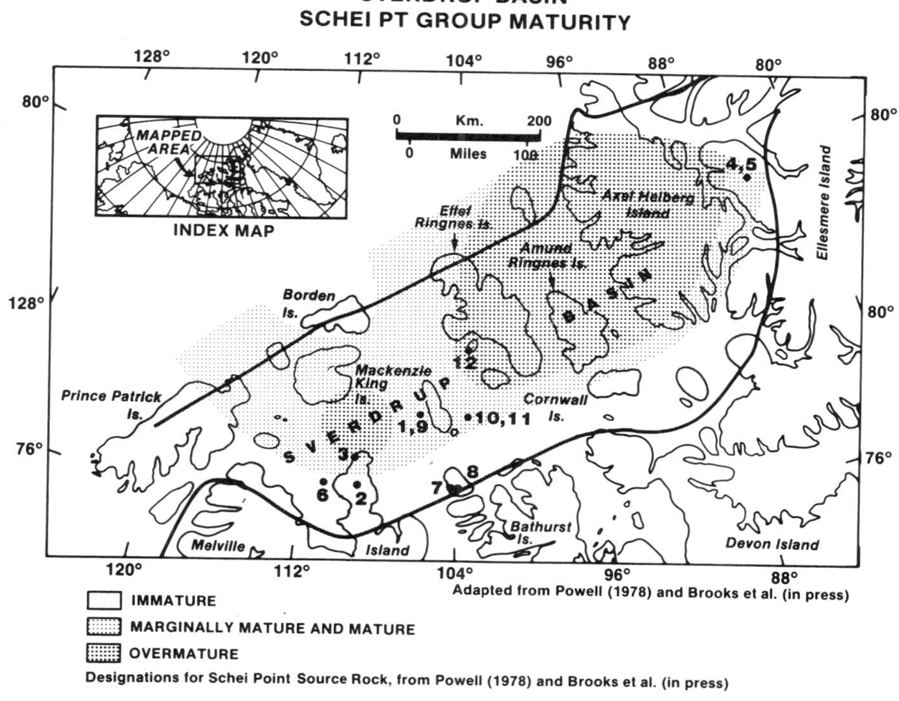

Figure 14.15 Map of Sverdrup Basin highlighting the maturity distribution in the Schei Point Group. Oil sample locations as shown (Table 14.1). Schei Point maturity data are adapted from Powell (1978) and Brooks et al. (in press).

a presumed area of high maturity resulting from the occurrence of salt diapirism (Brooks et al., in press). However, the low maturity oil 12 is reservoired immediately above an overmature portion of the Schei Point/Blaa Mountain (compare Fig. 14.15 with Fig. 14.2 and 14.3). Likewise, oil 7, an overmature oil, is found on the basin flank, while oil 10, the most immature oil of the entire sample set, is located in the vicinity of mature Schei Point source material.

The differences between present-day source maturity and present-day oil maturity can be interpreted in several ways. Initially, it is possible that the molecular differences previously attributed to maturity differences are in fact due to source differences. However, this possibility is discounted based on the previously-described complete consistency among a large set of molecular measurements conventionally used as maturity tools. Furthermore, known differences in source organic facies, as discussed in conjunction with Figure 14.9, do not correspond with parameters considered to represent molecular maturity. It is, therefore, assumed that the systematic molecular differences discussed earlier are predominantly maturity-induced.

A more reasonable alternative is that the geographic discrepancy between oil and source rock maturities is misleading. Two possibilities exist. First, the maps of Powell (1978) and Brooks et al. (in press) may not be sufficiently accurate for such a comparison. This is a real possibility, particularly when we consider the potentially confusing effects of salt diapirism and igneous instrusions on past heat flows in the basin. Localized maturation caused by anomalous thermal conditions could have created miniature zones of mature Schei Point in unknown (i.e., undrilled) locations throughout the basin (Stuart-Smith and Wennekers, 1977), although the volumetric significance of these effects is unclear.

A second reason that the discrepancy between oil and source maturity could be misleading derives from the documented uplift that has occurred along the margins of the Sverdrup Basin, and the ensuing possibility of remigration of reservoired hydrocarbons. Several workers (e.g., Snowdon and Roy, 1975) have documented uplift in the basin following maximum subsidence. This uplift probably occurred during early Tertiary time (Brooks et al., in press), *after* peak generation from the Schei Point (which occurred during late Cretaceous through early Tertiary time; Fischer et al., 1980). The uplift would have effectively quenched further generation from this source rock. Redistribution of migrated hydrocarbons during and after this uplift would have displaced reservoired oil, disturbing a geographical correlation between source and reservoir maturity that may once have existed.

The occurrence of this early Tertiary uplift also suggests that the oil maturation levels measured in this study are probably a function of the Schei Point source shale maturity at the time of expulsion. Indeed, change in molecular ratios which occurred post-trapping should not be significant in this basin, insomuch as present-day reservoir depths (Table 14.1) are considerably less than maximum burial depths of the Schei Point Group (Snowdon and Roy, 1977; Fischer et al., 1980). Differing oil maturities must, therefore, be a product of source maturity at the time of expulsion, and are perhaps complicated by redistribution of oil during a remigration episode postuplift. The imprint of source maturity is favored as a cause for present-day oil maturity, insomuch as some of the molecular maturity trends noted for this oil set have also been observed in Schei Point source shales (Brooks et al., in press).

This conclusion is supported by examining oil pairs produced from the same well. Oils 1 and 9 are from 2097 m and 2104 m in the Cisco B-66 well, whereas oils 10 and 11 are from 1769 m and 1792 m in the MacLean I-72 well (Table 14.1). The oils in each of these pairs, despite being reservoired within only 7 m and 23 m of one another (respectively), have widely different triaromatic steroid hydrocarbon chain length ratios (0.76 and 0.55 for the Cisco B-66 well, and 0.20 and 0.53 for the MacLean I-72 well; Table 14.2). Similar differences were noted by Brooks et al. (in press) for sterane epimer ratios of two oils from the Balaena D-58 well in the basin: $20S/20R$-$5\alpha,14\alpha,17\alpha$-ethylcholestane ratios are significantly *lower* in the oil at 1672 m than in the oil at 405 m. It would require a complex scenario of postmigration and posttrapping subsidence and subsequent uplift to account for the maturity levels of these Sverdrup oils if the maturity was a product of thermal conditions in the reservoir. It is considered more likely that these reservoirs filled at different times in geologic history, times when their source facies were at different levels of thermal maturity, and/or when redistribution of migratable hydrocarbons played a role.

Although this geologic explanation of a wide range of single-family oil maturities appears reasonable, the relatively uncommon occurrence of such large maturity ranges for commonly sourced oils of a single basin presumably requires unusual generative conditions. Such conditions may be attained by (1) locally pressurizing the source beds in such a way that expulsion is delayed as maturity advances continue (a "pressure-cooker" effect), or (2) selectively maturing the source bed(s) in a nonregional fashion by, for example, the introduction of heat from local occurrences of igneous intrusions, or heat flow anomolies caused by the presence of salt diapirs (Brooks et al., in press). Either possibility could yield discontinuous source bed maturity trends, and thus discontinuous oil maturity patterns. Both possibilities have been recognized in the Sverdrup Basin (Stuart-Smith and Wennekers, 1977).

CONCLUSIONS

Examination of the elemental concentrations, isotopic ratios and molecular distributions of 12 Sverdrup Basin oils/condensates suggests the following conclusions. (1) Most oils in Jurassic and Triassic reservoirs are commonly sourced from shales of the Triassic Schei Point Group. (2) Geochemical differences among the oils are a result of (a) organic facies variations within the Schei Point, and (b) differences in thermal maturity of the source at the time of expulsion. (3) The oils/condensates vary widely in thermal maturity, as monitored by internally consistent changes in the following ratios: $C_{21}/(C_{21} + C_{28})$-$20R$-triaromatic steroid hydrocarbons, rearranged/regular steranes, 18α(H)-trisnorneohopane/17α(H)-trisnorhopane, C_{23}-tricyclic terpane/hopane, n-C_{20}/n-C_{29}, and n-C_{29}/(1/2)(n-C_{28} + n-C_{30}). These maturity differences are also reflected in API gravity variations. (4) The lack of complete correspondence between oil/condensate maturity and present-day thermal maturity of the Schei Point source shales is attributed to a combination of local thermal anomalies and postgenerative remigration due to regional uplift.

These conclusions indicate that each of the major Sverdrup Basin reservoirs filled at a different time, and that significant increases in maturity may have resulted from delayed expulsion of oil/condensate from overmature source rocks. In a geologic context, this implies that knowledge of timing is critical to exploration in this region: The results of this study provide constraints on models developed to understand subsidence, petroleum generation and migration, subsequent uplift, and remigration of mature oils and thermal condensates in the Sverdrup Basin.

Acknowledgments

I am indebted to P.W. Brooks (Institute of Sedimentary and Petroleum Geology, Calgary) for providing me with the Sverdrup Basin sample set. I thank Z.A. Wilk for providing the GC-MS data, S.A. Bharvani and G. Marcinko for laboratory and data handling assistance, B.W. Bromley and G.G. Rinaldi for helpful advice, and G. Stadnicky and R. Davis for educating me on the geology of the Sverdrup Basin. Reviews by Dr. J. Zumberge and one anonymous reviewer have improved the original manuscript. Finally, I thank Unocal Science & Technology management for providing an environment where this type of research is possible, and for allowing its publication.

REFERENCES

Abbott, G.D. and Maxwell, J.R. (1988) Kinetics of the aromatization of rearranged ring-C monoaromatic steroid hydrocarbons. *Org. Geochem. 13*, 881–885.

Alexander, R., Kagi, R.I., Rowland, S.J., Sheppard, P.N., and Chirila, T.V. (1985) The effects of thermal maturity on distributions of dimethylnaphthalenes and trimethylnaphthalenes in some Ancient sediments and petroleums. *Geochim. Cosmochim. Acta 49*, 385–395.

Balkwill, H.R. and Fox, F.G. (1982) Incipient rift zone, western Sverdrup Basin, Arctic Canada. In *Arctic Geology and Geophysics* (eds. A.F. Embry and H.R. Balkwill) *Can. Soc. Petr. Geol.* Memoir 8, pp. 171–187.

Beach, F., Peakman, T.M., Abbott, G.D., Sleeman, R., and Maxwell, J.R. (1989) Laboratory thermal alteration of triaromatic steroid hydrocarbons. *Org. Geochem. 14*, 109–111.

Brooks, P.W., Embry, A.F., Goodarzi, F., and Stewart, R. (in press) Geochemical studies of the Sverdrup Basin (Arctic Islands), Part 1: Organic geochemistry and biological marker geochemistry of Schei Point Group (Triassic) and recovered oils. *Bull. Can. Petr. Geol.*

CURIALE, J. A. (1986) Origin of solid bitumens, with emphasis on biological marker results. *Org. Geochem. 10*, 559–580.

CURIALE, J. A. (1989) Integrated biological marker correlations among mature oils of the Sverdrup Basin. Division of Petroleum Chemistry Preprint, American Chemical Society (Dallas, Tx, April 9–14, 1989).

CURIALE, J. A., (in press). Applications of organic geochemical methods for hydrocarbon exploration. In *Organic Geochemistry* (eds. M. Engel and S. Macko) Vol. 6 of Topics in Geobiology, Plenum Press, New York.

CURIALE, J.A., CAMERON, D., and DAVIS, D.V. (1985) Biological marker distribution and significance in oils and rocks of the Monterey Formation, Calif. *Geochim. Cosmochim. Acta 49*, 271–288.

CURIALE, J.A., LARTER, S.R., SWEENEY, R.E., and BROMLEY, B.W. (1989) Molecular thermal maturity indicators in oil and gas source rocks. In *Thermal History of Sedimentary Basins* (eds. N.D. Naeser and T.H. McCulloh) pp. 53–72. Springer-Verlag.

FISHER, M.J., BARNARD, P.C., and COOPER, B.S. (1980) Organic maturation and hydrocarbon generation in the Mesozoic sediments of the Sverdrup Basin, Arctic Canada. In Fourth International Palynology Conference, Luknow, India 2, 581–588.

HENAO-LONDONO, D. (1977) A preliminary evaluation of the Arctic Islands. *Bull. Can. Petr. Geol. 25*, 1059–1084.

KRUGE, M.A., HUBERT, J.F., AKES, J., and MERINEY, P. (1989) Extended tricyclic terpanes: Molecular markers in Lower Jurassic synrift lacustrine black mudstones of the Hartford Basin, Connecticut. Abstract. *Bull. Amer. Assoc. Petr. Geol. 73*, 375.

LEENHEER, M.J. and ZUMBERGE, J.E. (1987) Correlation and thermal maturity of Williston Basin crude oils and Bakken source rocks using terpane biomarkers. In Williston Basin: Anatomy of a Cratonic Oil Province (ed. M.W. Longman), Rocky Mountain Association of Geologists, Denver, pp. 287–298.

LEWAN, M.D., BJOROY, M., and DOLCATER, D.L. (1986) Effects of thermal maturation on steroid hydrocarbons as determined by hydrous pyrolysis of Phosphoria Retort Shale. *Geochim. Cosmochim. Acta 50*, 1977–1987.

LIN, D.-P., LITORJA, L.A., and ABBAS, N.M. (1989) Determination of pentacyclic triterpanes in Arabian crude oil using gas chromatography/high resolution mass spectrometry. *Fuel 68*, 257–259.

MACKENZIE, A.S. (1984) Applications of biological markers in petroleum geochemistry. In *Advances in Petroleum Geochemistry* (eds. J. Brooks and D. Welte) Vol. 1, pp. 115–214. Academic Press, London.

MACKENZIE, A.S., REN-WEI, L., MAXWELL, J.R., MOLDOWAN, J.M., and SEIFERT, W.K. (1983) Molecular measurements of thermal maturation of Cretaceous shales from the Overthrust Belt, Wyoming, USA. In *Advances in Organic Geochemistry—1981* (eds. M. Bjoroy et al.) pp. 496–503.

MACKENZIE, A.S., DISKO, U., and RULLKÖTTER, J. (1983) Determination of hydrocarbon distributions in oils and sediment extracts by gas chromatography-high resolution mass spectrometry. *Org. Geochem. 5*, 57–63.

MACKENZIE, A.S. and MAXWELL, J.R. (1981) Assessment of thermal maturation in sedimentary rocks by molecular measurements. In *Organic Maturation Studies and Fossil Fuel Exploration* (ed. J. Brooks) pp. 239–254. Academic Press.

MACKENZIE, A.S., HOFFMAN, C.F., and MAXWELL, J.R. (1981) Molecular parameters of maturation in the Toarcian shales, Paris Basin, France–III. Changes in aromatic steroid hydrocarbons. *Geochim. Cosmochim. Acta 45*, 1345–1355.

NAESER, N.D. and MCCULLOGH, T.H. (1989) *Thermal History of Sedimentary Basins—Methods and Case Studies*. Springer-Verlag, New York.

POWELL, T.G. (1978) An assessment of the hydrocarbon source rock potential of the Canadian Arctic Islands. *Geol. Surv. Can.* Paper 78-12, 82 pp.

PROCTOR, R.M., TAYLOR, G.C., and WADE, J.A. (1984) Oil and natural gas resources of Canada. *Geol. Surv. Can.* Paper 83-31.

RADKE, M. (1987) Organic geochemistry of aromatic hydrocarbons. In *Advances in Petroleum Geochemistry* (eds. J. Brooks and D. Welte) pp. 141–208. Academic Press.

SAKATA, S., SUZUKI, N., and KANEKO, N. (1988) A biomarker study of petroleum from the Neogene Tertiary sedimentary basins in Northeast Japan. *Geochem. J. 22*, 89–105.

SEIFERT, W.K. and MOLDOWAN, J.M. (1986) Use of biological markers in petroleum exploration. In *Biological Markers in the Sedimentary Record* (ed. R.B. Johns) pp. 261–290. Elsevier.

SNOWDON, L.R. and ROY, K.J. (1975) Regional organic metamorphism in the Mesozoic strata of the Sverdrup Basin. *Bull. Can. Petr. Geol. 23*, 131–148.

STUART-SMITH, J.H. and WENNEKERS, J.H.N. (1977) Geology and hydrocarbon discoveries of Canadian Arctic Islands. *Bull. Amer. Assoc. Petr. Geol. 61*, 1–27.

TRETTIN, H.P. and HILLS, L.V. (1967) Triassic "Tar Sands" of Melville Island, Canadian Arctic Archipelago. In *Seventh World Petroleum Congress Proc. 3, 773–787.*

15

Reversed-Phase High-Performance Liquid Chromatography of Metalloporphyrins

Christopher J. Boreham

Abstract. High-resolution reversed-phase HPLC of Ni and VO porphyrins is an effective technique for obtaining detailed structural information needed for many organic geochemical applications. Overall, VO porphyrins show a smaller elution dependence on structural variations than do Ni porphyrins. In the former, the C_{28}–C_{32} ETIO series overlaps the C_{30}–C_{32} DPEP series, while they are completely separated as their Ni complexes. Separation within a pseudohomologous series of VO porphyrins is largely governed by the length of the alkyl substituent, t_R H < t_R methyl < t_R ethyl, while the elution order is reversed for the last two within the Ni series.

In any analytical scheme where demetallation and remetallation are used for porphyrin identification, there is the real prospect of selective decomposition. The relative proportions of ETIO and DPEP structures vary slightly as a result of switching the coordinating species while structures with more than one ring, especially fused rings, are preferentially decomposed.

INTRODUCTION

The usefulness of petroporphyrins as maturity indicators (Corwin, 1960; Didyk et al., 1975; Mackenzie et al., 1980; Barwise and Park, 1983; Barwise and Roberts, 1984; Baker and Louda, 1986; Sundararaman et al., 1988), in oil-oil (Hohn et al., 1982; Shi et al., 1982; Baker and Louda, 1986) and oil-source (Taguchi, 1975) correlation studies and in determining organic inputs and depositional environments (Moldowan et al., 1986; Hayes et al., 1987; Takigiku, 1987), lies in an ability to obtain pure fractions and to separate complex mixtures of structurally distinct species. The detailed resolution required for these applications has largely been met by techniques developed during the past decade using high-performance liquid chromatography (HPLC) (Hajibrahim et al., 1978; Mackenzie et al., 1980; Sundararaman, 1985; Barwise et al., 1986; Chicarelli et al., 1986, Sundararaman et al., 1988). Many of the earlier separations involved the normal-phase HPLC on 5 μm and more recently 3 μm silica of free-base porphyrins (Hajibrahim et al., 1978; Mackenzie et al., 1980; Barwise et al., 1986; Chicarelli et al., 1986) produced by acid-induced demetallation of the metalloporphyrins; usually nickel (II) (Ni) and vanadyl (VO) porphyrins. The advantages of this approach are that high purity with respect to porphyrin content can be readily achieved and the total metalloporphyrin content is analyzed in a single HPLC separation. The main disadvantage is that some porphyrin decomposition invariably occurs and it has been questioned whether there is preferential destruction of various structural groups (Sundararaman, 1985); HPLC of the intact metalloporphyrins eliminates this uncertainty.

Recently, high-resolution reversed-phase HPLC (3 μm, C_{18} bonded phase) has been used to analyze VO porphyrins (Sundararaman, 1985; Sundararaman et al., 1988). Although absolute structural information was not reported, the potential to resolve pseudohomologous series and possible structural and positional isomers was demonstrated. Reversed-phase HPLC has also been the method of choice for analysis of Ni porphyrins, albeit mainly as a tool for the purification of single species for ^{1}H NMR characterization. Many unprecedented porphyrin structures have been revealed in this way. With the most recent report on the structure-elution characteristics of Ni porphyrins using high-resolution reversed-phase HPLC (Boreham and Fookes, 1989), methodology is now available to analyze sedimentary porphyrins with the two most common coordinated metals.

This paper further defines the structure-elution characteristics of VO porphyrins under reversed-phase HPLC. Coupled with the previous study on the Ni analogues, it examines whether or not structural integrity is maintained through the combined processes of demetallation and remetallation.

EXPERIMENTAL

The 35 standard porphyrins examined in the previous study on the structure-elution characteristics of Ni porphyrins (Boreham and Fookes, 1989) are shown in Figure 15.1. From this standard set, various VO porphyrins were prepared. The standard

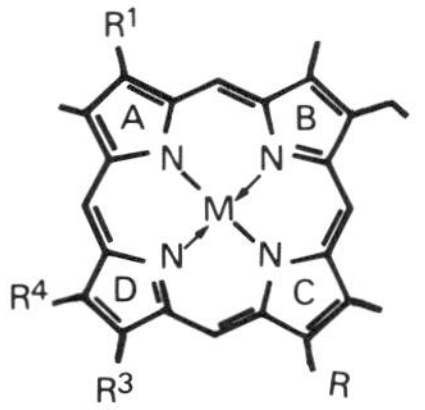

1a: $R^4 = CH_3$, $R^{1-3} = CH_2\,CH_3$ (C_{32})
1b: $R^3 = CH_3$, $R^{1,2,4} = CH_2\,CH_3$ (C_{32})
1c: $R^{1,4} = CH_3$, $R^{2,3} = CH_2\,CH_3$ (C_{31})
1d: $R^{2,4} = CH_3$, $R^{1,3} = CH_2\,CH_3$ (C_{31})
1e: $R^{1,2,4} = CH_3$, $R^3 = CH_2\,CH_3$ (C_{30})
1f: $R^1 = H$, $R^4 = CH_3$, $R^{2,3} = CH_2\,CH_3$ (C_{30})
1g: $R^2 = H$, $R^4 = CH_3$, $R^{1,3} = CH_2\,CH_3$ (C_{30})
1h: $R^1 = H$, $R^{2,4} = CH_3$, $R^3 = CH_2\,CH_3$ (C_{29})
1i: $R^2 = H$, $R^{1,4} = CH_3$, $R^3 = CH_2\,CH_3$ (C_{29})
1j: $R^{1,2} = H$, $R^4 = CH_3$, $R^3 = CH_2\,CH_3$ (C_{28})
1k: Unknown mixture of C_{27} Ni Etio

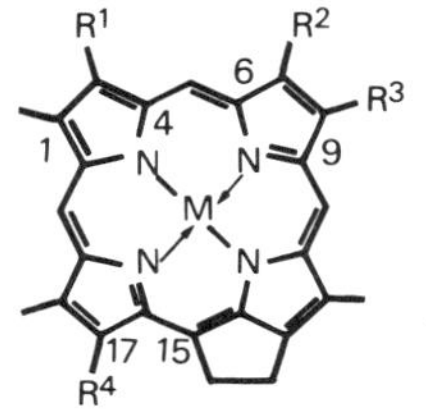

2a: $R^2 = CH_3$, $R^{1,3} = CH_2\,CH_3$, $R^4 = CH_2\,CH_2\,CH_3$ (C_{33})
2b: $R^2 = CH_3$, $R^{1,3,4} = CH_2\,CH_3$ (C_{32})
2c: $R^{1,2} = CH_3$, $R^{3,4} = CH_2\,CH_3$ (C_{31})
2d: $R^{2,3} = CH_3$, $R^{1,4} = CH_2\,CH_3$ (C_{31})
2e: $R^2 = H$, $R^{1,3,4} = CH_2\,CH_3$ (C_{31})
2f: $R^1 = H$, $R^2 = CH_3$, $R^{3,4} = CH_2\,CH_3$ (C_{30})
2g: $R^3 = H$, $R^2 = CH_3$, $R^{1,4} = CH_2\,CH_3$ (C_{30})
or $R^2 = H$, $R^3 = CH_3$, $R^{1,4} = CH_2\,CH_3$ (C_{30})
2h: $R^4 = H$, $R^2 = CH_3$, $R^{1,3} = CH_2\,CH_3$ (C_{30})

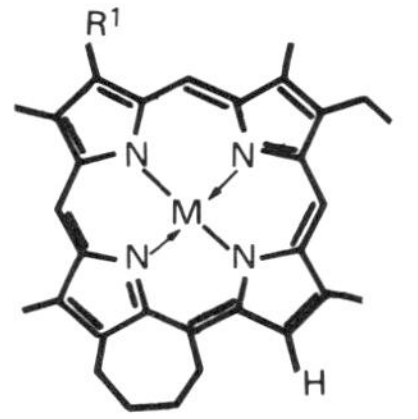

3a: $R^1 = CH_2\,CH_3$ (C_{32})
3b: $R^1 = CH_3$ (C_{31})
3c: $R^1 = H$ (C_{30})

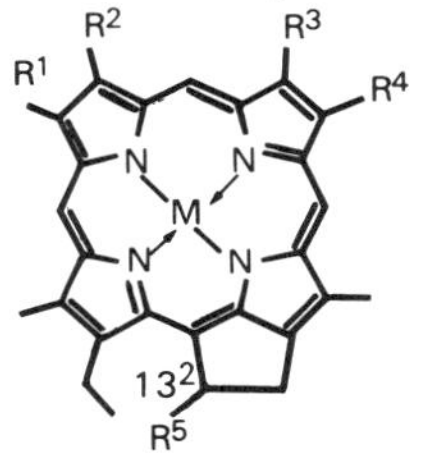

4a: $R^{1,2,3,4}$, 2 x CH_3, 2 x $CH_2\,CH_3$, $R^5 = CH_3$ (C_{33})
4b: $R^{1,2,3,4}$ 3 x CH_3, $CH_2\,CH_3$, $R^5 = CH_3$ (C_{32})
4c: $R^{1,2,3,4}$ H, 2 x CH_3, $CH_2\,CH_3$, $R^5 = CH_3$ (C_{31})
4d: $R^{1,3} = CH_3$, $R^{2,4} = CH_2\,CH_3$, $R^5 = OH$ (C_{32})

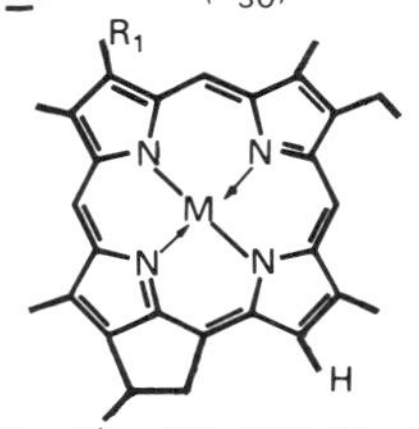

5a: $R^1 = CH_2\,CH_3$ (C_{31})
5b: $R^1 = H$ (C_{29})

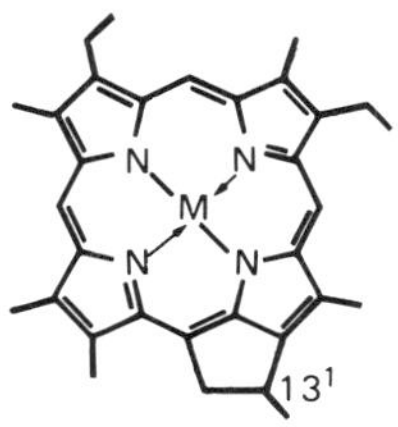

6: (C_{33})

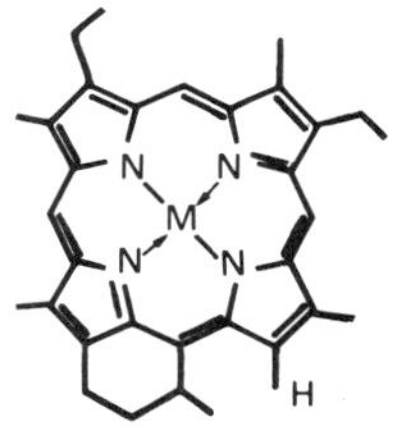

7: (C_{32})

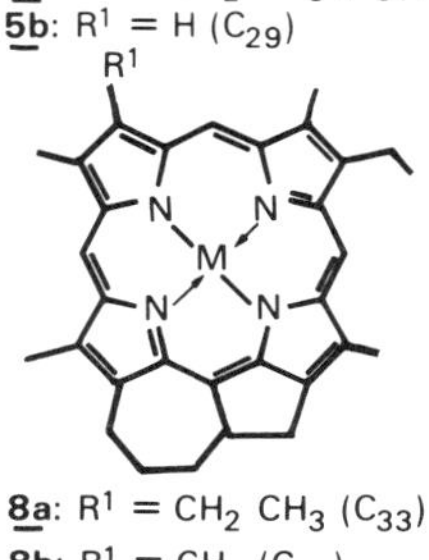

8a: $R^1 = CH_2\,CH_3$ (C_{33})
8b: $R^1 = CH_3$ (C_{32})

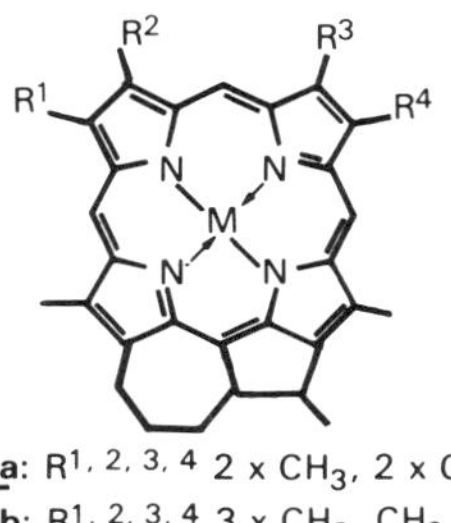

9a: $R^{1,2,3,4}$ 2 x CH_3, 2 x $CH_2\,CH_3$ (C_{34})
9b: $R^{1,2,3,4}$ 3 x CH_3, $CH_2\,CH_3$ (C_{33})

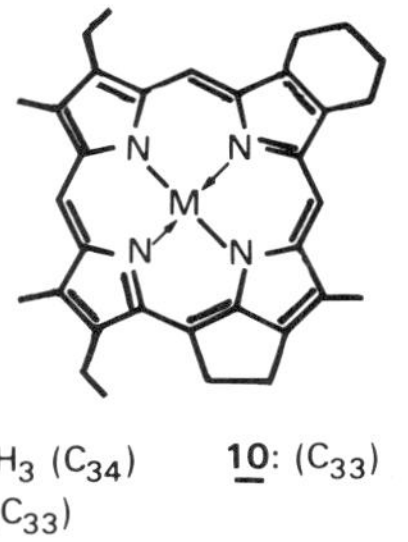

10: (C_{33})

16–3/270–1

Figure 15.1 Structures of standard metalloporphyrins. M = Ni for each species. M = VO for selected species, see text.

Ni porphyrin was demetallated in methanesulphonic acid following a modification (refluxing benzene for 45 min under argon) to the original procedure (Erdman, 1965). VO inserted in the resulting free-base porphyrin as follows (Erdman et al., 1956): excess vanadyl sulphate. $5H_20$ and the free-base porphyrin (50:1 w/w) were refluxed in glacial acetic acid/pyridine (2:1 v/v; 1 ml/μmole porphyrin) for 2 to 3 hours under argon. Subsequent column chromatographic clean-up (on silica using a petroleum ether—dichloromethane gradient, VO porphyrin eluting within the 30:70 mixture) gave the pure vanadyl porphyrin (purity determined by extinction coefficients and mass spectrometry).

HPLC analysis was performed using the same system as previously reported (Boreham and Fookes, 1989); Waters equipment employing three pumps, U6K injector and a 490 multiwavelength detector. The metalloporphyrin elution profile was visualized by monitoring the adsorbance at λ = 400 nm for Ni and λ = 406 nm or 415 nm for VO as well as simultaneously absorbance-ratio recording at 400 nm/550 nm and 415 nm/570 nm for Ni and VO, respectively. Separation was achieved on a precolumn (Guard-Pak μ-Bondapak C_{18}; Waters) and two columns (Nova Pak C_{18} 4 μ, each 150 × 3.9 mm i.d.; Waters) connected in series. Mobile phase conditions for Ni porphyrins were as described previously (Boreham and Fookes, 1989) and employed either isocratic flow in methanol (0.2% pyridine) when standard mixtures were analyzed to obtain absorbance-ratio values or a gradient profile from methanol (0.2% pyridine) to methanol/acetonitrile (85:15, 0.2% pyridine) at 1 ml/min for sedimentary porphyrin mixtures. VO porphyrins were analyzed either with the same columns as mentioned under isocratic flow at 1 ml/min in methanol/acetonitrile/water (45:45:10) (Sundararaman, 1985) or using a three-column combination (two 250 × 4.6 mm i.d., 150 × 4.6 mm i.d., 3 μ C_{18} Hypersil) under isocratic flow at 0.8 ml/min in methanol/acetonitrile/water (47.5:47.5:5) (Sundararaman et al., 1988). Column efficiencies were 32,000 plates/m for Ni and 50,000 plates/m for VO using identical columns.

Ni and VO porphyrins extracted from two horizons from the marine Toolebuc Formation, Australia were processed. The Ni porphyrins (200 μg) were demetallated with methanesulphonic acid, as mentioned, or sulphuric acid/trifluoroacetic acid/1,2-dithioethane [1:9 (1 ml): 10 μl] (Battersby et al., 1983). In the latter, reaction time was 5 minutes at room temperature and the reaction was done with and without addition of chloroform (100 μl, to assist in porphyrin solubility), and either in air or under a blanket of argon. Following neutralization (sat. $NaHCO_3$) and extraction into $CHCl_3$, the free-base porphyrins were purified on silica ($CHCl_3$ eluent). VO porphyrins (200 μg) were demetallated with methanesulphonic acid [0.5 ml in refluxing benzene (1 ml), 90 min] and the free-base porphyrins isolated as mentioned. Nickel was inserted into the free-base porphyrins using Ni $(acac)_2$ in refluxing benzene for 5 to 6 hours. (Verne-Misner et al., 1986; Boreham and Fookes, 1989).

RESULTS AND DISCUSSION

Structure-Elution Characteristics of Ni and VO Porphyrins

In comparing the elution characteristics of Ni (Boreham and Fookes, 1989), VO (Sundararaman, 1985), and free-base (Barwise et al., 1986; Chicarelli et al., 1986) porphyrins, it is evident that VO porphyrins behave differently from Ni or free-base porphyrins. With the last two, the ETIO members (C_{28}–C_{32}) were completely separated from the DPEP (C_{30}–C_{32}) series while for VO porphyrins there was marked overlap between the two structures. Furthermore, the elution order (retention times, t_R) within pseudohomologous series for Ni and free-base porphyrins was t_R H < t_R ethyl < t_R methyl while for VO porphyrins the t_R increased as the length of the alkyl substituent increased. For example, in the VO pseudohomologous series *1j*, *1i*, and *1g*; *1j*, *1h*, and *1f*; *1g*, *1d*, and *1a*; *1f*, *1c*, and *1a*; *2f*, *2c*, and *2b* the elution order is t_R H < t_R methyl < t_R ethyl (Figs. 15.2b and c). The overlap between the members of the ETIO and DPEP series, and the slightly shorter t_R for C_{32} ETIO, *1a* compared with C_{32} DPEP, *2b* is in accordance with previous observations (Sundararaman, 1985). Structural isomers are readily separable; the VO butanoporphyrins *3a* and *3b* are widely separated from their respective DPEP isomers, *2b* and *2c*. Thus structural isomers with fully alkylated substituents are retained longer than those with β-hydrogens (t_R *2b* > t_R *3a*; t_R *2c* > t_R *3b*; t_R *1e* > t_R *1f*, and *1g*). Separation of positional isomers is more difficult for VO species; coelution occurs for the VO C_{30} and C_{31} ETIO isomers (*1f* and *1g*; *1c* and *1d*, respectively; Fig. 15.2b) while these are partially resolved as their Ni analogues (Fig. 15.2d). The C_{29} ETIO isomers, *1h* and *1i* are not resolved as either metallo-species. The ring-D β-H VO C_{30} DPEP, *2h* is partially resolved from the ring-A isomer, *2f* (Fig. 15.2c) while these two positional isomers are completely resolved in their Ni form (Fig. 15.2d). For the VO porphyrins definite improvements in resolution are achieved using the longer 65 cm column combination. The C_{30} (t_R *1g* < t_R *1f*) and C_{31} (t_R *1d* < t_R *1c*) ETIO isomers are now resolved to the same extent and have the same elution order as were their Ni counterparts on the shorter column combination (Boreham and Fookes, 1989; and Fig. 15.2d). Furthermore, *8b* is resolved from and elutes after *2b*. However, some negative aspects do occur, namely, *1a* now coelutes with *2c*.

In the application of the HPLC technique to the analysis of sedimentary Ni porphyrins, absorbance-ratio recording was indispensible in the identification of different structural types (Boreham and Fookes, 1989). Systematic changes in the absorbance spectrum resulted in large variations in absorbance ratios. For example, the ratio of the absorbance at 400 nm to the absorbance at 550 nm for the Ni complexes within the structure *1*, *2*, and *3* ranged from 1.4–1.8, 3.3–3.6, and 9.2–9.5, respectively (Boreham and Fookes, 1989). These wavelengths corresponded to the Soret maxima (to within 1 nm) of the structure with the longest wavelength (i.e., *3* in the structures in Fig. 15.1) ratioed against the α-band maxima (to within

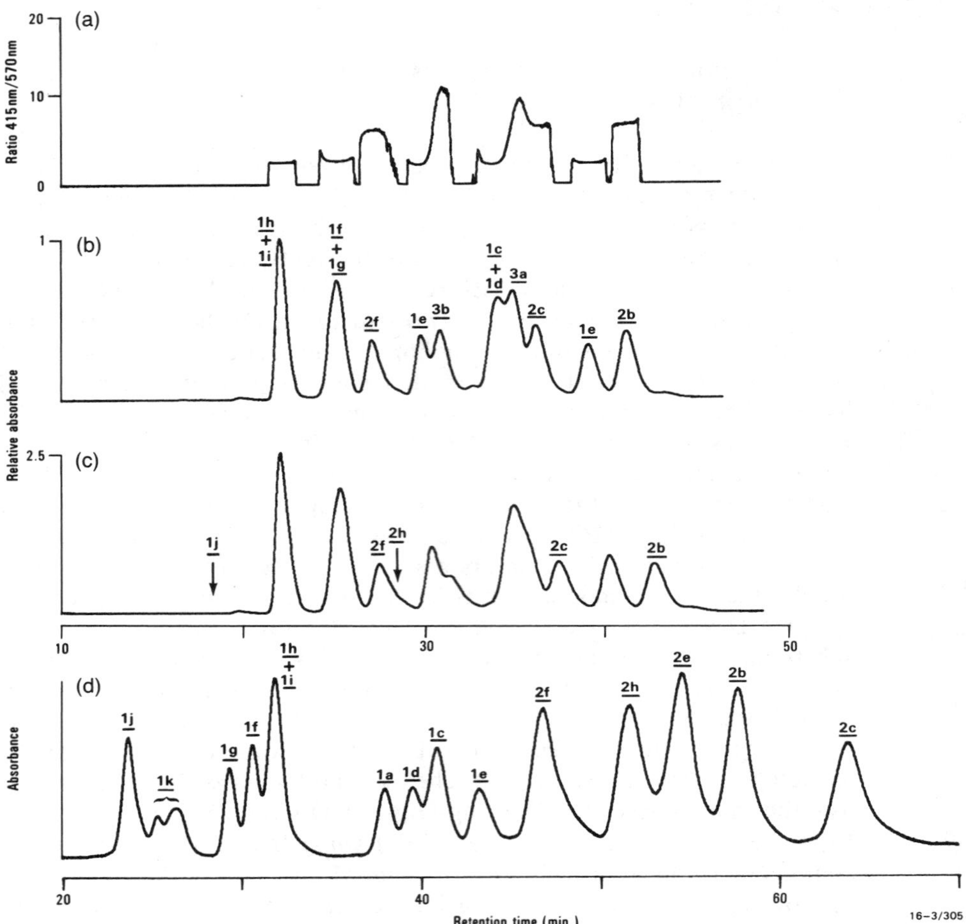

Figure 15.2 HPLC traces of synthetic mixtures of metalloporphyrins. (a) Absorbance-ratio trace, 415 nm/570 nm, corresponding to absorbance traces (b) and (c). (b) Absorbance trace at 415 nm for VO porphyrins. (c) Absorbance trace at 406 nm for VO porphyrins [same mixture as (b)]. (d) Absorbance trace at 400 nm for Ni porphyrins (isocratic, 1 ml/min., methanol/0.2% pyridine).

1 nm) of the structure with the shortest wavelength (*1*). Application of this criteria to the VO porphyrins resulted in a similar ratio trend. Hence, the absorbance-ratio of 415 nm/570 nm for VO *1a*, *1c–1j*; *2b–2c*, *2f*, *2h*; and *3a–3b* is in the range 2.5–3.4; 6.8–7.5; and 14.0–14.5, respectively (Fig. 15.2a). The absorbance-ratios measured on structures *5a*, *7*, *8a*, *9a*, and *10* are 8.0, 9.3, 9.8, 9.6, and 8.0, respectively.

The relative response for various structural types can be selectively modified depending on the monitoring wavelength. For example, monitoring at 415 nm, the

VO butanoporphyrins (*3a–b*, Soret maxima at 414 nm) are easily distinguished in both the absorbance and absorbance-ratio traces (Figs. 15.2b and a, respectively). However monitoring at 406 nm (Fig. 15.2c), the absorbance response for *3a–b* has decreased by half while that for all VO ETIO in mixture (*1*, Soret maxima at 407 ± 1 nm) and all VO DPEP in mixture (*2*, Soret maxima at 411 ± 1 nm) has increased approximately 2.5 and 1.3 fold, respectively, for the same absolute concentrations as in Fig. 15.2b. The overall effect is that there is a positive identification of the butano structures (*3a–b*) from the absorbance-ratio trace (Fig. 15.2a) but poor visualization via the absorbance trace (Fig. 15.2c). Therefore, depending on the structure of interest, its response can be effectively enhanced simply by monitoring at the appropriate wavelength.

Effects of Demetallation/Remetallation

Demetallation of metalloporphyrins is an effective way of rapidly upgrading the purity of a porphyrin mixture by exploiting the polarity differences on silica supports between the metalloporphyrins and their free-bases. Nonporphyrin material which often comprises the bulk of the crude metalloporphyrin fraction is effectively removed. Subsequent remetallation can then be used to convert an initial paramagnetic metalloporphyrin (e.g., VO) into monomeric diamagnetic species (e.g., Ni, Zn), more amenable to further characterization [e.g., ^{1}H NMR, HPLC (Ni complex)]. Recoverable yields, however, are of the order of 70 to 80 percent for each step (Sundararaman, 1985; Boreham, unpublished results). Although, in a qualitative sense, this loss may be tolerated, quantitatively it may be of significance if decomposition of certain structural types is selective. An improved understanding of the origins of petroporphyrins, and of their affinities to definite precursor chlorophylls or hemes, will enhance the use of petroporphyrins in depositional reconstructions. The relative and absolute concentrations of various structural types will be relevant factors, and therefore, the ramifications of demetallation/remetallation is an important consideration.

From the results of this study, changes to porphyrin compositions attributed individually to demetallation and remetallation cannot be completely determined. However, the effect of varying the demetallating conditions can be assessed to some extent, since the remetallating conditions are the same in all cases. Demetallation of Ni porphyrins with methanesulphonic acid results in very little change to the overall Ni porphyrins in sample 3910 although there is a slight decrease in the relative concentration of Ni C_{32} DPEP, *2b* (Fig. 15.3b) compared with the untreated Ni porphyrins (Fig. 15.3c). Mild selective oxidation/decomposition of this species may have occurred because the reaction conditions were not absolutely oxygen-free. Using the H_2SO_4/TFA/1,2-dithioethane method, artifact peaks can and do occur (e.g., Fig. 15.3a; this peak is eliminated when the reaction is performed under argon). There is also a major loss of Ni C_{31} DPEP (*2c*) which occurs under all the conditions examined for reasons which are presently unclear, although some loss through precipitation may have occurred since *2c* is one of the least soluble

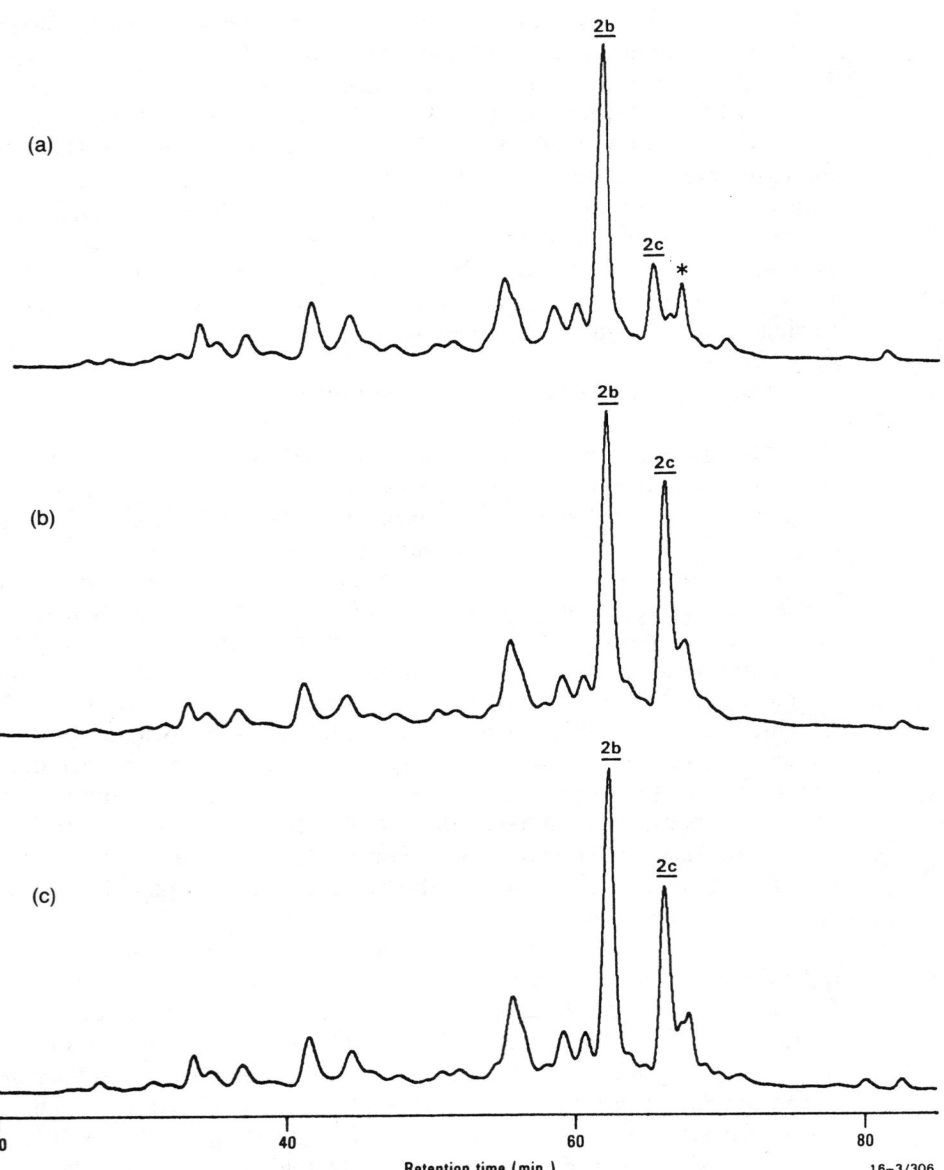

Figure 15.3 HPLC trace of total Ni porphyrins from Toolebuc Formation, sample 3910. (a) Ni porphyrins from trace (c) demetallated with H_2SO_4/TFA/1,2-dithioethane and remetallated with $Ni(acac)_2$ (* artifact). (b) Ni porphyrins of trace (c) demetallated with methanesulphonic acid and remetallated with $Ni(acac)_2$. (c) Untreated Ni porphyrins.

Ni porphyrins. In the cases where the untreated VO porphyrins (Fig. 15.4a) are reanalyzed as Ni porphyrins (Fig. 15.4b), some variations were seen across structural boundaries. In order to compare relative abundances, peak areas first must be adjusted using a normalizing factor equal to the absorbance of the Soret maxima for each structure divided by the absorbance at the monitoring wavelength (this assumes that extinction coefficients for each structural type are equal; for the Ni

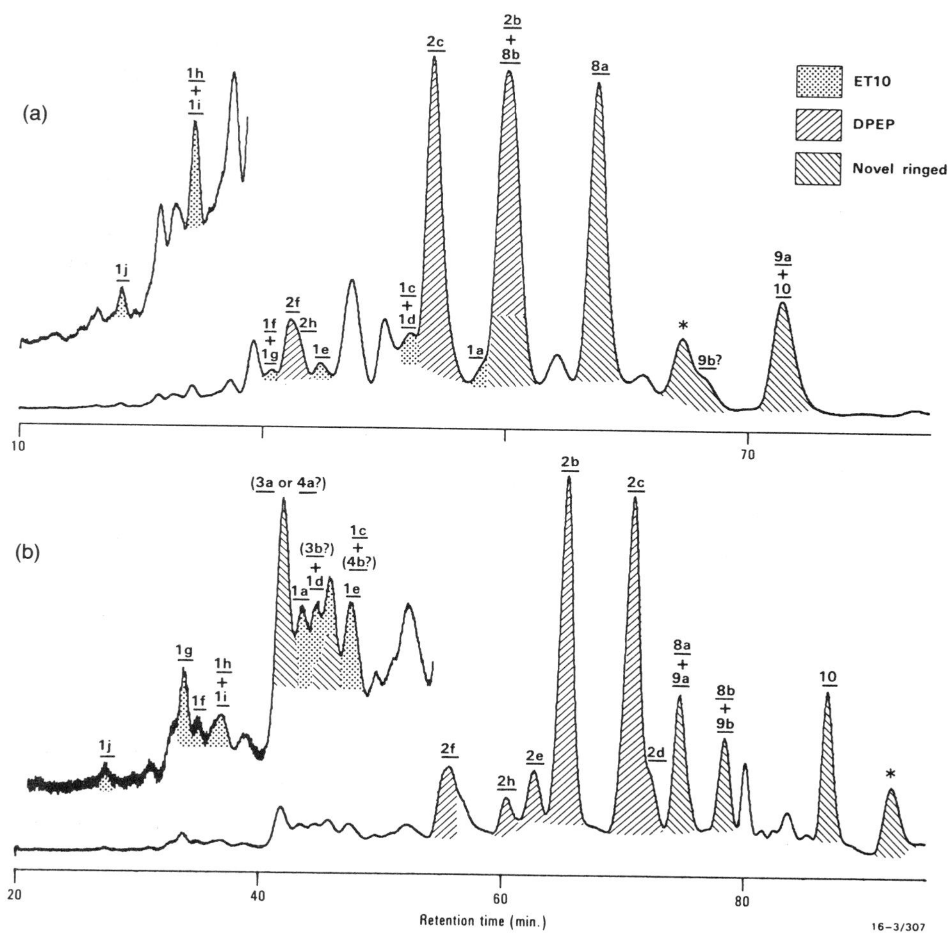

Figure 15.4 HPLC trace of total VO porphyrins from Toolebuc Formation, sample 4257. (a) Untreated VO porphyrins (absorbance at 415 nm). (b) VO porphyrins from trace (a) demetallated with methanesulphonic acid, Ni inserted, then analyzed as Ni porphyrins (absorbance at 400 nm). Peak * is tentatively assigned to the C_{32} ring-A methyl pseudohomolog of *10* on the basis of elution order and similar absorbance-ratio.

porphyrins Soret extinction coefficients for the structures *1–10* are within 10 percent (Popp et al., 1991), while the VO porphyrins are assumed to behave similarly). With this adjustment, relative concentrations of ETIO and DPEP porphyrins in Figure 15.4 remain little affected after demetallation followed by remetallation, while large changes occurred in the fused 7/5 ringed porphyrins *8* and *9* where there is a combined 50 percent loss.

CONCLUSIONS

Reversed-phase HPLC on 3 μ or 4 μm C_{18} bonded silica is an effective method for the detailed analysis of Ni and VO porphyrins. Structural types are readily separated. The addition of a five-membered ring of the DPEP type leads to further retention compared with the ETIO series. For the VO porphyrins this increased retention is much less than the Ni counterparts resulting in overlap of the C_{28}–C_{32} ETIO series with the C_{30}–C_{32} DPEP series for the former while complete separation occurs in the latter. Within a pseudohomologous series the elution order is t_R H < t_R methyl < t_R ethyl and t_R H < t_R ethyl < t_R methyl for VO and Ni, respectively. Structural isomers with β-H substituents show shorter retention times compared with their fully alkylated isomers for both metalloporphyrins. Using the same columns (4 μ), separation of positional isomers is more readily achieved as the Ni complex.

Demetallation has minimal effect on the relative proportions of ETIO and DPEP structures when methanesulphonic acid is used (Ni and VO). However, when a sulphuric acid/trifluoroacetic acid/1,2-dithioethane mixture (Ni) is used, an increased decomposition of DPEP structures can occur. Structures with more than one ring, especially strained fused rings, are the most susceptable to selective decomposition.

Acknowledgments

Mr. Algis Juodvalkis is thanked for his assistance with the HPLC analysis. C. J. B. publishes with the permission of the Director, Bureau of Mineral Resources, Geology and Geophysics.

REFERENCES

Baker, E.W. and Louda, J.W. (1986) Porphyrins in the geological record. In *Biological Markers in the Sedimentary Record* (ed. R.B. Johns) Elsevier, New York. pp. 125–225.

Barwise, A.J.G. and Park, P.J.D. (1983) Petroporphyrin fingerprinting as a geochemical marker. In *Advances in Organic Geochemistry* (eds. M. Bjorøy et al.) John Wiley, London. pp. 668–674.

Barwise, A.J.G. and Roberts, I. (1984) Diagenetic and catagenetic pathways for porphyrins in sediments. In *Advances in Organic Geochemistry, 1983* (eds. P.A. Schenck, J.W. Leeuw, and G.W.M. Lijmbach). *Org. Geochem. 6*, 167–176.

Barwise, A.J.G., Evershed, R.P., Wolfe, G.A., Eglinton, G., and Maxwell, J.R. (1986) High-performance liquid chromatographic analysis of free-base porphyrins I. An improved method. *J. Chromatogr. 368*, 1–9.

Battersby, A.R., Jones, K., and Snow, R.J. (1983) New methods of demetallating tetrapyrolic metallo-macrocycles. *Angew. Chem., Int. Ed. 22*, 734–736.

Boreham, C.J. and Fookes, C.J.R. (1989) Separation of nickel (II) alkyl porphyrins by reversed-phase high-performance liquid chromatography-methodology and application. *J. Chromatogr. 467*, 195–208.

Chicarelli, M.I., Wolff, G.A., and Maxwell, J.R. (1986) High-performance liquid chromatographic analysis of free-base porphyrins II. Structure effects and retention behavior. *J. Chromatogr. 368*, 11–19.

Corwin, A.H. (1960) Petroporphyrins. *Proc. 5th World Pet. Congr.* New York, N.Y. *paper V-10*, pp. 119–129.

Didyk, B.M., Alturki, Y.I.A., Pillinger, C.T., and Eglinton, G. (1975) Petroporphyrins as indicators of geothermal maturation. *Nature 256*, 563–565.

Erdman, J.G., Ramsey, V.G., Kalenda, N.W., and Hanson, W.E. (1956) Synthesis and properties of porphyrin vanadium complexes. *J. Amer. Chem. Soc., 78*, 5844–5847.

Erdman, J.G. (1965) *U.S. Patent 3 190*, 829.

Hayes, J.M., Takigiku, R., Ocampo, R., Callot, H.J., and Albrecht, P. (1987) Isotopic compositions and probable origins of organic molecules in the Eiocene Messel shale. *Nature 329*, 48–51.

Hajibrahim, S.K., Tibbetts, P.J.C., Watts, C.D., Watts, J.R., Eglinton, G., Colin, H., and Guiochon, G. (1978) Analysis of carotenoid and porphyrin pigments of geochemical interest by high-performance liquid chromatography. *Anal. Chem. 50*, 549–553.

Hohn, M.E., Hajibrahim, S.K., and Eglinton, G. (1982) High-pressure liquid chromatography of petroporphyrins: Evaluation as a geochemical fingerprinting method by principal components analysis. *Chem. Geol., 37*, 229–237.

Mackenzie, A.S., Quirke, J.M.E., and Maxwell, J.R. (1980) Molecular parameters of maturation in the Toarcian shales, Paris Basin, France -II. Evolution of metalloporphyrins. In *Advances in Organic Geochemistry, 1979* (eds. J.R. Maxwell and A.C. Douglas), Pergamon Press, Oxford. pp. 239–248.

Moldowan, J.M., Sundararaman, P., and Schoell, M. (1986) Sensitivity of biomarker properties to depositional environment and/or source input in the Lower Toarcian of SW-Germany. In *Advances in Organic Geochemistry 1985*, Part II: *Molecular and General Organic Geochemistry* (eds. D. Leythaeuser and J. Rullkötter). *Org. Geochem. 10*, 915–926.

Popp, B.N., Hayes, J.M., and Boreham, C.J. (1991) Microscale determination of the spectral characteristics and carbon-isotopic compositions of geoporphyrins. *Chem. Geol.*, submitted.

Shi, J., Mackenzie, A.S., Alexander, R., Eglinton, G., Gower, A.P., Wolff, G.A., and Maxwell, J.R. (1982) A biological marker investigation of petroleum and shales from the Shengli oil field, the Peoples Republic of China. *Chem. Geol. 35*, 1–35.

Sundararaman, P. (1985) High-performance liquid chromatography of vanadyl prophyrins. *Anal. Chem. 57*, 2204–2206.

Sundararaman, P., Biggs, N.R., Reynolds, J.C., and Petzer, J.C. (1988) Vanadyl porphyrins, indicators of kerogen breakdown and generation of petroleum. *Geochim. Cosmochim. Acta 52*, 2337–2341.

Taguchi, K. (1975) Geochemical relationships between Japanese Tertiary oils and their source rocks. *Proc. 9th World Pet. Congress 2*, Applied Science, London. pp. 193–194.

Takigiku, R. (1987) Isotopic and molecular indicators of origins of organic matter in sediments, *Ph.D. Thesis*, Indiana Uni., Bloomington, USA.

Verne-Misner, J., Ocampo, R., Callot, H.J., and Albrecht, P. (1986) Molecular fossils of chlorophyll c of the 17-nor-DPEP series. Structure determination, synthesis, geochemical significance. *Tetrahedron Lett. 27*, 5257–5260.

16

Comparison of Natural and Laboratory Simulated Maturation of Vanadylporphyrins

Padmanabhan Sundararaman

Abstract. There is a close similarity between vanadylporphyrin distributions of bitumens extracted from core and cuttings from the Monterey Formation of increasing maturity and extracts of an immature Monterey rock pyrolyzed at increasing temperatures. Both in the natural maturation sequence and laboratory pyrolysis experiments, the Porphyrin Maturity Parameter (PMP) increases gradually during early stages of maturation. This is followed by a rapid increase during the main phase of oil generation.

INTRODUCTION

Vanadylporphyrins occur as an extremely complex mixture, consisting of homologous series of structural isomers. The two major structural isomers are generically referred to as DPEP and ETIO, based on the presence of one or no exocyclic ring in the porphyrin structure (Fig. 16.1). The ratio of DPEP/ETIO porphyrins has been observed to decrease with an increase in maturity (Mackenzie et al., 1980; Louda and Baker, 1981; Barwise and Park, 1983; Barwise and Roberts, 1984; Barwise, 1987). The change in the DPEP/ETIO ratio during maturation is attrib-

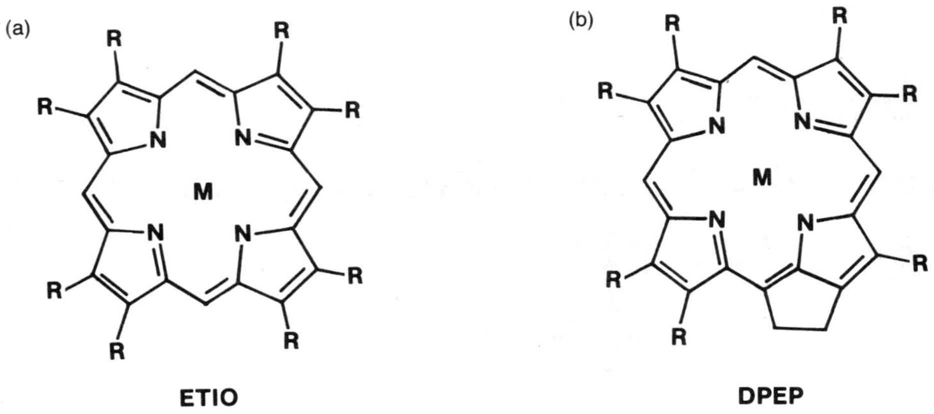

(a) ETIO (b) DPEP

Figure 16.1 Generalized structures of DPEP and ETIO porphyrins (R = alkyl substituents).

utable to a decrease in the relative concentration of C_{31} and C_{32} DPEP and a concomitant increase in the concentration of C_{29} and C_{28} ETIO porphyrins (Sundararaman et al., 1988). This observation has led to the quantification of maturation using a Porphyrin Maturity Parameter PMP = $C_{28}E/(C_{28}E + C_{32}D)$, where $C_{28}E$ and $C_{32}D$ are ETIO porphyrins with 28 carbons and DPEP porphyrins with 32 carbons, respectively. In this chapter, I report the results of laboratory pyrolysis maturation experiments which closely parallel the changes observed in porphyrin distribution with burial depth.

EXPERIMENTAL

The complete details of vanadylporphyrin isolation, purification, and HPLC analysis are described in previous publications (Sundararaman et al., 1988; Sundararaman, 1985).

Briefly, the bitumens and pyrolysates were separated on an alumina column into saturate/aromatic and polar fractions. The polar fractions which contain the vanadylporphyrins were further purified on methanesulfonic acid bonded silica using methylene chloride as eluent. The purified vanadylporphyrins were chromatographed using 3-μm Hypersil C_{18} columns (two 4.6 mm × 25 cm and one 4.6 mm × 15 cm in series). The mobil phase was 47.5% CH_3OH, 47.5% CH_3CN, and 5% H_2O. The flow rate was 0.8 ml/min. The eluent was monitored with a diode array detector set at 406 nm.

SAMPLES

Hydrous pyrolysis experiments were performed on organic-rich core material (1651.7–1653.5 m; 5419–5425 ft) from the Siliceous member of the Monterey Formation from a well located in the onshore Santa Maria Basin (Peters et al., 1990). For the natural maturation sequence, bitumens extracted from the Monterey Formation of a well from offshore California were analyzed. The type of organic matter in both sets of samples is marine Type II amorphous and is uniform throughout the sampled interval.

HYDROUS PYROLYSIS

The hydrous pyrolysis technique used was identical to that described by Lewan et al., (1979). The hydrous pyrolysis experiments were carried out in one-liter stainless steel reactors. The reactors were loaded with 400 g of homogenized rock and 260 mL of water. Sealed reactors were pressurized with helium and heated to selected temperatures in the range of 260–340°C for 72 hours. Products were separated after cooling the vessels (Peters et al., 1990).

RESULTS AND DISCUSSION

Vanadylporphyrin distributions of bitumens extracted from Monterey source rocks of increasing maturity from a single well show systematic changes (Fig. 16.2a). The most significant changes are observed in peaks eluting between 30 and 90 minutes. At early stages of maturation (shallower depths, 6990′–7880′) the concentrations of $C_{32}D$ and $C_{31}D$ decrease systematically with increase in depth, relative to concentrations of $C_{28}E$ and $C_{29}E$. Changes are also observed in the peaks eluting between $C_{29}E$ and $C_{31}D$ (between 45 and 60 min), however, due to considerable overlap of peaks in this region of the chromatograms, no systematic trend could be observed. During advanced stages of maturation (8194′ and 8610′) the relative concentrations of $C_{28}E$ and $C_{29}E$ decrease and those of earlier eluting peaks (between 30 and 38 min) increase. These earlier eluting peaks are low molecular weight (carbon number) ETIO porphyrins.

The vanadylporphryin distributions of artificially matured samples (Fig. 16.2b) parallel the maturity trend observed in the well data (Fig. 16.2a). With increase in pyrolysis temperature, the relative concentrations of $C_{31}D$ and $C_{32}D$ decrease and those of $C_{28}E$ and $C_{29}E$ increase. At higher pyrolysis temperatures (330°C and 340°C), the relative concentrations of $C_{28}E$ and $C_{29}E$ decrease and those of earlier eluting peaks (between 30 and 36 min) increase. Some minor differences between the natural maturation and the laboratory simulated maturation are worth noting. The relative concentration of $C_{32}D$ is higher than that of $C_{31}D$ during early stages of natural maturation, whereas in the laboratory pyrolysis the relative con-

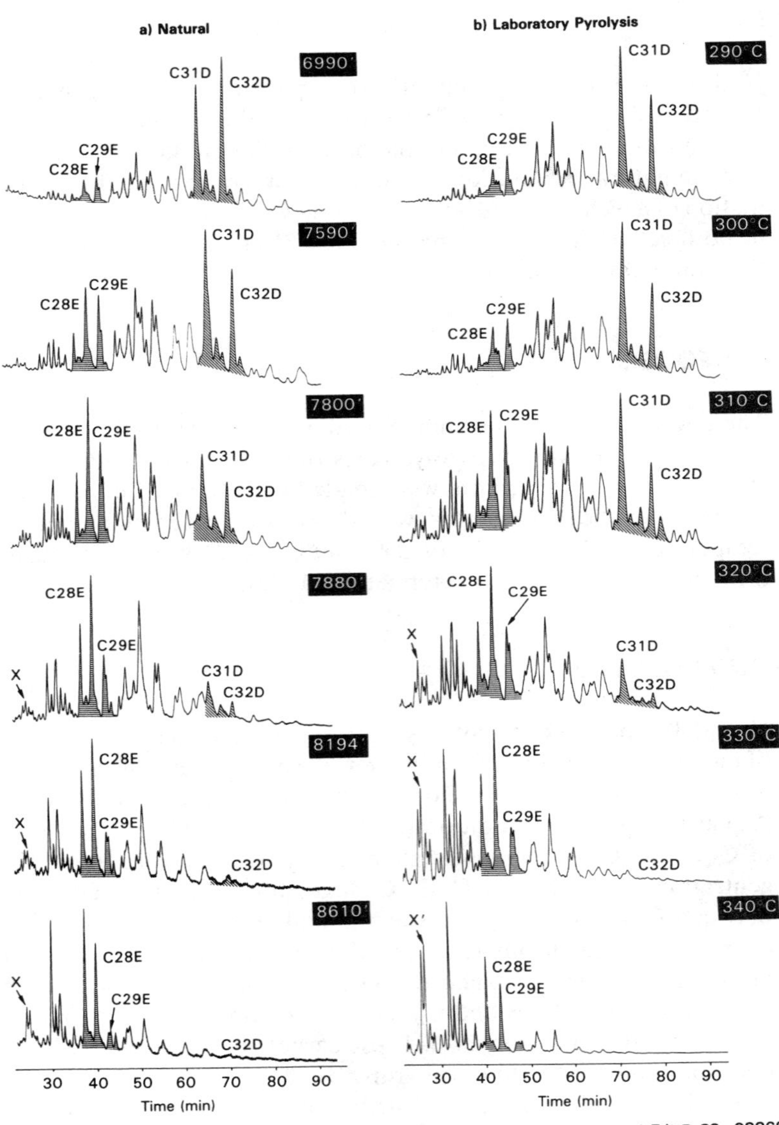

Figure 16.2 HPLC of vanadylporphyrins of bitumens extracted from (a) an immature source rock pyrolyzed at different temperatures for 72 hours and (b) source rocks of increasing maturity from a single well.

Table 16.1 PMP DATA FOR NATURAL AND LABORATORY PYROLYSIS SAMPLES

	Natural Maturation				Artificial Maturation		
	PMP		PI	HI		PMP	HI
Depth (ft)	$C_{28}E/(C_{28}E + C_{32}D)$	T_{max}	$S_1(S_1 + S_2)$	S_2/TOC	Pyrolysis Temperature	$C_{28}E/(C_{28}E + C_{32}D)$	S_2/TOC
6000	0.18	422	0.07	—	260	0.11	496
6600	0.18	422	0.04	—	270	0.08	484
6617	0.16	424	0.11	447	280	0.15	483
6810	0.10	421	0.11	—	290	0.16	353
6990	0.13	—	—	—	300	0.30	263
7410	0.39	—	—	—	310	0.59	229
7590	0.45	—	—	—	320	0.74	158
7800	0.70	429	0.10	—	330	0.99	112
7880	0.90	434	0.23	331	340	0.99	35
8010	0.89	433	0.18	—			
8197	0.94	433	0.50	240			
8200	0.92	432	0.19	—			
8202	0.94	432	0.25	—			
8610	1.00	437	0.27	—			
9000	0.83	436	0.33	161			
9000	1.00	435	0.31	—			
9210	1.00	439	0.28	—			
9390	1.00	438	0.30	—			
9390	1.00	437	0.27	—			
9600	1.00	438	0.28	—			
9600	1.00	440	0.28	—			
11057	1.00	441	0.35	—			

centrations are reversed. At advanced stages of maturation (330°C and 340°C) the relative concentration of the peak eluting around 20 minutes (marked X in Fig. 16.2) is much higher in the laboratory pyrolysis experiments (Fig. 16.2b) compared to natural maturation (Fig. 16.2a).

The Porphyrin Maturation Parameter PMP = $C_{28}E/(C_{28}E + C_{32}D)$ increases gradually during early stages of maturation both in natural samples and laboratory pyrolysis samples (Table 16.1 and Fig. 16.3). This is followed by a rapid increase, corresponding to the main stage of oil generation in the natural samples based on other parameters such as Rock-Eval production index, T_{max}, and vitrinite reflectance data. In the laboratory pyrolysis experiments, the rapid change corresponds to a significant breakdown of kerogen and the generation of extractable material (Sundararaman et al., 1988).

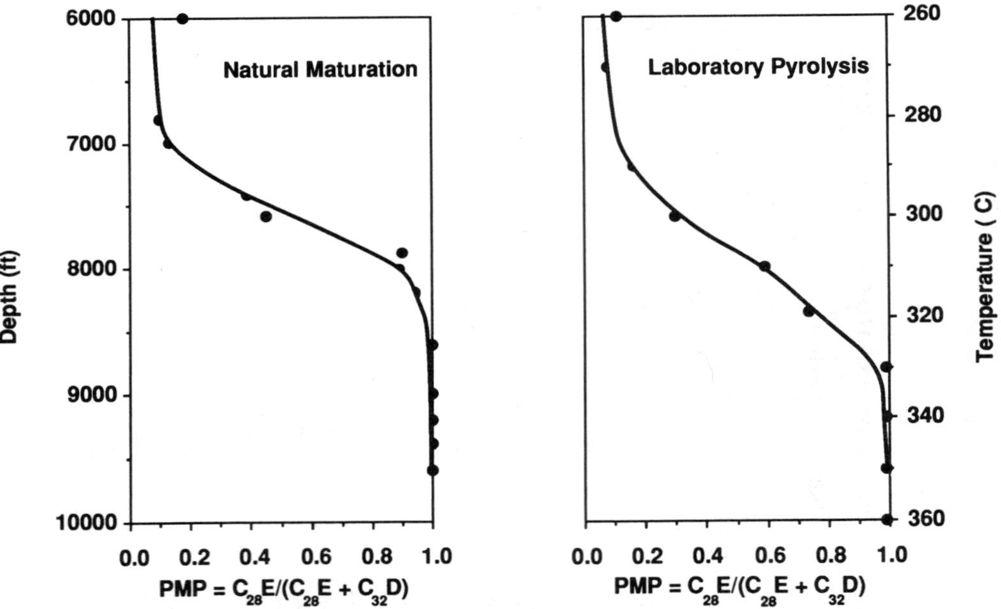

Figure 16.3 Porphyrin Maturity Parameter (PMP) plotted against (a) depth for a series of source rocks of increasing maturity and (b) pyrolysis temperature of an immature source rock pyrolyzed at different temperatures.

CONCLUSIONS

Maturation, as a function of burial and by hydrous pyrolysis, results in a decrease in the relative concentrations of $C_{32}D$ and $C_{31}D$ porphyrins and a concomitant increase in the $C_{28}E$ and $C_{29}E$ porphyrins. At more advanced stages of maturation, the relative concentration of $C_{28}E$ and $C_{29}E$ porphyrins decreases and that of lower molecular weight porphyrins increases. Changes in vanadylporphyrin distributions of samples from the subsurface and hydrous pyrolysis parallel one another as a function of maturity. These changes have been quantified through the use of PMP.

Acknowledgments

I thank J. J. Jaime for porphyrin analyses, R. F. Dias and K. E. Peters for the pyrolysis experiments, and Chevron Oil Field Research Company for permission to publish this work.

REFERENCES

Barwise, A.J.G. (1987) Mechanism involved in altering Deoxophylloerythroetioporphyrin-Etioporphyrin ratios in sediments and oils. In *Metal Complexes in Fossil Fuels*. Geochemistry, Characterization and Processing; A.C.S. Symposium Series 344 (eds. R.H. Filby and J.F. Branthaver), pp. 100–109.

Barwise, A.J.G. (1987) Mechanism involved in altering Deoxophylloerythroetioporphyrin-Etioporphyrin ratios in sediments and oils. In *Metal Complexes in Fossil Fuels*. Geochemistry, Characterization and Processing; A.C.S. Symposium Series 344 (eds. R.H. Filby and J.F. Branthaver), pp. 100–109.

Barwise, A.J.G. and Roberts, I. (1984) Diagenetic and catagenic pathways for porphyrins in sediments. In *Advances in Organic Geochemistry 1983* (eds. P.S. Schenk, J.W. de Leeuw, and G.W.M. Lijmbach) *Organic Geochemistry 6*. Pergamon Press, Oxford. pp. 167–176.

Lewan, M.D., Winters, J.C., and McDonald, J.H. (1979) Generation of oil-like pyrolyzates from organic-rich shales. *Science 203*, pp. 897–899.

Louda, J.W. and Baker, E.W. (1981) Geochemistry of tetrapyrrole, carotenoid and perylene pigments in sediments from the San Miguel Gap (Site 467) and Baja California borderland (Site 471), Deep Sea Drilling Project, Leg 63. Init. Repts. DSDP 63, 785–818.

Mackenzie, A.S., Quirke, J.M.E., and Maxwell, J.R. (1980) Molecular parameters of maturation in the Toarcian Shales, Paris Basin, France-II. Evolution of metalloporphyrins. In *Advances in Organic Geochemistry 1979* (eds. J.R. Maxwell and A.G. Douglas). Pergamon Press, Oxford. pp. 239–248.

Peters, K.E., Moldowan, J.M., and Sundararaman, P. (1990) Effects of hydrous pyrolysis on biomarker thermal maturity parameters: Monterey Phosphatic and Siliceous members. *Org. Geochem., 15*, 249–265.

Sundararaman, P., Biggs, W.R., Reynold, J.G., and Fetzer, J. (1988) Vanadylporphyrins, indicators of kerogen breakdown and generation of petroleum. *Geochim. Cosmochim. Acta, 52*, 2337–2341.

Sundararaman, P. (1985) High-performance liquid chromatography of vanadylporphyrins. *Anal. Chem. 57*, 2204–2206.

17

In Vitro Biodegradation of Steranes and Terpanes: A Clue to Understanding Geological Situations[1]

Patricia Chosson, Jacques Connan, Daniel Dessort, and Colette Lanau

Abstract. Seventy-three pure strains of bacteria, selected from international collections among Pseudomonadaceae (33) and Actinomycetaceae (40), were screened for their capacity to degrade steranes and terpanes contained in total alkanes isolated from various crudes (West Rozel oil, Utah; Vic-Bilh, Aquitaine Basin, France). Reproducible in vitro, the biodegradation of steranes was observed with seven strains, the most noticeable effects occurred with gram-positive strains belonging to *Nocardia* and *Arthrobacter* genera.

Regular steranes and their methylated homologues are biodegraded to various degrees. The rate of biodegradation of regular steranes was found to vary in the sequence $\alpha\alpha\alpha$R (5α(H),14α(H),17α(H)20*R*-sterane) > $\alpha\beta\beta R$ (5α(H),14β(H),17β(H)20*R*-sterane) > $\alpha\beta\beta S$ > $\alpha\alpha\alpha S$ with $C_{27} > C_{28} > C_{29}$. 17α(H),21β(H)-hopanes are also degraded but without marked changes in terpane patterns. Slight variations in terpane ratios, however, indicate that the rate of biodegradation of $\alpha\beta$-hopanes is 22*R* > 22*S* and $C_{35} > C_{30}$. Results on sterane biodegradation were obtained using

[1]Given as oral presentation at the 197th American Chemical Society Meeting in Dallas, April 1989. This chapter is dedicated to the memory of W.K. Seifert who did pioneering work to promote biomarker chemistry in oil exploration.

branched and cyclic alkane fractions of natural samples as well as pure sterane mixtures.

In vitro biodegradation of steranes and terpanes reproduces in-reservoir biodegradation under geological conditions. No demethylated hopanes were produced in our laboratory experiments. Literature data, complemented by the results of this laboratory study, allow us to propose that demethylated hopanes are not formed through in-reservoir biodegradation of αβ-hopanes but are preexisting structures, concentrated in severely biodegraded oils by selective removal of regular hopanes. Biodegradation of steranes and hopanes enhances other minor families of polycyclic terpanes, the occurrence of which varies according to the organic facies of the source rocks. These families are, consequently, highly diagnostic of specific environmental conditions and may be used to reconstruct paleoenvironments of the source rocks of severely biodegraded crudes.

INTRODUCTION

The biodegradation of steranes and terpanes in crude oils in the reservoir has been observed and described by numerous authors (Reed, 1977; Seifert and Moldowan, 1979; Rullkötter and Wendisch, 1982; McKirdy et al., 1983; Volkman et al., 1983; Connan, 1984; Sandstrom and Philp, 1984; Seifert et al., 1984; Brooks et al., 1988; Jiang Zhusheng et al., 1988; Zhang Dajing et al., 1988).

Several researchers have attempted to reproduce these field observations by in vitro biodegradation studies. Crude oils investigated were inoculated with microorganisms either in pure or in mixed cultures. Most of these laboratory studies failed (Rubinstein et al., 1977; Connan et al., 1980; Teschner and Wehner, 1985).

A unique set of experiments, performed by Goodwin et al. (1983), provided evidence that in vitro biodegradation of steranes and terpanes was possible; however, the reported results were obtained only once and were not further investigated. More recently, Brakstad and Grahl-Nielsen (1988) confirmed the in vitro biodegradability of steranes by studying the weathering of North Sea crudes (whole oil + sea water, 0.5 to 12.5 months in tanks open to the air).

A better understanding of the biodegradation of these polycyclic alkanes appears very important in petroleum geochemistry, because these biomarker classes are currently used in oil-to-oil and oil-to-source rock correlation as well as for maturity assessment (Mackenzie, 1984). Therefore, our goal was to degrade steranes and terpanes in the laboratory under controlled and reproducible conditions in order to answer several questions: What kind of microorganisms degrade steranes and triterpanes? What is the sequence of the removal of various classes of steranes? To what extent is it possible to metabolize steranes and terpanes in the laboratory? Goodwin et al. (1983) have already demonstrated that regular steranes are more easily degradable than rearranged steranes, as stated earlier by Seifert and Moldowan (1979) and that C_{27} steranes are removed at a higher rate than C_{29} steranes. Among αβ-hopanes, the rate of biodegradation is $22R > 22S$ and $C_{35} > C_{34} >$

$C_{33} > C_{32} > C_{31} > C_{30} > C_{29}$. Tricyclic terpanes were not found to be biodegraded under laboratory conditions.

In the laboratory experiments, we used branched and cyclic alkanes from various oils (West Rozel crude, Great Salt Lake, Utah; Vic-Bilh oil, Aquitaine Basin, S.W. France) as well as pure compounds (5α(H),14α(H),17α(H)20*R*-cholestane; 5α(H),14α(H),17α(H)20*S*-cholestane; 5α(H),14β(H),17β(H)20*R*-cholestane). Seventy-three bacteria belonging to Pseudomonadaceae (33) and Actinomycetaceae (40) were classified according to their capacity to degrade natural sterane and terpane mixtures. Biodegradation of steranes and terpanes was achieved with *Nocardia*, *Arthrobacter* and *Mycobacterium*. Kinetics of sterane consumption are discussed by referring to geological settings.

MATERIALS AND METHODS

Branched and Cyclic Alkanes from Crude Oil

Basic geochemical data of West Rozel and Vic-Bilh oils are listed in Table 17.1. The Rozel oil comes from the West Rozel No. 3 well, drilled in the heavy-oil deposit, Great Salt Lake, Utah (Bortz, 1987). The Rozel Point oil, cropping out in the neighborhood of the West Rozel field, was extensively studied for organic sulfur compounds (Schmid, 1986; Sinninghe Damste et al., 1987; Sinninghe Damste and De Leeuw, 1987). The West Rozel oil, chosen here for its unusual richness in regular steranes (Fig. 17.1), allows an easy GC recognition of biodegradation features during the screening experiments. The West Rozel oil analyzed depicts an *n*-alkane pattern with a striking even predominance (n-C_{16}, n-C_{18}, . . . , n-C_{26}). According to Bortz (1987), this oil may not be fully representative of the West Rozel field because of the biodegradation. The West Rozel oil is sulfur-rich (12%, Table 17.1). The Vic-Bilh oil (Fig. 17.1) is typical of the sulfur-rich crude oils (Table 17.1) from the Aquitaine Basin that originate from Barremo-Jurassic source rocks. According to its geochemical properties, the Vic-Bilh oil is comparable to the Pécorade oil previously used in our in vitro experiments with *Pseudomonas oleovorans* (Connan et al., 1980).

Branched and cyclic alkanes were prepared by applying a 5 Å molecular sieve treatment to remove *n*-alkanes. Branched and cyclic alkanes of Vic-Bilh and West Rozel oils were chosen for the great difference in relative concentrations of steranes and terpanes (TT/ST = 4.9 and 0.14, respectively, Table 17.1).

Microorganisms

Microorganisms were selected from international collections, except for *Nocardia species* SEBR 16 that we isolated earlier from a soil sample. The following genera were examined: *Nocardia* (25), *Mycobacterium* (11), *Corynebacterium* (1), *Arthrobacter* (2), *Protoaminobacter* (1) and *Pseudomonas* (33).

Table 17.1 BASIC GEOCHEMICAL DATA OF CRUDE OILS USED IN LABORATORY EXPERIMENTS

						Gross Composition C_{15+} Fraction					C_{15+} Alkanes			
Well	Depth	Age of Reservoir	API Gravity (15°C)	δ^{13} C(‰/PDB)	δ^{34} S (‰/CDT)	Alkanes %	"Aromatics" %	Resins %	Asphaltenes %	Sulphur %	Prist/n − C_{17}	Phyt/n − C_{18}	Prist/Phyt	TT/ST
West Rozel N° 3	2354 feet	Upper Pliocene	6	−25.8	−10.8	2.0	33.2	51.5	13.3	12	1.39	2.63	0.24	0.14
Vic Bilh N° 2	2201 2234 Meters	Barremian	26.7	−23.4	−1.0	30.2	44.5	22.4	2.9	3.4	0.11	0.18	0.69	4.9

Abbreviations:

PR/n-C_{17} pristane/n-C_{17} alkane
PH/n-C_{18} phytane/n-C_{18} alkane
PR/PH pristane/phytane
TT/ST triterpane to sterane ratio: ratio of m/z 191 integral (C_{27}–C_{35}) to m/z 217 integral (C_{27}–C_{29})

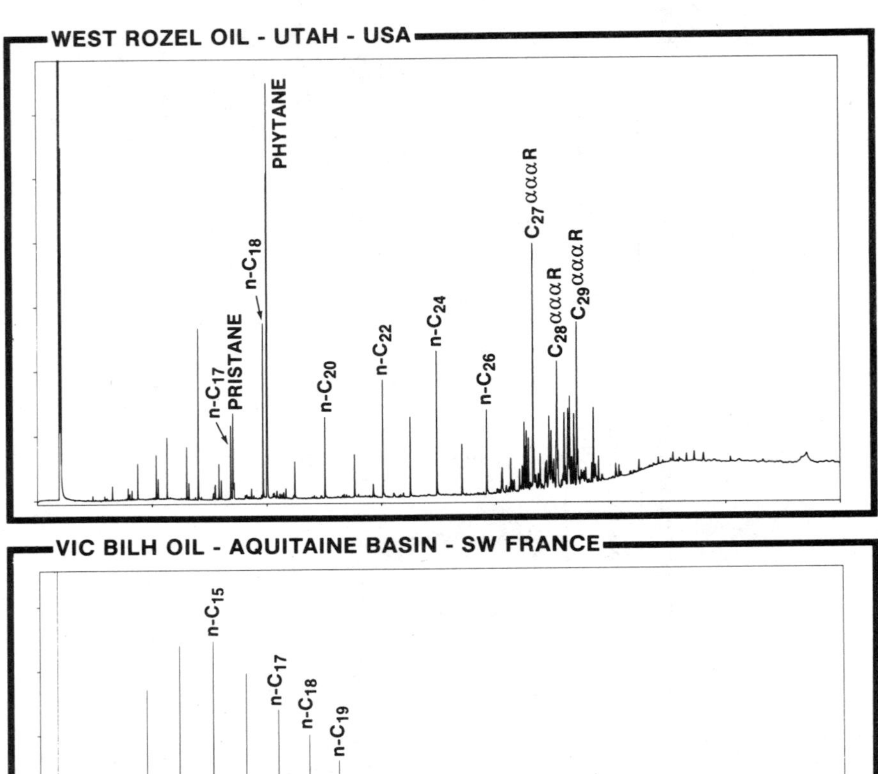

Figure 17.1 Gas chromatograms of total alkanes from West Rozel oil (Utah) and Vic-Bilh oil (Aquitaine Basin, S.W. France).

Analytical Procedures

Liquid cultures were grown in 250 ml Erlenmeyer flasks using 40 ml of sterile nutrient medium with the following concentrations (in g liter^{-1}, distilled water): $(NH_4)_2SO_4$: 2; $CaCl_2,2H_2O$: 0.01; $FeSO_4,7H_2O$: 0.01; K_2HPO_4: 2; $MgSO_4,7H_2O$: 0.2; glycerol: 10; yeast extract: 20; pH adjusted to 6.7. Flasks were incubated at

28°C on a rotary shaker (220 rpm) for 48 h (first stage of culture, 2.5% inoculation) and 24 h (second stage of culture, 5% inoculation).

Branched and cyclic alkanes (70 μg in 100 μl of decalin) or reference sterane compounds (Chiron, 8 μg per sterane in 200 μl of decalin) were added to the subculture broth (10^9 cells ml^{-1}). After various periods of incubation (from 1 to 15 days), one flask was removed, whole subculture was acidified then extracted with dichloromethane. In kinetic studies, an internal standard (5α-androstane, 8 μg in 100 μl of decalin) was added prior to the extraction step. The organic layer was dried with Na_2SO_4 and evaporated under vacuum. The organic residue was subsequently chromatographied over a silica gel column using *n*-hexane. Branched and cyclic alkanes were analyzed by GC (column:50 m × 0.21 mm i.d., coated with OV-1, film thickness: 0.11 μm; temperature program: 80 to 300°C at 1.6°C min^{-1}; H_2 as carrier gas) and computerized GC-MS (column 60 m × 0.25 mm i.d., coated with DB_5, film thickness: 0.10 μm; temperature program: 40 to 130°C (9°C min^{-1}) and 130 to 300°C (2.5°C min^{-1})) using a Finnigan 4500 quadrupole mass spectrometer (EI 70 eV, emission current: 200 μA) equipped with an Incos data system.

RESULTS OF IN VITRO EXPERIMENTS

When carried out according to Goodwin et al. (1983), all of our laboratory experiments were negative within a 10-day incubation time. These failures led us to choose the strategy consisting of screening pure bacterial strains inoculated with branched and cyclic alkanes from crude oils and with pure steranes. In the experimental set up proposed (Fig. 17.2), growth culture is reproducible, incubation periods are short and analytical methods are simple.

Biodegradation Efficiency of Various Gram-Positive Bacteria

Among the 73 aerobic bacteria tested, 7 gram-positive species (Table 17.2) were clearly positive for the degradation of steranes and 2 of them could affect terpanes. Gram-negative bacteria screened were all found to be unable to achieve such a biodegradation. The present result partially explains why we did not succeed in biodegrading the Pecorade steranes and terpanes using *Pseudomonas oleovorans* (Connan et al., 1980).

Within this set of experiments, the biodegradation of steranes was recognized by changes in sterane patterns due to the selective removal of 5α(H),14α(H), 17α(H)-steranes with the 20*R* configuration. All of the 7 strains of *Nocardia*, *Arthrobacter*, and *Mycobacterium* strains are able to consume these steranes within 8 days. *Nocardia globerula* ATCC 21506 appears to be the least effective bacterium whereas *Nocardia species* SEBR 16 provided the most extensive biodegradation.

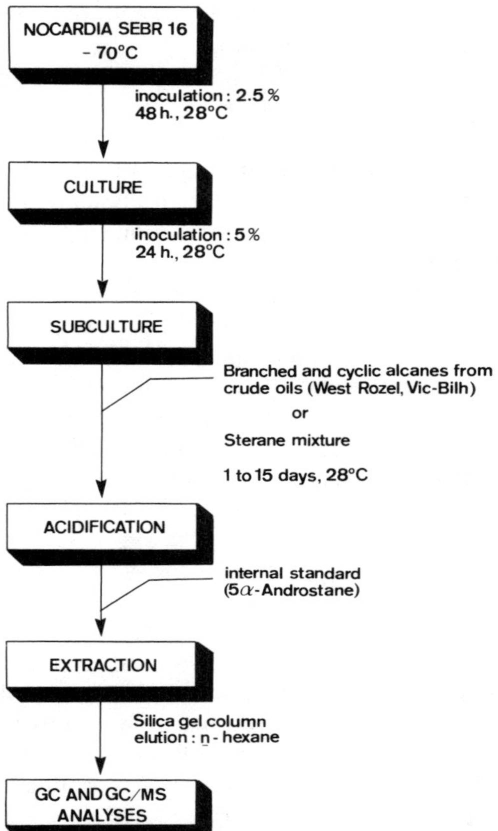

Figure 17.2 Flowchart of experimental set up.

This bacterial strain, originally isolated from a soil sample, was identified as a *Nocardia* by referring to Bergey's manual (1986). The cell wall, according to its composition, belonged to type IV and contained mycolic and tuberculostearic acids. The biochemical characteristics studied were not sufficient for a species assignment.

Step-by-Step Biodegradation of the West Rozel Oil Steranes by Nocardia Species SEBR 16

Nocardia species SEBR 16, the most efficient bacterium in this work in biodegrading steranes, was used to study the sequential degradation of various stereoisomers present in branched and cyclic alkanes from West Rozel oil. Changes in the regular sterane, methylsterane and triterpane composition were monitored by computerized-GC-MS analysis. Figure 17.3 depicts the step-by-step evolution of regular steranes (m/z 217) with time of incubation ranging from 2 to 15 days.

Table 17.2 CAPACITY OF 7 GRAM-POSITIVE BACTERIA TO BIODEGRADE STERANES IN BRANCHED AND CYCLIC ALKANES FROM CRUDE OILS (8 DAYS)

Microorganisms International Reference	Microorganisms Sanofi Elf Biorecherches Reference	Origin of Branched and Cyclic Alkanes	Steranes						Triterpanes			
			C_{27}		C_{28}		C_{29}					
			αααR	αββR	αααR	αββR	αααR	αββR	Tm	Ts	C_{29}αβH	C_{30}αβH
Nocardia globerula ATCC 21506	Nocardia SEBR 666	Vic Bilh Oil	+	−	++	−	+	−	−	−	−	−
Nocardia globerula ATCC 21505	Nocardia SEBR 772	Vic Bilh Oil	+++	−	+++	−	+	−	−	−	−	−
Nocardia species ATCC 19170	Nocardia SEBR 773	Vic Bilh Oil	+++	−	+++	−	+++	−	−	−	−	−
Mycobacterium species ATCC 29472	Mycobacterium SEBR 763	Vic Bilh Oil	+++	−	+++	−	++	−	−	−	−	−
Arthrobacter simplex ATCC 13260	Arthrobacter SEBR 770	Vic Bilh Oil	+++	−	+++	−	+++	−	−	−	−	−
Nocardia species	Nocardia SEBR 16	West Rozel Oil	+++	+	+++	+	+++	+	−	+	+?	−
Nocardia erythropolis ATCC 4277	Nocardia SEBR 769	West Rozel Oil	+++	−	+++	−	++	−	−	+	+?	−

Abbreviations in Figure 17.3. T_s: 18α(H)-22,29,30-trisnorneohopane; T_m: 17α(H)-22,29,30-trisnorhopane; C_{29}αβH: 17α(H),21β(H)-norhopane; C_{30}αβH: 17α(H),21β(H)-hopane.

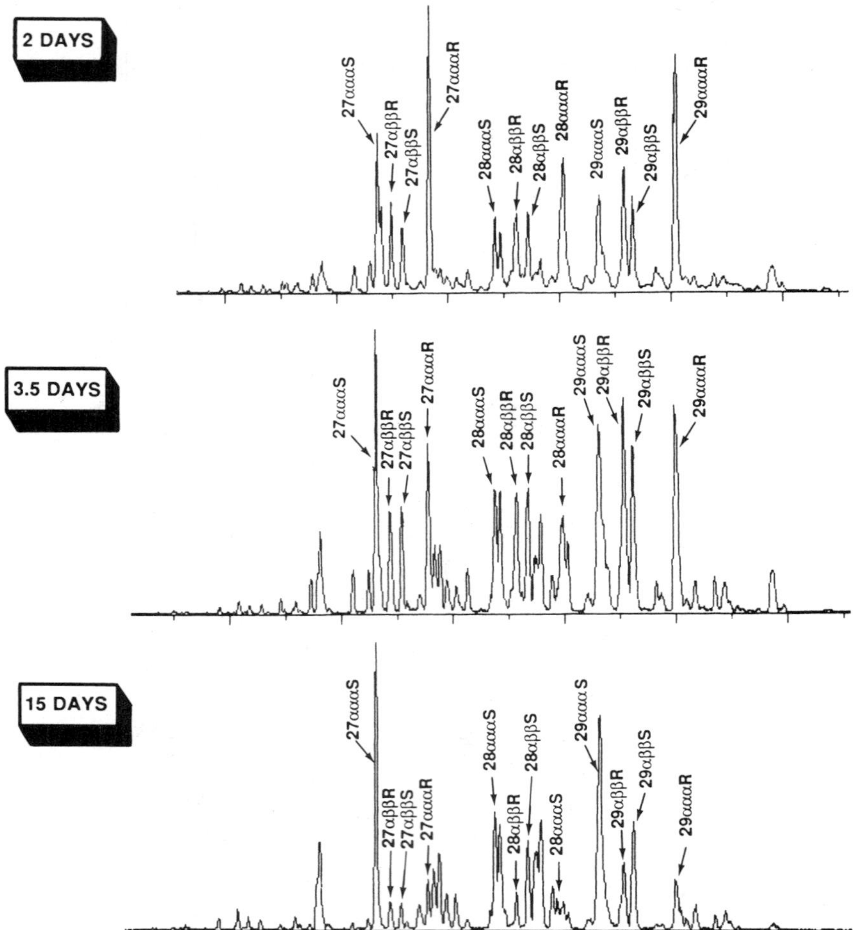

Figure 17.3 Evolution of sterane patterns (*m/z* 217) during in vitro biodegradation of branched and cyclic alkanes from West Rozel oil by *Nocardia species* SEBR 16. 27ααα*R*: 5α(H),14α(H),17α(H)20*R*-cholestane; 27ααα*S*; 5α(H),14α(H), 17α(H)20*S*-cholestane; 27αββ*S*; 5α(H),14β(H),17β(H)20*S*-cholestane.

Nocardia species SEBR 16 preferentially degrades the 20*R* isomers of 5α(H),14α(H),17α(H)-steranes (C_{27}, C_{28}, C_{29}) which are drastically reduced after 3.5 days of incubation (Table 17.3). The 5α(H),14α(H),17α(H)20*S*-steranes were most resistant to bacterial attack which resulted in the increase of their relative concentration during the experiment. The 5α(H),14β(H),17β(H)-steranes are also attacked as shown by the significant decrease of C_{27}αββ(*R* + *S*) steranes (Fig. 17.3). These thermodynamically more stable steranes are removed at a slower rate, however, than their αααR-sterane counterpart (Fig. 17.4a). Among the αββ-ster-

Table 17.3 STEPWISE CHANGES IN STERANE RATIOS DURING IN VITRO BIODEGRADATION OF BRANCHED AND CYCLIC ALKANES FROM WEST ROZEL OIL BY *NOCARDIA SPECIES* SEBR 16

SEBR. Ref. / Days of Incubation \ Sterane Ratios	SNEA(P) Ref.	C_{27} Regular Steranes				C_{28} Reg. Ste.		C_{29} Regular Steranes				αββ Steranes		
		27ααS / 27ααR	27ββS / 27ααR	27ββS / 27ααS	27ββR / 27ββS	28ααR / 28ααS	28ββS / 28ααS	29ααS / 29ααR	29ββS / 29ααR	29ββS / 29ααS	29ββR / 29ββS	27 (%)	28 (%)	29 (%)
B/C Alkanes of West Rozel Oil	2325	0.28	0.11	0.41	1.29	3.10	1.10	0.30	0.22	0.76	1.37	27	28	45
2 Days	B11986	0.52	0.25	0.48	1.20	2.00	1.10	0.40	0.34	0.87	1.32	32	39	39
3.5 Days	B11987	1.86	0.62	0.33	0.97	0.80	1.00	0.92	0.94	0.90	1.30	25	29	46
15 Days	B11989	38.18	2.39	0.07	1.00	0.20	0.60	4.07	2.02	0.50	0.62	13	41	46

Abbreviations in Figure 17.3. 28αααR: 5α(H),14α(H),17α(H)20*R*-24-methylcholestane; 29αααR: 5α(H),14α(H),17α(H)20R-24-ethyl-cholestane; percent 27αββ: percent 27αββ *R* + *S* in total αββ steranes.

anes, the 20*R* isomers are preferentially removed as illustrated by the data presented in Figure 17.3 and the biodegradation efficiency varies according to the sequence $C_{27} > C_{28} > C_{29}$. Figure 17.4b shows sterane ratios in a diagram currently used to assess maturity and migration of oils and source rocks (Seifert and Moldowan, 1981; Mackenzie, 1984).

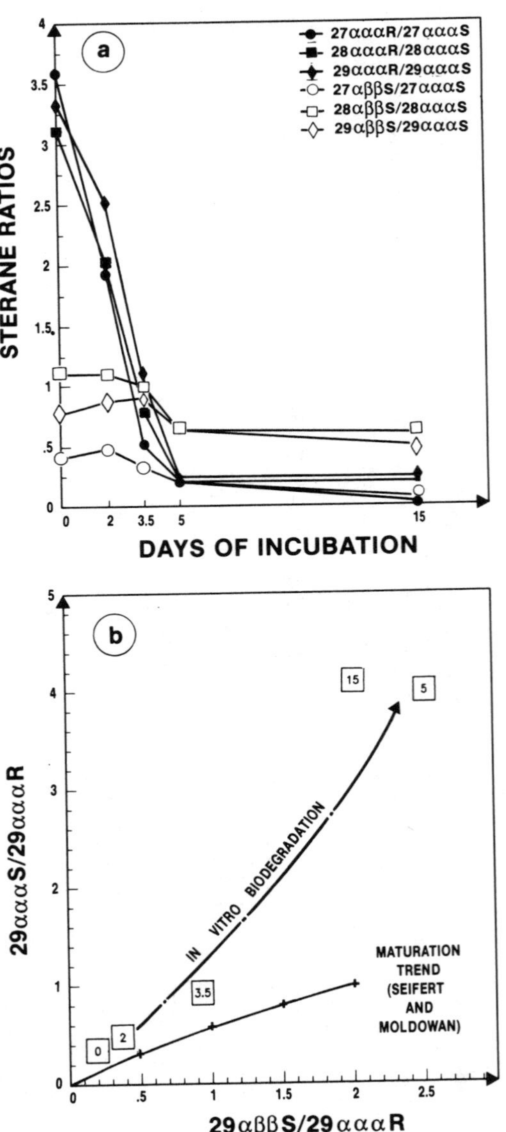

Figure 17.4 Evolution of sterane and terpane ratios during in vitro biodegradation of branched and cyclic alkanes from West Rozel oil by *Nocardia species* SEBR 16. Abbreviations in Tables 17.3 and 17.4.

Evolution of methylsteranes (m/z 231), methylated in position 2, 3, or 4, is presented in Figure 17.5. When increasing incubation, the sequence, as well as the intensity in the decrease of the relative concentrations of methylsteranes, mimics the evolution of regular steranes (Fig. 17.3). Based on the hypothesis that methylsteranes display a behavior similar to that of regular steranes, the most likely chemical structures are proposed. These structure assignments were confirmed by comparing the m/z 231 and 232 mass fragmentograms.

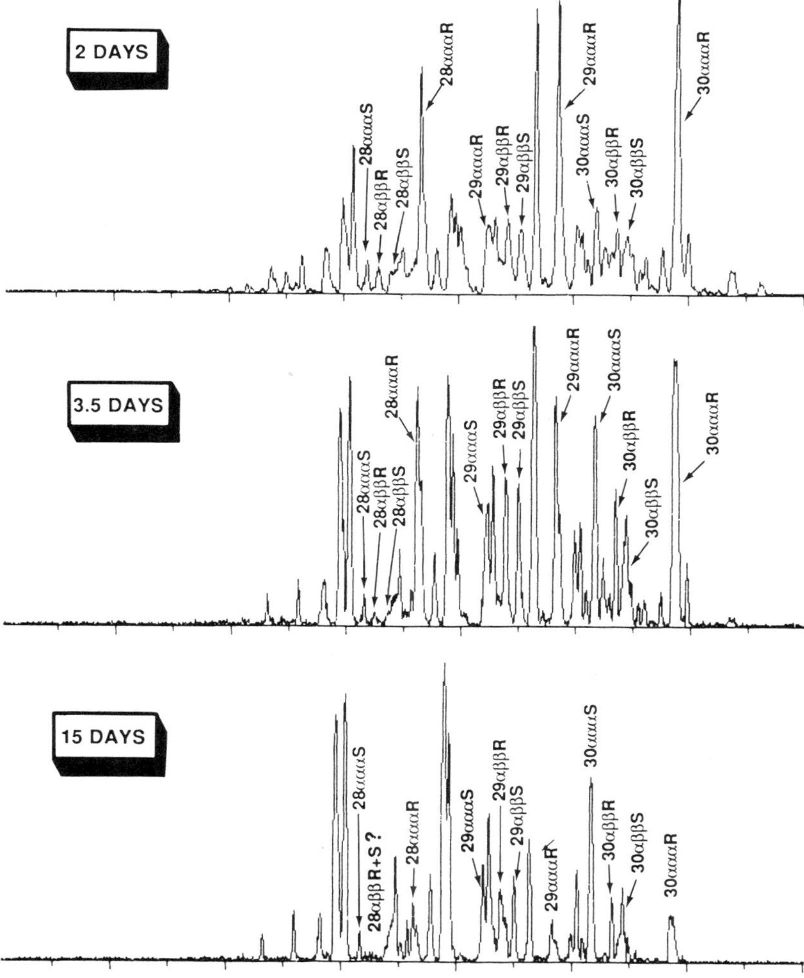

Figure 17.5 Evolution of methylsteranes patterns (m/z 231) during in vitro biodegradation of branched and cyclic alkanes from West Rozel oil by *Nocardia species* SEBR 16. 28αααS: methyl-5α(H),14α(H),17α(H)20*S*-cholestane; 28αββS: methyl-5α(H),14β(H),17β(H)20*S*-cholestane. Methyl group is presumed at position 4.

Table 17.4 STEPWISE CHANGES IN STERANE AND TERPANES RATIOS DURING IN VITRO BIODEGRADATION OF BRANCHED AND CYCLIC ALKANES FROM WEST ROZEL OIL BY *NOCARDIA SPECIES* SEBR 16

SEBR. Ref. Days of Incubation \ Terpane Ratios	SNEA (P) Ref.	TT/ST	Σ Steranes + Methylsteranes + Triterpanes			Triterpanes				Steranes/Terpanes		
			ST %	TT %	MST %	29αβH/30αβH	% 22S/C_{32}αβH	GAMM./C_{30}αβH	35αβHR/35αβHS	27αααR × 10/29αβH	28αααR × 10/29αβH	29αααR × 10/29αβH
B/C Alkanes of West Rozel Oil	2325	0.14	63	9	28	0.19	62	0.70	0.46	278	170	253
2 Days	B11986	0.15	61	9	29	0.18	57	0.70	0.44	212	192	160
15 Days	B11989	0.68	41	28	29	0.17	57	0.77	0.38	Low	Low	12

Abbreviations:

TT/ST	triterpane (C_{27}–C_{35}) to sterane (C_{27}–C_{29}) ratio
ST %	percent steranes (C_{27}–C_{29})
TT %	percent terpanes (C_{27}–C_{35})
MST %	percent methylsteranes (C_{28}–C_{30})
29αβH/30αβH	17α(H),21β(H)-norhopane/17α(H),21β(H)-hopane
%22*S*	22S-17α(H),21β(H)-bishomohopane/22S-17α(H),21β(H)-bishomohopane + 22R-17α(H),21β(H)-bishomohopane
GAMM./C_{30}αβH	gammacerane/17α(H),21β(H)-hopane
35αβH*R*/35αβH*S*	17α(H),21β(H)-C_{35} hopane (22*R*)/17α(H),21β(H)-C_{35} hopane (22*S*)

Mass fragmentograms, diagnostic of terpanes (m/z 191), do not show obvious changes among triterpanes. Several ratios (Table 17.4) suggest, however, that the biodegradation of terpanes indeed occurs. It seems that C_{30} $\alpha\beta$ hopane is more quickly removed than gammacerane as was also seen in geological samples (Zhang Dajiang et al., 1988) and that the 22*R* isomer of C_{35} $\alpha\beta$ hopanes is preferentially degraded as was previously found by Goodwin et al. (1983). The preliminary data indicate that between 25 and 50 percent of C_{35} $\alpha\beta$-hopanes, gammacerane and C_{30} $\alpha\beta$-hopane have disappeared when 5α(H),14α(H),17α(H)20*R*-steranes have been completely metabolized. This terpane removal does not significantly affect the distribution pattern and contrasts with sterane degradation where the 20*R* isomers are selectively attacked.

Step-by-Step Biodegradation of a Mixture of Pure Steranes by Nocardia Species SEBR 16

A mixture of three steranes (5α(H),14α(H),17α(H)20*R*-cholestane; 5α(H),14α(H),17α(H)20*S*-cholestane; 5α(H),14β(H),17β(H)20*R*-cholestane) were inoculated with *Nocardia species* SEBR 16 under experimental conditions previously described (Fig. 17.2). 5α-Androstane, added prior to the extraction step (Fig. 17.2), was used as internal standard for quantitation. The results (Fig. 17.6 and Table 17.5) clearly show a sequential biodegradation of the three stereoisomers in a way that fully confirms the stepwise evolution recorded in the West Rozel oil experiment. 5α(H),14α(H),17α(H)20*R*-cholestane was attacked first as shown in the data of the one-day experiment and had completely disappeared after 5 days. The biodegradation of 5α(H),14β(H),17β(H)20*R*-cholestane started at the same time but at a slower rate (Table 17.5). During the whole experiment, 5α(H),14α(H),17α(H)20*S*-cholestane remains undegraded (Fig. 17.6).

DISCUSSION OF DATA FROM LABORATORY EXPERIMENTS WITH RESPECT TO FIELD OBSERVATIONS

Principle Results of Laboratory Experiments

All positive experiments carried out with pure strains (*Nocardia*, *Anthrobacter*, *Mycobacterium*), branched and cyclic alkanes of various oils (Vic-Bilh, Aquitaine Basin; Pacassa 3 and Camaro E1, Angola, unpublished; West Rozel, USA) or pure sterane mixtures led us to converging observations.

- Regular steranes and their methylated homologues are biodegraded to various degrees, the rate of biodegradation being $\alpha\alpha\alpha R > \alpha\beta\beta R > \alpha\beta\beta S > \alpha\alpha\alpha S$ with $C_{27} > C_{28} > C_{29}$ (at least for $\alpha\alpha\alpha R$ and $\alpha\beta\beta R + S$),
- $\alpha\beta$ hopanes are biodegraded without marked changes in terpane patterns.

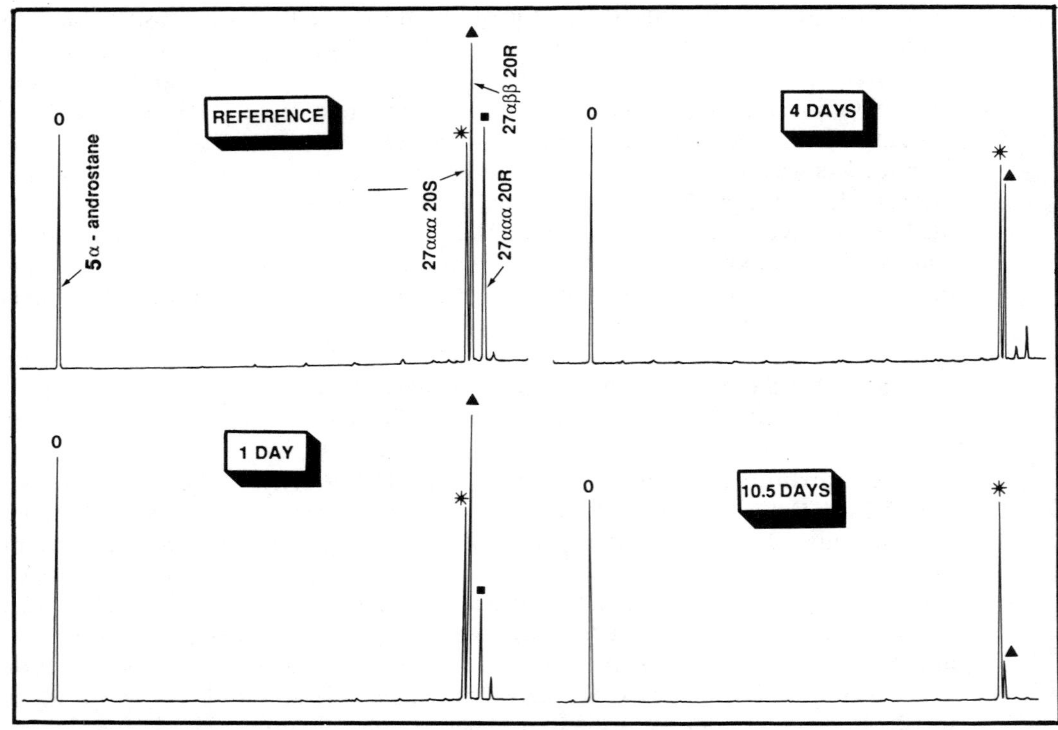

Figure 17.6 Gas chromatograms of an authentic standard mixture of steranes (27ααα*R*, 27ααα*S*, 27αββ*R*) during in vitro biodegradation by *Nocardia species* SEBR 16. Incubation time from 1 to 10.5 days.

Table 17.5 IN VITRO BIODEGRADATION OF A REFERENCE STERANE MIXTURE BY *NOCARDIA SPECIES* SEBR 16: CHANGES IN STERANE RATIOS

Days of Incubation	27αααS / And.	27αββR / And.	27αααR / And.	27αββR / 27αααS	27αααR / 27αααS
0					
Reference	1.15	1.59	1.16	1.37	1.00
1	0.91	1.29	0.42	1.42	0.46
2	0.86	1.24	0.29	1.43	0.34
3	0.89	0.87	0.08	0.97	0.09
4	0.94	0.79	0.05	0.84	0.05
5	0.79	0.42	0.01	0.53	0.01
6	0.78	0.33	0.01	0.42	0.02
7.5	0.95	0.33	0.06	0.35	0.06
8	1.20	0.21	0.02	0.17	0.02
10.5	1.11	0.18	0.01	0.16	0.01

Abbreviations in Figure 17.3, and: 5α-Androstane (internal standard).

However, incipient variations of terpane ratios indicate that the rate of biodegradation of αβ hopanes is 22*R* > 22*S* and $C_{35} > C_{30}$. Gammacerane appears to be more resistant than αβ hopanes. Homologous series of hopanes demethylated in C-10 position and demethylated tricyclic terpanes were not recorded in the present laboratory experiments.

Comparison Between Laboratory Data and Field Observations

Laboratory simulations. A review of the published literature (Table 17.6) clearly indicates that the selective biodegradation of C_{27}–C_{29} 5α(H),14α(H),17α(H)-steranes with the biological 20*R* configuration has been quoted by Rullkötter and Wendisch (1982), McKirdy et al. (1983), Volkman et al. (1983), Sandstrom and Philp (1984), Seifert et al. (1984), Landais and Connan (1986), and Zhang Dajing et al. (1988). In their pioneering work, Seifert and Moldowan (1979) proposed a reverse order of the attack of certain diastereoisomers when they noticed the preferential loss of the C_{27} 13β(H),17α(H)20*S*-diasteranes. Of particular interest are the results of Volkman et al. (1983) on Australian crude oils (Paleozoic Barrow and Dampier subbasins, Western Australia). In this geological setting, the sequence of removal of regular steranes ααα*R* > αββ*R* + *S* > ααα*S* fully agrees with what we observed in our West Rozel oil experiments. Careful examination of the starting material reveals that the original characteristics of sterane distribution patterns are highly variable: mainly regular steranes (West Rozel, Vic-Bilh; oils studied by Seifert and Moldowan, 1984), regular steranes with subordinate diasteranes (McKirdy et al., 1983; Goodwin et al., 1983, laboratory experiments; Landais and Connan, 1986; Zhang Dajiang et al., 1988), comparable levels of diasteranes and regular steranes (Goodwin et al. 1983, field data). Furthermore, the relative concentrations of C_{27}-, C_{28}-, and C_{29}-steranes exhibit a fairly large range: occurrence of significant amounts of C_{27}-, C_{28}-, and C_{29}-steranes in Vic-Bilh and West Rozel, predominent C_{28}-and C_{29}-steranes in Karamay oils (McKirdy et al., 1983; Landais and Connan, 1986; Zhang Dajiang et al., 1988). The initial sterane distribution patterns may certainly influence the authors' conclusions. For instance, when Zhang Dajiang et al. (1988) concluded that the sequential removal of ααα*R*-steranes began with C_{29}- and C_{28}-homologues, it is likely that this was influenced by the fact that C_{27} regular steranes occurred in much smaller amounts. A similar difficulty was encountered in the Lodève Basin (Landais and Connan, 1986) where C_{27} regular steranes are lacking.

Despite the difficulties in applying the West Rozel results to various case histories, it seems realistic to use them as guidelines to understand some geological situations described in the literature.

Table 17.6 BIODEGRADATION OF STERANES AND TERPANES UNDER GEOLOGICAL AND LABORATORY CONDITIONS

	Order of Bacterial Attack →				Effect of Severe Biodegradation
	Regular Steranes	αβ-Hopanes	Diasteranes	Other Terpanes	
Seifert and Moldowan 1979	28-29S > 28-29R ?		27S > 27R 27 > 29	Tricyclic terpanes	Formation of demethylated hopanes
Rullkötter and Wendisch 1982	28αββR > 28αββS 28αααR > 29αααS 29αββR > 29αββS 27 > 28-29	29 > 30-32			Formation of demethylated hopanes via oxygenated intermediates Formation of 25-normoretanes Formation of onoceranes?
McKirdy *et al.* 1982	αααR > αααS αββR > αββS	αβ-Hopanes	Diasteranes		
Volkman *et al.* 1983	αααR > αββR + S > αααS 100%* > 50%* > 10%* *% removal		27R > 27S 27 > 29		Formation of demethylated hopanes via hopane conversion
Goodwin *et al.* 1983	Labo. 27 > 28 > 29	22R > 22S			No demethylated hopanes
	Field	35 > 34 > ··· > 30 > 29 αβ-Hopanes	27 > 28 > 29 Diasteranes		Concentration of demethylated hopanes and of normoretanes

Sandstrom and Philp 1984	29αααR > 29 αααS		Diasteranes	Tricyclic Terpanes	Demethylated hopanes present
Seifert *et al.* 1984	αααR > αααS 29αααR > 29αααS + 29αββR-S 28αααR > 28αααS + 28αββR-S for αααS: 27 > 28 > 29	27-29 > 30-35	Diasteranes	Gammacerane	No demethylated hopanes. Concentration of homologous series of normal side chain 17α-hopanes
Landais and Connan 1986	αααR > αααS 29 ≅ 28	22R > 22S + Degradation Tetrac. Terpanes (24/4)		—Tricyclic terpanes including extended members (C31-C40) concentrated —Gammacerane	No demethylated hopanes
Zhang Dajiang *et al.* 1988	αααR > αααS + αββR + S 29 > 28 > 27	αβ-Hopanes	Diasteranes	Pregnanes/Tricyclics Gammacerane	Formation of demethylated hopanes Concentration of 9,10-secosteranes and 8,14-secohopanes
This report field data (Aquitaine Basin)	27 > 28 > 29	αβ-Hopanes + Tetracyclic Terpanes (24/4)		Methyl αβ-Hopanes Hexahydrobenzohopanes Tricyclic Terpanes Pregnanes	No demethylated hopanes
This report Laboratory (West Rozel)	αααR > αββR > αββS > αααS 27 > 28 > 29 for αααR and αββR+S	22R > 22S 35 > 30			No demethylated hopanes

Review of literature data giving removal sequence among steranes and terpanes.

Laboratory Simulation: Some Discrepancies with Geological Case Histories

The preferential removal of regular ααα-steranes with the biological 20*R* configuration is reproduced by the incubation of branched and cyclic alkanes from Vic-Bilh oil with *Arthrobacter simplex* SEBR 770. A similar evolution occurs with methylated steranes; this allows a 20*R* assignment of configuration to the degradable components. No marked changes appear in the terpane pattern or in terpane ratios, despite the fact that αβ-hopanes may have decreased somewhat in concentration, as suggested by quantitative data (unpublished). The Aquitaine Basin oils, for example, the Vic-Bilh oil have biomarker distribution significantly different from that of West Rozel oil (TT/ST = 5 instead of 0.1 in West Rozel oil). Nevertheless, the overall mechanisms of sterane biodegradation appear to be the same in both oils. *Arthrobacter simplex* SEBR 770, used in laboratory experiments, was found to be less active than *Nocardia species* SEBR 16, because αββ*R*- and αββ*S*-steranes were lightly attacked by the former.

As typical natural asphalts from the Aquitaine Basin, samples from well Pécorade 26 were chosen. This well has reached the main producing zone of the Pécorade oil field, and a representative oil sample from 2671 to 2790 m depth, was analyzed as a reference. Asphalts in Cenomanian horizons at 1520 and 1550 m are biodegraded counterparts of the oil accumulated at depth. This oil is not biodegraded and similar to those previously used in our in vitro experiments with *Pseudomonas oleovorans* (Connan et al., 1980).

At 1550 m, biodegradation is extremely severe: C_{27}–C_{29} regular steranes are almost completely removed (Fig. 17.7), whereas αβ-hopanes and tetracyclic terpanes (24/4) are extensively depleted. A preferential attack of 20*R*-steranes and 20*R*-methylsteranes (Fig. 17.8) was found neither in Aquitaine Basin samples nor in the bitumens from Iraq. In Iraq (Connan, 1988), the most commonly biodegradation sequence observed is: $C_{27}\alpha\alpha\alpha(R + S) + C_{27}\alpha\beta\beta(R + S) > C_{28}\alpha\alpha\alpha(R + S) + C_{28}\alpha\beta\beta(R + S) > C_{29}\alpha\alpha\alpha(R + S) + C_{29}\alpha\beta\beta(R + S) > C_{21} + C_{22}$. Such a sequence is the same as in Goodwin et al.s' (1983) laboratory experiments with whole oil samples inoculated 11.5 months with a mixed population of aerobic bacteria, molds, and yeasts. The oil used by Goodwin et al. (1983), unknown in origin, shows a sterane pattern comparable to those of our crude oils in France (Aquitaine Basin) and in Iraq. The selective hopane destruction caused a corresponding selective enrichment of bacterially resistant tricyclic terpanes (23/3 to 26/3), αβ-methylhopanes and hexahydrobenzohopanes (Fig. 17.9). In this asphalt as in many other natural asphalts from the Aquitaine Basin, demethylated hopanes were not detected. The same situation has been described for archeological bitumens collected in ancient cities, mainly in Mesopotamia. Ranging from mildly to severely biodegraded, these samples never contain demethylated hopanes (Connan, 1988).

Apparently, the experimental approach applied by Goodwin et al. (1983)

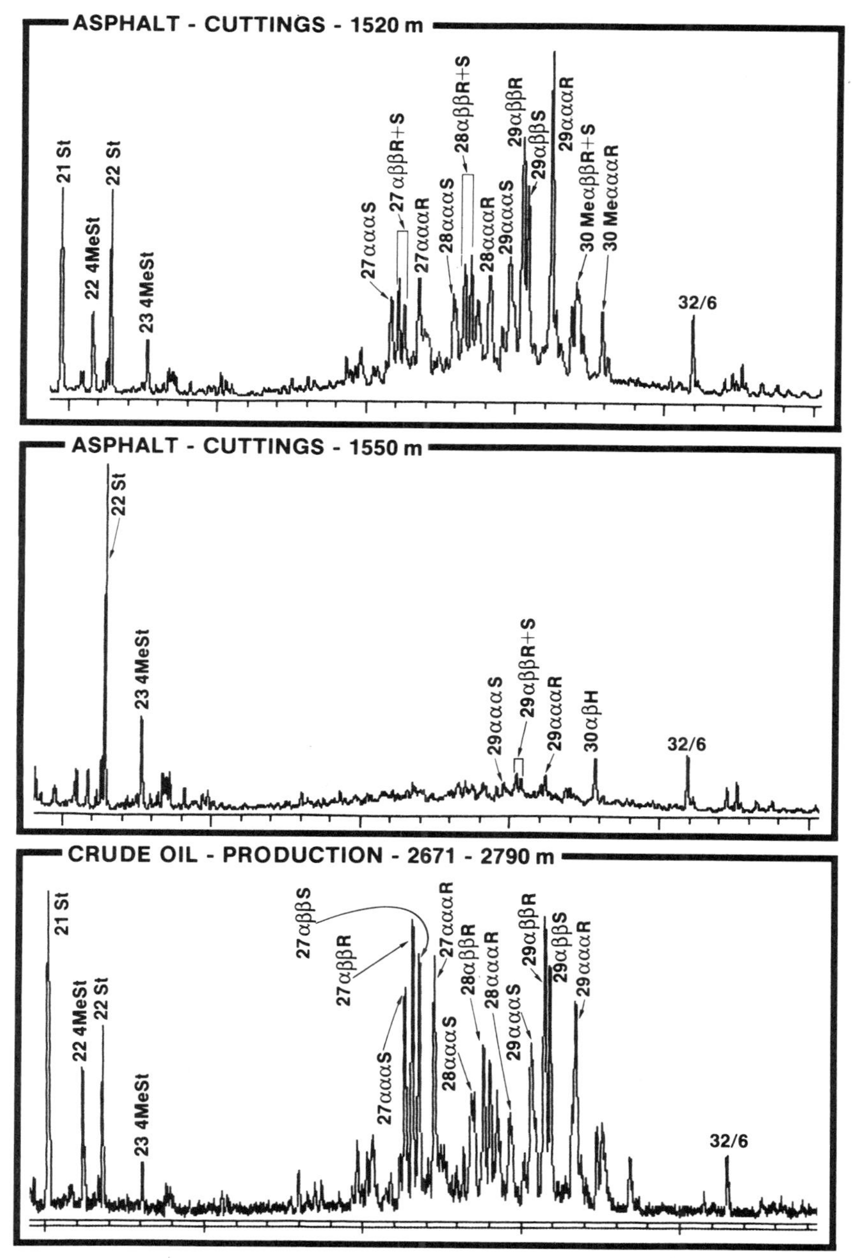

Figure 17.7 Fragmentogram, *m/z* 217, showing biodegradation of steranes in geological samples: a case history in crude oils of the Pécorade 26 well in the Aquitaine Basin (S.W. France). Abbreviations in Table 17.3.

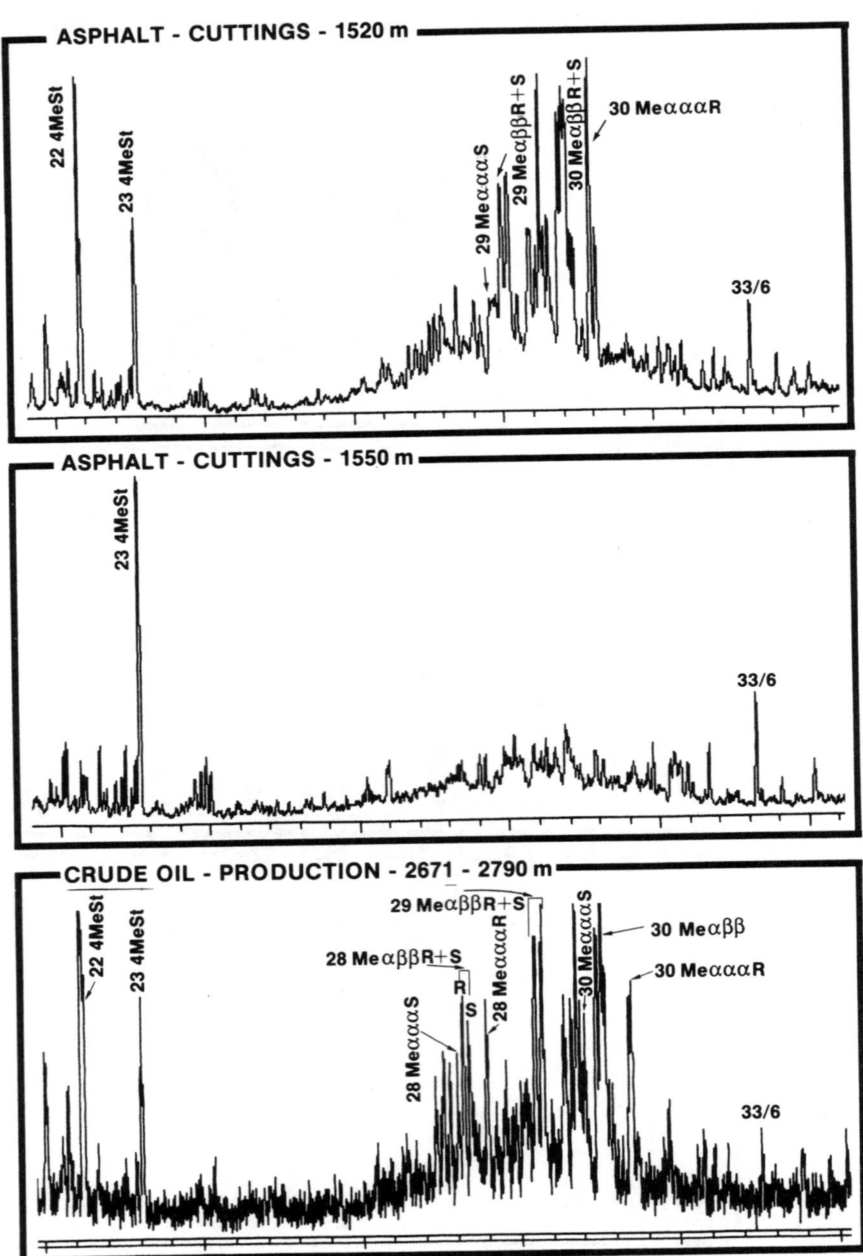

Figure 17.8 Fragmentogram, *m/z* 231, showing biodegradation of methylsteranes in geological samples: a case history in crude oils of the Pécorade 26 well in the Aquitaine Basin (S.W. France). 30 Meααα*R*: methyl-5α(H),14α(H),17α(H)20*R*-ethyl-cholestane.

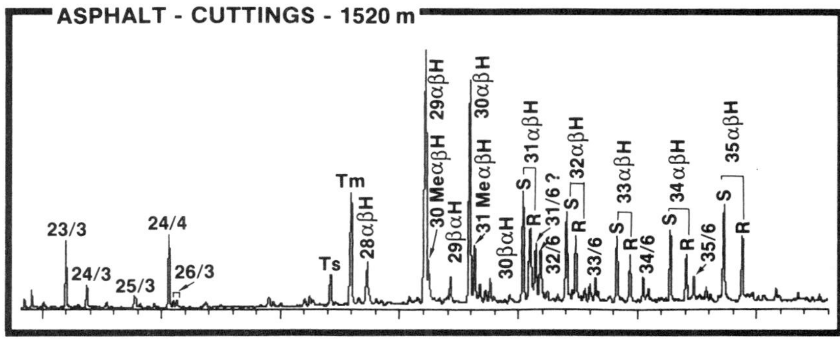

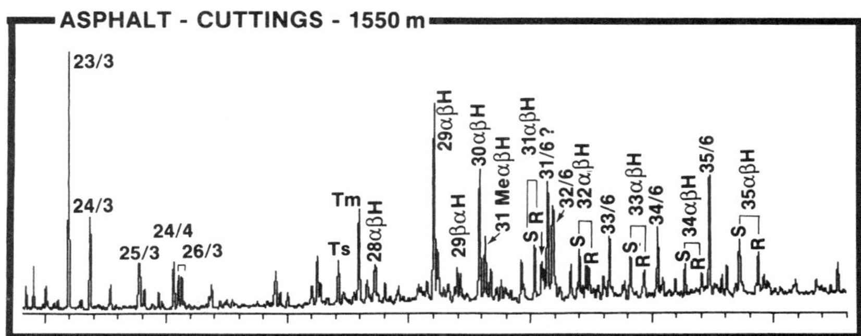

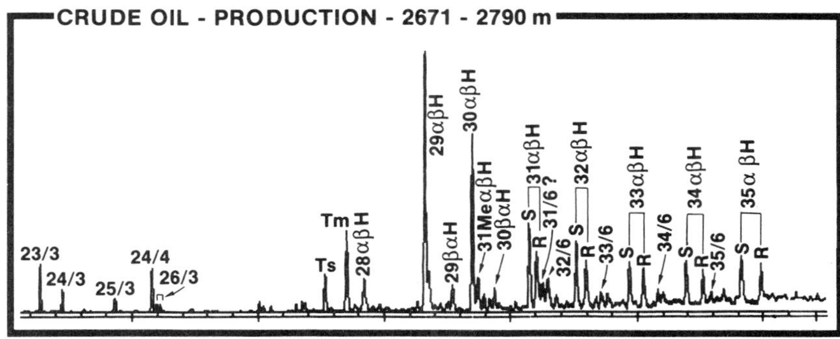

Figure 17.9 Fragmentogram, *m/z* 191, showing biodegradation of terpanes in geological samples: a case history in crude oils of the Pécorade 26 well in the Aquitaine Basin (S.W. France).

reproduces quite well natural biodegradation and weathering effects observed in asphalts of the Aquitaine Basin and in archeological samples of Iraq.

Demethylated Hopanes: Neogenesis or Relative Enrichment of Existing Components?

Summary of literature data. Severe biodegradation of steranes and αβ hopanes reveals detectable amounts of various new families of polycyclic alkanes. The basic question is if these structures are products of polycyclic alkane precursors metabolized by bacteria or preexisting minor compounds enhanced by selective removal of more readily biodegradable families (e.g., steranes and αβ-hopanes).

Among the classes of polycyclic alkanes detected in extremely biodegraded crudes are secohopanes (Restle, 1983), hexahydrobenzohopanes (Connan and Dessort, 1987), gammacerane, demethylated hopanes, onoceranes, and 30-nor-17α(H)-hopanes (Table 17.5), tricyclic terpanes (Reed, 1977) including the extended members, neohopanes. Of particular interest are demethylated hopanes whose identification has been preferentially made in crude oils with *n*-alkanes either present or absent. The occurrence of hopanes demethylated at C-10 (25-nor-17α(H)-hopanes, Rullkötter and Wendisch, 1982; Trendel et al., 1990) is regarded as a key indication of paleobiodegradation in a seemingly nonbiodegraded oil containing *n*-alkanes (Volkman et al., 1983). According to several authors (Seifert and Moldowan, 1979; Volkman et al., 1983; Philp, 1983) 25-norhopanes would result from biotransformation of αβ-hopanes during in-reservoir biodegradation of crude oils.

To substantiate this hypothesis, two main arguments may be quoted.

1. Demethylated hopanes and, to some extent, demethylated tricyclic terpanes (Howell et al., 1984) were preferentially identified in severely biodegraded crude oils (Table 17.5),
2. Demethylated hopanes, frequently recognized among residual alkanes of severely biodegraded oils, have not been detected so far in pyrolysates of their asphaltenes obtained at various temperatures (300 to 370°C; Cassini and Eglinton, 1986). Similar conclusions may also be extended to other biomarker classes. For example, 25,28,38-trisnorhopane, 28,30-bisnorhopane, and 18α(H)-oleanane (Cassini and Eglinton, 1986; Fowler and Brooks, 1987; Jones et al., 1987) have not been detected in pyrolysates of asphaltenes, of NSO compounds (Jones et al., 1987), and of kerogen (Moldowan et al., 1984; Noble et al., 1985). These data are generally interpreted by assuming that the precursors of the above-mentioned biomarkers enter the sedimentary record as free lipids which are not subsequently incorporated into the polymeric network of asphaltenes, resins or kerogens. In the case of demethylated hopanes, this supports the hypothesis that demethylated hopanes are genetically related to hopanes via in-reservoir biodegradation.

There are several arguments, however, to argue against this hypothesis.

1. Demethylated hopanes, as well as demethylated tricyclic terpanes, may possibly be indigenous source rock constituents as was mentioned by Philp (1983) and Howell et al. (1984). The occurrence of 25-norhopanes in association with 28,30-bisnorhopane and 25,28,30-trisnorhopane has indeed been reported by Noble et al. (1985) in some Western Australian shales. Noble et al. (1985) proposed that their incorporation into the sediment as alkanes is brought about by reworking of biodegraded petroleum deposits. Therefore, they ascribed a fossil origin to these polycyclic alkanes when incorporated into the sediment.
2. Demethylated hopanes were not formed in detectable amounts during biodegradation under laboratory conditions (Goodwin et al., 1983; later this chapter). In that respect, it seems very likely that the microbial degradation of αβ-hopanes does not proceed directly to 25-norhopanes but may involve oxygenated intermediates, as was observed for main classes of hydrocarbons (Connan, 1984). A preliminary ongoing study, undertaken to study the metabolic pathways of sterane consumption, fully substantiates this idea.
3. Demethylated hopanes are not recorded everywhere (Table 17.5). Severely biodegraded oils in some sedimentary basins, in which particular environmental conditions prevailed, are completely devoid of demethylated hopanes. Such a situation was noticed in the U.S. Gulf Coast (Seifert and Moldowan, 1979), Switzerland (Goodwin et al., 1983; Connan, 1984), in the Aquitaine Basin (this report and unpublished data), in Greece (Seifert et al., 1984) and in Iraq (Connan, 1988).

Demethylated Hopanes: Preexisting Enriched Biomarkers During In-Reservoir Biodegradation

In an attempt to define the most likely origin of demethylated hopanes, we carried out a statistical treatment of our biomarker data bank. The screening of 1650 biomarker analyses on crude oils and rock extracts, on a worldwide basis, leads us to identify 216 samples in which the C_{29} demethylated hopane was detected.

Among these 216 samples, more than 60 are rocks in which the extractable organic matter is undoubtedly indigenous. This set of rocks comprises some well-known source rocks of prolific petroleum provinces in the world, for example, the Melania, Kissenda, and Crabe formations in Gaboon, the Pointe Noire and Pointe Indienne marls in Congo, and the Kimmeridge clays of the North Sea (U.K. and Norway). An example of the biological marker distribution of such a source rock is given in Figure 17.10. The organic matter, located in a shaly formation equivalent to the Pointe Noire marls, is immature (Thermal Alteration Index = 2.5; low isomerization in steranes) and the rock is not impregnated. It contains a significant

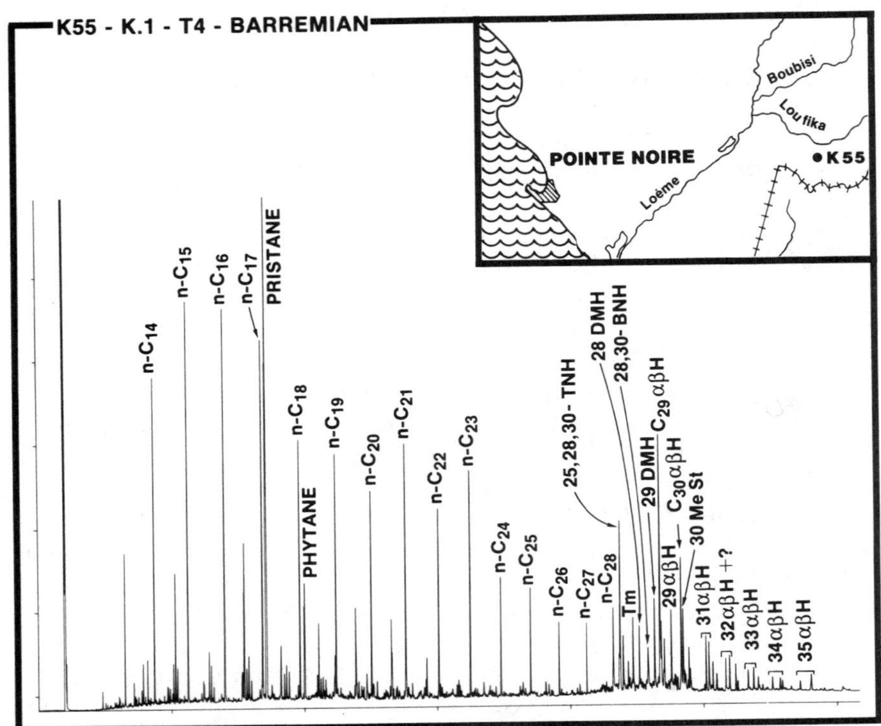

Figure 17.10 Gas chromatogram of total alkanes from an immature source rock in onshore Congo. 28 DMH: 25,30-bisnor-17α(H),21β(H)-hopane; 29 DMH: 25-nor-17α(H),21β(H)-hopane; 28,30-BNH: 28,30-bisnorhopane; 25,28,30-TNH: 25,28,30-trisnorhopane.

amount of 25,28,30-trisnorhopane, 28,30-bisnorhopane and C_{28} and C_{29} 25-nor-αβ-hopanes (Fig. 17.10). The Congo sample, however, is not a typical example. Generally, when they are present in source rocks extracts, C_{28} and C_{29} demethylated hopanes are minor in comparison to C_{29}αβ-hopane (average C_{29} demethylated hopane to C_{29} αβ-hopane ratio = 0.15). The common occurrence of the C_{29} demethylated hopane as a minor component of the terpanes and, therefore, of the alkanes, does explain why this molecular compound is difficult to detect. In fact, it is probably present in many rock extracts and pyrolysates, but at a very low level, which explains why this compound and other family members are not more often recorded in rock samples. As far as pyrolysates are concerned, one may recall that these compounds are diluted by prominent families of alkylated series (*n*-, iso-, anteiso-, cyclohexyl-, cyclopentylalkanes) and, therefore, are difficult to identify. Furthermore, review of samples in which demethylated hopanes were detected shows that their occurrence seems to be restricted to shaly sediments (partly oxic?) with significant input of clastic material. As mentioned, the detection of demethylated hopanes in source rocks is presently restricted to organic matter-rich shales which were not deposited under strongly anoxic conditions. Demethylated hopanes were not found so far in source rocks from evaporitic and hypersaline deposits. They have not even been identified in severely biodegraded oils from these basins. For instance, they have never been observed in asphalts and source rocks from the Barremo-Jurassic of the Aquitaine Basin, in crude oils from Sicily or in archeological bitumens from Iraq (Connan, 1988). Their occurrence, in the sedimentary record in relation to particular environmental conditions, substantiates the idea that demethylated hopanes are indigenous biomarkers of source rocks. Like αβ-hopanes, they may derive from demethylated hopanoidal precursors which occur early under particular depositional conditions, excluding the extremely reducing ones of hypersaline systems. Inspection of our crude oil data bank confirms that demethylated hopane concentration is significantly enhanced in severely biodegraded oils.

Enhanced Concentration of Bacterially Resistant Polycyclic Alkanes: A Clue to Recognition of Palaeobiodegradation

The behavior of demethylated terpanes towards biodegradation strongly resembles that of other minor terpane families, for example, secohopanes, secosteranes, hexahydrobenzohopanes, 30-nor-17α(H)-hopanes, onoceranes, and tricyclic terpanes including the extended members because all of these polycyclic alkanes are also concentrated and sometimes even detected only when bacterial degradation of crude oils occurs. Of particular interest are the practical consequences of these results. They provide diversified guidelines to assess palaeobiodegradation phenomena in various basins in the world. In fact, highly abnormal amounts of any of the bacterially resistant polycyclic alkanes may be indicative of paleobiodegradation in the history of an oil. To confirm such a conclusion, it is necessary to calibrate the biomarker concentration levels on related source rocks. Abnormal amounts of demethylated hopanes in *n*-alkane-rich crudes are a good indicator of palaeobiodegradation. For instance, in the Angola offshore basin, we have explained an abnormal concentration of tricyclic terpanes in crude oil as being due to a palaeobiodegradation phase (Connan et al., 1987). Similarly, abnormal amounts of secohopanes in Gaujacq oil (Aquitaine Basin) reflect a palaeobiodegradation that has taken place as a consequence of diapiric tectonism.

CONCLUSIONS

Steranes and terpanes were biodegraded in the laboratory under reproducible conditions using seven gram-positive strains of *Nocardia*, *Arthrobacter*, and *Mycobacterium*. *Nocardia species* SEBR 16 provided the most extensive biodegradation within an 8-day period whereas 33 species of *Pseudomonas* were unable to achieve such a degradation.

Significant biodegradation has been observed either on branched and cyclic alkanes of various oils (West Rozel, Utah; Vic-Bilh, Aquitaine Basin, France) or on pure sterane mixtures (5α(H),14α(H),17α(H)20*R*-cholestane; 5α(H),14α(H), 17α(H)20*S*-cholestane; 5α(H),14β(H),17β(H)20*R*-cholestane).

Regular steranes and their methylated homologues are biodegraded to various degrees, the rate for regular steranes being $\alpha\alpha\alpha R > \alpha\beta\beta R > \alpha\beta\beta S > \alpha\alpha\alpha S$ with $C_{27} > C_{28} > C_{29}$ at least for $\alpha\alpha\alpha R$ and $\alpha\beta\beta R + S$.

αβ-Hopanes were also slightly biodegraded as was indicated by incipient variations of terpane ratios, the rate being $22R > 22S$ and $C_{35} > C_{30}$.

In vitro biodegradation of steranes and terpanes under laboratory conditions leads to biomarker distribution which also have been observed in geological settings. Laboratory experiments provide reliable methods to understand in-reservoir biodegradation of polycyclic hydrocarbons.

No demethylated hopanes were detected in our laboratory experiments using West Rozel and Vic-Bilh oils, that is, generated under highly reducing conditions in hypersaline systems.

Demethylated hopanes (25-norhopanes) appear as not generated during biodegradation of αβ-hopanes in the reservoir. These preexisting minor compounds have probably been concentrated by selective removal of other more readily degradable polycyclic alkanes, namely, steranes and terpanes. Demethylated hopanes were found in more than 60 rock extracts. Their abnormal concentration in crude oils is a reliable indicator of palaeobiodegradation. Biodegradation of hopanes and steranes enhances the concentration of other minor polycyclic alkanes, that is, secohopanes, hexahydrobenzohopanes, secosteranes, tricyclic terpanes, and 30-nor-17α(H)-hopanes. These polycyclic alkanes are not ubiquitous but are very diagnostic of particular environmental conditions (demethylated hopanes mostly in shaly sediments, hexahydrobenzohopanes in evaporitic carbonates or marls). Their abnormal enrichment in *n*-alkane rich crudes may be regarded as a palaeobiodegradation indicator.

Acknowledgments

We are grateful to F. Bernard for his continuous support of this project over the years and to P. Maldonado for his interest and encouragement. We thank the Société Nationale Elf-Aquitaine for its financial support and permission to publish.

REFERENCES

Bergey (1986) *Manual of Systematic Bacteriology* (ed. J.P. Butler) *2*, Williams and Wilkins. pp. 1458–1471.

Bortz, L.C. (1987) Heavy-oil deposit, Great Salt Lake, Utah. In *Exploration for Heavy*

Crude Oil and Natural Bitumen (ed. R.R. Meyer), Am. Assoc. Pet. Geol. Studies in Geology N° 25 pp. 555–563, Tulsa.

BRAKSTAD, F. and GRAHL-NIELSEN, O. (1988) Identification of weathered oils. *Marine poll. Bull. 19*, 319–324.

BROOKS, P.W., FOWLER, M.G., and MACQUEEN, R.W. (1988) Biological marker and conventional organic geochemistry of oil sands/heavy oils, Western Canada Basin. *Org. Geochem. 12*, 519–538.

CASSANI, F. and EGLINTON, G. (1986) Organic geochemistry of Venezuelan extra-heavy oils 1. Pyrolysis of asphaltenes: A technique for correlation and maturity evaluation of crude oils. *Chem. Geol. 56*, 167–183.

CONNAN, J., RESTLE, A., and ALBRECHT, P. (1980) Biodegradation of crude oil in Aquitaine Basin. In *Advances in Organic Geochemistry-1979* (eds. A.G. Douglas and J.R. Maxwell) Pergamon Press. pp. 1–17.

CONNAN, J. (1984) Biodegradation of crude oils in reservoirs. In *Advances in Petroleum Geochemistry* (eds. J. Brooks and D.H. Welte) *1*, Academic Press. pp. 299–335.

CONNAN, J. and DESSORT, D. (1987) Novel family of hexacyclic hopanoid alkanes (C_{23}–C_{35}) occurring in sediments and oils from anoxic paleoenvironments. *Org. Geochem. 11*, 103–113.

CONNAN, J., LEVACHE, D., SALVATORI, T., RIVA, A., BURWOOD, R., and LEPLAT, P. (1987) Petroleum geochemistry in Angola: An AGELFI study—Abstract of 13th International Meeting on Organic Geochemistry, Venice, September 21–25.

CONNAN, J. (1988) Quelques secrets des bitumes archéologiques de Mésopotamie révélés par les analyses de géochimie organique pétrolière. *Bull. Centres Rech. Explor.-Prod. Elf-Aquitaine 12*, 759–787.

FOWLER, M.G. and BROOKS, P.W. (1987) Organic Geochemistry of Western Canada Basin tar sands and heavy oils. 2 Correlation of tar sands using hydrous pyrolysis of asphaltenes. *Energy & Fuels*, 459–467.

GOODWIN, N.S., PARK, P.J.D., and RAWLINSON, A.P. (1983) Crude oil biodegradation under simulated and natural conditions. In *Advances in Organic Geochemistry-1981* (eds. M. Bjorøy et al.) John Wiley & Sons. pp. 650–658.

HOWELL, V.J., CONNAN, J., and ALDRIDGE, A.K. (1984) Tentative identification of demethylated terpanes in non biodegraded and slightly biodegraded crude oils from the Los Llanos Basin, Colombia. *Org. Geochem. 6*, 63–92.

JIANG, ZHUSHENG, PHILP, R.P., and LEWIS, C.A. (1988) Identification of novel bicyclic alkanes from steroid precursors in crude oils from Kelamayi oil field in China. *Geochim. Cosmochim. Acta 52*, 491–498.

JONES, D.M., DOUGLAS, A.G., and CONNAN, J. (1987) Hydrocarbon distributions in crude oil asphaltene pyrolyzates. 1. Aliphatic compounds. *Energy & Fuels* 468–476.

LANDAIS, P. and CONNAN, J. (1986) Source rock potential and oil alteration in the uraniferous basin of Lodève (Hérault, France). *Sci. Geol. Bull. 39*, 293–314.

MACKENZIE, A.S. (1984) Application of biological markers in petroleum geochemistry. In *Advances in Petroleum Organic Geochemistry* (eds. J. Brooks and D.H. Welte) Academic Press. pp. 115–124.

MCKIRDY, D.M., ALDRIDGE, A.K., and YPMA, P.J.M. (1983) A geochemical comparison of some crude oils from Pre-Ordovician carbonate rocks. In *Advances in Organic Geochemistry-1981* (eds. M. Bjorøy et al.) John Wiley & Sons. pp. 99–107.

MOLDOWAN, J.M., SEIFERT, W.K., ARNOLD, E., and CLARDY, J. (1984) Structure proof and significance of stereoisomeric 28,30-bisnorhopanes in petroleum and petroleum source rocks. *Geochim. Cosmochim. Acta 48*, 1651–1661.

NOBLE, R., ALEXANDER, R., and KAGI, R.J. (1985) The occurrence of bisnorhopane, trisnorhopane and 25-norhopanes as free hydrocarbons in some Australian shales. *Org. Geochem. 8*, 171–176.

PHILP, R.P. (1983) Correlation of crude oils from the San Jorges Basin, Argentina. *Geochim. Cosmochim. Acta 47*, 267–275.

REED, W.E. (1977) Molecular compositions of weathered petroleum and comparison with its possible source. *Geochim. Cosmochim. Acta 41*, 237–247.

RESTLE, A. (1983) Etude de nouveaux marqueurs biologiques dans les pétroles biodégradés: cas naturels et simulation in vitro. Thèse Docteur ès Sciences. Université Louis Pasteur, Strasbourg.

RUBINSTEIN, I., STRAUSZ, O.P., SPYCKERELLE, C., CRAWFORD, R.J., and WESTLAKE, D.W.S. (1977) The origin of the oil sand bitumens of Alberta: A chemical and a microbiological simulation study. *Geochim. Cosmochim. Acta 41*, 1341–1353.

RULLKÖTTER, J. and WENDISCH, D. (1982) Microbial alteration of 17α(H)-hopanes in Madagascar asphalts: Removal of C-10 methyl group and ring opening. *Geochim. Cosmochim. Acta 4*, 1545–1553.

SANDSTROM, M.W. and PHILP, R.P. (1984) Biological marker analysis and stable carbon isotopic composition of oil seeps from Tonga. *Chem. Geol. 43*, 167–180.

SCHMID, J.C. (1986) Marqueurs biologiques soufrés dans les pétroles. Thèse de Doctorat. Université Louis Pasteur, Strasbourg.

SEIFERT, W.K. and MOLDOWAN, J.M. (1979) The effect of biodegradation on steranes and terpanes in crude oils. *Geochim. Cosmochim. Acta 43*, 111–126.

SEIFERT, W.K. and MOLDOWAN, J.M. (1981) Paleoreconstruction by biological markers. *Geochim. Cosmochim. Acta 45*, 738–794.

SEIFERT, W.K., MOLDOWAN, J.M., and DEMAISON, G.J. (1984) Source correlation of biodegraded oils. *Org. Geochem. 6*, 633–643.

SINNINGHE DAMSTE, J.S. and DE LEEUW, J.W. (1987) The origin and fate of isoprenoid C_{20} and C_{15} sulphur compounds in sediments and oils. *Intern. J. Environ. Anal. Chem. 28*, 1–19.

SINNINGHE DAMSTE, J.S., DE LEEUW, J.W., KOCK-VAN DALEN, A.C., DE ZEEUW, M.A., DE LANGE, F., RIJPSTRA, W.I.C., and SCHENCK, P.A. (1987) The occurrence and identification of series of organic sulfur compounds in oils and sediments extracts. I. A study of Rozel Point Oil (USA). *Geochim. Cosmochim. Acta 51*, 2369–2391.

TESCHNER, M. and WEHNER, H. (1985) Chromatographic investigations on biodegraded crude oils. *Chromatographia 20*, 407–416.

TRENDEL, J.M., GUILHEM, J., CRISP, P., REPETA, D., CONNAN, J., and ALBRECHT, P. (1990) Identification of two C-10 demethylated C_{28} hopanes in biodegraded petroleum. *J. Chem. Soc., Chem. Commun., 5,* 424–425.

VOLKMAN, J.K., ALEXANDER, R., KAGI, R.I., and WOODHOUSE, G.W. (1983) Demethylated hopanes and their application in petroleum geochemistry. *Geochim. Cosmochim. Acta 47*, 785–794.

ZHANG DAJIANG, HUANG DIFAN, and LI JINCHAO (1988) Biodegraded sequence of Karamay oils and semi-quantitative estimation of their biodegraded degrees in Junggar Basin, China. In *Advances in Organic Geochemistry-1987* (eds. L. Mattavelli and L. Novelli) Pergamon Press. pp. 295–302.

18

Biological Markers in Petroleum Asphaltenes: Possible Mode of Incorporation

Sylvie Trifilieff, Odette Sieskind, and Pierre Albrecht

Abstract. *n*-Heptane precipitated asphaltenes from sulfur-rich crude oils have been submitted to oxidative degradation with ruthenium tetroxide. The carboxylic acids formed during this procedure are mostly comprised of monocarboxylic acids, usually dominated by linear components (C_9–C_{30}). Linear dicarboxylic acids (C_{11}–C_{26}) and some benzene polycarboxylic acids also appear in the more mature petroleums.

The structures of the branched and cyclic acids and their carbon number distributions (e.g., acyclic isoprenoid acids, C_{15}–C_{21}; hopanoic acids, C_{28}–C_{36}; 3β-carboxysteranes, C_{28}–C_{30}) are consistent with a selective oxidation of aromatic subunits and concomittant liberation of alkyl moieties attached to them. These results indicate that the corresponding alkyl chains may have become incorporated into the asphaltenes by Friedel-Crafts-type reactions operating on a portion of the alcohols or olefins during diagenesis or early maturation. A simulation experiment confirmed such a possibility in the case of the steroids.

However, oxidation of possible biologic-sourced alkyl aromatic molecules present in asphaltenes and cleavage of other types of functions could also explain the formation of some oxidation products.

INTRODUCTION

Although petroleum asphaltenes have been the subject of many studies, their structural elucidation is still far from clear (Speight, 1984 and 1985). A very general picture developed by Yen (1974) views the asphaltenes as being composed of aromatic subunits cross-linked by aliphatic chains. Very few well-defined structural elements of asphaltenes have been established and their mode of linkage to the macromolecular network is unclear (e.g., Rubinstein et al., 1979, and references therein).

In this study, we have submitted *n*-heptane precipitated asphaltenes from sulfur-rich crude oils to an oxidative degradation with ruthenium tetroxide (Carlsen et al., 1981), a method previously used for the oxidation of coals (Stock and Tsé, 1983) and various petroleum fractions (Mojelsky et al., 1985). This method, which has been checked on model compounds, typically degrades aromatic subunits and eventually liberates alkyl moieties attached to them as carboxylic acids bearing one extra carbon in the form of a carboxylic acid function. Our results show that the asphaltenes yield substantial amounts of linear, isoprenoid, and polycyclic acids. Furthermore, the carbon-number distributions of the branched and cyclic acids give hints as to a possible mode of incorporation of some biological markers into the asphaltene macromolecules.

SAMPLES STUDIED

The asphaltenes of four crude oils precipitated with *n*-heptane have been studied by oxidative degradation (Table 18.1). The four oils are sulphur rich (6–15%) and are from source rocks that were deposited in carbonate evaporitic environments. The study will be illustrated by results obtained from Lameac and Rozel Point samples. The data from Bati Raman and Boscan show great similarities with those from Lameac (Trifilieff, 1987).

Table 18.1 CRUDE OILS SELECTED FOR OXIDATION OF THEIR nC_7-ASPHALTENES WITH RUTHENIUM TETROXIDE (% OF ACIDS IS IN ALL CASES RELATIVE TO ASPHALTENES)

Crude Oil	Origin	Age	Asphaltenes %	% of Acids from Oxidation of Asphaltenes	% of Free Acids from Crude Oil
Bati Raman	Turkey	Cretaceous	19	3.6	0.3
Boscan	Venezuela	Cretaceous	12	4.5	0.7
Lameac	France (Aquitaine Basin)	Cretaceous	10	7.4	0.3
Rozel Point	USA (Utah)	? Seep Oil	30	2.9	0.4

EXPERIMENTAL PART

General

Mass spectra were obtained by electron impact (70 eV) on an LKB 9000S GC-MS system equipped with a PDP 11E/10 computer. GC conditions were the following: fused silica columns (SE 30 25 m × 0.3 mm), 100–290°C/min, 3°C/min, 3 ml helium/min.

Gas-chromatographic analyses were carried out on a Carlo Erba 4160 gas chromatograph equipped with glass columns (25–40 m, 0.3 mm) coated with apolar (SE 54, OV 101) or polar phases (OV 1701, Superox). Hydrogen was used as a carrier gas (2–3 ml/min).

Liquid-chromatographic separations were made on silica-gel columns (40–63 μm, MERCK) or thin layer plates (MERCK $60F_{254}$) of varying thickness (0.25–2.00 mm).

Precipitation of Asphaltenes

The asphaltenes of the crude oils were obtained following the method developed by Speight (1984). The viscosity of the oils was reduced by slightly heating them on a water bath. To 50 g of petroleum were added 2 liters of *n*-heptane while stirring. After 12 hours of decantation at room temperature, the precipitated asphaltenes were separated from the maltenes by filtration on sintered glass under argon. The asphaltenes were subsequently redissolved in a minimal amount of toluene (10 ml/g of asphaltenes) and reprecipitated by adding a large excess of *n*-heptane (500 ml/g of asphaltenes). After 12 hours of decantation at room temperature, the asphaltenes were again separated by filtration (as mentioned) and stored in glass vessels under argon.

Degradation of Asphaltenes with Ruthenium Tetroxide

The oxidation of asphaltenes was carried out following a method adapted from the procedure described by Carlsen et al. (1981) in which RuO_4 is used as a catalyst in the presence of CH_3CN and excess $NaIO_4$.

Typically, 1 g of asphaltenes was dissolved in 10 ml of CCl_4 and 10 ml of CH_3CN; 20 ml of doubly distilled water, 5 g of $NaIO_4$ and 50 mg of $RuCl_3$ were added, and the mixture was stirred for 24 hours at room temperature under argon.

The mixture was subsequently filtered over sintered glass (No. 4). The filtrate was acidified to pH 1 and extracted three times with chloroform. The solid residue was dissolved or suspended in water, acidified and the mixture was also extracted with chloroform. The chloroform phases were assembled and the solvent was evaporated giving the total organic extract. The latter was redissolved in a small amount of chloroform and methanol and esterified with diazomethane (Boer and Baker, 1963). The esterified mixture was separated on a silica-gel column (200 g) and chromatographed (methylene chloride) to give a fraction of medium polarity containing both the methylesters of mono and dicarboxylic acids, which were further purified by thin-layer chromatography (TLC) (CH_2Cl_2 eluent). In one case (Rozel Point), the branched and cyclic mono- and diacid methyl esters were obtained by removal of the linear components by urea adduction (Dastillung, 1976).

Test Experiments

In order to check whether free acids are coprecipitated in the preparation procedure of asphaltene, the latter were dissolved in chloroform and treated with diazoethane (Arndt, 1943). The resulting mixture was then divided into two aliquots. The first one was treated by the ruthenium tetroxide oxidation procedure described, leading to a mixture of ethyl and methyl esters. The second aliquot was separated by column chromatography over silica gel into a nonpolar and a polar fraction by elution with CH_2Cl_2 and $CHCl_3/MeOH/H_2O$ (65:25:4 by volume, respectively). The nonpolar fraction was further purified by SiO_2 TLC and the zone corresponding to ethyl esters of mono and dicarboxylic acids was separated. The polar fraction was further studied in order to check if some coprecipitated acids eventually escaped esterification with diazoethane. It was, therefore, separated on a silica-gel column impregnated with KOH (McCarthy and Duthie, 1962). After elution of the neutrals with ether, the acidic fraction was eluted with 2 to 10 percent of formic acid in ether and esterified with diazomethane. The zone corresponding to mono and dicarboxylic acids was separated by TLC.

The experiments showed that in all cases free coprecipitated acids were present in small amounts, but negligible in comparison with those formed on oxidation with ruthenium tetroxide.

Separation of the Free Acids Present in the Crude Oils

The crude oils (15–50 g) were chromatographed on KOH impregnated silica gel following the method of McCarthy and Duthie (1962), as indicated. The acid fraction was esterified with diazomethane. Purification of the ester fraction by SiO_2 TLC yielded the mono and diacids as their methylesters.

Ruthenium Tetroxide Oxidation of Model Compounds

1-phenyloctadecane and 3,7,11,15-tetramethyl-1-phenylhexadecane were prepared by synthesis starting from octadecan-1-ol and phytol, respectively (Trifilieff, 1987).

Oxidation of 1-phenyloctadecane with ruthenium tetroxide under the same conditions as those used for the oxidative degradation of the asphaltenes yielded nonadecanoic acid as the major compound (68%) along with octadecanoic acid (11%) and heptadecylphenylketone (5%). Oxidation of 3,7,11,15-tetramethyl-1-

phenylhexadecane yielded 4,8,12,16-tetramethyl heptadecanoic acid as the major compound (55%) and small amounts of phytanoic acid (5%).

3β-phenylcholestane was prepared by synthesis starting from cholestan-3β-ol (Sieskind, unpublished results). Treatment of 3β-phenyl cholestane with ruthenium tetroxide under the same conditions as mentioned essentially yielded 3β-carboxycholestane (60%), which was identified by comparison with the synthetic reference compound that follows.

Synthesis of 3-Carboxycholestanes

3α- and 3β-carboxycholestanes were prepared by synthesis starting from cholestan-3-one as described elsewhere (Trifilieff et al., in preparation).

RESULTS

Ruthenium tetroxide oxidation according to the method developed by Carlsen et al. (1981) generally degrades aromatic entities (unless they are deactivated by electron withdrawing groups) and liberates the alkyl moieties attached to them as carboxylic acids bearing one extra carbon coming from the oxidative degradation of the aromatic system.

Application of this method, in which ruthenium tetroxide is used in catalytic amounts in the presence of acetonitrile, to coals and coal macerals indeed produces series of linear mono- and dicarboxylic acids, mostly with short chains, as well as mixture of aromatic acids. The latter are presumably formed by the breakdown of the aromatic subunits occurring in the coal macromolecules (Stock and Tsé, 1983; Stock and Wang, 1985 and 1986; Blanc and Albrecht, 1990). Similar products have also been formed from various petroleum fractions, including C_5-asphaltenes (Mojelsky et al., 1985).

Oxidation of Model Compounds

In order to assess the limits of this oxidation reaction under our conditions, we have tested it on two alkylbenzene model compounds, 1-phenyloctadecane and 3,7,11,15-tetramethyl-1-phenylhexadecane (see Experimental Part section).

Oxidation of 1-phenyloctadecane produces the expected compound, nonadecanoic acid, in relatively high yield (68%), but also yields minor amounts of stearic acid (11%) and of heptadecylphenyl ketone (5%). Likewise, degradation of 3,7,11,15-tetramethyl-1-phenylhexadecane generates the expected compound, 4,8,12,16-tetramethylheptadecanoic acid, as the dominant product (55%); however, a minor amount of phytanic acid (5%) is also formed. In both cases, the formation of minor side-products appears to be due to oxidation at the benzylic position, followed by a Bayer-Villiger type of reaction and hydrolysis of the phenyl ester under the conditions used.

General Aspects; Linear Acids Formed from Asphaltenes

Ruthenium tetroxide oxidation of the C_7-asphaltenes, obtained from four selected crude oils, produces between 2.9 and 7.4 percent of acids (Table 18.1).

Oxidation of the asphaltenes from Rozel Point crude oil yields almost exclusively a mixture of monocarboxylic acids (Fig. 18.1) dominated by linear chains (C_9–C_{28}). However, acyclic isoprenoid acids and polycyclic acids are also present in substantial amounts. The linear acids display a maximum around C_{15} and show a significant odd carbon-number predominance between C_{13} and C_{23}. If these acids are formed by the mechanism described, this would mean that alkyl chains with an even predominance are attached to aromatic moieties in the asphaltene macromolecules. In comparison, the total alkane distribution of the immature petroleum is dominated by acyclic isoprenoid and steroid structures and is dissimilar to the acids formed by asphaltene oxidation (Fig. 18.2). Remarkably, the *n*-alkanes show an even predominance between C_{18} and C_{26}, a feature that is often observed in petroleums and sediments deposited in evaporitic carbonate environments (e.g., Connan et al., 1986 and references therein). The free *n*-alkanes may be formed by reduction of predominantly even linear acids (or alcohols), leading to a predominance of even carbon numbered linear alkanes. Based on the asphaltene-oxidation results, the same even carbon numbered chains may be incorporated by attachment to aromatic subunits into the asphaltenes.

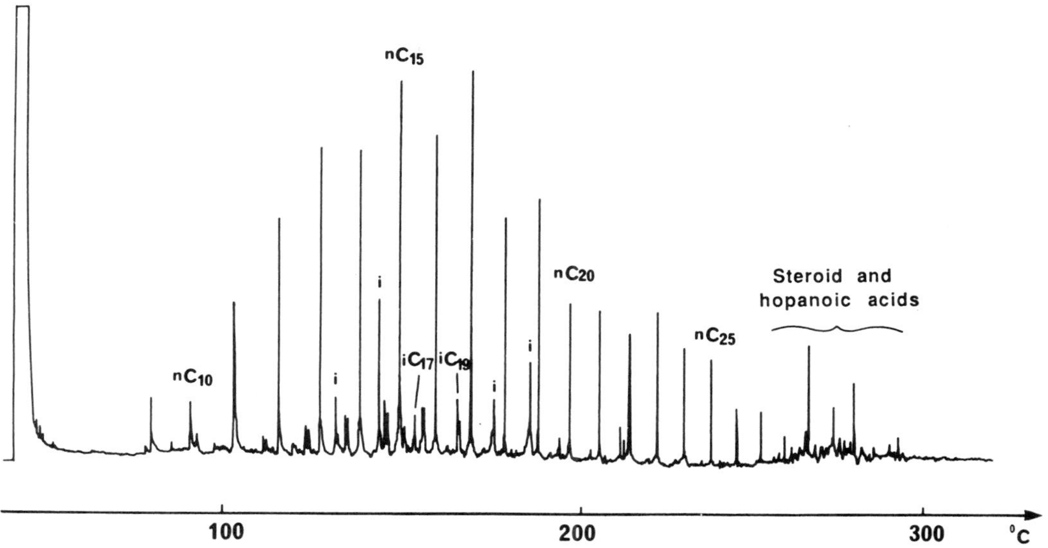

Figure 18.1 Gas chromatogram of the methyl esters of the acids obtained on RuO_4 oxidation of the asphaltenes from Rozel Point crude oil. Conditions: SE 54, 35 m × 0.32 mm, 40–300°C, 4°C/min. Carbon numbers are for the acids. *n* = linear monocarboxylic; *i* = isoprenoid monocarboxylic.

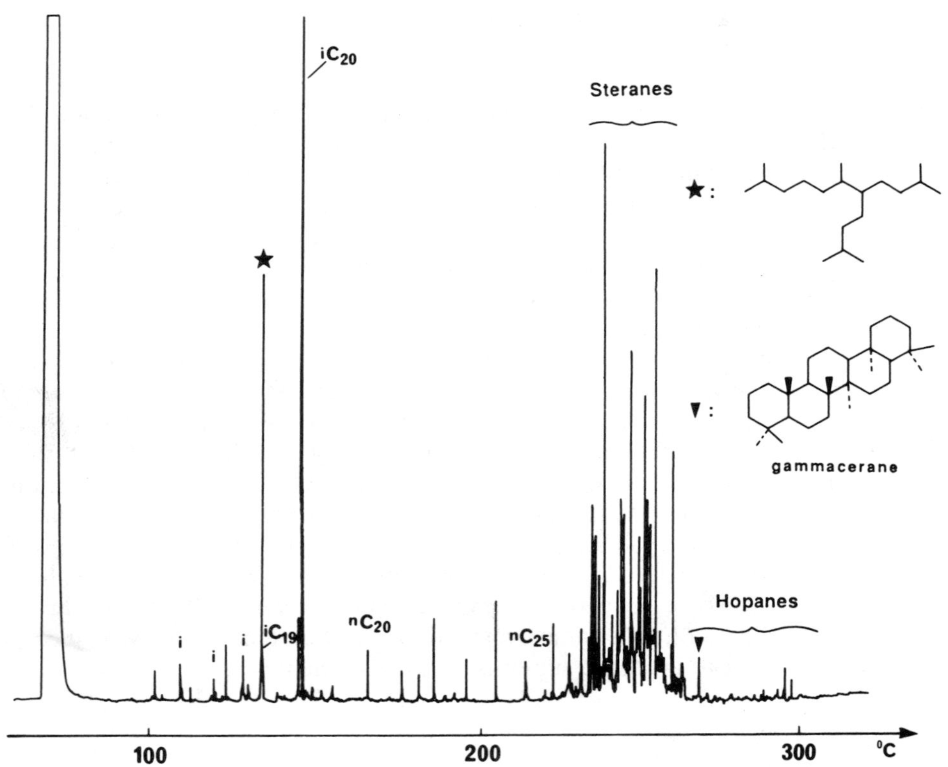

Figure 18.2 Gas chromatogram of the total alkanes from Rozel Point crude oil. Conditions: SE 54, 35 m × 0.32 mm, 40–100°C, 10°C/min., 100–300°C, 4°C/min. n = linear; i = isoprenoid.

The more mature Lameac crude oil shows a very different picture (Fig. 18.3). In this case, a higher proportion of acids is obtained from the asphaltenes (Table 18.1). Their distribution is again dominated by linear chains (C_9–C_{34}) with a maximum around C_{12}–C_{15} but, unlike in the case of Rozel Point, it is smooth and shows no predominance. α,ω-Dicarboxylic acids are also present (C_{11}–C_{26}) but do not show any predominance either. Acyclic isoprenoid acids and polycyclic acids can be observed, although in much lesser amounts than for Rozel Point crude oil. Finally, the small amounts of benzene polycarboxylic acids that are detected could represent end products of the oxidation of some aromatic subunits occurring in the asphaltenes. High proportions of these compounds are obtained from the oxidation of coals, which are usually enriched in aromatic structures (Stock and Wang, 1986; Blanc and Albrecht, 1990). In contrast, the gas chromatogram of the total alkanes of the same oil (Fig. 18.4) displays a distribution typical of petroleums from evaporitic carbonate environments: slight even carbon number predominance of n-alkanes (C_{16}–C_{30}), low pristane/phytane ratio, significant contribution of penta-

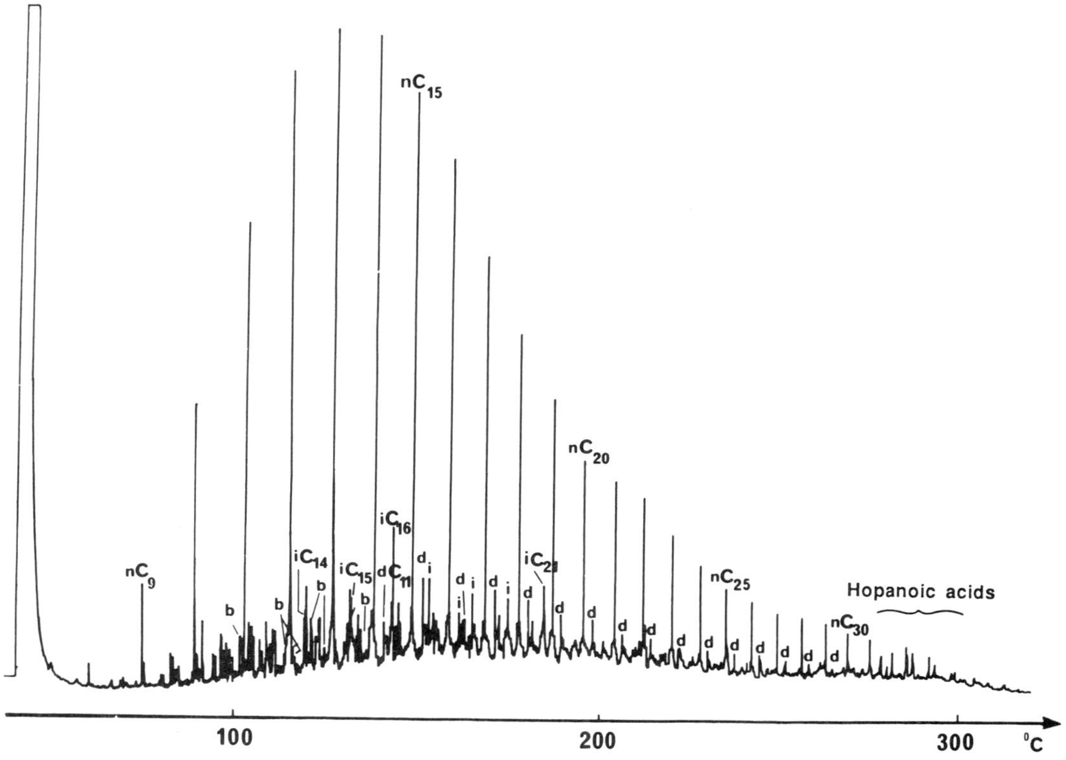

Figure 18.3 Gas chromatogram of the methyl esters of the acids obtained on RuO_4 oxidation of the asphaltenes from Lameac crude oil. Conditions as in Figure 18.1. Carbon numbers are for the acids. n = linear monocarboxylic; i = isoprenoid monocarboxylic; d = linear dicarboxylic; b = benzene polycarboxylic.

cyclic hopanoids (see below) (Palacas, 1983; Rullkötter et al., 1984; Hussler, 1985; Connan et al., 1986; Didyk et al., 1978).

The acids obtained from oxidation of asphaltenes of the two other crude oils (Bati Raman and Boscan) show great similarities with those from Lameac and, therefore, will not be discussed any further.

Branched and Cyclic Acids from Oxidation of Asphaltenes

The branched and cyclic acids from the oxidation of the asphaltenes of Rozel Point crude oil were obtained by urea adduction of the straight-chain components. They are dominated by a series of regular acyclic isoprenoid acids (C_{15}–C_{21}) (Eglinton et al., 1966; Douglas et al., 1970; Spyckerelle, 1973) and several families of polycyclic acids (Fig. 18.5; see Appendix).

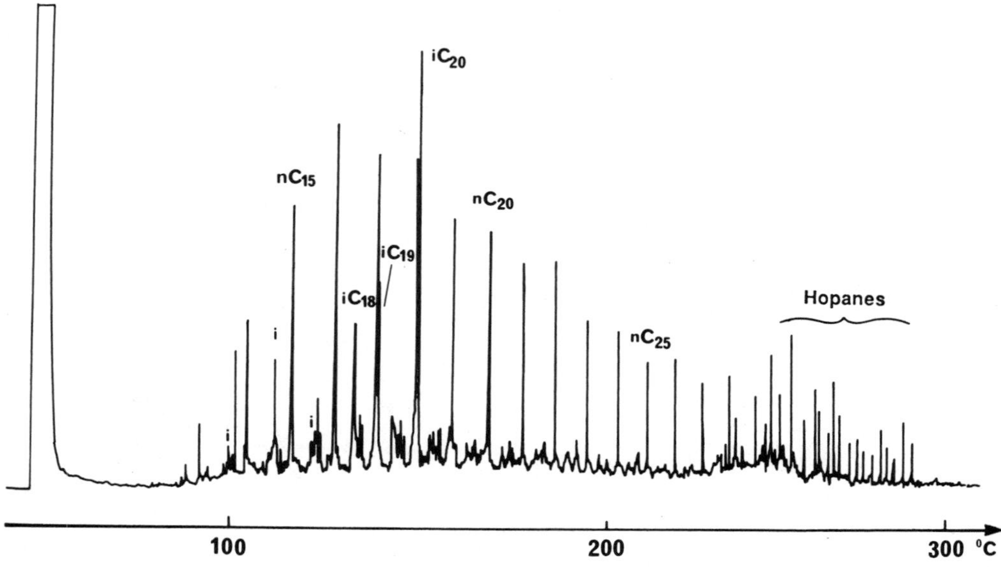

Figure 18.4 Gas chromatogram of the total alkanes from Lameac crude oil. Conditions as in Figure 18.1. *n* = linear; *i* = isoprenoid.

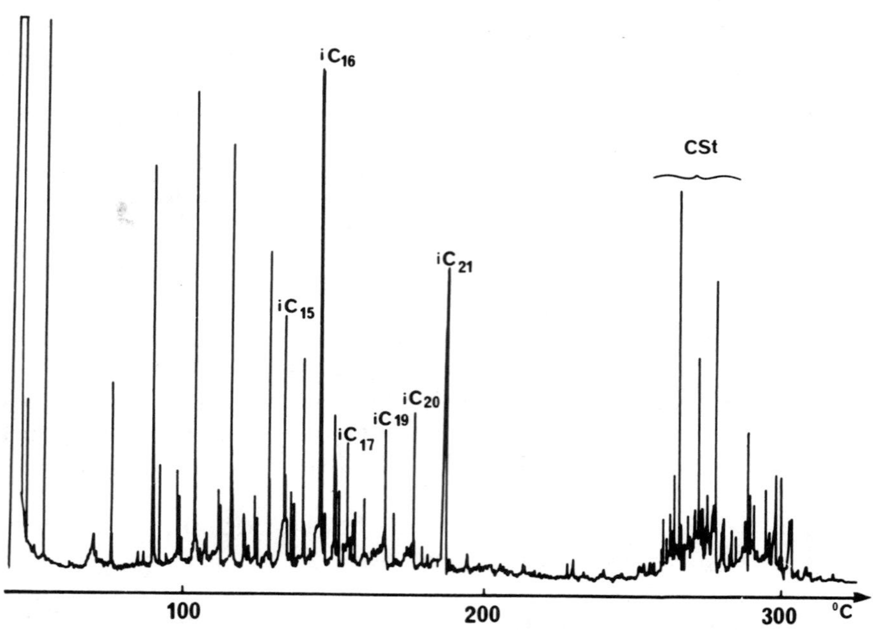

Figure 18.5 Gas chromatogram of the branched and cyclic acids (as their methyl esters) obtained on RUO_4 oxidation of the asphaltenes from Rozel Point crude oil. Conditions as in Figure 18.1. *i* = isoprenoid monocarboxylic; CSt = 3β-carboxysteranes (C_{28}–C_{30}). Short-chain linear acids are incompletely removed by urea adduction.

The most abundant isoprenoid acids are the C_{16} and C_{21} members, which probably form by oxidative cleavage of farnesyl and phytyl chains bound to an aromatic unit in the asphaltene molecules.

More intriguing is the prominent occurrence of a series of steroid acids, which have been conclusively identified by comparison with a C_{27} synthetic standard as 3β-carboxycholestanes (C_{28}–C_{30}) (Trifilieff et al., in preparation). These acids bear an extra carbon at position 3 in comparison with regular steroids and may arise from an oxidative cleavage of the corresponding steranes attached via position 3 to an aromatic subunit of the asphaltenes. The methyl esters of this new type of steroid acids are shown on the *m/z* 275 fragmentogram (Fig. 18.6), which is homologous to the *m/z* 217 fragmentogram of the steranes. As observed for the free steranes occurring in the alkane fraction of this oil (Fig. 18.7), the predominance of the biochemical stereoisomers (20*R* 5αH,14αH,17αH) reflects the high degree of immaturity of this oil.

Small amounts of carboxysteranes bearing an extra methyl group were also detected (C_{29}–C_{31}). Their mass spectra [*m/z* 444; 458; 472 ($M^{+\cdot}$ 70%); 289 (100%);

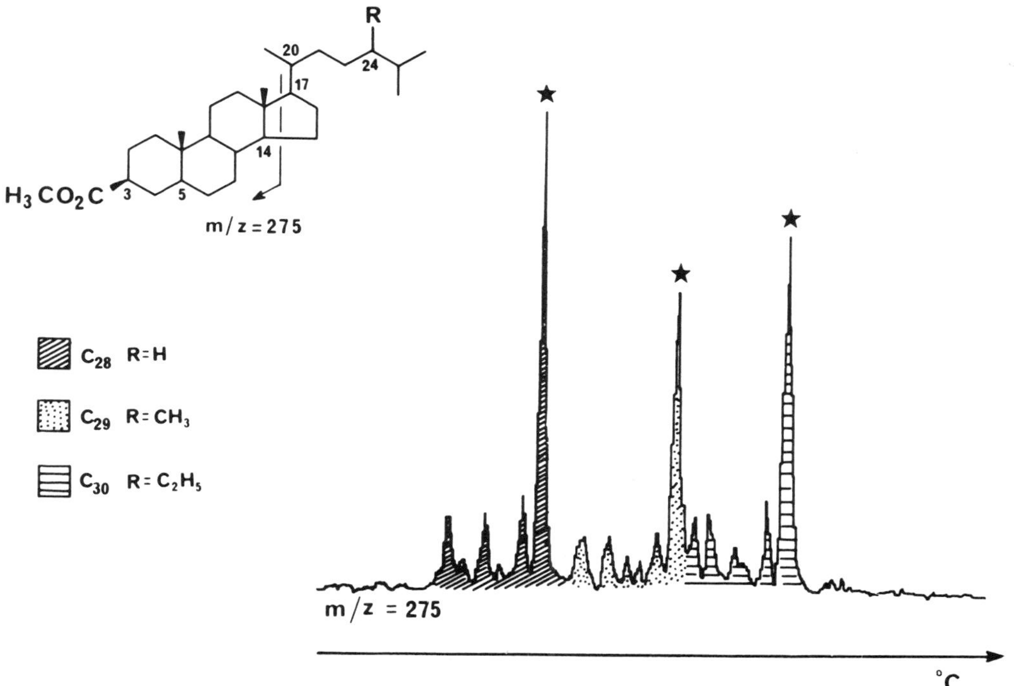

Figure 18.6 Mass fragmentogram *m/z* 275 showing the distribution of carboxysteranes (as their methyl esters) obtained on RuO_4 oxidation of the asphaltenes of Rozel Point crude oil. Conditions: SE 30, 25 m × 0.30 mm, 100–290°C, 3°C/min. * = 20*R* 3β-carboxysteranes.

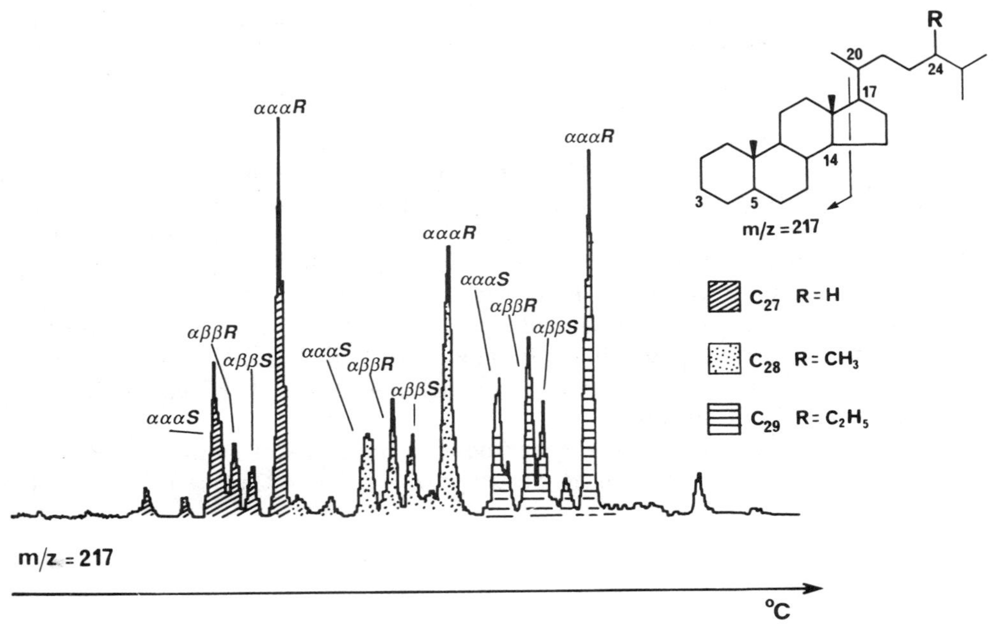

Figure 18.7 Mass fragmentogram *m/z* 217 showing the distribution of steranes from Rozel Point crude oil. Conditions as in Figure 18.6.

221 (35%)] indicate that they could correspond to 3β-carboxysteranes with an extra methyl group on ring A or B, perhaps at position 4. However, their precise structure has not yet been elucidated.

Besides the 3β-carboxysteranes, a series of hopanoic acids (Ensminger et al., 1974; Van Dorsselaer, 1975; Jaffé et al., 1988) is also present in the oxidation products of the Rozel Point asphaltenes (Fig. 18.8). They are essentially composed of 17αH,21βH isomers and dominated by the C_{32} homologues (22*R* and 22*S*). Their carbon-number distribution ranges from C_{30} to C_{36}. The existence of the C_{36} homologue is particularly striking because it may arise from oxidation of a C_{35} hopanoid moiety attached to an aromatic subunit in the asphaltenes. The lower homologues may also ultimately originate from the incorporation of shorter chain hopanoids into the asphaltenes.

In contrast, the *m/z* 191 fragmentogram obtained from the alkane fraction of the oil shows a very different distribution in which the C_{30} and C_{35} 17αH,21βH hopanes are predominant, along with gammacerane. The hopanoic acids obtained by oxidation are apparently more immature than the corresponding alkanes, based on the ratios of the 22*R* and 22*S* diastereoisomers and the occurrence of small amounts of a 17βH,21βH isomer. The lower maturity of the asphaltene-oxidation products is confirmed by steroid acids formed on oxidation, which show a more immature distribution compared to free steranes occurring in the oil (Figs. 18.6

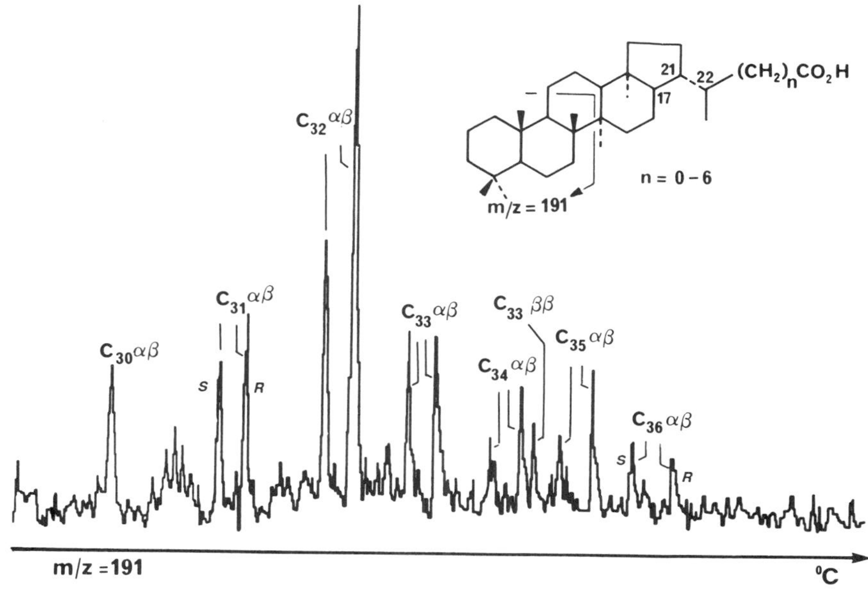

Figure 18.8 Mass fragmentogram *m/z* 191 showing the distribution of hopanoic acids (as their methyl esters) obtained upon RuO_4 oxidation of the asphaltenes of Rozel Point crude oil. Conditions as in Figure 18.6. αβ: 17αH,21βH; ββ: 17βH,21βH; *S*, *R*: configurations at C-22.

and 18.7). It seems that these molecules are relatively protected towards isomerization as compared with their free counterparts due to their incorporation in the macromolecular matrix of the asphaltenes, which is in agreement with previous observations (e.g., Rubinstein et al., 1979).

The branched and cyclic acids obtained upon RuO_4 oxidation of the Lameac crude oil asphaltenes also show a series of acyclic isoprenoid acids (C_{15}–C_{21}) (Fig. 18.3). Although the acids are less abundant than in the case of the Rozel Point sample, it is quite remarkable that the series also culminates at C_{21}, which again may correspond to the oxidation of a phytyl moiety attached to an aromatic ring in the asphaltenes. Steroid acids appear to be absent. In contrast, C_{28}–C_{36} hopanoic acids are abundant with the C_{32} homologues predominating (Fig. 18.9). In the hopanoic acids liberated from the Lameac asphaltenes, thermodynamic equilibrium seems to have been reached at position 22. Free hopanes contained in the corresponding oil show a distribution ranging from C_{27} to C_{35}, are dominated by the C_{30} αβ homologue, and display thermodynamic equilibrium at position 22 in the higher homologues.

In all cases, the amounts of free acids obtained from the crude oils are one order of magnitude less than those obtained by oxidation of the asphaltenes with ruthenium tetroxide (Table 18.1). Furthermore, they generally show different dis-

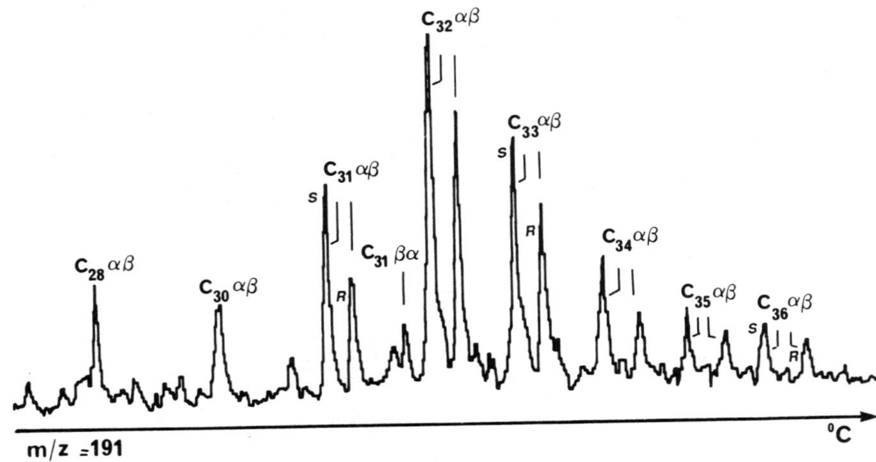

Figure 18.9 Mass fragmentogram *m/z* 191 showing the distribution of hopanoic acids (as their methyl esters) obtained on RuO_4 oxidation of the asphaltenes of Lameac crude oil. Conditions as in Figure 18.6. αβ: 17αH,21βH; *S*, *R*: configurations at C-22.

tributions which leave no ambiguity as to the origin of the acids from oxidation of asphaltenes rather than from liberation of acids coprecipitated with the asphaltenes (Trifilieff, 1987).

DISCUSSION

Given the specificity of RuO_4 oxidation on our model compounds, most of the acids derived from crude oil asphaltenes are likely to arise from the oxidation of alkyl moieties attached to an aromatic subunit. In each case, the extra carboxylic carbon comes from the oxidation of the aromatic system (Fig. 18.10).

There are several clear arguments in favor of this hypothesis: (1) the carbon numbers of the major acyclic isoprenoids acids (C_{16} and C_{21}), the largest hopanoic acid (C_{36}), and the major steroid acid series (C_{28}–C_{30}); (2) the odd predominance of the *n*-alkanoic acids in the case of Rozel Point asphaltene-oxidation products; and (3) the position of the carboxylic group adjacent to the carbon bearing a functionality in the most likely biological precursors (e.g., phytol, bacteriohopanetetrol, sterols). The existence of 3β-carboxysteranes among the Rozel Point asphaltene-oxidation products is particularly striking in this respect.

However, there are quite a few differences between the oxidation products obtained for the asphaltenes of Rozel Point and Lameac crude oils. For example, the acids formed by oxidation of the Rozel Point asphaltenes are highly dominated by monocarboxylic acids, whereas the Lameac oxidation products contain significant amounts of α,ω-dicarboxylic acids. This suggests that long alkyl chains could

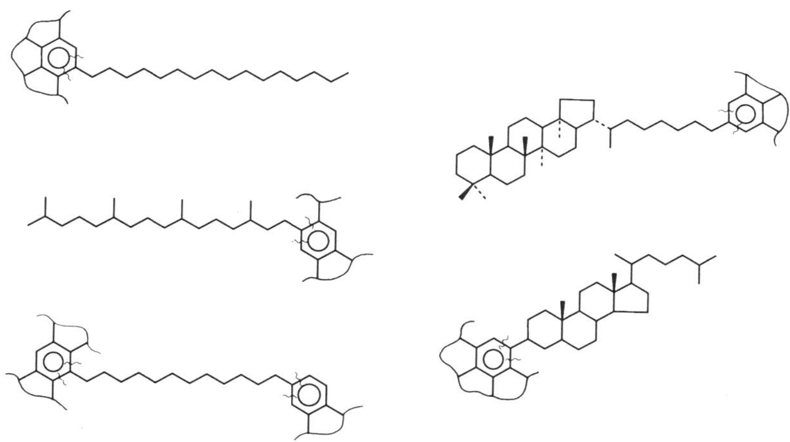

Figure 18.10 Structural entities possibly occurring in C_7-asphaltenes of sulfur-rich crude oils and from which linear and polycyclic acids may be cleaved off by treatment with ruthenium tetroxide.

act as a cross-link between aromatic subunits in the Lameac asphaltenes. Furthermore, the presence of some benzene polycarboxylic acids in the Lameac asphaltene-degradation products suggests that its aromatic subunits might be more extended than those within the asphaltenes of Rozel Point.

Finally, our results show that the structures of the acids obtained upon oxidation of the asphaltenes reflect the degree of maturity of a crude oil. Therefore, these acids may be used as maturity parameters, along with other molecules occurring in the alkane, aromatic, and polar fractions. They could become particularly useful in the case of biodegraded petroleums in which many of the "free" molecules are altered by the biodegradation process (Seifert and Moldowan, 1979).

One basic question which one might raise at this stage is to understand how the alkyl structures became linked into the asphaltene macromolecules. In this respect, a possible explanation has been furnished by previous results obtained in our laboratory in the case of steroids. When cholestanol is heated under mild conditions (refluxed in benzene for 1 hr) in the presence of K-10 montmorillonite clay, a good yield (40%) of 3β-phenylcholestane is formed via a Friedel-Crafts-type reaction operating on the alcohol. The reaction is surprisingly regiospecific, since no 2-phenylcholestane is observed (Sieskind, unpublished results). Oxidation of 3β-phenylcholestane with ruthenium tetroxide under the same conditions used in the current study produces the corresponding 3β-carboxycholestane in good yield.

This result suggests that stanols and eventually other alcohols (hopanols, isopranols) may be linked to aromatic molecules or subunits in the asphaltenes (or in any other fraction) via Friedel-Crafts-type reactions, occurring during diagenesis or early maturation (Fig. 18.11). Acid catalysis may be provided by certain clay

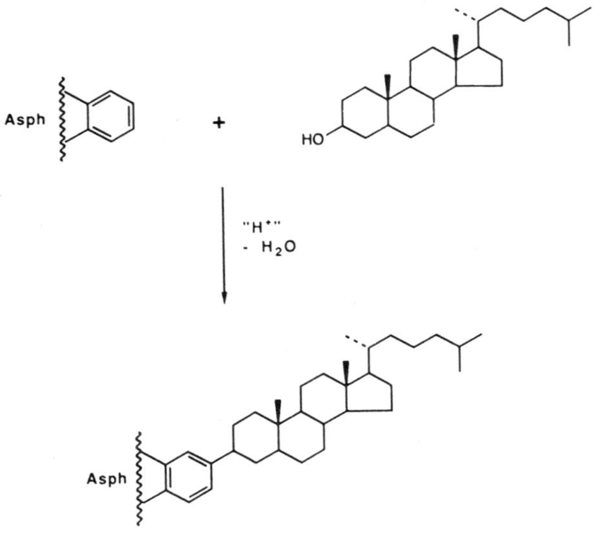

Asph
+
HO
"H⁺"
- H_2O
Asph

Figure 18.11 Possible mode of incorporation of steroids into asphaltenes via a Friedel-Crafts-type alkylation operating on cholestan-3β-ol.

minerals or by acid sites present in the organic matter. By such a mechanism, bacteriohopanetetrol, the likely bacterial precursor of most sedimentary hopanoids (Ourisson et al., 1984), may become attached to an aromatic system, with the last four carbons of the side chain forming an extra aromatic ring by loss of two molecules of water (Fig. 18.12). Alternatively, RuO_4 oxidation of a compound already containing an aromatic ring, such as a thienyl hopane which occurs in sulfur rich sediments and crude oils (Valisolalao et al., 1984), could be incorporated into the asphaltenes by reacting with an alcohol or olefinic group existing in the asphaltenes. Ruthenium tetroxide oxidation is likely to also liberate a C_{32} hopanoic acid from such an entity (Fig. 18.12). Chromanes (Sinninghe Damsté et al., 1987) could likewise become incorporated into the asphaltenes by a Friedel-Crafts-type reaction and liberate C_{21} isoprenoid acid by oxidation.

In the case of straight chains, the mechanism would *a priori* not favor attachment via the terminal, less stable carbocation. We have indeed confirmed this point with a linear primary alcohol under the same conditions as used for cholestanol. Contrary to our expectation, the formation of 1-phenylalkane was, however, not negligible (Sieskind, unpublished results).

Another possible explanation is the existence in the asphaltenes of alkyl aromatic subunits directly inherited from biological material. Such molecules are not known at the present time in the case of steroids or hopanoids. However, in the case of the acyclic isoprenoids and tocopherols, which have been identified in sediments (Brassell et al., 1983), or ubiquinones, which are widespread in living organisms (Harwood and Russell, 1983) may provide adequate starting materials susceptible to incorporation into the asphaltenes. In particular, tocopherol-type molecules may liberate a C_{21} isoprenoid acid under our oxidation procedures.

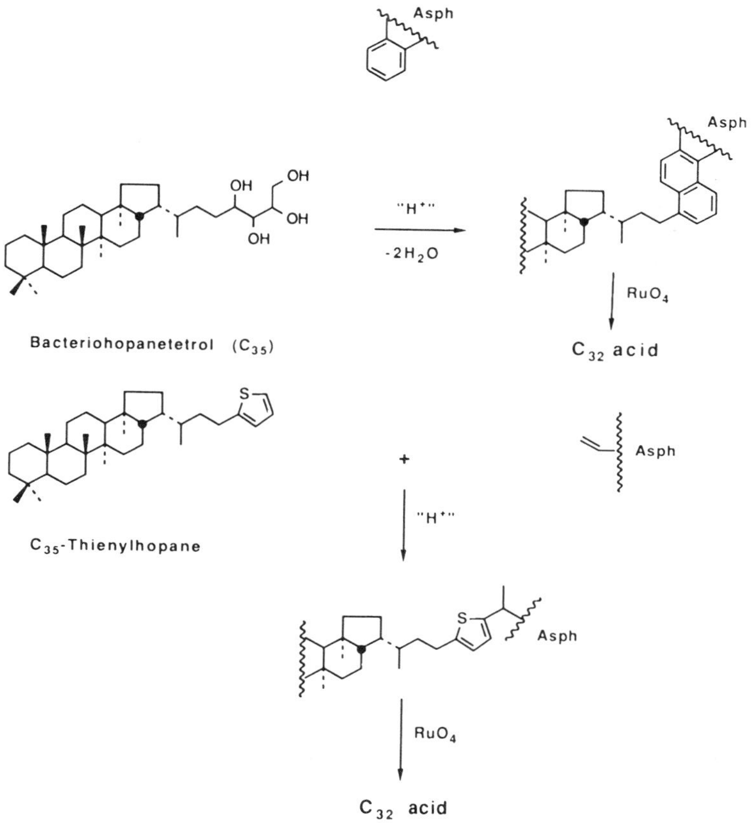

Figure 18.12 Possible mode of incorporation of hopanoid skeletons into asphaltenes via Friedel-Crafts-type reactions. Alcohols or carboxylic acid functions occurring in asphaltenes may constitute other reactive species.

Finally, it is possible that RuO_4 oxidation may cleave other functionalities, a point that has not been checked carefully so far. Preliminary data obtained in our laboratory have shown that, following oxidation of ethers to esters, these products are to a certain extent hydrolyzed to the corresponding acids in the conditions of the oxidation reaction (C. Reiss and P. Albrecht, unpublished results). It cannot be excluded that some of the long chains or other alkyl moieties might be liberated from the asphaltenes via such a pathway. The recent finding of significant amounts of 3β-carboxysteranes as free acids in marls from evaporitic series does suggest that they may be at least partly linked into the asphaltenes as the corresponding esters (Dany et al., 1990). This point is now being checked with specific reagents. Furthermore, given the importance of sulfur in the cross-linking of asphaltenes from sulfur-rich crude oils, we are also checking the behavior of alkyl sulfides under the conditions of the oxidation reaction.

CONCLUSIONS

Oxidative degradation of *n*-heptane precipitated asphaltenes from sulfur-rich crude oils leads to a mixture of acids representing between 2.9 and 7.4 percent of the original asphaltene weight. In all cases, monocarboxylic acids are by far the major components and are usually dominated by linear compounds (C_9 to around C_{30}). In the more mature petroleums, long-chain α,ω-dicarboxylic acids (C_{11}–C_{26}) are also present in substantial amounts, along with some benzene polycarboxylic acids.

The branched and cyclic acids are dominated by a series of regular acyclic isoprenoid acids (C_{15}–C_{21}) in which the C_{16} and C_{21} homologues are usually predominant. Hopanoic acids are also present with a carbon range extending to C_{36}. Furthermore, 3β-carboxysteranes are major components of the acids formed on oxidation of the asphaltenes of Rozel Point crude oil.

The structures of the identified compounds and their carbon number distributions appear consistent with a specific oxidation process preferentially degrading aromatic subunits occurring in the asphaltene macromolecules and consequently liberating aliphatic moieties attached to them in the form of carboxylic acids bearing one extra carbon. The presence of α,ω-dicarboxylic acids in the Lameac asphaltene-oxidation products suggests that long alkyl chains could act as a cross-link between aromatic subunits. The occurrence in the same oil of small amounts of benzene polycarboxylic acids implies that the aromatic entities occurring in the asphaltenes could be more extended than in the case of Rozel Point. Furthermore, our results clearly demonstrate that the acids formed on oxidation of the asphaltenes reflect in their structures the degree of maturity of a crude oil. They may be potentially useful as maturity parameters in the case of biodegraded petroleums.

Our results give strong indications, especially in the case of the steroids, that the alkyl moieties could have become linked into the asphaltene macromolecules via Friedel-Crafts-type reactions operating on alcohols (or olefins?) during diagenesis or early maturation. Acid catalysis for these reactions may occur on certain clay minerals or acid sites in the organic matter.

Alternative explanations for the oxidation products include the incorporation into the asphaltenes of alkylaromatic molecules of direct biological origin and the cleavage of other types of functions during the oxidative degradation process, a point which is now being checked.

Acknowledgments

We thank the Institut Français du Pétrole for a research fellowship (S.T.) and financial support; G. Ryback (Shell, Sittingbourne, U.K.) for a sample of Rozel Point seep oil; D. Durrenberger, R. Hueber, G. Teller, and P. Wehrung for their collaboration in the mass spectral studies.

APPENDIX

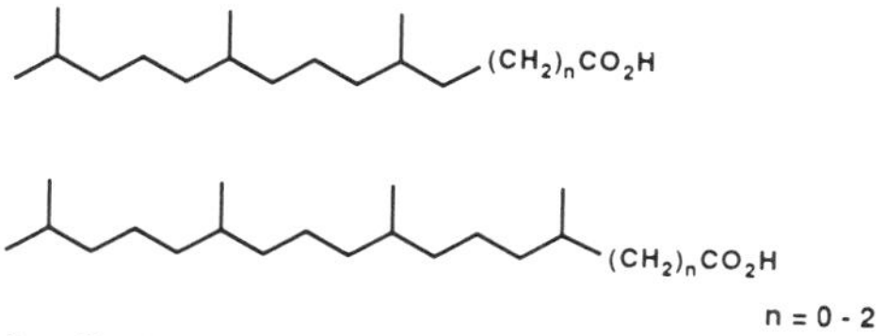

Acyclic isoprenoid acids (C_{15}-C_{21})

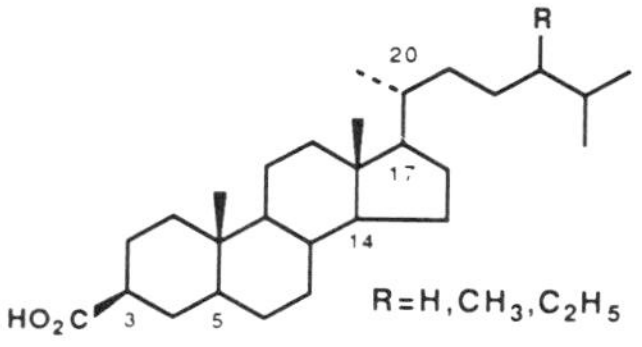

3β-carboxysteranes (C_{28}-C_{30})

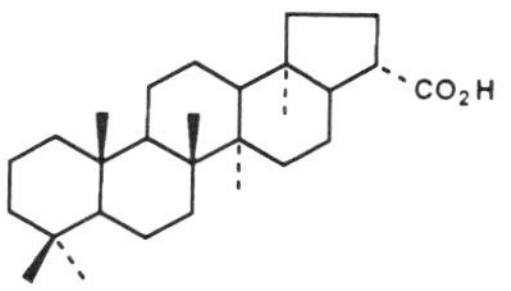

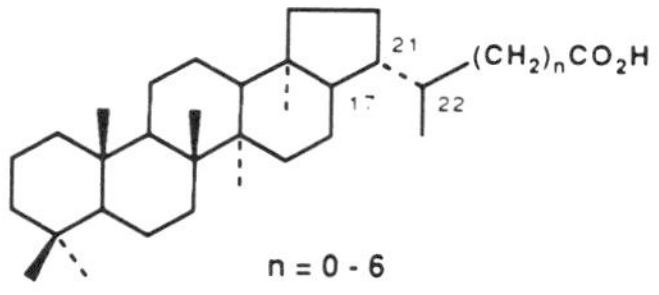

Hopanoic acids (C_{28}-C_{36})

Figure 18.13 Branched and cyclic biological markers occurring in the products of oxidation of C_7-asphaltenes from sulfur rich crude oils with ruthenium tetroxide.

REFERENCES

ARNDT, F. (1943) *N*-nitrosoethylurea. *Organic Synthesis 2*, 461–462.

BLANC, P. and ALBRECHT, P. (1990) Molecular markers in bitumen and macromolecular matrix of coals. Their evaluation as rank parameters. In *Advanced Methodologies in Coal Characterization*, (ed. H. Charcosset) Elsevier, Amsterdam, pp. 53–82.

BOER, T.J. and BAKER, H.J. (1963) Diazomethane. *Organic Synthesis 4*, 250–253.

BRASSELL, S.C., EGLINTON, G., and MAXWELL, J.R. (1983) The geochemistry of terpenoids and steroids. *Biochem. Soc. Trans. 11*, 575–586.

CARLSEN, H.J., KATSUKI, T., MARTIN, U.S., and SHARPLESS, K.B. (1981) A greatly improved procedure for ruthenium tetraoxide catalyzed oxidations of organic compounds. *J. Org. Chem. 46*, 3936–3938.

Connan, J., Bouroullec, J., Dessort, D., and Albrecht, P. (1986) The microbial input in carbonate-anhydrite facies of a sabkha paleoenvironment from Guatemala: A molecular approach. In *Advances in Organic Geochemistry 1985*; *Org. Geochem. 10*, 29–50.

Dany, F., Riolo, J., Trendel, J.M., and Albrecht, P. (1990) 3-Carboxysteranes, a novel family of fossil steroids. *J. Chem. Soc., Chem. Commun.*, in press.

Dastillung, M. (1976) Lipides de sédiments récents. Thèse de Doctorat ès-Sciences, Université Louis Pasteur, Strasbourg, France.

Didyk, B.M., Simoneit, B.R.T., Brassell, C.M., and Eglinton, G. (1978) Organic geochemical indicators of paleoenvironmental conditions of sedimentation. *Nature 272*, 216–222.

Douglas, A.G., Douraghi-Zadeh, K., Eglinton, G., Maxwell, J.R., and Ramsay, J.N. (1970) Fatty acids in sediments including the Green River shale (Eocene) and Scottish Torbanite (Carboniferous). In *Advances in Organic Geochemistry 1966*, (eds. G.D. Hobson and G.C. Speers) pp. 315–334.

Eglinton, G., Douglas, A.G., Maxwell, J.R., Ramsay, J.N., and Staellberg-Stenhagen, S. (1966) Occurrence of isoprenoid fatty acids in the Green River shale. *Science 153*, 1133–1135.

Ensminger, A., Van Dorsselaer, A., Spyckerelle, C., Albrecht, P., and Ourisson, G. (1974) Pentacyclic triterpenes of the hopane type as ubiquitous geochemical markers: Origin and significance. In *Advances in Organic Geochemistry 1973*, (eds. B. Tissot and F. Bienner) *Technip*, Paris, pp. 245–260.

Harwood, J.L. and Russell, N.J. (1983). *Lipids in plants and microbes.* George Allen and Unwin, London.

Hussler, G. (1985) Marqueurs géochimiques en séries carbonatées. Thèse de Doctorat ès-Sciences, Université Louis Pasteur, Strasbourg, France.

Jaffé, R., Albrecht, P., and Oudin, J.L. (1988) Carboxylic acids as indicators of oil migration: II. Case of the Mahakam Delta, Indonesia. *Geochim. Cosmochim. Acta 52*, 2599–2607.

McCarthy, R.D. and Duthie, A.H. (1962) A rapid quantitative method for the separation of free fatty acids from other lipids. *J. Lipid Res. 3*, 117–119.

Mojelsky, T.W., Montgomery, D.S., and Strausz, O.P. (1985) Ruthenium (VIII) catalyzed oxidation of high molecular weight components of Athabasca oil sand bitumen. *AOSTRA J. Res. 2*, 131–137.

Ourisson, G., Albrecht, P., and Rohmer, M. (1984) The microbial origin of fossil fuels. *Sc. Amer. 251*, 44–51.

Palacas, J. (1983) Carbonate rocks as sources of petroleum: geological and chemical characteristics and oil-source correlations. Proceedings of the 11th World Petroleum Congress, Vol. 2, Wiley, Chichester, pp. 31–43.

Rubinstein, I., Spyckerelle, C., and Strausz, O.P. (1979) Pyrolysis of asphaltenes: A source of geochemical information. *Geochim. Cosmochim. Acta 43*, 1–6.

Rullkötter, J., Aizenshtat, Z., and Spiro, B. (1984) Biological markers in bitumen and pyrolysates of Upper Cretaceous bituminous chalks from the Ghareb Formation (Israel). *Geochim. Cosmochim. Acta 48*, 151–157.

SEIFERT, W.K. and MOLDOWAN, J.M. (1979) The effect of biodegradation on steranes and terpanes in crude oils. *Geochim. Cosmochim. Acta 43*, 111–126.

SINNINGHE DAMSTÉ, J.S., KOCK-VAN DALEN, A.C., DE LEEUW, J.W., SCHENCK, P.A., SHENG, G., and BRASSELL, S.C. (1987) The identification of mono-, di- and trimethyl 2-methyl-2-(4,8,12-trimethyltridecyl) chromans and their occurrence in the geosphere. *Geochim. Cosmochim. Acta 51*, 2393–2400.

SPEIGHT, J.G. (1984) The chemical nature of petroleum asphaltenes. In *Caractérisation des huiles lourdes et des résidus pétroliers*, Symposium International, Lyon 1984, (ed. B.P. Tissot) Technip, Paris, pp. 32–41.

SPEIGHT, J.G., WERNICK, D.L., GOULD, K.A., OVERFIELD, R.E., RAO, B.M.L., and SAVAGE, D.W. (1985) Molecular weight and association of asphaltenes: A critical review. *Rev. Inst. Fr. Pétrole 40*, 51–61.

SPYCKERELLE, C. (1973) Constituants organiques d'un sédiment crétacé. Thèse de Doctorat de Spécialité, Université Louis Pasteur, Strasbourg, France.

STOCK, L.M. and TSÉ, K. (1983) Ruthenium tetroxide catalyzed oxidation of Illinois *n*° 6 coal and some representative hydrocarbons. *Fuel 62*, 974–976.

STOCK, L.M. and WANG, S.H. (1985) Ruthenium tetroxide catalyzed oxidation of Illinois *n*° 6 coal. The formation of volatile monocarboxylic acids. *Fuel 64*, 1713–1717.

STOCK, L.M. and WANG, S.H. (1986) Ruthenium tetroxide catalyzed oxidation of coals. The formation of aliphatic and benzene carboxylic acids. *Fuel 65*, 1552–1562.

TRIFILIEFF, S. (1987) Etude de la structure des fractions polaires de pétroles (résines et asphaltènes) par dégradations chimiques sélectives. Thèse de Doctorat ès-Sciences, Université Louis Pasteur, Strasbourg, France.

VALISOLALAO, J., PERAKIS, N., CHAPPE, B., and ALBRECHT, P. (1984) A novel sulfur containing C_{35} hopanoid in sediments. *Tetrahedron Lett. 25*, 1183–1186.

VAN DORSSELAER, A. (1975) Triterpènes de sédiments. Thèse de Doctorat ès-Sciences, Université Louis Pasteur, Strasbourg, France.

YEN, T.F. (1974) Structure of petroleum asphaltenes and its significance. *Energy Sources 1*, 447–463.

19

Source Correlation and Maturity Assessment of Select Oils and Rocks from the Central Adriatic Basin (Italy and Yugoslavia)

J. Michael Moldowan, Cathy Y. Lee, Padmanabhan Sundararaman, Tito Salvatori, Andja Alajbeg, Bogdan Gjukić, Gerard J. Demaison, Nacer-Eddine Slougui, and David S. Watt

Abstract. The central part of the Adriatic basin contains several areas, both onshore and offshore, where crude oils have been encountered, in subsurface Mesozoic carbonate reservoirs, and also as surface seepages.

Two families of oils have been identified in the Central Adriatic basin. Both families have been related to carbonate or carbonate-rich source rocks.

1. A Liassic-Triassic family of oils present mainly on the Italian side of the basin.
2. A Cretaceous family of oils present mainly on the Yugoslavia side of the basin. These are almost exclusively found as surface oil seepages, near the Dalmatian Coast.

The oils and seeps in the basin were correlated using a multiparameter approach. Heavy biodegradation occurs in most of the seep oils, affecting the acyclic alkanes, steranes, and terpanes. The strongest support of oil source grouping is based on C_{27}–C_{28}–C_{29} monoaromatic- (MA) and C_{26}–C_{27}–C_{28} triaromatic- (TA) steroid distributions, carbon isotope ratios and, in the less biodegraded oils, C_{31}–C_{35} homohopane distributions and gammacerane indices.

Analysis of the parents (M+) of the *m/z* 231 daughter ion using GC-MSMS on a triple quadrupole mass spectrometer is introduced as the method of choice for measuring C_{26}–C_{27}–C_{28} TA-steroids distributions and to analyze C_{29} TA-steroids, analogues of "marine indicator" C_{30}-steranes. Steranes were analyzed by the M+ to *m/z* 217 parent to daughter transitions. Concentrations of C_{30}-steranes relative to C_{27}–C_{30} steranes vary inversely with C_{34} 17α(H)-homohopane concentrations. Low C_{30}-steranes and high C_{34} 17α(H)-homohopanes in the Cretaceous oil family probably indicate a restricted evaporitic source rock depositional environment.

Triassic oil to source correlations were aided by hydrous pyrolysis which permitted biomarker correlations between an artificially matured Liassic-Triassic source rock and some of the oils. All the oils evaluated by this study range widely in API gravity (from 5°API to 36°API) and sulfur content (from 9.5 to 1.7%). Although a few of these oils, particularly the seepage oils, are biodegraded, the major effect on oil gravity was found to be that of maturation. A correlation was clearly established between API gravity and the transformation ratio of MA-steroids to TA-steroids. This relationship was used to interpolate the original API gravity of heavily biodegraded oil seeps.

Biomarker maturity parameter studies were carried out on rock samples from some deep wells in an attempt to directly observe the onset of oil generation in the Mesozoic carbonate sequence. The aromatic steroid aromatization parameter [TA C_{28}/(TA C_{28} + MA C_{29})] and the porphyrin maturity parameter (PMP) [C_{28} Etio/(C_{28} Etio + C_{32} DPEP)] showed, at least in one well, that values comparable with mature oils were not reached until a depth of about 6 km. Such great depth is compatible with the low-geothermal gradients prevailing in this general area and with the burial history.

INTRODUCTION

There are many oil shows, seepage oils, and several commercial accumulations of oil in the central part of the Adriatic Basin (Jacob et al., 1983; Pieri and Mattavelli, 1986). They display great diversity in physical properties and chemical compositions such as API gravities ranging from 5 to 36 degrees, and sulfur concentrations from 2 to 11 percent. Unravelling the reasons for this diversity presents a challenge to the organic geochemist. It is clear from previous work (Seifert and Moldowan, 1986) that source, maturity, and biodegradation may affect such properties in oils.

This study attempts to tie the oils to specific source rocks and to understand differences in their history through geochemistry. Previous studies (e.g., Seifert et al., 1984; Rullkötter et al., 1985) have shown that the best way to correlate oils, particularly heavily biodegraded seepages, is through a combination of biological marker and stable isotope analyses. Together with geological and geophysical data, such information may be used to support a basin model with predictive value to the explorationist.

The locations of the samples available for this study are shown in Figure 19.1, and Table 19.1 gives additional identification information. The samples include

Figure 19.1 Adriatic Basin sample locations.

seepage oils and source rocks onshore and offshore Italy and surface seepages and source rocks at or near the central Dalmation coast, Yugoslavia, including the coastal islands (Jacob et al., 1983). One heavy oil, Ravni Kotari-3, produced during a deep test in an exploratory well in that area is also included.

Geological/Geographical Setting

The area of study in the central to southern part of the Adriatic basin is indicated on the map, Figure 19.1. Organic-rich source rocks containing sulfur-rich Type II kerogen from the Lower Liassic-Triassic occur within carbonate-rich outcrop sequences and cores both in Italy and Yugoslavia. Triassic cores plus outcrops in Italy have been analyzed, but are not described here in detail. Cretaceous organic-rich rocks with Type II kerogen were also sampled in Yugoslavia. However, the areal stratigraphic distribution of the source beds and what controls it is still unclear. Indications are that at least some of the Lower Liassic-Triassic source rocks were deposited in intraplatform, possibly hypersaline, lagoons where anoxic conditions developed because of chemical stratification. The carbonate rocks of the Liassic-Triassic platform facies are generally without source potential because they were deposited under oxic open marine conditions. However, some deep-sea basins have also been found to be barren of source rocks. Furthermore, we suspect thin Cretaceous source beds found on the Italian side of the Adriatic basin, for

Table 19.1 SAMPLE IDENTIFICATION AND DEGREE OF DESTRUCTION BY BIODEGRADATION, ADRIATIC

Group No.	Country	Type	Well/Name or Code	Reservoir Age[b]	Depth, M	Biodeg.[a] Ranking	Paraffins[c] %	Paraffins[c] Destruction	Isoprenoids Destruction	Steranes Destruction	Hopanes Destruction	GC-MS No.
IA	Italy	Oil	Katia	Olig-Cret	1780–1797	0	14.4	None	None	None	None	563
IA	Italy	Oil	Santa Maria	Eo-U. Cret	2295–2337	0	14.8	None	None	None	None	561
IA	Italy	Seepage	Tocca da Casauria	N/A	Surface	5	0	Total	Total	None	None	666
IA	Italy	Oil	Alanno-39	Miocene	561–567	3	3.7	Partial	Partial?	None	None	694
IA	Italy	Oil	Aquila	Paleo-U. Cret	3866–3924	0	13.4	None	None	None	None	560
IB	Italy	Oil	Rovesti	L. Cret	2529–2547	2	7.0	Partial	None	None	None	562
IC	Italy	Oilsoak rock	Filletino	Triassic	Surface	5	0	Partial	Partial	None	None	796
ID	Yugoslavia	Asphalt veins	Glamoč	Triassic	Surface	0	14.0	None	None	None	None	701
II	Yugoslavia	Oil	Ravni Kotari-3	L. Cret	2713–2967	0	>15	None	None	None	None	558
II	Yugoslavia	Oilsoak rock	Rošca	U. Cret	Surface	9	0	Total	Total	Total	Partial	702
II	Yugoslavia	Asphalt	Vrgorac	U. Cret	Surface	8	0	Total	Total	Total	Partial	646
II	Yugoslavia	Asphalt	Vinišće	U. Cret	Surface	5	0	Total	Total	None	None	645
II	Yugoslavia	Oilsoak rock	Škrip	U. Cret	Surface	8	0	Total	Total	Total	Partial	556
II	Yugoslavia	Oilsoak rock	Okruglica	U. Cret	Surface	9	0	Total	Total	Total	Partial	557
III	Yugoslavia	Oilsoak rock	Palanka	U. Jurassic	Surface	9	0	Total	Total	Total	Partial	700

[a]1–3 Light, 4–5 Moderate, 6–7 Heavy, 8–9 Very Heavy (See Table 19.2 for definitions.)

[b]For surface seepages, age of stained rock.

[c]See Figure 19.2.

example, the Bonarelli layer, may have been preferentially deposited along the slopes where the impinging oxygen-minimum layer allowed better preservation of organic matter, particularly at times of oceanic anoxic events (Cenomanian-Turonian and Santonian). Such a depositional setting has been documented in Central Italy by Arthur and Premoli (1982). The thin Cretaceous organic-rich "fish beds" of the Dalmatian Coast (Šebečić, 1981) were deposited at the same time, but also partly in association with intraplatform basin developments.

The top of the oil generation window is thought to be depressed from 4500 to 6000 m or more by the *abnormally low-geothermal gradients* prevailing in the Adriatic basin. Gradients as low as 10°C/km (0.55°F/100 ft) are common in the carbonate bank area. Elsewhere, gradients are somewhat higher, but it appears that in most parts of the basin the oil generation window can only affect the Triassic.

Experimental

Experimental procedures and conditions have been published previously (Seifert et al., 1983; Sundararaman, 1985; Moldowan et al., 1985; Moldowan and Fago, 1986). Conditions peculiar to given analyses are provided as footnotes to figures.

OIL TO OIL CORRELATIONS

Biodegradation

Biodegradation is a factor that must be considered in any oil study of this type where API gravities range from a low of 5.1° (immature or biodegraded?) to a high of 36.5° (mature or less biodegraded?). Biodegradation destroys *n*-paraffins, isoprenoids, steranes, and hopanes in sequential order (Seifert et al., 1984; Connan, 1984).

Table 19.1 lists the samples and their biodegradation characteristics. The biodegradation ranking is based on the system shown in Table 19.2 which was derived from our earlier reports (Seifert and Moldowan, 1979; Seifert et al., 1984).

Although the reservoired oil samples display a wide range of API gravities (5.1–36.5°API, Table 19.4), only two oils, Rovesti and Alanno-39, show light biodegradation (Table 19.1). The heavy oils Katia (14.3°API) and Ravni Kotari-3 (5.1°API) show no sign of biodegradation. This observation is similar to that shown for Monterey sourced oils of California (Orr, 1986)—that biodegradation is not the principal factor determining oil gravity. Also, like the Monterey oils, the Adriatic basin oils contain high concentrations of sulfur (Table 19.4). It is the sulfur cross-linking of the hydrocarbon chains (Rullkötter et al., 1985; Schmid et al., 1987) which probably determines the API gravity of these oils.

Except for Tocca da Casauria, the seepage oils were isolated by washing oil-stained or oil-soaked rocks. Several seeps from Yugoslavia show very heavy biodegradation with removal of C_{27}-to C_{30}-steranes, 17α(H)-hopanes (Ranking 8, Table

Table 19.2 BIODEGRADATION RANKING SUMMARY[a]

Extent	Ranking	Indicator (Definition)
Light	1	Lower homologs of *n*-paraffins depleted
Light	2	General depletion of *n*-paraffins
Light	3	Only traces of *n*-paraffins remain
Moderate	4	No *n*-paraffins, acyclic isoprenoids intact
Moderate	5	Acyclic isoprenoids absent
Heavy	6	Steranes partly degraded
Heavy	7	Steranes degraded, diasteranes intact
Very Heavy	8	Hopanes partly degraded
Very Heavy	9	Hopanes absent, diasteranes attacked
Severe	10	C_{26}–C_{29} aromatic steroids attacked

[a]In cases where 25-desmethylhopanes are formed, hopane degradation may occur as early as Ranking 6, that is, simultaneously with or possibly before sterane degradation, but after removal of acyclic isoprenoids. No 25-desmethylhopanes were observed in this study.

19.1) and in some cases, C_{27}- to C_{30}-diasteranes (Ranking 9, Table 19.1). All heavily degraded seeps show the type of biodegradation similar to that previously reported for an oil from the U.S. Gulf Coast (Seifert and Moldowan, 1979) and several seepage oils from Western Greece (Seifert et al., 1984) where 17α(H)-hopanes are removed without the appearance of 17α(H)-25-desmethylhopanes. This type of biodegradation has also been demonstrated in the laboratory (Goodwin et al., 1983). Numerous examples of very heavy biodegradation with the occurrence of 17α(H)-25-norhopanes have also been reported (Reed, 1977; Seifert and Moldowan, 1979; Rullkötter and Wendisch, 1982; Philp, 1983; Volkman et al., 1983; Goodwin et al., 1983).

Very heavily biodegraded seepage oils may have had longer exposure to aerobic conditions than those showing light to moderate biodegradation, which could indicate that the less biodegraded seeps are fresher or more active. On the other hand, differences in local conditions such as climate or ecology at the seepage site might influence the extent of biodegradation. In addition, microbial activity in the seep may have occurred during secondary migration or in a subsurface reservoir before reaching the surface.

Seeps showing severe biodegradation with evidence of C_{26}–C_{29} MA- and TA-steroid depletion (Wardroper et al., 1984) are not found in this study. Thus MA- and TA-steroid ratios can be used in correlation and maturity assessment in the Adriatic basin.

The oil from Rovesti-1 (2529–2597 m) shows *n*-paraffin depletion (Table 19.1) compared to the oil produced from a shallower horizon (2385–2406, Fig. 19.2) which is probably due to light biodegradation. Microbial activity in the reservoir is generally thought to be quenched by elevated temperatures, perhaps above 66°C (Philippi, 1977), however the Rovesti-1 reservoir temperature is only 49°C.

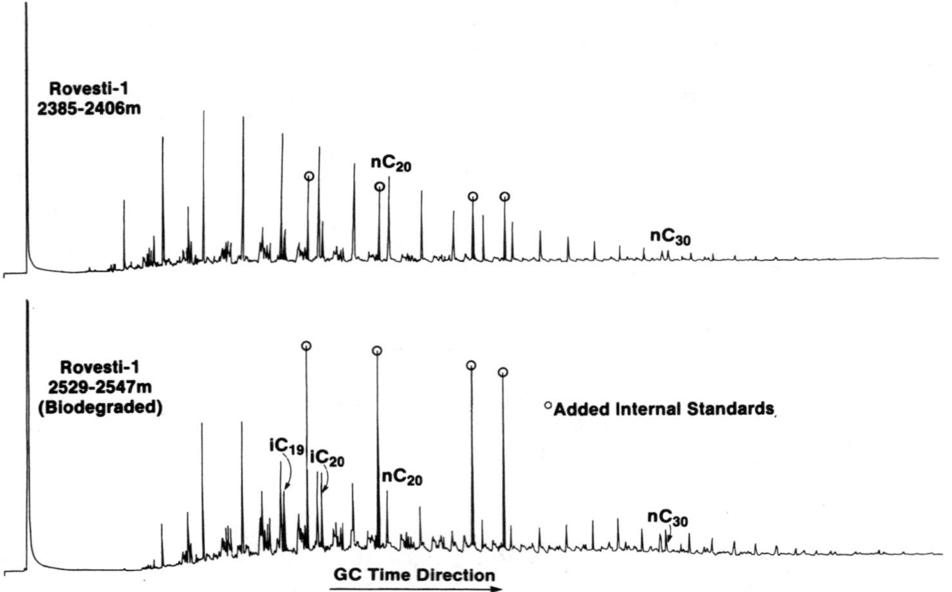

Figure 19.2 Capillary gas chromatography of saturate fractions shows biodegradation of an oil at great depth in the reservoir. OV 101—12 m fused silica capillary column (Hewlett Packard) with H_2 carrier gas, 80–325°C at 10°C/min, Hewlett-Packard 5880A gas chromatograph with split injector ratio set at 100:1. Internal standards are 3-methylheptadecane, 3-methylnonadecane, 2-methyldocosane and 3-methyltricosane (Chemical Samples Co.) added from a standard solution by volume to allow calculation of percent paraffins, percent isoprenoids (See Table 19.1).

Source Grouping

The parameters found most useful in source correlation of the oils and seeps are presented in Table 19.3 and Figures 19.3, 19.4, and 19.6. Based on stable carbon isotopes (Table 19.3), the oils can be classified into at least two major groups, low ratios (δC^{13} = -30.35 to -26.73) and high ratios (δC^{13} = -23.54 to -21.58), although the spreads are large enough to permit additional groups or subgroups. A recent investigation of oils from the Po Basin, Italy, showed similar ranges in those oils (Riva et al., 1986). Some of the spreads in the $\delta^{13}C$ concentrations, at least in the seeps, may be caused by the removal of components by biodegradation. Removal of the saturates, which tend to be isotopically lighter than the aromatics usually results in isotopically heavier biodegraded crudes (Fuex, 1977; Stahl, 1978; Schoell, 1984).

The MA-steroids and TA-steroids, unaffected by biodegradation are, potentially, the most applicable biomarkers to source correlation. The distribution of

Table 19.3 ADRIATIC BASIN SOURCE PARAMETERS

Group No.	Identification	Gammacerane[a] Index	TA-Steroids[b] C_{26}-*S*/C_{28}-*S*	C_{27}-*R*/C_{28}-*R*	Dia.[c]/Reg.	iC_{19}/iC_{20}	$\delta^{13}C$ per mil[d]
IA	Katia	2.6	0.25	0.54	0.04	0.67	−28.31
IA	Santa Maria-3	3.1	0.10	0.76	0.02	0.77	−28.99
IA	Casauria	2.9	0.25	0.62	0.01	B	−28.30
IA	Alanno	2.4	0.12	0.54	0.04	0.72	−27.39
IA	Aquila	3.3	0.09	0.84	0.17	0.73	−27.60
IB	Rovesti	1.8	0.10	0.99	0.35	0.77	−27.77
IC	Filletino	6.6	0.47	0.74	0.00	B	−28.40
ID	Glamoč	0.9	0.14	0.77	0.08	0.40	−30.35
II	Ravni Kotari-3	2.3	0.44	0.53	0.00	0.51	−22.61
II	Rošca	B	0.21	0.40	B	B	−21.58
II	Vrgorac	B	0.15	0.24	B	B	−23.05
II	Vinišće	1.5	0.40	0.50	0.01	B	−23.54
II	Škrip	B	0.32	0.42	B	B	−22.88
II	Okruglica	B	0.20	0.46	B	B	−22.74
III	Palanka	B	0.76	0.88	B	B	−26.73

[a]Gammacerane Index = Gammacerane/17α(H)-Hopane (× 10).

[b]*S* and *R* indicate 20*S* and 20*R*, respectively. Measurements from GC-MS *m/z* 231 chromatograms.

[c]13β,17α(H)-Diacholestanes (20*S* + 20*R*)/ααα- + ββα-cholestanes (20*S* + 20*R*).

[d]Measured on the whole oil versus Pee Dee Belemnite standard. B = Biodegraded components in parameter.

C_{27}–C_{28}–C_{29} MA-steroids shown in Figure 19.3 does not clearly resolve the two main groups as shown by carbon isotopes. The seepage oils, Vrgorac and Vinišće, from the isotopically high group fall near the isotopically low oils. However, one seepage oil, Palanka, is widely separated from the others and is designated as Group III in Table 19.3. In retrospect, Palanka could also be considered in its own class isotopically. Thus the major groups (I–III) in Table 19.3 are indicated.

The distributions of TA-steroids measured by MID (*m/z* 231) GC-MS and by GC-MSMS ($M^{+\cdot} \rightarrow m/z$ 231) are shown by the ratios C_{26} 20*S*/C_{28} 20*S* and C_{27} 20*R*/C_{28} 20*R* (Table 19.3) and by the triangular diagram (Fig. 19.4), respectively. Due to the coelution of C_{26} 20*R* and C_{27} 20*S* in the GC-MS analysis of fragment *m/z* 231, it is not possible to show TA-steroids in a ternary diagram using (MID) GC-MS and the values within Groups I and II (Table 19.3) are widely variable providing only a weak basis to group the samples. However, TA-steroid analysis by GC-MSMS ($M^{+\cdot} \rightarrow m/z$ 231) separates C_{26} 20*R* and C_{27} 20*S* on different chromatograms (Fig. 19.5). The C_{26}–C_{27}–C_{28} TA-steroid distributions by GC-MSMS (Fig. 19.4) distinguish the Group II from the Group I samples, the latter showing a tendency toward higher C_{27} TA-steroids. There is an analogous tendency toward higher C_{28} MA-steroid concentrations in the Group I oils relative to the Group II oils (Fig. 19.3) in support of the proposal that TA-steroids are catagenic daughters

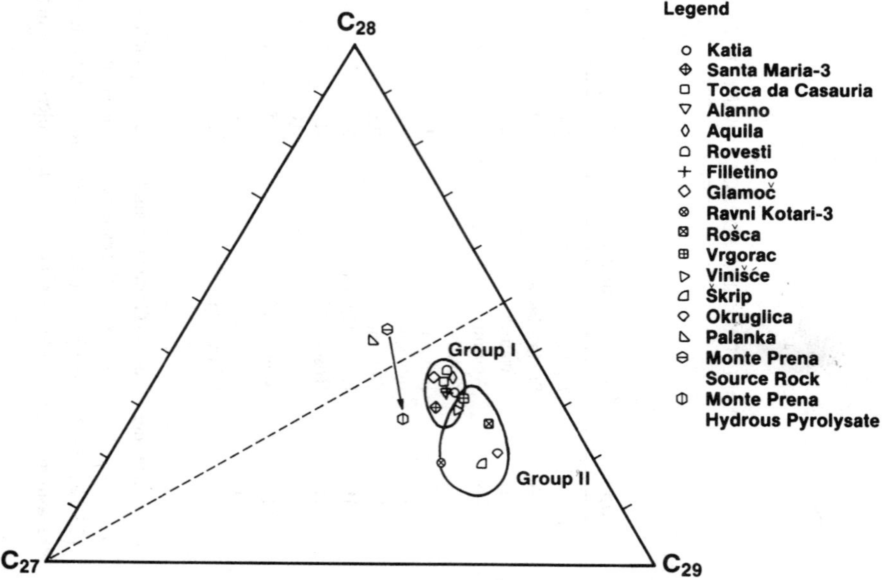

Figure 19.3 C_{27}–C_{28}–C_{29} Monoaromatic steroid distribution diagram shows overlap of Groups I and II, discrimination of Group III (Palanka). MA-steroids for each carbon number are the sum of $5\alpha + 5\beta(20S + 20R)$ plus $(10\beta \rightarrow 5\beta)$ CH_3-rearranged $(20S + 20R)$ (Moldowan and Fago, 1986). Monte Prena bitumen extract before and after hydrous pyrolysis is shown with arrow.

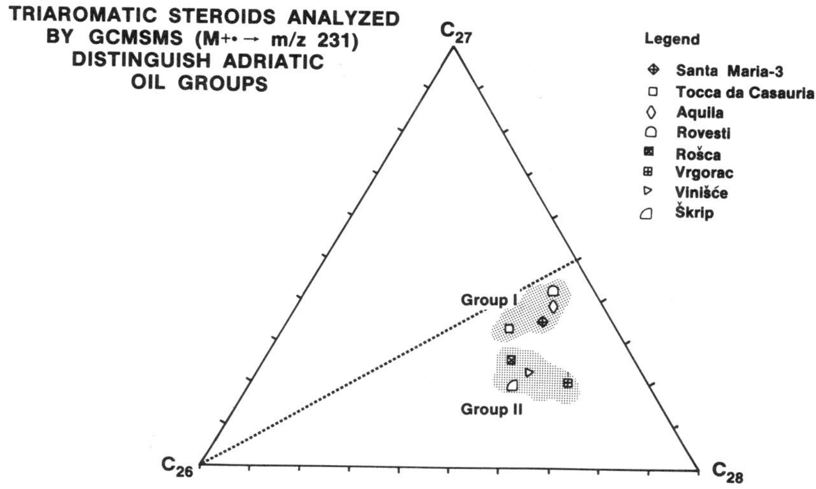

Figure 19.4 C_{26}–C_{27}–C_{28} Triaromatic steroid distribution diagram distinguishes Group I and Group II oils and seeps. Data were obtained using parent ion analysis of $M^+ \rightarrow 231$ on the aromatic fractions of the oils (Fig. 19.5). Oils not shown were not analyzed. Each carbon number is the sum of the 20*S* and 20*R* epimers. The 24*R* and 24*S* epimers were not separated (see Gallegos and Moldowan, 1991).

of MA-steroids (Mackenzie et al., 1982). The relationship between Figures 19.3 and 19.4 falls short of being a 1 to 1 correspondence, however, which may indicate different rates of conversion for C_{27}, C_{28}, and for C_{29} MA-steroids to C_{26}, C_{27}, and C_{28} TA-steroids, respectively; or it may reflect differences in rates of conversion for regular and rearranged MA-steroids to TA-steroids (Moldowan and Fago, 1986).

The ratios of Ni/(Ni + V) porphyrins were measured for all the samples in Table 19.1 and found to be near zero, that is, ~100% vanadyl dominated. This would tend to support an origination of the oils from organic matter deposited in a highly reducing marine ecosystem with a high bacterial sulfate reduction activity (Lewan, 1984; Moldowan et al., 1986a).

Further evidence for these groupings may be obtained from the homohopane distributions (Fig. 19.6) in the oils and in those seeps where the homohopanes have survived biodegradation. The two Yugoslavia samples having homohopanes, Ravni Kotari-3 and Vinišće, show a strong preservation of the C_{34} member and some preservation of the C_{35} member (Fig. 19.6a). These two oils are assumed to be representative of Group II. The five Italian oils composing Group IA show less preferential preservation of the C_{34} member (Fig. 19.6a) but equivalent C_{35} preservation to the Group II oils. Among the Group I oils, the Rovesti oil shows a relatively depressed C_{34} concentration (Fig. 19.6b). This difference in Rovesti may be an indication of a slightly different depositional environment for its source rock and it is subgrouped IB. The seepage oil from Filletino, Group IC, differs from

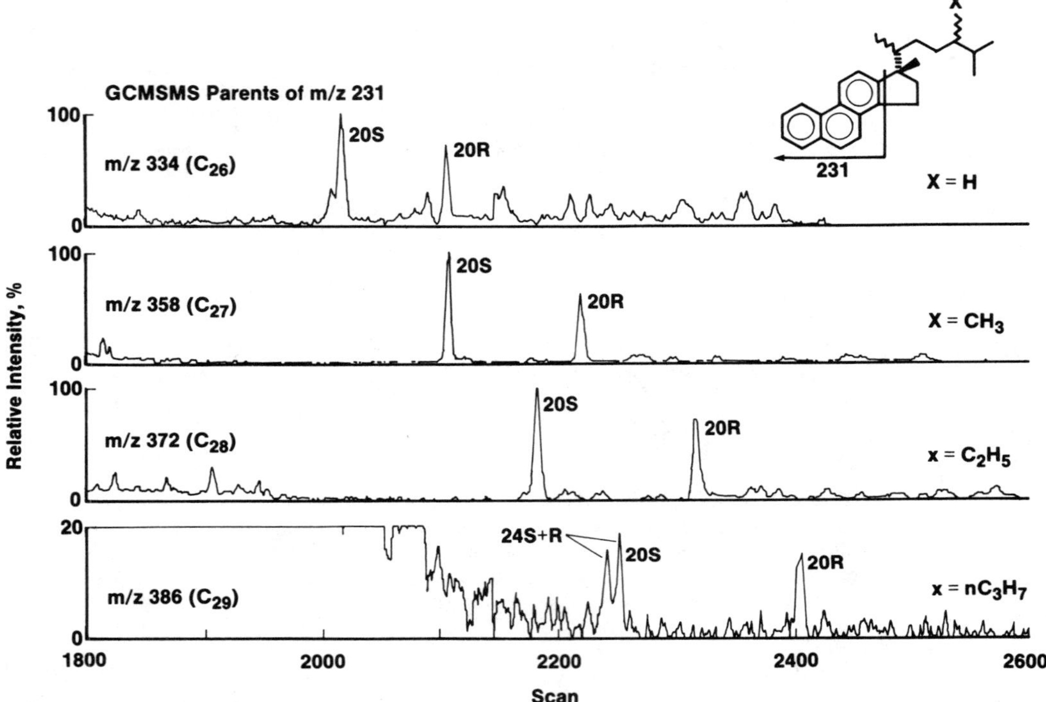

Figure 19.5 Example of a parent ion analysis for *m/z* 231 daughters (Rovesti Oil) using collision-activated decomposition (GC-MSMS). Experiments were performed on aromatic oil fractions using a Finnigan MAT TSQ 70 system scanning at 1.5 sec intervals, a 60 m J and W Scientific DB-1 capillary column, H_2 carrier gas, programmed 150–320°C at 2°C/minute. The overall sensitivity of the method is somewhat less than regular MID GC-MS of *m/z* 231 due to low stability of TA-steroid molecular ions. However, the method offers improved selectivity for analysis of minor components such as the C_{29} TA-steroids that could not be detected in these oils using regular (MID) GC-MS *m/z* 231. The C_{29} TA-steroids were identified by coinjection of a mixture of triaromatic steroids prepared from 24-*n*-propylcholestane by catalytic dehydrogenation isomerization using a Pd/C catalyst (Seifert et al., 1983).

Group IA in showing a significant preservation of C_{32}, C_{34}, and C_{35} 17α(H)-homohopanes relative to C_{33}. The Glamoč seep oil from Yugoslavia shows an entirely different homohopane distribution, with C_{33} and C_{35} preservation relative to C_{32} and C_{34}, and accordingly is Group ID.

Homohopane distributions appear to be both a function of bacterial source input (Ourisson et al., 1984) and depositional environment (e.g., Demaison et al., 1984). Moldowan et al. (1986b) observed a prominent C_{33} 17α(H)-homohopane 22*S* + 22*R* doublet in a series of Lower Toarcian shales. Its fluctuation appeared independent of other indications of paleoenvironmental changes in the cores (i.e., pH/Eh) and probably reflects changes in the bacterial ecology during sedimentation. On the other hand, concentrations of the C_{35} 17α(H)-homohopanes increased during periods of anoxia relative to the C_{31}-homologues.

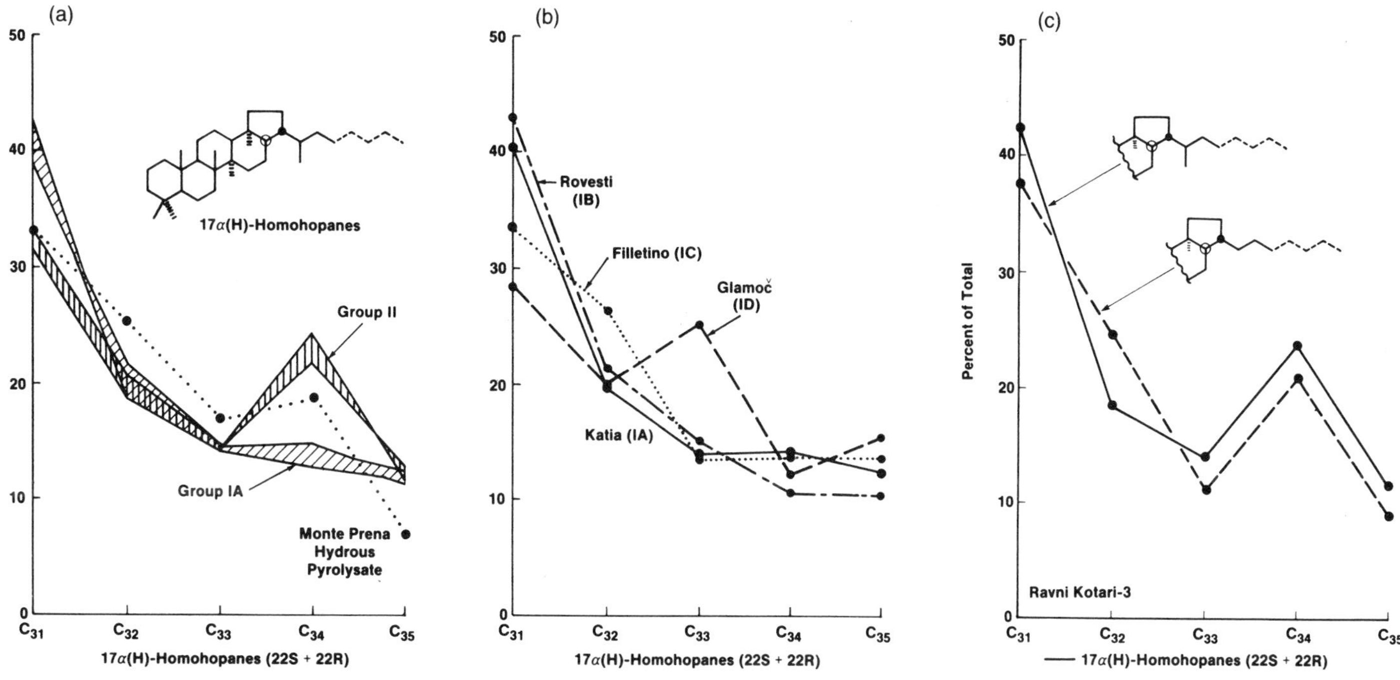

Figure 19.6 Homohopane distributions as an aid in source correlation. (a) Comparison of major Adriatic oil groups and an organic rich Triassic source rock. (b) Individual variations among Group I subgroups. (c) Extended 30-nor-17α(H)-homohopane distributions are comparable to 17α(H)-homohopane distributions in Adriatic Oils.

The seep oil from Filletino (Group IC) shows prominent C_{34}- and C_{35}-homohopanes similar in homohopane distribution to those reported by Connan et al. (1986) in evaporitic source rocks from Guatemala and by Palacas et al. (1984) in carbonate source rocks from the South Florida basin. This Filletino seep also shows an unusually high gammacerane index (6.57, Table 19.3) which has been noted as a possible indicator of hypersaline evaporitic source rock deposition (Moldowan et al., 1985): the gammacerane possibly being derived from tetrahymanol in lower aquatic organisms (Venkatesan, 1989; ten Haven et al., 1989). The high C_{34} predominance in the homohopanes of the Group II oils may also indicate an evaporitic source rock from a restricted paleoenvironment for those oils, while the relatively less pronounced C_{34} homohopanes in the Group I oils could indicate a less restricted carbonate source environment of deposition.

It has been demonstrated that bacteriohopanes (Ourisson et al., 1984) bind with sulfur, probably formed via bacterial sulfate reduction (P. Albrecht, personal communication). Hopanes liberated from the complex show a predominant C_{35} homologue typical of many oils from anoxic marine source rocks. Lack of such C_{35}-homohopane preservation in petroleums derived from sediments deposited in freshwater lakes, or in oxic to suboxic marine basins, may indicate a lack of sulfate reduction. In these cases, hopanes might only be attached without sulfur to the kerogen (Trifilieff et al., 1991). Preservation of intermediate homohopane homologues (C_{32}, C_{33}, and C_{34}) may indicate partial, possibly bacterially mediated, oxidation typical of mildly suboxic depositional environments (Demaison et al., 1984) (Fig. 19.7).

The paleoenvironmental conclusions based on 17α(H)-homohopane distributions are also supported by other parameters in Table 19.3. Moldowan et al. (1986a) found a relationship between pristane/phytane ratios and diasterane/regular

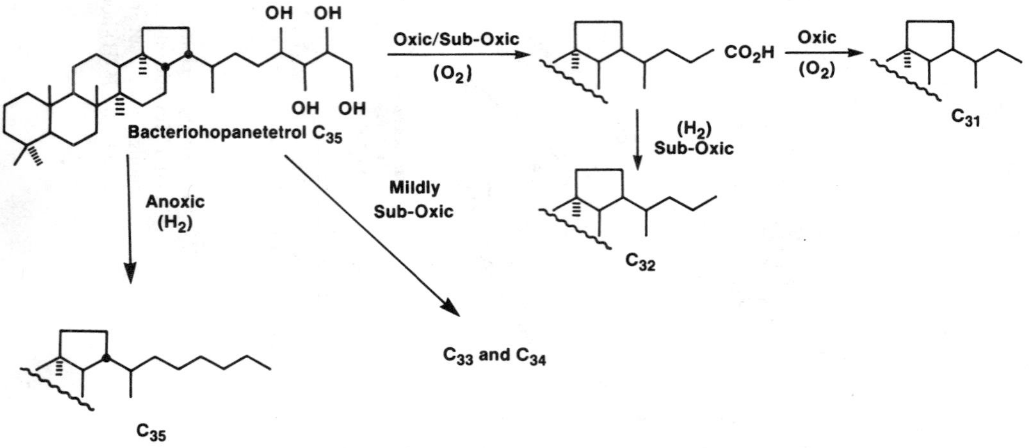

Figure 19.7 Homohopane diagenesis in saline basins is mediated by depositional environment.

sterane ratios in a sequence of L. Toarcian shales. Both parameters increase in sections of core where the organic matter was exposed to higher oxidation during deposition. In this study, the samples with the greatest C_{34} or C_{35} 17α(H)-homohopane preservation (Ravni Kotari-3, Vinišće, Glamoč, and Filletino) also show very low diasteranes and Ravni Kotari-3 and Glamoč also show a relatively low-pristane/phytane ratio. The covariation of these biomarker parameters suggests that Ravni Kotari-3, Vinišće, and Filletino were generated from source rocks deposited in an anoxic to very mildly suboxic paleoenvironment compared to a suboxic one for the samples of Group IA. As will be discussed, Glamoč and Aquila are the most mature oils and the elevations in their diasterane/regular sterane ratios could be reflecting higher maturity. Rovesti, on the other hand, is less mature. Its elevated diasterane/regular sterane ratio plus lack of preferential C_{34} 17α(H)-homohopane preservation indicates a stronger suboxic source rock paleoenvironment of deposition than the oils of Group IA.

All unbiodegraded to moderately biodegraded oils and seeps show C_{30}-steranes of the type previously reported as marine indicators (Moldowan et al., 1985). These C_{30}-steranes are relatively less abundant in the oils showing the most predominant C_{34}-homohopanes (Fig. 19.8). Both high C_{34}-homohopanes and low C_{30}-steranes in oils have been associated with evaporitic marine sediments (Connan et al., 1986, and Mello et al., 1988, respectively).

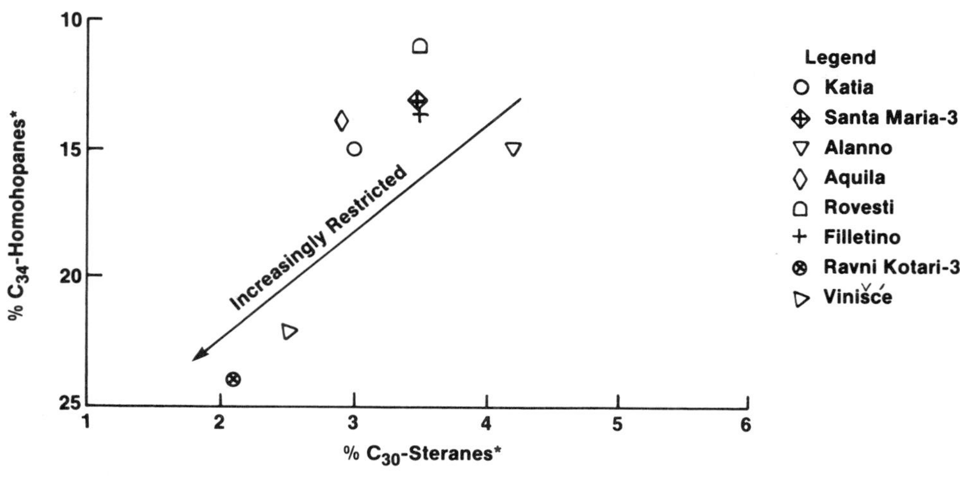

Figure 19.8 C_{30}-steranes and C_{34}-homohopanes respond to restricted marine depositional environments.

Extended 17α(H)-30-Norhopane Series

A series of hopanoid compounds showing the absence of side-chain branching was suggested in a study of biodegraded oils from Western Greece (Seifert et al., 1984). The metastable reaction monitoring (MRM) GC-MS method (Fig. 19.9) facilitates the observation of the suspected unbranched (normal) side-chain hopanoids. Authentic 29,30-bisnor-17α(H)-hopane (C_{28}), 30-nor-29-methyl-17α(H)-hopane (nC_{30}) and 30-nor-29-pentyl-17α(H)-hopane (nC_{34}) were used in (MRM) GC-MS coinjection experiments to make the identifications. The Santa Maria-3 oil sample shown in Figure 19.9 is unbiodegraded. Indeed, all the oil and seepage samples in this study showed this series of compounds. Thus, the extended 17α(H)-30-norhopane series occurs independent of biodegradation and is more biodegradation resistant than 17α(H)-hopanes.

The 17α(H)-30-norhopane series commences at C_{34}. The C_{34} member has the same side-chain length as the C_{35}-hopane and, therefore, its precursor could substitute for bacteriohopanetetrol as a bacterial cell membrane component (Rohmer et al., 1979). The absence of 17α(H)-29,30-bisnorhopane (C_{28}) in many crudes has been rationalized on the basis of requiring cleavage of two carbon-carbon bonds at positions 22–29 and 22–30 of the hopane side chain (Kimble et al., 1974; Mol-

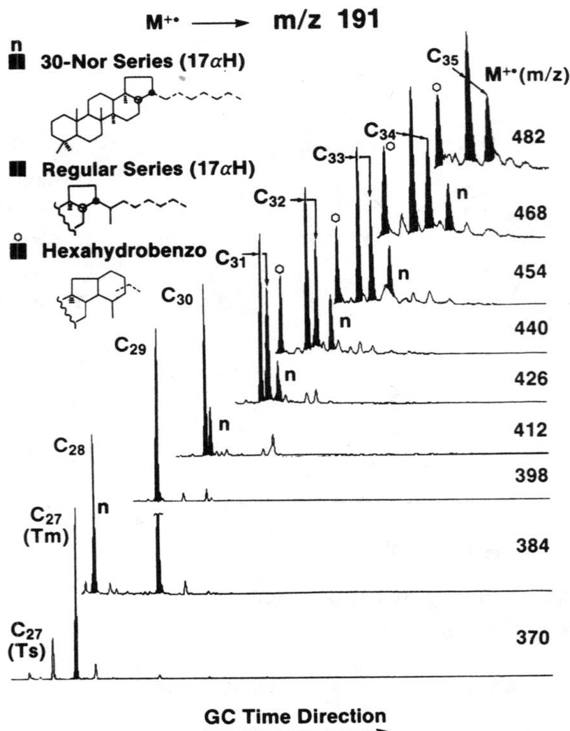

Figure 19.9 Metastable Reaction Monitoring GC-MS is applied to hopane compound class analysis, Santa Maria-3 oil, saturate fraction. Parent ($M^{+\cdot}$) → daughter (*m/z* 191) transitions monitored using B^2/E linked scans (Haddon, 1979). VG Micromass 7070H mass spectrometer using selective ion monitoring on Finnigan INCOS data system. J and W Scientific DB-1 capillary column using H_2 carrier gas in a HP 7620A gas chromatograph scanning at 1.5 sec intervals, programmed 150–320°C at 2°C/minute.

dowan et al., 1984). Thus, 17α(H)-29,30-bisnorhopane may be related to extended 30-norhopanoid precursors by cleavage of only one carbon-carbon bond at C-22 providing a rationale for its limited occurrence.

Distributions of extended C_{30}–C_{34} 30-nor-17α(H)-homohopanes (Fig. 19.6c) were plotted and compared to those of C_{31}–C_{35} 17α(H)-homohopanes. The similar distributions found for the two series suggests (1) similar bacterial precursor compound(s) except for the presence or absence of a C-30 methyl group, (2) that the geochemistry of the two series is similar, and (3) that the more biodegradation resistant 30-nor-17α(H)-homohopane series can substitute for the 17α(H)-homohopane series in correlating heavily biodegraded oils at least to rank 8 (Table 19.2). The seep oils from Škrip (Rank 8) and Vinišće show a predominent C_{33} 30-norhomohopane comparable to Ravni Kotari-3 (Fig. 19.6c), but those from Rošca and Okruglica (Rank 9) showed decreased C_{32}–C_{34} 30-norhopanes, possibly due to biodegradation. Distributions of C_{30} to C_{34} 30-norhopanes for the Group IA oils were consistent and comparable to those of the C_{31} to C_{35} 17α(H)-hopanes.

Maturity

For a maturity assessment, it is first best to compare those samples of the same source grouping because those oils should have the most similar combination of source input plus source rock catalytic effect. Parameters used for maturity correlation are listed in Table 19.4. TA-C_{28}/(TA-C_{28} + MA-C_{29}) (aromatization) and C_{28} Etio/(C_{28} Etio + C_{32} DPEP) PMP are the two most powerful parameters because they are largely independent of source input (Mackenzie et al., 1981, 1982; Moldowan et al., 1986a; Raedeke et al., 1987; Sundararaman, Moldowan, and Seifert, 1988; Sundararaman et al., 1988) and resistant to biodegradation (Seifert et al., 1984; Seifert and Moldowan, 1986). In fact, the heavily biodegraded oils in this study show no destruction of these biomarkers. Also, because of the calcareous source rocks in this basin, any catalytic effect of clays on the T_m/T_s ratio (Seifert and Moldowan, 1978) should be suppressed making it a useful maturity parameter. Based on those parameters, maturity levels have been assigned to the samples in this study. The levels, given in Table 19.4, range from mature to immature.

Immature oils, typified by all the seeps and oils from Yugoslavia except Glamoč and by the seep from Filletino, Italy, are bitumen-like products expelled at a low level of maturation. They probably result from breaking of sulfur-sulfur and sulfur-carbon bonds in the sulfur-rich kerogens of organic-rich source rocks (Rullkötter et al., 1985; Orr, 1986). Immature oils are characterized by high viscosity and low-API gravity.

The "early mature" oils (Table 19.4) are probably generated from source rocks at the beginning of the main phase of oil generation with the cracking of some carbon-oxygen and highly substituted carbon-carbon bonds. These oils tend to be medium-API gravity oils of possible economic interest. The mature oils come from a source rock in the main phase of oil generation by breaking of carbon-

Table 19.4 MATURITY RELATIONSHIPS OF ADRIATIC OILS AND SEEPS

Maturity Level[b]	Group No.	Well	$TA\text{-}C_{28}/$[a] $(TA\text{-}C_{28} + MA\text{-}C_{29})$	(PMP) $C_{28}E/(C_{29}E + C_{32}D)$	Tm/Ts	C_{29}-Steranes[c] S/(S+R)	$\beta/(\beta + \alpha)R$	% Sulfur	°API Gravity
2	IA	Katia	0.53	0.49	4.18	0.44	0.60	6.2	14.3
2+	IA	Santa Maria-3	0.80	0.45	4.00	0.50	0.67	5.5	23.0
2+	IA	Alanno	0.78	0.42	2.63	0.44	0.61	4.8	23.1(B)
3	IA	Aquila	0.87	0.67	2.70	0.49	0.64	1.7	36.5
3	IB	Rovesti	0.83	0.64	2.20	0.49	0.65	2.7	24.2(B)
1	IC	Filletino	0.47	0.00	28.18	0.49	0.63	11.5	B
3	ID	Glamoč	0.90	–	3.56	0.48	0.63	2.1	–
1	II	Ravni Kotari-3	0.09	–	62.00	0.52	0.69	9.4	5.1
1	II	Rošca	0.23	0.26	B	B	B	7.4	B
1	II	Vrgorac	0.29	0.00	B	B	B	6.8	B
1	II	Vinišće	0.32	0.00	18.76	0.46	0.63	7.0	B
1	II	Škrip	0.29	0.28	B	B	B	6.2	B
1+	II	Okruglica	0.57	0.29	B	B	B	6.4	B
1	III	Palanka	0.43	0.16	B	B	B	7.6	B

[a]MA-Steroid Aromatization Parameter $TA\text{-}C_{28} = 20S + 20R$; $MA\text{-}C_{29} = 5\alpha + 5\beta\ (20S + 20R)$.

[b]1 = Immature, 2 = Early Mature, 3 = Mature.

[c]Data from Metastable Reaction Monitoring GC-MS: $S/(S + R) = \alpha\alpha\alpha 20S/(\alpha\alpha\alpha 20S + \alpha\alpha\alpha 20R)$, $\beta/(\beta + \alpha)R = \alpha\beta\beta 20R/(\alpha\beta\beta 20R + \alpha\alpha\alpha 20R)$; B = Biodegradation.

carbon bonds in the kerogen. They tend to be high-gravity oils of potentially high economic value.

The API gravity is an important property for evaluation of a prospect as it determines the oil's producibility, all other factors being equal. API gravity generally correlates inversely to sulfur content, especially in high sulfur crudes (Hughes et al., 1985; Orr, 1986), and it is not surprising that the Adriatic oils show such a relationship as demonstrated in Figure 19.10.

The percentage of sulfur in these reservoired oils appears to depend on two factors. Primarily, it depends on the depositional environment of the source rock (Moldowan et al., 1985). Thus, oils from source rocks (particularly carbonate and evaporitic) deposited in an iron deficient, anoxic, aqueous environment that supports sulfate-reducing bacteria, tend to be high in sulfur (Tissot and Welte, 1984). The second factor is maturation of the oil which decreases the sulfur content. A

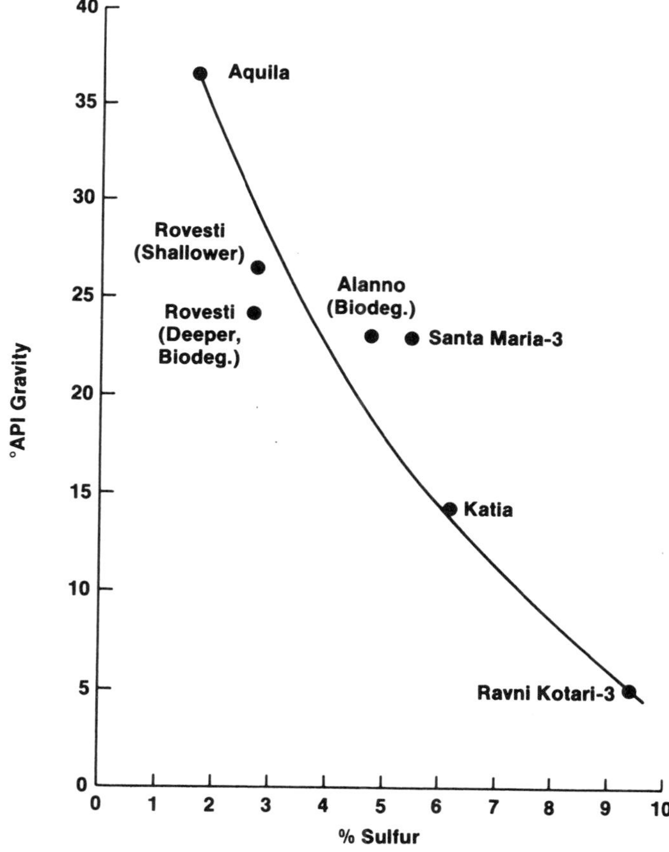

Figure 19.10 Sulfur concentration correlates with API gravity in Adriatic Basin oils.

third possible factor, biodegradation in the reservoir appears to have only a minor influence on sulfur content and API gravity of the Adriatic oils. This is illustrated by the data points for the lightly biodegraded (Table 19.1) Alanno oil and the deeper Rovesti oil in Figure 19.10, which show the same °API gravity to percent sulfur relationship as the unbiodegraded ones. If we assume in the case of the Adriatic oils that the depositional environmental effect on sulfur richness is largely uniform, then the relative API gravities of the oils should be largely maturity-dependent.

For the Adriatic oils, we established a systematic relationship between API gravity and the aromatization parameter (Fig. 19.11). The curve is obtained by using only the oils with known API gravity (unbiodegraded or lightly biodegraded).

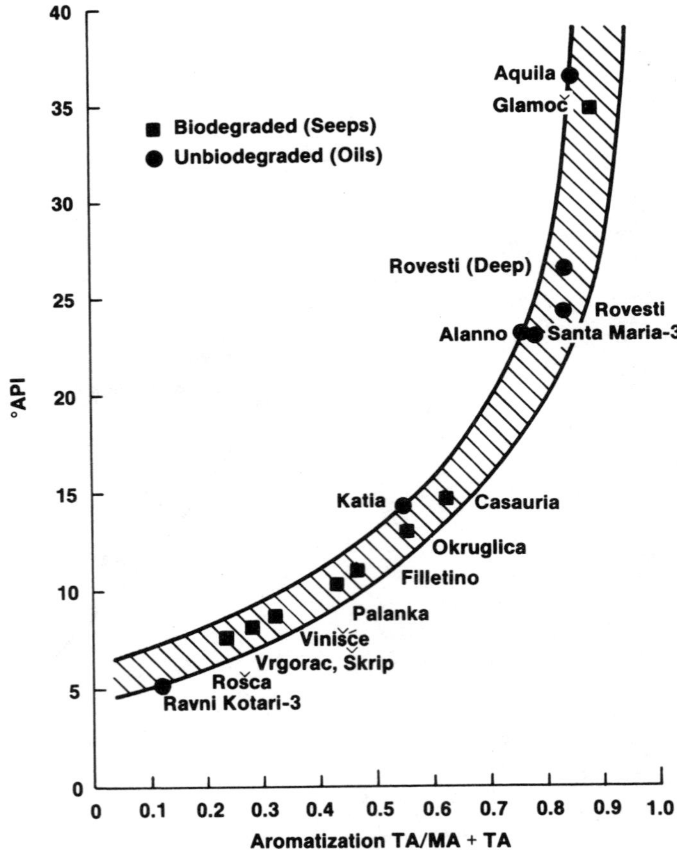

Figure 19.11 The MA-steroid aromatization parameter (Table 19.3) shows a relationship with API gravity in Adriatic Basin oils. The curve is prepared using the oils (known °API gravity). Seeps are added according to their MA-steroid aromatization values implying projection of API gravities.

The curve is then used to determine API gravity of the biodegraded seeps through their aromatization ratios as if they were unbiodegraded. The method seems most accurate at gravities below 25°API where the aromatization scale ranges from 0.1 to 0.8. Above 25°API, there is only a small increase in aromatization, from 0.8 to 0.9. The results in Figure 19.11 indicate that if any of the seeps, except Glamoč, were found in the reservoir without heavy biodegradation, their API gravities would be $\leq$ 14. Glamoč would be a high-gravity oil, $>$ 25°API.

SOURCE ROCK-TO-OIL CORRELATION BY HYDROUS PYROLYSIS

A selection of source rocks covering the Mesozoic era, mostly outcrops from the Yugoslavian study area, were evaluated by standard geochemical techniques and compared with the oils and seep oils by isotopes and biological markers. The generation of the Group II oil and seepages from Cretaceous age carbonates was indicated by correlation with an Upper Cretaceous outcrop.

The correlation of the Group I oils and seeps was more problematic because the oils occur in reservoir rocks having a wide range of ages (Table 19.1). The source rock candidates were selected from deep cores or outcrops, but all were immature. Previous attempts to correlate immature source rocks with mature oils have generally yielded mixed results because of inherent differences in biomarker composition between immature bitumen and kerogen generation products (Eglinton and Douglas, 1988). The differences probably arise because biomarkers in immature bitumens are diagenesis products of trapped free lipids of the contributing organisms, while biomarkers generated from the kerogen represent structural components of cell membranes (Chappe et al., 1980). Thus kerogen and immature bitumen biomarker distributions may differ greatly, just as bound and free lipid distributions may differ in organisms (e.g., Goodwin, 1973).

An organic rich Triassic source rock outcropping near Monte Prena, Italy, was selected to demonstrate the efficacy of immature source rock-to-oil correlation using hydrous pyrolysis. The carbon isotopic ratio of Monte Prena's bitumen (-29.7 per mil) indicates a potential correlation with the Group IA oils (range -28.99 to -27.39, Table 19.3). The richness of the rock is indicated by a total organic carbon content of 22 percent. Rock-Eval analysis showed it to be of oil prone Type II kerogen (Espitalie et al., 1977) with hydrogen and oxygen indices of 638 (mg HC/g TOC) and 45 (mg CO_2/g TOC), respectively. Its T_{max} value of 423°C indicates immaturity and its production index 0.03 ($S_1/S_1 + S_2$) shows that it has not generated significantly (Peters, 1986). Monte Prena's immaturity is confirmed by biomarker ratios measured on the bitumen. The ratio of $20S/(20S + 20R)$ C_{29}-steranes was only 0.04 compared with low maturity oils showing ratios of 0.23 and up (Seifert and Moldowan, 1981). The ratio $22S/(22S + 22R)$ 17α(H)-bishomohopanes (C_{32}) was 0.19, far below the threshold for oil generation, probably ≥ 0.50 (Seifert and Moldowan, 1980, 1986; Mackenzie et al., 1981). TA-steroids and Etio-porphyrins were virtually absent, and 17β,21β(H)-hopane was present, consistent with immaturity.

The immature Monte Prena source rock was pyrolyzed under hydrous conditions (Lewan et al., 1986) in a pressure vessel at 320°C for 72 hours. The oil generated from that experiment is compared by GC-MS analysis with a Group IA oil, Katia (Fig. 19.12). The results of this sterane (*m/z* 217) analysis show a dramatic shift from the immature extract pattern, which is heavily skewed toward C_{29}, to the mature oil pattern which contains considerable C_{27} and C_{28} components. The pattern of the C_{27}–C_{29} 5α,14α,17α(H),20*R*-steranes (ααα-20*R*) indicated by the dotted line graph for the Monte Prena hydrous pyrolysate is nearly identical to that for Katia in Figure 19.12.

Although the hydrous pyrolysis experiment has unlocked the source-sensitive C_{27}-C_{28}-C_{29}-sterane ratios of the kerogen-bound steranes, sterane epimerization has taken a different pathway in the pyrolysate. The ααα-20*S* C_{29} component is much stronger relative to ααα-20*R* C_{29} in the pyrolysate than in Katia oil, while the αββ C_{29} epimers are in lower concentrations relative to the ααα C_{29} epimers. This supports the concept that the equilibration kinetics are slower for the formation

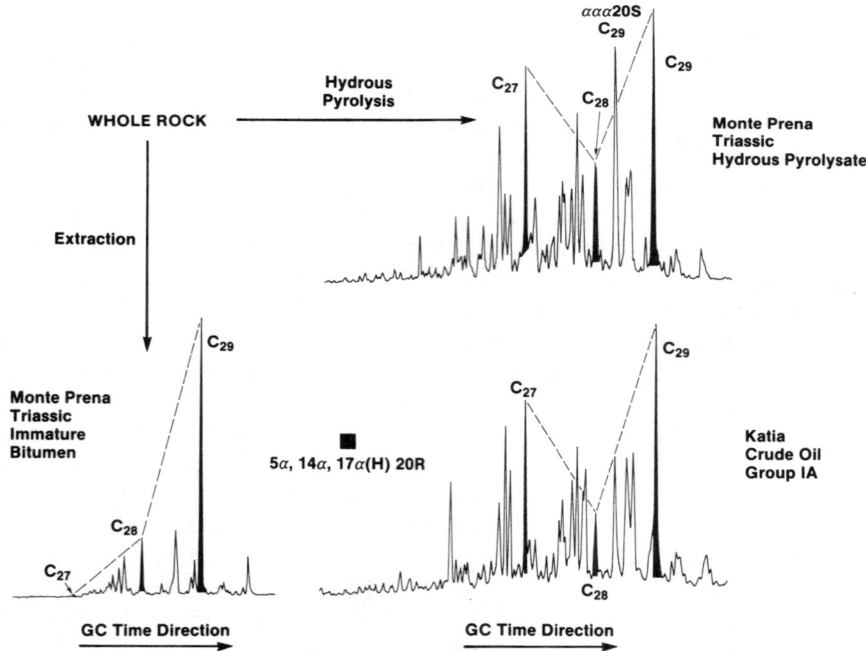

Figure 19.12 Comparison of sterane (*m/z* 217) fragmentograms of saturate fractions from a hydrous pyrolysis experiment and an oil. Selective ion monitoring GC-MS experiment using GC conditions as specified in Figure 19.6. (a) Mostly 5β,14α,17α(H)-C_{29}(20*R*). (b) Mixture 5β,14α,17α(H)-C_{29}(20*R*) + 5α,14β,17β(H)-C_{29}(20*S* + 20*R*). (c) Mostly 5α,14β,17β(H)-C_{29}(20*S* + 20*R*).

of the αββ (20*S* and 20*R*) components which are more effective maturity indicators at higher thermal maturation levels than the ααα-20*S* isomers (Sundararaman, Moldowan, Seifert, 1988).

The distribution of C_{27}-C_{28}-C_{29}-MA-steroids can be useful for source correlation (Seifert et al., 1983; 1984) contributing information independent of the steranes. This is probably due to MA-steroid formation from different precursor sterols, for example, having a double bond in the side chain (Moldowan et al., 1986a; Riolo et al., 1986). The distribution of C_{27}-C_{28}-C_{29} MA-steroids was compared between the Monte Prena hydrous pyrolysate oil and its immature bitumen (Fig. 19.3). In the ternary diagram, a shift in MA-steroid distribution is documented. The MA-steroids in the hydrous pyrolysate are in juxtaposition to the Group I oils, indicating a possible correlation.

The extended 17α(H)-homohopanes were a minor component of the *m*/*z* 191 fragmentogram of Monte Prena bitumen compared with 17β(H)-hopanes and moretanes. The hydrous pyrolysate, however, yielded a meaningful 17α(H)-homohopane distribution shown in Figure 19.6a. Although the 17α(H)-homohopane distribution differs from that of the Group IA oils, a preferential preservation of the 17α(H)-tetrahomohopane (C_{34}) homologue is apparent in some of the Group IA oils (Fig. 19.6a). The high C_{34} homohopane preservation in Monte Prena probably indicates a more reducing depositional environment than its lateral counterparts which sourced the Group IA oils.

It is also worth mentioning that the moretane/hopane ratios (C_{29}–C_{31}) were 0.10 in the pyrolysate compared with 0.05–0.06 in the oils. This apparent preservation of moretane stereochemistry in the kerogen has been demonstrated using several pyrolysis techniques (Seifert and Moldowan, 1980; Soldan and Cerqueira, 1986; Tannenbaum et al., 1986) even when the rock is already mature.

The hydrous pyrolysate *m*/*z* 191 fragmentogram shows a significant gammacerane peak yielding a gammacerane index of 4.0. This value is comparable to the Group IA oils (range 2.4 to 3.3, Table 19.3).

MATURITY AND PRESERVATION OF ORGANIC MATTER IN SEDIMENTS

Vitrinite reflectance (R_o) has been the standard way to measure a maturation profile in well cuttings and the onset of oil generation is generally assumed to be R_o = 0.50% (Teichmüller, 1971; Dow, 1977). This method has been supplemented by Rock-Eval measurements such as T_{max} (Espitalie et al., 1977; Peters, 1986) which generally marks oil generation at 435°C. However, both of these methods often are not sensitive enough to define the depth of onset of oil generation.

The geochemical log from Brač-1 well (Fig. 19.1) shows the samples to contain low total organic carbon (0.1–0.4% TOC). However, the organic matter in those carbonates/anhydrites yields slightly elevated hydrogen indices, typically 200–300 by Rock-Eval analysis. Also monitored in the Brač-1 well cuttings were: (1) the conversion of MA-steroids to TA-steroids, (2) the conversion of 20*R* to 20*R* + 20*S* C_{29}-steranes, (3) the conversion of 22*R* to 22*R* + 22*S* C_{32}-homohopanes, (4) the relative proportions of a C_{28} Etio and a C_{32} DPEP petroporphyrin, and (5) changes in the substitution patterns of methylphenanthrenes. The parameters are referred to as MA-steroid aromatization (TA/MA + TA), C_{29}-sterane isomerization, C_{32}-homohopane isomerization, the porphyrin maturity parameter (PMP), and the methylphenanthrene index (MPI-1). These measurements were compared with vitrinite reflectance and Rock-Eval T_{max}.

Using these parameters an attempt was made to determine the depth at which the thermal alteration of organic matter has reached the level necessary to have generated petroleum. This is not to imply that petroleum could actually be generated from these lean carbonate/anhydrite rocks. In similar studies these methods have been applied to source rocks and shaly sequences (Mackenzie and McKenzie, 1983; Hong et al., 1986), but not to lean carbonate/anhydrite sequences.

Figure 19.13 shows *m*/*z* 191 and *m*/*z* 217 chromatograms from the saturate fractions of two limestones (732 i and 732 k) and an anhydrite (732 m). The chromatograms are representative of the sterane and terpane distributions seen in the Brač-1 well with respect to diasterane concentrations and homohopane distributions. That is, diasterane concentrations range from low to moderate (732 m to i, respectively). A slight preferential preservation of C_{34} in the C_{31} to C_{35}-17α(H)-homohopane envelope of doublets is observed in sample 732 i, while in sample 732 k the C_{34} epimers are strongly preserved. In anhydrite sample 732 m, the C_{35}-17α(H)-homohopanes predominate, although biomarker distributions in other anhydrites near the bottom of Brač-1 more closely resemble sample 732 k (Fig. 19.13). Samples 732 i and k resemble biomarker distributions resulting from mildly suboxic to anoxic paleoenvironments of deposition in marine carbonate source rocks (Palacas et al., 1984) which show varying amounts of diasteranes and C_{32} to C_{34} 17α(H)-homohopane preservation (see Fig. 19.6 and discussion). Higher diasterane concentrations and lower homohopane preservation, as in sample 732 i, probably indicates more oxidation (Moldowan et al., 1986a). The depositional environment for evaporite sample 732 m was the most highly reducing of the three samples. The biomarker distributions in sample 732 k are typical of most in the Brač-1 well. Figure 19.13 also shows a typical limestone sample (757 y) from the Famoso-1 well (Fig. 19.1). Relatively high-diasterane concentrations, plus 17α(H)-hopane distribution patterns which drop steeply from C_{31} to C_{35}, are indicative of its oxygenated environment of deposition, in contrast to the more reducing paleoenvironments of Brač-1.

Thus, in spite of the low TOC values, the H/C ratios plus the sterane and terpane fingerprints of Brač-1 suggest that high quality, possibly Type II, kerogen is preserved in sediments throughout the section. Such sediments must have been deposited under oxygen-deficient bottom waters with only low levels of organic productivity. In contrast to the usual case for organic lean carbonates where organic matter is strongly oxidized, that in the Brač-1 well appears to have been preserved in mildly suboxic to anoxic paleoenvironments (Demaison et al., 1984).

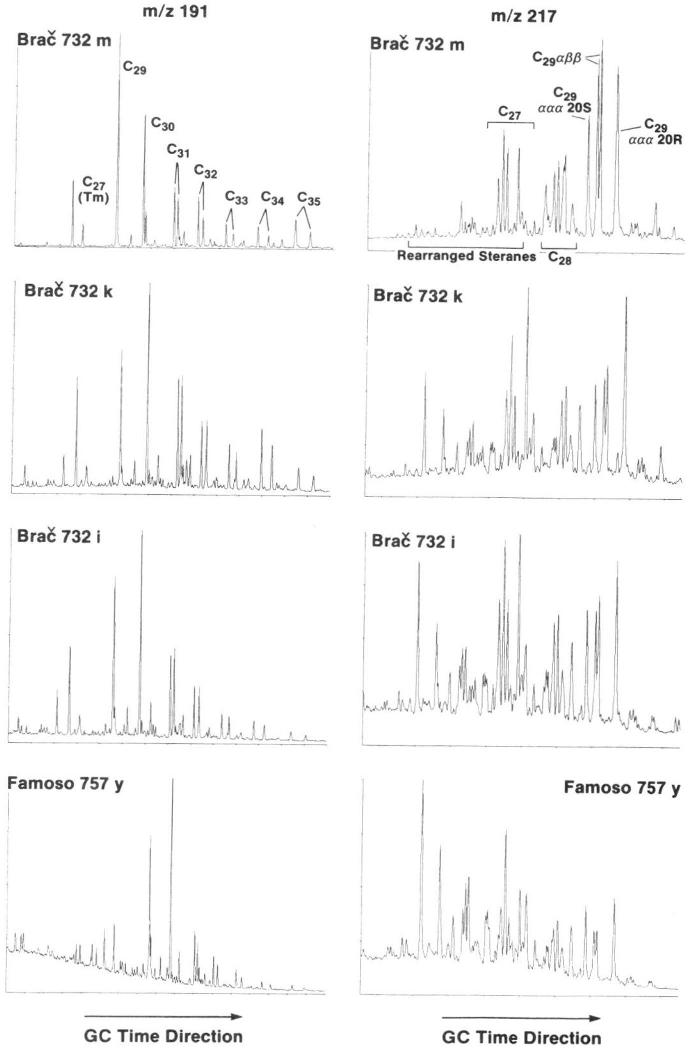

Figure 19.13 Terpane (*m/z* 191) and sterane (*m/z* 217) fingerprints indicate depositional environments of sediments in Brač-1 and Famoso-1 wells. Samples correspond to points in Figure 19.14(a) where depths are indicated. Selective ion monitoring using GC conditions as specified in Figure 19.6.

The values for aromatization and PMP (Fig. 19.14) are nearly constant between 2000 and 5000 m with a sharp increase below 5000 m. The MPI-1 values fluctuate greatly between 2000 and 5000 m indicating a potential problem of sensitivity to subtle changes in organic facies. Recent work on the phenanthrene-based

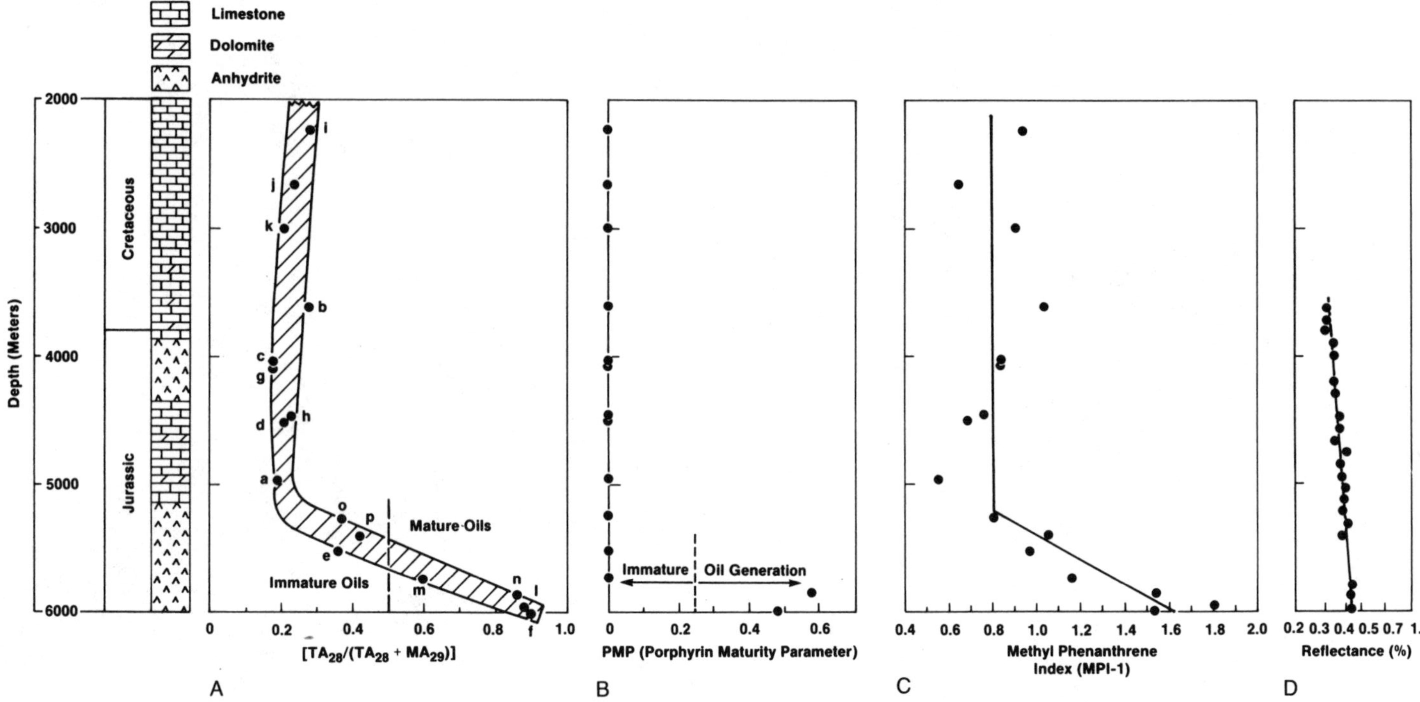

Figure 19.14 Maturation/generation parameters applied versus depth of cuttings in Brač-1 well (Fig. 19.1).

maturity parameters indicated variability in Type II kerogens (Radke et al., 1982, 1986; Radke and Welte, 1983). Nevertheless, there also appears to be a systematic increase in MPI-1 near the bottom of Brač-1. In detail, the sharp increase in each parameter takes place at a slightly different depth: 4965–5265 m for aromatization (Fig. 19.14a), between 5735–5855 m for PMP (Fig. 19.14b) and 4965–5525 m for MPI-1 (Fig. 19.14c).

Aromatization data for the oils within the Adriatic Basin in Table 19.4 indicate that the mature oils with API gravities > 14° have aromatization ratios > 0.5. In the Brač-1 well sediments, these aromatization values lie below 5550 m. PMP values (Table 19.4) for mature Adriatic oils are > 0.4; they are reached below 5800 m in the Brač-1 well.

The correspondence between oil maturities and the maturities of bitumen extracts of the sediments in Brač-1 cannot be coincidental. Generation of mature bitumen at the bottom of Brač-1 indicates the onset of the proper maturation level for the generation of petroleums at depths between 5500–5800 m. In this depth interval, the vitrinite reflectance is only 0.45 percent (Fig. 19.14d), a value which is well below that typical of the beginning of the principal zone of oil generation in shales, generally above 0.5 percent R_o (Tissot and Welte, 1984). Likewise, the T_{max} values from Rock-Eval analysis are low, reaching an average value of only about 430°C at the bottom of Brač-1. The beginning of the oil generation window is generally considered to be T_{max} = 435°C (Peters, 1986). It should be emphasized that this analysis applies to the Upper Jurassic sequence at the bottom of Brač-1 (Fig. 19.14). The result may be different for other kerogens as found in Liassic-Triassic source rocks in Italy and southern Yugoslavia and Upper Cretaceous source rocks in Yugoslavia.

The isomerization parameters for steranes and homohopanes for the Brač-1 well level off near equilibrium values (Seifert and Moldowan, 1986) below 3000 m. Between 3000 and 6000 m $\alpha\alpha\alpha$ C_{29}-sterane isomerization, $20R/(20S + 20R)$, ranges unsystematically between 0.46 and 0.50. This range is comparable to that shown in the oils (Table 19.4). Actually, equilibrium of the $\alpha\alpha\alpha$ C_{29} $20R/(20R + 20S)$ sterane ratio at petroleum generative temperatures has been shown (Seifert and Moldowan, 1981) and calculated using the Allinger MM2 force field (van Graas et al., 1982) to be in the 0.53–0.55 range. Relatively shallow C_{29}-sterane isomerization contrasting with aromatization only at great depth in the Brač-1 well is consistent with the hypothesis of Mackenzie et al. (1982) that sterane isomerization is more time-dependent (Mesozoic-long time) while aromatization is more temperature-dependent (low-geothermal gradient–low temperature).

CONCLUSIONS

1. Two major families of oils were found in the Central Adriatic Basin as determined by carbon isotope ratios, aromatic steroids, and 17α(H)-homohopane distributions. They are probably related to carbonate-rich Cretaceous and Liassic-Triassic source rocks.

2. A series of C_{28}–C_{34} 17α(H)-30-norhopanes was indicated, using the metastable reaction monitoring GC-MS technique and coelution of authentic standards, to be present in all the oils and seep oils from the Adriatic regardless of degree of biodegradation.
3. The distribution of C_{31} to C_{35} 17α(H)-homohopanes in oils and rocks may be used to monitor depositional environment.
4. Biodegradation of an oil in the reservoir may occur at 2400 m under a sufficiently low-geothermal gradient.
5. Hydrous pyrolysis of an immature source rock is an effective tool to correlate its biomarkers with an oil, by revealing the character of the biomarkers in the kerogen.
6. The porphyrin maturity parameter (PMP) may be used to identify the onset of the zone of oil generation in a well.
7. The key factor determining the API gravity of the oils is probably sulfur cross-linking. In high-sulfur crudes, the percent sulfur is related to maturity which can be measured by certain biomarker maturity parameters. The MA-steroid aromatization parameter $[TA_{28}/(TA_{28} + MA_{29})]$ was found to vary systematically with API gravity and may be used as a predictor for API gravity in heavily biodegraded oils and seeps as if they were unbiodegraded.
8. The monoaromatic steroid aromatization parameter may be used to confirm the onset of oil generation maturity in sediments by comparison with early generated oils, assuming that source rocks for the oils had kerogen with similar generation kinetics to the sediments.
9. High diasterane/sterane ratios in bitumens extracted from limestone show that clay catalysis is not the only factor regulating diasterane formation.
10. Parent ions of *m/z* 231 recorded by GC-MSMS provide improved C_{26}-C_{27}-C_{28} TA-steroid distributions compared to MID GC-MS (*m/z* 231). The analysis is a sensitive detector for C_{29} TA-steroids, analogues of "marine indicator" C_{30}-steranes indicated by coelution studies using triaromatized 24-*n*-propylcholestane.

Acknowledgments

The authors wish to dedicate this chapter to the memory of Dr. Wolfgang K. Seifert (1931–1985) who was project leader during the early stages of this work. E. J. Gallegos provided assistance in GC-MS instrumentation. Geochemical data were provided by the Organic Geochemistry Group of COFRC, La Habra, Calif. P. Albrecht provided a sample of 29,30-nor-17α(H)-hopane and helpful discussions. L. A. Wraxall, F. J. Fago, P. Novotny, and M. Pena provided technical assistance. B. Alpern provided vitrinite reflectance data. Appreciation to L. Novelli and L. Mattavelli of AGIP for their collaborative efforts. Thanks to M. Schoell for carefully reviewing the manuscript. The authors thank D. W. Lewis and R. H. Sheppard of COPI and F. L. Campbell of COFRC for their support, and the managements of COFRC, COPI, AGIP, and INA for permission to publish.

REFERENCES

ARTHUR, M.A. and PREMOLI, S.I. (1982) Development of wide-spread organic-rich strata in the Mediterranean Tethys. In *Nature and Origin of the Cretaceous Carbon-rich Facies* (eds. S.O. Schlanger and M.B. Cita), Academic Press. pp. 7–54.

CHAPPE, B., MICHAELIS, W., and ALBRECHT, P. (1980) Molecular fossils of archaebacteria as selective degradation products of kerogen. In *Advances in Organic Geochemistry 1979* (eds. A.G. Douglas and J.R. Maxwell), Pergamon Press, Oxford. pp. 265–274.

CONNAN, J. (1984) Biodegradation of crude oils in reservoirs. In *Advances in Petroleum Geochemistry* Vol. 1 (eds. J. Brooks and D. Welte), Academic Press, London. pp. 299–335.

CONNAN, J., BOUROULLEC, J., DESSORT, D., and ALBRECHT, P. (1986) The microbial input in carbonate-anhydrite facies of a sabkha palaeoenvironment from Guatemala: A molecular approach. In *Advances in Organic Geochemistry 1985* (eds. D. Leythaeuser and J. Rullkötter), Pergamon Journals, Oxford. pp. 29–50.

DEMAISON, G., HOLCK, A.J.J., JONES, R.W., and MOORE, G.T. (1984) Predictive source bed stratigraphy; a guide to regional petroleum occurrence. In *Proc. 11th World Petr. Congr.* 2, PD1, John Wiley, London. pp. 17–29.

DOW, W.G. (1977) Kerogen studies and geological interpretations. *J. Geochem., Explor.* *7*, 79–99.

EGLINTON, T.I. and DOUGLAS, A.G. (1988) Quantitative study of biomarker hydrocarbons released from kerogens during hydrous pyrolysis. *Energy and Fuels 2*, 81–88.

ESPITALIE, J., MADEC, M., TISSOT, B. and LEPLAT, P. (1977) Source rock characterization method for petroleum exploration. *Offshore Technology Conference*, OTC 2935, Houston, Texas, May 2–5, 1977, pp. 439–444.

FUEX, A.N. (1977) The use of stable carbon isotopes in hydrocarbon exploration. *J. Geochem. Explor. 7*, 155–188.

GALLEGOS, E.J. and MOLDOWAN, J.M. (1991) The effect of injection hold time on GC resolution and the effect of collision gas on mass spectra in geochemical "biomarker" research. Chapter 9 in this text.

GOODWIN, T.W. (1973) Comparative biochemistry of sterols in eukaryotic microorganisms. In *Lipids and Biomembranes of Eukaryotic Microorganisms* (ed. J.A. Erwin), Academic Press, New York. pp. 1–40.

GOODWIN, N.S., PARK, P.J.D., and RAWLINSON, T. (1983) Crude oil biodegradation. In *Advances in Organic Geochemistry 1981* (eds. M. Bjorøy et al.), John Wiley, Chichester. pp. 650–658.

VAN GRAAS, G., BAAS, J.M.A., DE GRAAF, V., and DE LEEUW, J.W. (1982) Theoretical organic geochemistry. I. The thermodynamic stability of several cholestane isomers calculated by molecular mechanics. *Geochim. Cosmochim. Acta 46*, 2399–2402.

HADDON, W.F. (1979) Computerized mass spectrometry linked scan system for recording metastable ions. *Anal. Chem. 51*, 983–988.

TEN HAVEN, H.L., ROHMER, M., RULLKÖTTER, J., and BISSERET, P. (1989) Tetrahymanol, the most likely precursor of gammacerane, occurs ubiquitously in marine sediments. *Geochim. Cosmochim Acta 53*, 3073–3079.

HONG, Z.-H., LI, H.-X., RULLKÖTTER, J., and MACKENZIE, A.S. (1986) Geochemical application of sterane and triterpane biological marker compounds in the Linyi Basin. In *Advances in Organic Geochemistry 1985* (eds. D. Leythaeuser and J. Rullkötter), Pergamon Journals, Oxford. pp. 433–439.

HUGHES, W.B., HOLBA, A.G., MÜLLER, D.E., and RICHARDSON, J.S. (1985) Geochemistry of greater Ekofisk crude oils. In *Geochemistry in Exploration of the Norwegian Shelf* (ed. B.M. Thomas), Graham and Trotman. pp. 75–92.

JACOB, H., JENKO, K., and SPAIĆ, V. (1983) Dispersed solid to semisolid natural oils and bitumens of the Yugoslav Adriatic coastal area (outer Dinarides). *NAFTA 34*(12), pp. 693–700, Zagreb (in Croatian with maps).

KIMBLE, B.J., MAXWELL, J.R., PHILP, R.P., EGLINTON, G., ALBRECHT, P., ENSMINGER, A., ARPINO, P., and OURISSON, G. (1974) Tri- and tetraterpenoid hydrocarbons in the Messal Oil Shale. *Geochim. Cosmochim. Acta 38*, 1165–1181.

LEWAN, M.D. (1984) Factors controlling the proportionality of vanadium to nickel in crude oils. *Geochim. Cosmochim. Acta 48*, 2231–2238.

LEWAN, M.D., BJORØY, M., and DOLCATER, D.L. (1986) Effects of thermal maturation on steroid hydrocarbons as determined by hydrous pyrolysis of Phosphoria Retort Shale. *Geochim. Cosmochim. Acta 50*, 1977–1987.

MACKENZIE, A.S. and MCKENZIE, D. (1983) Isomerization and aromatization of hydrocarbons in sedimentary basins formed by extension. *Geol. Mag. 120*, 417–470.

MACKENZIE, A.S., LEWIS, C.A., and MAXWELL, J.R. (1981) Molecular parameters of maturation in the Toarcian shales, Paris Basin France—IV. Laboratory thermal alteration studies. *Geochim. Cosmochim. Acta 45*, 2369–2376.

MACKENZIE, A.S., LAMB, N.A., and MAXWELL, J.R. (1982) Steroid hydrocarbons and the thermal history of sediments. *Nature 295*, 223–226.

MELLO, M.R., TELNAES, N., GAGLIANONE, P.C., CHICARELLI, M.I., BRASSELL, S.C., and MAXWELL, J.R. (1988) Organic geochemical characterization of depositional palaeoenvironments of source rocks and oils in Brazilian marginal basins. In *Advances in Organic Geochemistry 1987* (eds. L. Mattavelli and L. Novelli). *Org. Geochem. 13*, 31–45.

MOLDOWAN, J.M. and FAGO, F.J. (1986) Structure and significance of a novel rearranged monoaromatic steroid hydrocarbon in petroleum. *Geochim. Cosmochim. Acta 50*, 343–351.

MOLDOWAN, J.M., SEIFERT, W.K., ARNOLD, E., and CLARDY, J. (1984) Structure proof and significance of stereoisomeric 28,30-bisnorhopanes in petroleum and petroleum source rocks. *Geochim. Cosmochim. Acta 48*, 1651–1661.

MOLDOWAN, J.M., SEIFERT, W.K., and GALLEGOS, E.J. (1985) Relationship between petroleum composition and depositional environment of petroleum source rocks. *Bull. Am. Assoc. Pet. Geol. 69*, 1255–1268.

MOLDOWAN, J.M., SUNDARARAMAN, P., and SCHOELL, M. (1986a) Sensitivity of biomarker properties to depositional environment and/or source input in the Lower Toarcian of SW-Germany. In *Advances in Organic Geochemistry 1985* (eds. D. Leythaeuser and J. Rullkötter) Pergamon Journals, Oxford, pp. 915–926.

MOLDOWAN, J.M., SUNDARARAMAN, P., and SCHOELL, M. (1986b) The effects of depositional environment on biomarker diagenesis. Symp. on Org. Geochem. of Petr. and Source Rocks. Division of Geochemistry, 191st ACS National Meeting, New York, April 13–18, 1986.

ORR, W.L. (1986) Kerogen/asphaltene/sulfur relationships in sulfur-rich Monterey oils. In *Advances in Organic Geochemistry 1985* (eds. D. Leythaeuser and J. Rullkötter), Pergamon Journals, Oxford. pp. 499–516.

OURISSON, G., ALBRECHT, P., and ROHMER, M. (1984) The microbial origin of fossil fuels. *Sci. Am. 251*, 44–51.

PALACAS, J.G., ANDERS, D.E., and KING, J.D. (1984) South Florida Basin—A prime example of carbonate source rocks in petroleum. In *Petroleum Geochemistry and Source Rock Potential of Carbonate Rocks (AAPG Studies in Geology No. 18)* (ed. J.G. Palacas), Am. Assoc. Pet. Geol., Tulsa. pp. 71–96.

PETERS, K.E. (1986) Guidelines for evaluating petroleum source rock using programmed pyrolysis. *Bull. Am. Assoc. Pet. Geol. 70*, 318–329.

PHILIPPI, G.T. (1977) On the depth, time and mechanism of origin of the heavy to medium-gravity naphthenic crude oils. *Geochim. Cosmochim. Acta 41*, 33–52.

PHILP, R.P. (1983) Correlation of crude oils from San Jorges Basin, Argentina. *Geochim. Cosmochim. Acta 47*, 267–275.

PIERI, M. and MATTAVELLI, L. (1986) Geologic framework of Italian petroleum resources. *Bull. Am. Assoc. Pet. Geol. 70*, pp. 103–130.

RADKE, M. and WELTE, D.H. (1983) The methylphenanthrene index (MPI). A maturity parameter based on aromatic hydrocarbons. In *Advances in Organic Geochemistry 1981* (eds. M. Bjorøy et al.), John Wiley, Chichester. pp. 504–512.

RADKE, M., WELTE, D.H., and WILLSCH, H. (1982) Geochemical study on a well in the Western Canada Basin: Relation of the aromatic distribution pattern to maturity of organic matter. *Geochim Cosmochim. Acta 46*, 1–10.

RADKE, M., WELTE, D.H., and WILLSCH, H. (1986) Maturity parameters based on aromatic hydrocarbons: Influence of the organic matter type. In *Advances in Organic Geochemistry 1985* (eds. D. Leythaeuser and J. Rullkötter).

RAEDEKE, L.D., SUNDARARAMAN, P., and MOLDOWAN, J.M. (1987) Comparison of porphyrin and hydrocarbon data with other indicators of maturity and oil generation. Abstracts, American Association of Petroleum Geologists Convention, June 1987, Los Angeles, Calif.

REED, W.E. (1977) Molecular compositions of weathered petroleum and comparison with its possible source. *Geochim. Cosmochim. Acta 41*, 237–247.

RIOLO, J., HUSSLER, G., ALBRECHT, P. and CONNAN, J. (1986) Distribution of aromatic steroids in geological samples: Their evaluation as geochemical parameters. In *Advances in Organic Geochemistry 1985* (eds. D. Leythaeuser and J. Rullkötter), Pergamon Journals, Oxford. pp. 981–990.

RIVA, A., SALVATORI, T., CAVALIERE, R., RICCHIUTO, T., and NOVELLI, L. (1986) Origin of oils in Po Basin, Northern Italy. In *Advances in Organic Geochemistry 1985* (eds. D. Leythaeuser and J. Rullkötter), Pergamon Journals, Oxford. pp. 391–400.

ROHMER, M., BOUVIER, P., and OURISSON, G. (1979) Molecular evolution of biomembranes: Structural equivalents and phylogenetic precursors of sterols. *Proc. Natl. Acad. Sci. 76*, 847–851.

RULLKÖTTER, J. and WENDISCH, D. (1982) Microbial alteration of 17α(H)-hopane in Madagascar asphalts: Removal of C-10 methyl group and ring opening. *Geochim. Cosmochim. Acta 46*, 1543–1553.

RULLKÖTTER, J., SPIRO, B., and NISSENBAUM, A. (1985) Biological marker characteristics of oils and asphalts from carbonate source rocks in a rapidly subsiding graben, Dead Sea, Israel. *Geochim. Cosmochim. Acta. 49*, 1357–1370.

SCHMID, J.C., CONNAN, J., and ALBRECHT, P. (1987) Occurrence and geochemical significance of long-chain dialkylthiocyclopentanes. *Nature 329*, 54–56.

SCHOELL, M. (1984) Stable isotopes in petroleum research. In *Advances in Petroleum Geochemistry* Vol. 1 (eds. J. Brooks and D. Welte), Academic Press, London. pp. 215–245.

ŠEBEČIĆ, B. (1981) Chemical analytical assessment of the source rock potential of bituminous rocks from the outer Dinarides (Yugoslavia). *NAFTA 33*(10), pp. 537–544, Zagreb (in Croatian with maps).

SEIFERT, W.K. and MOLDOWAN, J.M. (1978) Applications of steranes, terpanes and monoaromatics to the maturation, migration and source of crude oils. *Geochim. Cosmochim. Acta 42*, 77–95.

SEIFERT, W.K. and MOLDOWAN, J.M. (1979) The effect of biodegradation on steranes and terpanes in crude oils. *Geochim. Cosmochim. Acta 43*, 111–126.

SEIFERT, W.K. and MOLDOWAN, J.M. (1980) The effect of thermal stress on source rock quality as measured by hopane stereochemistry. In *Advances in Organic Geochemistry 1979* (eds. A.G. Douglas and J.R. Maxwell), Pergamon Press, Oxford. pp. 229–237.

SEIFERT, W.K. and MOLDOWAN, J.M. (1981) Paleoreconstruction by biological markers. *Geochim. Cosmochim. Acta 45*, 783–794.

SEIFERT, W.K. and MOLDOWAN, J.M. (1986) Use of biological markers in petroleum exploration. In *Biological Markers in the Sedimentary Record* (ed. R.B. Johns) Elsevier Science, Amsterdam. pp. 261–290.

SEIFERT, W.K., CARLSON, R.M.K., and MOLDOWAN, J.M. (1983) Geomimetic synthesis, structure assignment, and geochemical correlation application of monoaromatized petroleum steranes. In *Advances in Organic Geochemistry 1981* (eds. M. Bjorøy et al.) John Wiley, Chichester. pp. 710–724.

SEIFERT, W.K., MOLDOWAN, J.M., and DEMAISON, G.J. (1984) Source correlation of biodegraded oils. *Org. Geochem. 6*, 633–643.

SOLDAN, A.L. and CERQUEIRA, J.R. (1986) Effects of thermal maturation on geochemical parameters obtained by simulated generation of hydrocarbons. In *Advances in Organic Geochemistry 1985* (eds. D. Leythaeuser and J. Rullkötter), Pergamon Journals, Oxford. pp. 339–345.

STAHL, W.J. (1978) Source rock-crude oil correlation by isotopic type-curves. *Geochim. Cosmochim. Acta 42*, 1573–1577.

SUNDARARAMAN, P. (1985) High-performance liquid chromatography of vanadyl porphyrins. *Anal. Chem. 57*, 2204–2206.

SUNDARARAMAN, P., MOLDOWAN, J.M., and SEIFERT, W.K. (1988) Incorporation of petroporphyrins into geochemical correlation problems. In *Geochemical Biomarkers* (eds. T.F. Yen and J.M. Moldowan), Harwood Academic Publ., Chur, Switzerland. pp. 373–382.

SUNDARARAMAN, P., BIGGS, W.R., REYNOLDS, J.G., and FETZER, J.C. (1988) Vanadylporphyrins, indicators of kerogen breakdown and generation of petroleum. *Geochim. Cosmochim. Acta 52*, 2337–2341.

TANNENBAUM, E., RUTH, E., HUIZINGA, B.J., and KAPLAN, I.R. (1986) Biological marker distribution in coexisting kerogen, bitumen and asphaltenes in Monterey Formation diatomite, California. In *Advances in Organic Geochemistry 1985* (eds. D. Leythaeuser and J. Rullkötter), Pergamon Journals, Oxford. pp. 531–536.

Teichmüller, M. (1971) Anwendung kohlenpetrographischer methoden bei der Erdöl- und Erdgasprospektion. *Erdöl u. Kohle 24*, 69–76.

Tissot, B.P. and Welte, D.H. (1984) Petroleum Formation and Occurrence, Springer-Verlag, Berlin. pp. 1–699.

Trifilieff, S., Sieskind, O., and Albrecht, P. (1991) Biological markers in petroleum asphaltenes: Possible mode of incorporation. Chapter 18 in this text.

Venkatesan, M.I. (1989) Tetrahymanol: Its widespread occurrence and geochemical significance. *Geochim. Cosmochim. Acta 53*, 3095–3101.

Volkman, J.K., Alexander, R., Kagi, R.I., and Woodhouse, G.W. (1983) Demethylated hopanes in crude oils and their application in petroleum geochemistry. *Geochim. Cosmochim. Acta 47*, 785–794.

Wardroper, A.M.K., Hoffmann, C.F., Maxwell, J.R., Barwise, A.J.G., Goodwin, N.S., and Park, P.J.D. (1984) Crude oil biodegradation under simulated and natural conditions—II. Aromatic steroid hydrocarbons. *Org. Geochem. 6*, 605–617.

Index